Advanced Research in Biotechnology

Advanced Research in Biotechnology

Edited by **Joy Adam**

R CALLISTO
REFERENCE

New York

Published by Callisto Reference,
106 Park Avenue, Suite 200,
New York, NY 10016, USA
www.callistoreference.com

Advanced Research in Biotechnology
Edited by Joy Adam

International Standard Book Number: 978-1-63239-022-6 (Hardback)

Printed in the United States of America.

Contents

Preface

Biotechnology is the use of living organisms and systems to make or modify any product for specific use to improve the quality of human lives. The commercial production of enzymes like insulin, introduction of cry genes from Bacillus thuringiensis to cotton, or introducing disease-resistant genes to tobacco plants are all applications of biotechnology. While genetically modified crops have changed the agricultural scenario of the world, genetically modified organisms have enhanced the medical scenario. This has been possible because biotechnology permits for a single gene to be transferred to the plant or animal, one that is desired, by eliminating the undesired genes. Biotechnology is also used to produce various chemicals for DNA fingerprinting and gene therapy. Biotechnology is a boon for human beings which is being used to its potential these days.

Solid State Fermentation of a Raw Starch Digesting Alkaline Alpha-Amylase from Bacillus licheniformis RT7PE1 and Its Characteristics, Family-Specific Degenerate Primer Design: A Tool to Design Consensus Degenerated Oligonucleotides, Optimization of Fermentation Medium for the Production of Glucose Isomerase Using Streptomyces sp. SB-P1, Experimental Investigations on the Effects of Carbon and Nitrogen Sources on Concomitant Amylase and Polygalacturonase Production by Trichoderma viride BITRS-1001 in Submerged Fermentation and Involvement of the Ligninolytic System of White-Rot and Litter-Decomposing Fungi in the Degradation of Polycyclic Aromatic Hydrocarbons are significant themes which cover the advancements in biotechnology.

This book is the result of channelized energies of experts from the industry. I wish to thank them and the others involved in the completion of this book.

Editor

Production of Enzymes from Agroindustrial Wastes by Biosurfactant-Producing Strains of *Bacillus subtilis*

**Francisco Fábio Cavalcante Barros, Ana Paula Resende Simiqueli,
Cristiano José de Andrade, and Gláucia Maria Pastore**

Department of Food Science, Faculty of Food Engineering, University of Campinas, P.O. Box 6121, 13083-862 Campinas, SP, Brazil

Correspondence should be addressed to Francisco Fábio Cavalcante Barros; fabiobarros10@gmail.com

Academic Editor: Triantafyllos Roukas

Bacteria in the genus *Bacillus* are the source of several enzymes of current industrial interest. Hydrolases, such as amylases, proteases, and lipases, are the main enzymes consumed worldwide and have applications in a wide range of products and industrial processes. Fermentation processes by *Bacillus subtilis* using cassava wastewater as a substrate are reported in the technical literature; however, the same combination of microorganisms and this culture medium is limited or nonexistent. In this paper, the amylase, protease, and lipase production of ten *Bacillus subtilis* strains previously identified as biosurfactant producers in cassava wastewater was evaluated. The LB1a and LB5a strains were selected for analysis using a synthetic medium and cassava wastewater and were identified as good enzyme producers, especially of amylases and proteases. In addition, the enzymatic activity results indicate that cassava wastewater was better than the synthetic medium for the induction of these enzymes.

1. Introduction

The species of the genus *Bacillus* are known to be producers of enzymes of industrial interest. These bacteria are responsible for approximately 50% of the total enzyme market [1], which is estimated at 1.6 billion dollars. One of the primary advantages of using these species for enzyme production is that they are easily grown and maintained in the laboratory because they adapt to changes in the growing conditions that hinder the development and enzymatic synthesis of other microorganisms [2].

Among the different categories of enzymes, hydrolases are those with the largest industrial application, and among these, amylases, in particular alpha- and beta-amylases, have received special attention [3]. These enzymes catalyze the hydrolysis of starch and are produced by a wide range of microorganisms; however, for commercial applications, they are generally derived from *Bacillus* [1, 3–6], such as *B. licheniformis*, *B. stearothermophilus*, and *B. amyloliquefaciens*. These enzymes are applied to several industrial sectors, such as the food, fermentation, textiles, detergents, and paper industry [3, 5, 6]. The main amylases produced by *Bacillus* are resistant to heat, which is commercially important because numerous processes require high temperatures. Thus, the sensitivity to heat is no longer a limiting factor for their use [3].

Another relevant group is the proteases, which represent approximately 30% of the total sales of enzymes worldwide [7]. Proteases are predominantly applied to the food, textile, pharmaceutical, and detergent industries [1]. Some microorganisms produce low amounts of these enzymes, which impair their industrial application. However, in most cases, by adopting simple methods, such as the use of a specific and optimized medium, it is possible to increase production yields. The thermostable proteases produced by *Bacillus* spp. are among the most industrially important ones [7].

Finally, lipases, that is, enzymes that catalyze triacylglycerol hydrolysis, are widely used in organic chemistry due to their high specificity and selectivity [8]. Thus, they have received considerable attention because of their potential use in industrial processes [9], especially as biocatalysts. Among the reasons for the enormous potential of these enzymes are their high stability in organic solvents, the nonrequirement for cofactors, and broad substrate specificity [8]. However, these enzymes are only moderately stable at high temperatures where most industrial processes are performed. This drawback might be solved with the use of lipases produced by thermophilic microorganisms [8]. *Bacillus subtilis* secretes different lipases that vary according to growth and environmental conditions, pH, and amino acid supply [10].

The aim of this research project was to study the production of amylases, proteases, and lipases by *Bacillus subtilis* strains previously identified as biosurfactant producers, including the use of cassava wastewater as culture medium.

2. Materials and Methods

2.1. Microorganisms and Inoculum Preparation. The strains of *Bacillus subtilis* assessed in this study were ATCC 21332 from the *American Type Culture Collection* and LB2B, LB115, LB117, LB1a, LB5a, LB114, LB262, LB157, and LB2a from the culture collection of the Bioflavors Laboratory (DCA/FEA/Unicamp). All the strains were previously identified as producers of lipopeptide surfactants [11]. These cultures were maintained in inclined nutrient agar and refrigerated between 5 and 7°C.

For experiments in solid medium, the inoculum was prepared in Petri dishes containing nutrient agar (30°C, 24 h). The isolated colonies with specific growth characteristics (irregular shape, wavy edges, whitish, waxy, and flat top) were transferred using an inoculation needle to the solid media. In the liquid medium experiments the inoculum was produced as described by Barros et al. [12].

2.2. Culture Media. The culture media were classified into complex solids, synthetic liquids, and cassava wastewater. All media were maintained refrigerated until use.

2.2.1. Complex Solid Media. The complex solid media used were the following.

  (i) Differential agar for extracellular lipase-producing microorganisms as described by Lin et al. [13]. The composition percentages (w/v) in distilled water were olive oil 2.0, peptone 0.3, yeast extract 0.2, K_2HPO_4 0.2, $MgSO_4 \cdot 7H_2O$ 0.1, Na_2CO_3 0.1, agar 2.0, and rhodamine B 0.001.

  (ii) In order to test the production of extracellular protease, a medium was prepared as described by Giongo [14]. The concentration of each component $(g \cdot L^{-1})$ in distilled water was meat peptone 5.0, yeast extract 3.0, skimmed milk powder 10.0, and agar 12.0.

  (iii) To assess the presence of extracellular amylases, the medium described by Giongo [14] was adapted by the replacement of the skimmed milk powder with the same concentration of cassava starch.

2.2.2. Synthetic Liquid Media. The liquid media were prepared from a basal medium with the following composition $(g \cdot L^{-1}$ of distilled water): yeast extract 1.0, KH_2PO_4 2.0, $(NH_4)_2SO_4$ 5.0, sodium citrate 1.0, $MgSO_4 \cdot 7H_2O$ 0.2, $CaCl_2 \cdot 2H_2O$ 0.01, $FeCl_3$ 0.001, $MnSO_4 \cdot 7H_2O$ 0.001, and $ZnSO_4 \cdot 7H_2O$ 0.001. Supplements of carbon sources were added in a proportion of 1.0% (v/v) for liquids or (w/v) for solids according to the following description: olive oil for lipase, glucose for protease and cassava starch for amylase. After preparation, 50 mL of media was placed in 125 mL Erlenmeyer flasks and sterilized in an autoclave at 121°C for 20 min.

2.2.3. Cassava Wastewater. Cassava wastewater from a cassava flour factory was heated until boiling, cooled until it reached 5°C, and centrifuged at 5°C and at 1,600 ×g for 20 min. This process was aimed at the solubilization of starch, the removal of suspended solids, and the elimination of hydrogen cyanide. After this process, the resulting liquid was called treated cassava wastewater. For the experiments, 50 mL of this medium was placed into 125 mL Erlenmeyer flasks and sterilized in an autoclave at 121°C for 20 min.

2.3. Culture Selection in Solid Medium. Each culture was inoculated using an inoculation needle in Petri dishes containing the specific medium for each experiment. All cultures were incubated at 30°C for 72 h, and the colony diameter and halo formation were each measured every 24 h. An average colony diameter was obtained by measuring two perpendicular axes. The halos were calculated as the total hydrolysis diameter minus the colony diameter. All experiments were performed in duplicate and repeated two times.

2.4. Fermentative Process in Liquid Medium. The sterilized treated cassava wastewater and the synthetic media were inoculated and incubated at 30°C with agitation at 150 rpm for 60 h. An amount of 1 mL of inoculum was added to each glass flask containing 50 mL of media. Samples were aseptically obtained at regular intervals to determine the enzymatic activity.

2.5. Fermentative Process in Bioreactor. Fermentations were carried out in bench bioreactor (3.0 L) with 1.5 L of sterile treated cassava wastewater, which was added to the bioreactor vessel 100 mL of the inoculum. The apparatus is best described by Barros et al. [12]. The operating conditions were temperature, aeration, and stirring maintained at 30°C, 1 vvm, and 150 rpm, respectively. Samples were aseptically obtained at regular intervals to determine the enzymatic activity. The enzymatic activities were also performed in recovered foam simultaneously times to the culture medium.

2.6. Enzymatic Activity Measurement

2.6.1. Enzymatic Extract Preparation. Each sample was centrifuged at 3,900 ×g in an eppendorf centrifuge tube for 2 min to remove cells. The supernatant, called enzymatic extract, was used to determine the enzymatic activity.

2.6.2. Proteolytic Activity Measurement. The proteolytic activity was based on the capacity of extracts to promote casein hydrolysis. To assess this activity, 0.5 mL of enzymatic extract was added to a casein solution 1.2% (w/v) in phosphate buffer 0.2 M (pH 7.0). After mixing vigorously, the flasks were incubated in a water bath at 37°C for 30 min. After incubation, the reaction was stopped by adding 4 mL of acetate buffer 0.2 M (pH 4.0), cooled in an ice bath (0°C) and filtered with filter paper. After filtration, 1 mL of the liquid was removed and added to 3 mL of NaOH 1 M and 0.5 mL of the Folin-Ciocalteau reagent diluted 1:1 of distilled water. Then, the absorbance of samples was measured in a spectrophotometer at 660 nm. A blank measurement was determined by adding distilled water instead of enzymatic extract for each group of samples. The enzymatic activity was determined using a tyrosine standard curve, and one activity unity (U) was defined as 1 μmoL of tyrosine released per mL of enzyme per hour [15].

2.6.3. Amylolytic Activity Measurement.
The amylolytic activity was based on the extract capability to promote starch hydrolysis. To assess this activity, 0.5 mL of extract was added to 5 mL soluble starch 1.0% (w/v) in phosphate buffer 0.2 M (pH 7.0). Flasks were incubated at 37°C for 10 minutes. After incubation, the reaction was stopped by adding 5 mL HCl 0.1 N. Then, 0.5 mL of this solution was added to 5 mL iodide-iodate 5–0.5% (w/v) and diluted 1:9 in phosphate buffer 0.2 M pH 7.0. In parallel, a blank sample and a substrate blank were used. The blank reaction was obtained in a similar manner as the samples, differing only by the addition of distilled water instead of extract. The substrate blank consisted of 5 mL HCl 0.1 N, 5 mL phosphate buffer 0.2 M (pH 7.0), and 0.5 mL distilled water. The absorbance of the samples was read in a spectrophotometer at 580 nm, and their activities were determined using a starch standard curve. One enzymatic activity unity was defined as the reduction of 0.1 mg of starch in 10 minutes of reaction with 0.5 mL extract [4].

2.6.4. Lipolytic Activity Measurement.
The lipolytic activity was based on the extract capability to promote the hydrolysis of olive oil triacylglycerols. In this experiment, 1 mL of extract was added to flasks containing 4 mL phosphate buffer 0.2 M (pH 9.0), 1 mL CaCl$_2$ 110 mM, and 5 mL emulsion 25% (v/v) olive oil in Arabic gum 7% (w/v). The flasks were incubated at 37°C and 160 rpm for 20 min. After incubation, the reaction was stopped by adding 15 mL acetone:ethanol (1:1), and it was kept in a shaker for 5 more minutes. A blank sample was obtained in a similar manner as the extract samples, differing only by the use of distilled water instead of extract. The fatty acids released by the triacylglycerol hydrolysis were titrated with NaOH 0.05 M, and 20 μL phenolphthalein solution (0.5% in ethanol (w/v)) was used as indicator. One lipase unity (U) was defined as the amount of enzyme able to release one μmol of fatty acid per minute (μmoL·min^{-1}) under the conditions described above [16, 17].

3. Results

3.1. Culture Selection in Solid Medium.
All strains showed a halo formation in all experiments for all enzymes (Figures 1(a), 1(b), and 1(c)).

In the amylase production screening, the LB5a strain showed the largest halo at end of fermentation—even considering the considerable growth of the colony on the plate (12.0 mm). Its performance was substantially better than the second best strain, ATCC 21332, which showed an average halo diameter of 6.5 mm at 72 h (Figure 1(a)). For protease experiments (Figure 1(b)), the largest halos were produced by LB1a, which presented halo diameter of 24 mm, a significant vantage over the other strains at 72 h. All other strains showed values closer to 15 mm, except LB5a, which showed values slightly over 10 mm. For lipase, the largest halos were produced by LB5a strain, reaching a maximum halo diameter of 9.8 mm at 72 h. In some experiments there was a reduction in the halo size after 48 h (Figure 1(c)). The experiments are illustrated by Figure 2.

3.2. Enzymatic Activity in Synthetic Liquid Medium.
All results of enzymatic activity in synthetic liquid medium are shown in Figure 3, which show enzymatic activity as function of fermentation time.

The amylolytic enzymes showed an intense activity increase at the beginning of the experiment. The maximum activity values were approximately 16 U for both strains, but the apex was reached at different times: 10 and 16 hours for LB1a and LB5a, respectively. After few hours, there was a strong reduction of activity to levels ranging from 4 to near zero U for all strains (Figure 3(a)). Concerning the protease activity, there was a gradual rise in their levels during almost all the fermentation processes, a behavior that was similar for both strains. The enzymatic activity reached its peak at values of approximately 1.2 U and 1.3 U, for LB5a and LB1a, respectively. The activity peaks were attained simultaneously at approximately 48 h. However, LB1a maintained values near the maximum until 62 h. After reaching the maximum, there was an accentuated reduction of activity to levels close to the fermentation beginning (Figure 3(b)). The lipolytic activity was very low through all processes, showing only small variations (Figure 3(c)).

3.3. Enzymatic Activity in Cassava Wastewater.
All results of enzymatic activity in cassava wastewater medium are shown in Figure 4, which shows enzymatic activity as function of fermentation time.

For amylases the results were substantially high when compared with results reached in synthetic culture medium supplemented with soluble starch. In cassava wastewater medium, while LB1a reached maximum over 400 U, LB5a reached about 150 U. Regarding the kinetic behavior, both strains were similar to each other and when compared to the results obtained in the synthetic medium they showed a significant increase of activity in the first hours followed by a considerable reduction after a short time (Figure 4(a)).

The proteolytic activities achieved by both strains were higher throughout the process in cassava waste than in synthetic medium (Figure 4(b)). However, the maximum values for both strains were reached at similar periods to those found in the synthetic medium, that is, at approximately 50 h. Such as in the synthetic medium, LB1a also showed the highest peak in the wastewater. Comparing the curves

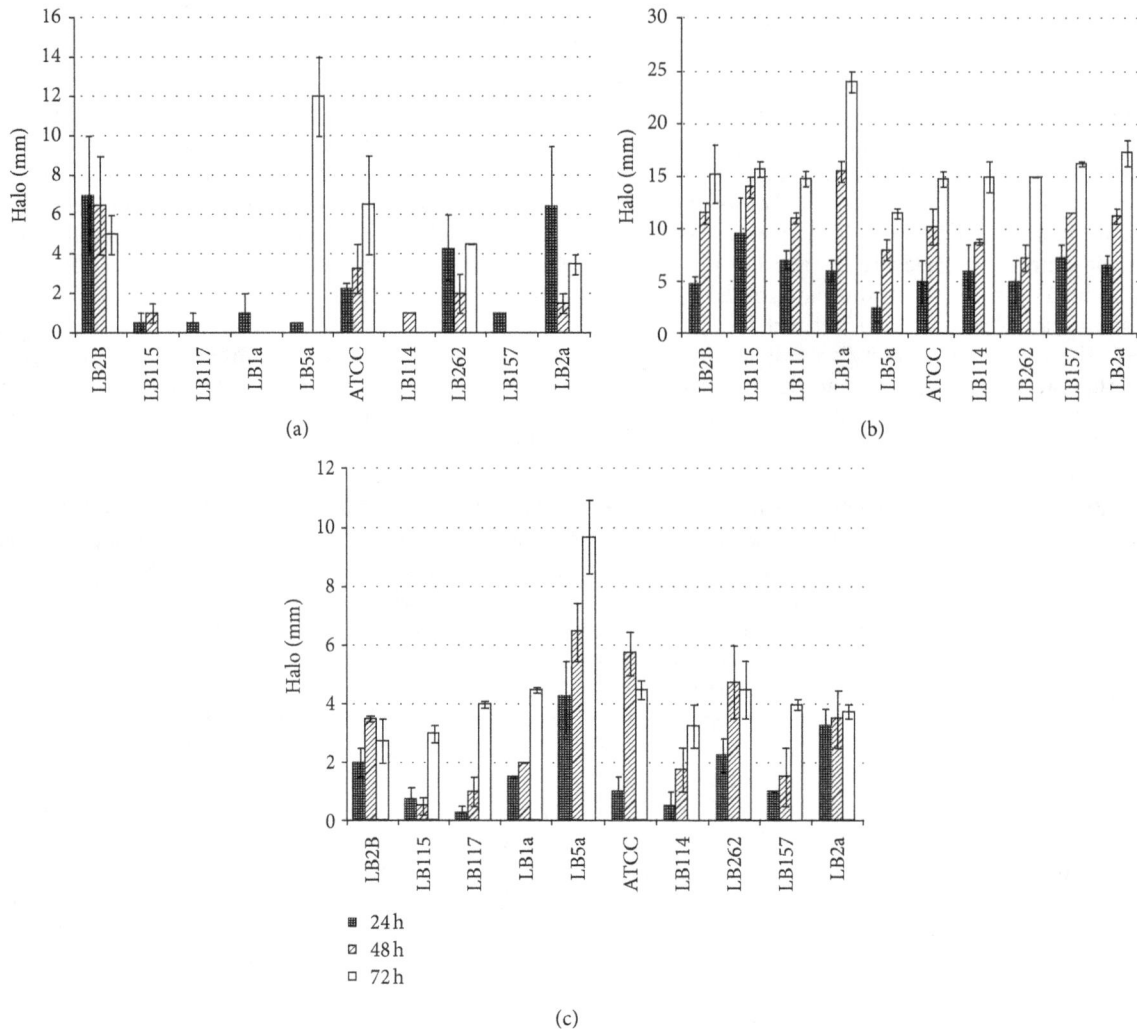

FIGURE 1: Average halo diameter in plates for LB1a and LB5a *Bacillus subtilis* strains: (a) amylase, (b) protease, and (c) lipase.

of both strains (Figure 4(b)), it is possible to see consistently higher results for LB1a than those found with LB5a during the whole process. Only close to the end of the fermentation process, after the peak was reached, the difference between the culture media became insignificant, especially due to the considerable reduction in the activity in both media.

The lipolytic activity values remained nearly constant for both strains throughout the studied period (Figure 4(c)). However, the activity in cassava wastewater was always lower than in the synthetic medium during the whole fermentation process for both microbial cultures. The reduction was more intense for LB5a, which showed approximately 50% decreased activity in the cassava wastewater when compared with the synthetic medium. In contrast, the reduction for LB1a oscillated at about 30%.

3.4. Enzymatic Activity in Bench Bioreactor. As seen in Figure 5, there is a strong increase in amylase activity in the culture medium at the first hours, from near to 30 U at 12 h to values around 120 U at 36 h. After that, the activity was gradually reduced until the end of fermentation. The

measurement of amylolytic enzymes in the foam recovered from bioreactor did not show significant difference between the values found in foam and culture medium. For protease, the curves started from 10 U at the fermentation initiation to values near 65 U between 48 and 60 h. After this maximum period, the value dropped to 35 U in 12 h. It was possible to observe after 36 h a significant difference between the enzymatic activity in the foam and the culture medium, where the values found in foam was always higher. For lipase, the values were very low since beginning and continued very low until the end of fermentation. There was no significant difference between foam and culture medium.

4. Discussion

4.1. Culture Selection in Solid Medium. The halo formation in all experiments indicated that all tested strains produce extracellular enzymes which were able to hydrolyze their respective substrates.

In general way, in amylase experiment the colonies displayed substantial development, which was higher than

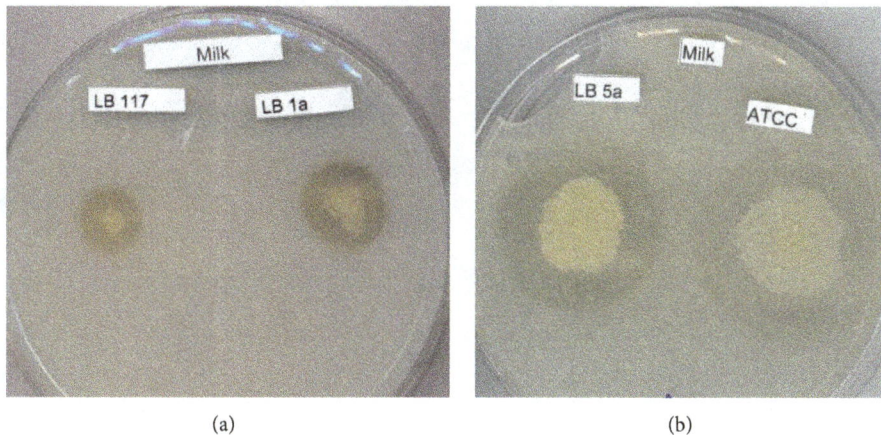

(a)   (b)

FIGURE 2: Milk agar plates after 24 h (a) and 72 h (b) inoculated with *Bacillus subtilis* strains LB117 and LB1a (a) and LB5a and ATCC21332 (b); the proteolysis halos are evident.

that found in the lipase and protease tests, thereby it was not possible to measure the halo of all strains at all times. The high speed of growth of the colonies could be explained by the fact that carbohydrates are the preferred carbon source for microorganisms of the genus *Bacillus* [12, 18]. Likewise, considering that the ultimate goal was the production of enzymes in cassava wastewater, the results of starch hydrolysis of the LB5a strain represented a critical finding. Considering the best results at 72 h, LB5a strain was selected to the following step.

The smallest protease halo shown by LB5a was opposite to the behavior observed in the amylase and lipase activity experiments (Figure 1). In the other hand, the LB1a strain, which had poor amylolytic and moderate lipolytic activities, showed the largest halo formation in the protease activity test. Considering the best results at 72 h, LB1a strain was selected for the activity experiments in liquid medium.

For lipase, the LB5a strain showed the best results in all the measurements (Figure 1(c)), which allow us to infer a good activity. This is in accordance with previous works that relate the use of solid media supplemented to emulsified triglycerides as standard methodology for the selection of lipase-producing microorganisms [17, 19]. In the other hand, although there was no strong positive relationship between the size of a halo in solid medium and the production of alkaline lipase in liquid medium, there is a greater tendency for microorganisms that showed a bigger halo to have a greater lipolytic activity in liquid medium [17]. Thus, the LB5a strain was selected for additional experiments. It is important to observe that the reduction of halo diameter after 48 h in some strains (Figure 1(c)) is related to faster growth of colonies rather than the formation of halos.

In addition, both selected strains revealed good biosurfactant production, in which the production from LB5a was higher than LB1a, as reported by Nitschke et al. [11]. The enzyme production coupled with biosurfactant could make these strains economically interesting, since they could be suitable for an integrated production process.

4.2. Enzymatic Activity in Synthetic Liquid Medium. All experiments aiming to detect enzymatic activity in synthetic liquid medium confirmed the data found in the solid medium screening. The presence of enzymatic activity for the two strains previously selected was confirmed.

The intense increase of amylolytic activities was similar to LB1a and LB5a (Figure 2(a)). These demonstrate that the studied bacteria were capable of using cassava starch as carbon source since it was the only C source available in the medium. Moreover, the production of enzymes occurred in the first hours of fermentation as expected. However, the different times found to the apex, 10 and 16 hours for LB1a and LB5a, respectively, showed that two strains present different abilities in the use of this substratum. Although the process happened more quickly in this experiment in comparison with other studies, the intense increase in activity at the start of the fermentation process is consistent with the data found in the collected works [6]. In this study the peak of enzymatic production ranged between 20 and 36 h, depending on the conditions used. The intense reduction of activity just after maximum could be interrelated to the depletion of the starch in the culture medium.

There is an apparent discrepancy of these results when compared with those found in solid medium experiments (Figure 1(a)) since LB1a growth produced only small amylase halos at 24 h, while no halos were visualized at 48 and 72 h. However, these data may not indicate a low enzymatic activity. This supposition is corroborated by the considerable culture growth on plates with starch-rich culture medium; thus, the colony spread faster than the hydrolysis halo. In other words, this evidence indicates that the cell development was strong because this medium provided better conditions to the microorganisms' survival and growth. This outcome demonstrates that both strains were highly adapted to the use of cassava starch as a carbon source.

Although the protease activity in solid medium was consistently higher for the LB1a, in liquid medium only at maximum activity there was a significant advantage of this strain.

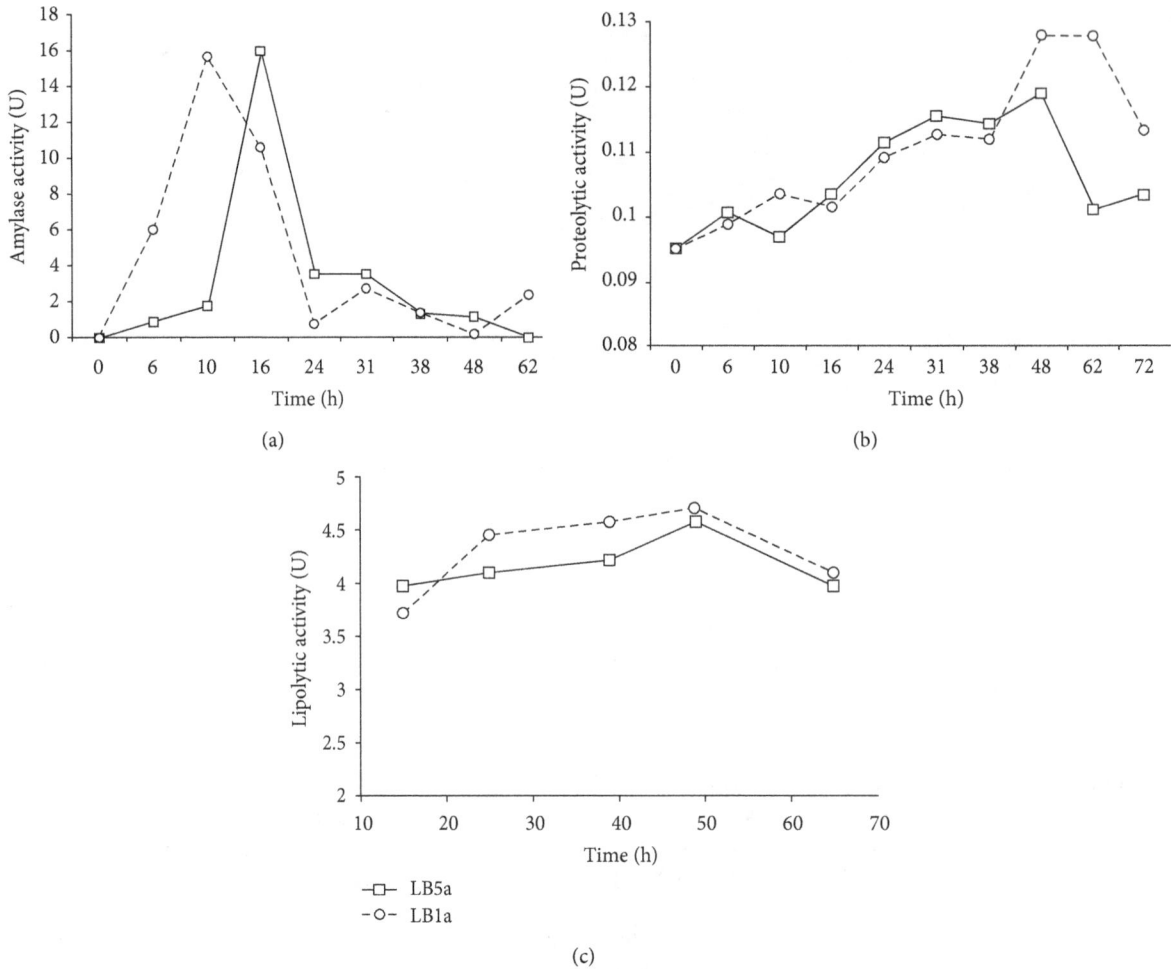

(a)

(b)

(c)

FIGURE 3: Enzymatic activity kinetics of LB5a and LB1a strains in a synthetic medium supplemented: (a) amylase, (b) protease, and (c) lipase.

Nevertheless, the activity can be considered low. The reduced activity throughout the process to both strains (Figure 2(b)) might be related to the fact that the cassava wastewater is not necessarily an inductor since the C source used was glucose and there was no protein substrate. Moreover, the strains were previously selected as good biosurfactant producers [11], and as the degradation of the biosurfactant produced by *Bacillus subtilis* 21332 at late fermentation time may have a direct liaison with the protease activity of the medium [18]; thus, the low activities found were probably a condition for their high biosurfactant production. This supposition is supported by the maintenance of the maximum activity for long periods at the end of fermentation, which is also related to the results obtained using the solid medium (Figure 1(b)). Finally, the accentuate decrease near the end of fermentation might be related to the total depletion of carbon sources and the beginning of the death phase of the cellular growth.

Very low levels of lipase activity even in the presence of inducer show that the lines are not good producer of this enzyme.

4.3. Enzymatic Activity in Cassava Wastewater. The production of biosurfactants by *Bacillus subtilis* in cassava wastewater was reported by Nitschke et al. [11], Barros et al. [12], and Barros et al. [20]. However, the use of the combination of these microorganisms and substrate for the enzyme production was not previously described. Thus, to evaluate this potential, the same strains were used in synthetic medium essays—LB1a and LB5a. As seen in all experiments, both strains were capable of producing extracellular enzymes in cassava wastewater. However, the use of a complex medium instead of a synthetic one affected the results significantly.

The amylose activity of cultures grown in cassava wastewater was higher than at in the synthetic medium (Figure 3(a)). The difference also was found when comparing the activity levels of the two strains. Hence, unlike the synthetic medium, in which similar values were found for both microorganisms, in cassava wastewater the LB1a strain showed values approximately 40 times higher than in the synthetic medium, while the LB5a values were only 10 times higher. The difference between peaks of enzymatic activity might be related to the results found in the synthetic media. For instance, LB5a showed short period at high level of activity in synthetic medium while LB1a presented a longer time at high levels (Figure 2(a)). Besides, in synthetic medium, the LB1a reached high activity levels before LB5a.

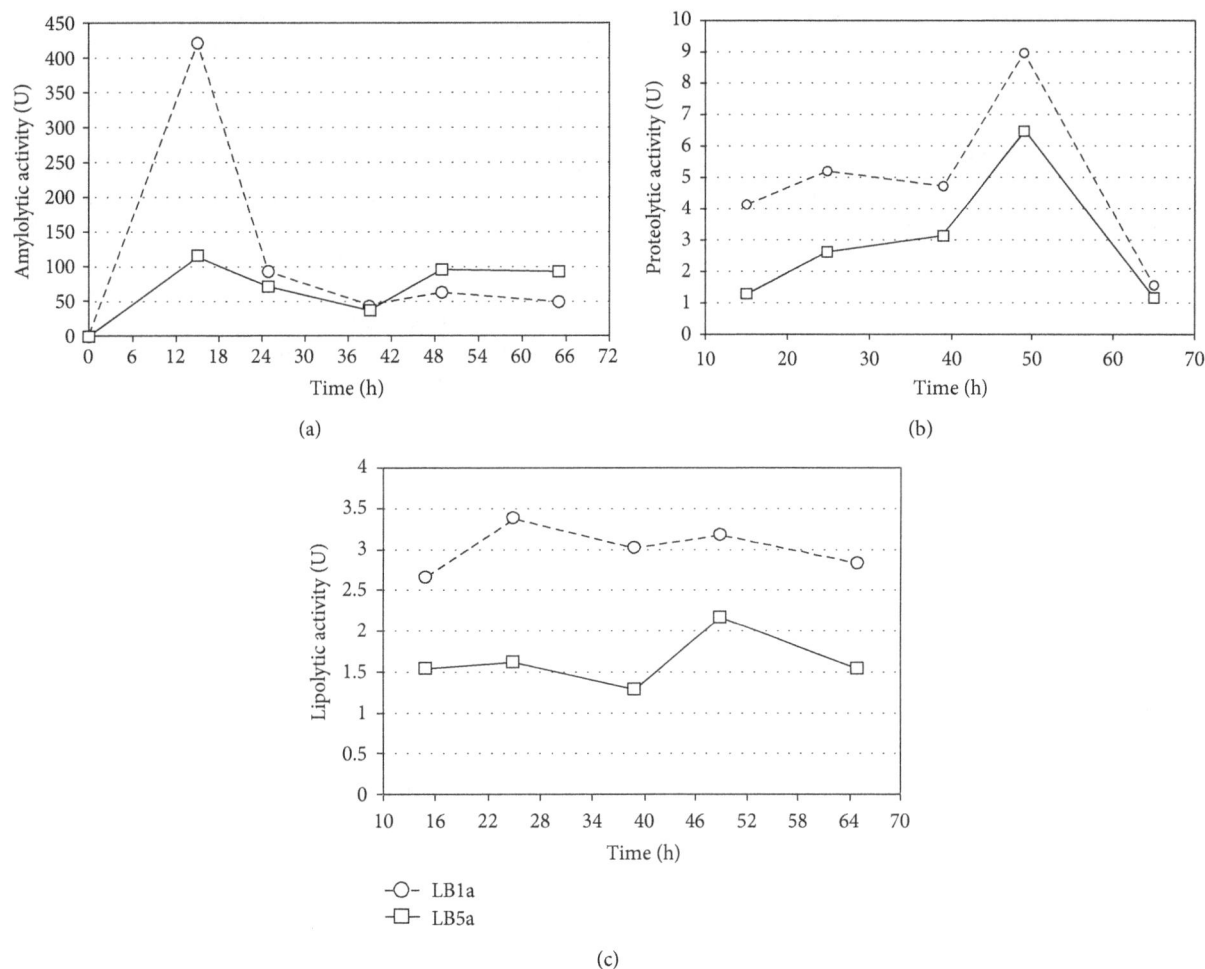

FIGURE 4: Enzymatic activity kinetics of LB5a and LB1a strains grown in glass flasks using cassava wastewater as culture medium: (a) amylase, (b) protease, and (c) lipase.

Cassava wastewater has a combination of sugars of high molecular weight, mainly soluble starch, and low molecular weight sugars, primarily saccharose, fructose, and glucose. The latter were preferentially consumed by the cultures. Consequently, after the depletion of the lower molecular weight sugars, the cultures started to produce amylases to provide glucose for their metabolism. One indication of this behavior was shown at previous experiments [12], which were was performed fermentations with the LB5a strain in cassava wastewater using a bioreactor. They detected a rise in reducing sugars concentration in the first 12 h of the fermentation, while the total sugars concentration had a sharp decline. The elevation of low molecular weight sugars is from the hydrolysis of starch.

The higher production of protease when compared with synthetic media could be explained by the waste composition, since it is very rich in important nutrients for *Bacillus* development [12]. The behavior of kinetics is very similar to that found in this work for synthetic medium.

Finally, the low lipolytic activity in both strains might be explained by the absence of an inductor in cassava wastewater beyond the facts already discussed earlier.

4.4. *Enzymatic Activity in Bioreactor.* The kinetic curves of amylase activity in bioreactor are consistent with literature data [4], where the peak of enzyme production is around 24 and 36 hours according to the variation in the conditions used, although the composition of the medium is different. The initial increase in amylase activity is obviously due to the presence of starch in the composition of the culture medium. Subsequently, starch hydrolysis released glucose to the microbial culture. Thus, this increase is not completely simultaneous with the exponential growth phase. Cell growth in the early hours of the fermentation is supported by the low molecular weight sugars like glucose, fructose, and sucrose. The effective use of glucose from starch hydrolysis by the microbial culture occured only after depletion of culture medium glucose content. Similar results are presented in cassava by Nitschke [18] and Barros et al. [12].

The enzyme activity also was measured in the foam removed from the fermenter. This aspect was considered important, since in this case, the primary recovery system of the biosurfactant is based on the principle of the bubble column, a method used for recovery and purification of surfactants [21] and enzymes [16]. Thus, different activities

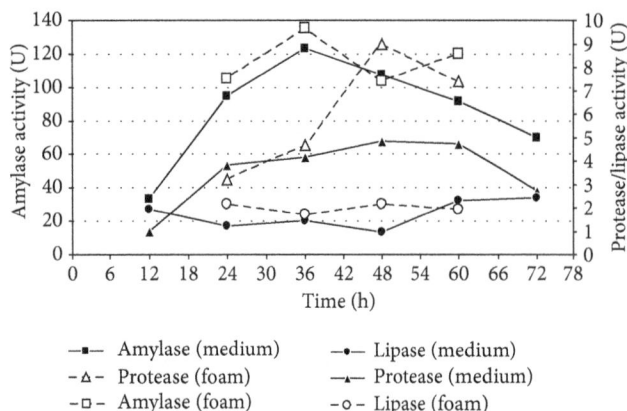

FIGURE 5: Enzymatic activity kinetics of LB5a and LB1a strains grown in bioreactor using cassava wastewater as culture medium: (a) amylase, (b) protease, and (c) lipase.

were expected between the activities found in the middle and in the foam. However, considering the activity found in the foam, despite having higher activities in most of sampling time (except for the last measurement); there, were no significant differences between the values found in the fermented medium. Thus, recovery was not detected for these categories of enzymes, because this recovery could be easily evidenced by a higher activity in the foam, which was not the case. In fact, there is no a similar behavior for all proteins recovery in the recovery using bubble column [21].

As in experiments in glass flasks, the increase of proteolytic activity was gradual, reaching a maximum between 48 and 60 hours. The low activity even at its apex may be related to the absence of protein that could be an inducer. Furthermore, as discussed before, the LB5a was previously selected as good biosurfactant-producing strain, which could be related to a low proteolytic activity. Although their values are increasing since the early hours, the proteolytic activities do not appear to be directly linked to microbial growth, because their maximum values are reached only about 24 hours after the end of exponential growth phase.

The proteases apparently are recovered by foam. In this case, the enzyme activities were significantly higher in the foam, especially at peak times, with values about twice greater in the foam than in the culture medium. Despite being significant, this enhancement factor was smaller than that for the surfactin, whose concentrations in the foam can reach up to 60 times the concentration in the culture broth. Regarding the lipase, as the flask experiments, the activity was low at all times, both in the culture medium and in the foam. There was no significant difference between the values of activity found in both.

5. Conclusion

Among the 10 strains of *Bacillus subtilis* tested for the production of enzymes, LB1a and LB5a were identified as potential sources of amylases and proteases. The production of these two enzymes in cassava wastewater was better than

that in the synthetic medium, showing the great potential of this agroindustrial residue for use as an alternative substrate. Similarly, the use of cassava wastewater as a substrate for enzyme production by *Bacillus* has not been reported in the literature until now. Additionally, the protease produced by the strain was carried by the foam produced during fermentation, which provides a simpler and economically feasible recovery.

Our results indicate the possibility of an integrated process for obtaining these enzymes and biosurfactants, which were also produced by these strains during the fermentation of cassava residue, that could greatly increase the economic viability of these strains.

References

[1] M. Schallmey, A. Singh, and O. P. Ward, "Developments in the use of *Bacillus* species for industrial production," *Canadian Journal of Microbiology*, vol. 50, no. 1, pp. 1–17, 2004.

[2] F. Kunst and G. Rapoport, "Salt stress is an environmental signal affecting degradative enzyme synthesis in *Bacillus subtilis*," *Journal of Bacteriology*, vol. 177, no. 9, pp. 2403–2407, 1995.

[3] Z. Konsoula and M. Liakopoulou-Kyriakides, "Co-production of α-amylase and β-galactosidase by *Bacillus subtilis* in complex organic substrates," *Bioresource Technology*, vol. 98, no. 1, pp. 150–157, 2007.

[4] M. J. Syu and Y. H. Chen, "A study on the α-amylase fermentation performed by *Bacillus amyloliquefaciens*," *Chemical Engineering Journal*, vol. 65, no. 3, pp. 237–247, 1997.

[5] A. Pandey, P. Nigam, C. R. Soccol, V. T. Soccol, D. Singh, and R. Mohan, "Advances in microbial amylases," *Biotechnology and Applied Biochemistry*, vol. 31, no. 2, pp. 135–152, 2000.

[6] C. E. S. Teodoro and M. L. L. Martins, "Culture conditions for the production of thermostable amylase by *Bacillus* sp," *Brazilian Journal of Microbiology*, vol. 31, pp. 298–302, 2000.

[7] W. C. A. Nascimento, C. R. Silva, R. V. Carvalho, and M. L. L. Martins, "Otimização de um meio de cultura para a produção de proteases por um *Bacillus* sp," *Ciência e Tecnologia de Alimentos*, vol. 27, pp. 417–421, 2007.

[8] J. R. Martín, M. Nus, J. V. S. Gago, and J. M. Sánchez-Montero, "Selective esterification of phthalic acids in two ionic liquids at high temperatures using a thermostable lipase of Bacillus thermocatenulatus: a comparative study," *Journal of Molecular Catalysis B*, vol. 52-53, pp. 162–167, 2008.

[9] E. Lesuisse, K. Schanck, and C. Colson, "Purification and preliminary characterization of the extracellular lipase of *Bacillus subtilis* 168, an extremely basic pH-tolerant enzyme," *European Journal of Biochemistry*, vol. 216, no. 1, pp. 155–160, 1993.

[10] T. Eggert, U. Brockmeier, M. J. Dröge, W. J. Quax, and K. E. Jaeger, "Extracellular lipases from *Bacillus subtilis*: regulation of gene expression and enzyme activity by amino acid supply and external pH," *FEMS Microbiology Letters*, vol. 225, no. 2, pp. 319–324, 2003.

[11] M. Nitschke, C. Ferraz, and G. M. Pastore, "Selection of microorganisms for biosurfactant production using agroindustrial wastes," *Brazilian Journal of Microbiology*, vol. 35, no. 1-2, pp. 81–85, 2004.

[12] F. F. C. Barros, A. N. Ponezi, and G. M. Pastore, "Production of biosurfactant by *Bacillus subtilis* LB5a on a pilot scale using cassava wastewater as substrate," *Journal of Industrial Microbiology and Biotechnology*, vol. 35, no. 9, pp. 1071–1078, 2008.

[13] S.-F. Lin, C.-M. Chiou, and Y.-C. Tsai, "Effect of triton X-100 on alkaline lipase production by *Pseudomonas pseudoalcaligenes* F-111," *Biotechnology Letters*, vol. 17, no. 9, pp. 959–962, 1995.

[14] J. L. Giongo, *Caracterização e aplicação de proteases produzidas por linhagens de Bacillus sp [thesis]*, Universidade Federal do Rio Grande do Sul, 2006.

[15] G. A. Macedo, G. M. Pastore, H. A. Sato, and Y. K. Park, *Bioquímica Experimental de Alimentos*, Editora Varela, São Paulo, Brazil, 2005.

[16] J. N. Dos Prazeres, J. A. B. Cruz, and G. M. Pastore, "Characterization of alkaline lipase from *Fusarium oxysporum* and the effect of different surfactants and detergents on the enzyme activity," *Brazilian Journal of Microbiology*, vol. 37, no. 4, pp. 505–509, 2006.

[17] J. N. Prazeres, *Produção, purificação e caracterização da lipase alcalina de Fusarium oxysporum [thesis]*, Universidade Estadual de Campinas, 2006.

[18] M. Nitschke, *Produção e caracterização de biossurfatante de Bacillus subtilis utilizando manipueira como substrato [thesis]*, Universidade Estadual de Campinas, 2004.

[19] F. Cardenas, E. Alvarez, M. S. De Castro-Alvarez et al., "Screening and catalytic activity in organic synthesis of novel fungal and yeast lipases," *Journal of Molecular Catalysis B*, vol. 14, no. 4–6, pp. 111–123, 2001.

[20] F. F. C. Barros, A. P. Dionísio, J. C. Silva, and G. M. Pastore, "Potential uses of cassava wastewater in biotechnological processes," in *Cassava: Farming, Uses, and Economic Impact*, C. M. Pace, Ed., pp. 33–54, Nova Science, New York, NY, USA, 2011.

[21] R. F. Perna, *Fracionamento de surfactina em coluna de bolhas e espuma [M.S. thesis]*, Universidade Estadual de Campinas, Campinas, Brazil, 2010.

Production of a Thermostable and Alkaline Chitinase by *Bacillus thuringiensis* subsp. *kurstaki* Strain HBK-51

Secil Berna Kuzu,[1] Hatice Korkmaz Güvenmez,[1] and Aziz Akin Denizci[2]

[1] *Biotechnology & Molecular Biology Division, Department of Biology, Cukurova University, 01330 Adana, Turkey*
[2] *Research Institute for Genetic Engineering and Biotechnology, The Scientific and Technical Research Council of Turkey (TUBITAK), Marmara Research Center Campus, Gebze-Kocaeli 41470, Turkey*

Correspondence should be addressed to Hatice Korkmaz Güvenmez, hkorkmaz@cu.edu.tr

Academic Editor: Triantafyllos Roukas

This paper reports the isolation and identification of chitinase-producing *Bacillus* from chitin-containing wastes, production of a thermostable and alkaline chitinasese, and enzyme characterization. *Bacillus thuringiensis* subsp. *kurstaki* HBK-51 was isolated from soil and was identified. Chitinase was obtained from supernatant of *B. thuringiensis* HBK-51 strain and showed its optimum activity at 110°C and at pH 9.0. Following 3 hours of incubation period, the enzyme showed a high level of activity at 110°C (96% remaining activity) and between pH 9.0 and 12.0 (98% remaining activity). Considering these characteristics, the enzyme was described as hyperthermophile-thermostable and highly alkaline. Two bands of the enzyme weighing 50 and 125 kDa were obtained following 12% SDS-PAGE analyses. Among the metal ions and chemicals used, Ni^{2+} (32%), K^+ (44%), and Cu^{2+} (56%) increased the enzyme activity while EDTA (7%), SDS (7%), Hg^{2+} (11%), and ethyl-acetimidate (20%) decreased the activity of the enzyme. *Bacillus thuringiensis* subsp. *kurstaki* HBK-51 is an important strain which can be used in several biotechnological applications as a chitinase producer.

1. Introduction

Chitin, a linear β-(1,4)-linked *N*-acetylglucosamine (GlcNAc) polysaccharide, is the main structure component of the fungal cell wall and the exoskeletons of invertebrates, such as insects and crustaceans. Chitinase plays a variety of important roles in these organisms ranging from nutrition to defense and control of ecdysis in insects. The importance of chitinases in many biological processes makes their inhibitors important targets for potential antifungal and insecticidal agents as well as antimalarial agents [1].

It is one of the most abundant naturally occurring polysaccharides and has attracted tremendous attention in the fields of agriculture, pharmacology, and biotechnology. Chitin is the second most abundant component of biomass on earth [2]. This linear polymer can be hydrolyzed by bases, acids or enzymes, such as lysozyme, some glucanases, and chitinase. Chitinases (EC 3.2.1.14), essential enzymes catalyzing the conversion of chitin to its monomeric or oligomeric components (low-molecular-weight products), have been found in a wide range of organisms, including bacteria, plants, fungi, insects, and crustaceans. Because chitin is not found in vertebrates, it has been suggested that inhibition of chitinases may be used for the treatment of fungal infections and human parasitosis [3]. Biological control, using microorganisms to repress plant disease, offers an alternative environmentally friendly strategy for controlling agricultural phytopathogens [4]. The production of inexpensive chitinolytic enzymes is an important element in the utilization of shellfish wastes that not only solves environmental problems but also promotes the economic value of marine products, so chitinases have been studied and purified from many microorganisms, and their enzymatic properties have been investigated [5]. Global annual recovery of chitin from the processing of marine crustacean waste is estimated to be around 37.300 metric tons [6].

2. Materials and Methods

2.1. Chemicals. Chitin-Azure was (Sigma-Aldrich), PMFS, Protein Marker (BSA), CBB R-250, DMSO and the others were purchased from Merck (Germany).

2.1.1. Isolation of Bacteria. Crabs, campus soil, and compost were mixed and held to approximately 30 days outside (exposure sun, rain or etc.). After this period, sterile distilled water was added to material and homogenized. 10 mL supernatant wae transferred to glass tube and incubated at 80°C for 10 min to eliminate vegetative forms of bacteria. 1 mL sample inoculated to 10 mL LB and incubated at 37°C for 6 hours and then sample was diluted $10^{-6} - 10^{-7}$ fold with freshly prepared LB, then streaked on N1 agar plate. After incubation at 37°C overnight, single colonies were selected (550 colonies in total) and stock cultures prepared [7].

2.2. Preparation of Colloidal Chitin. Preparation of colloidal chitin was performed from commercial chitin (C9752, Sigma-Aldrich Co, USA) according to Wen et al. [3] with minimal modifications. Fivegrams of chitin powder were added into 60 mL of concentrated HCL and left in refrigerator (at 4°C) overnight with vigorous stirring. The mixture was added to 2 litres of ice-cold Et-OH (95%) with rapid stirring and kept overnight at room temperature (at 22°C). The precipitate was collected by centrifugation at $6,000 \times g$ (4°C) for 25 min. The precipitate was washed with sterile distilled water until the colloidal chitin became neutral (pH 7.0) and then stored and at 4°C for further applications.

2.3. Screening Chitinase Producer Strains. All strains (total 550) were selected from soil including chitin wastes, tested one by one for chitinase activity on the CHDA (chitinase-detection agar). Chitinase producer strains were determined after 3–5 days at 37°C by visualizing the clear zone formed surrounding of the colonies on the CHDA plate. CHDA agar, used first detection of chitinase positive strains and preparation of stock culture, by adding 10% g of colloidal chitin and 20 g of agar in M9 medium containing (g/L); 0.65 $NaHPO_4$, 1.5 KH_2PO_4, 0.5 NH_4Cl, 0.12 $MgSO_4$, 0.005 $CaCl_2$, 0.25 NaCl, pH = 6.5 [3, 8].

2.4. Enzyme Production. HBK-51 strain (*Bacillus thuringiensis* subsp. *kurstaki*) was used as chitinase producer. 1 mL of HBK-51 strain fresh culture was inoculated into 10 mL of LB medium (containing 1% colloidal chitin as the sole source of C and N) and incubated overnight at 37°C with shaking (150 rev min^{-1}). The culture was transferred into a 2 litres glass bottle containing 500 mL M9-Medium which was supplemented with 1% colloidal chitin and incubated at 37°C on a rotary shaker at 150 rev min^{-1} for 3 days [3]. The culture was centrifuged for 20 min at $9000 \times g$, at 4°C and supernatant was separated from cell debris. Then

supernatant was filtered (with filter paper) and precipitated with Et-OH (ethyl-alcohol) [9]. All experiments were done with this enzyme preparation.

2.5. Purification of the Chitinase. Proteins in the culturesupernatant fluid were precipitated with Et-OH (ethyl-alcohol) at −33°C overnight. Et-OH (ethyl-alcohol) was added 70% of the original volume of supernatant. The precipitate was recovered by centrifugation ($12,000 \times g$ (4°C) for 30 min.), dissolved in a small amount of 0.1 M Na-phosphate buffer (pH 7.0), and stored at 4°C [10, 11].

2.6. Measurement of Enzyme Activity. Chitinase activity was measured with chitin-azure as a substrate [12]. Enzyme solution (0.5 mL) was added to substrate solution (0.5 mL), which contained 1.0% chitin azure in a sodium acetate buffer (100 mmol/L, pH 7.0), and the mixture was incubated at 50°C for 30 min. After centrifugation ($12,000 \times g$, 15 min, at 4°C), enzyme activity was measured at 595 nm absorbance using spectrophotometer. One unit of chitinase was the amount of enzyme that produced an increase of 0.01 in the absorbance [12, 13].

2.7. pH and Temperature Optima. The optimum pH and temperature were determined by incubation in a buffer at different pH values (3.0–10.0) and temperatures (30–120°C) under standard assay conditions using chitin-azure as the substrate. The buffers used were 50 mM sodium acetate buffer (pH: 3.0–6.0), 50 mM Tris-HCl buffer (pH: 7.0–9.0), and 50 mM Glycine-NaOH buffer (pH: 10.0–12.0). The optimum temperature for enzyme activity was measured under standard assay in the range of 30–120°C with intervals of 10°C [14].

2.8. Thermal Stability. The effect of the temperature on the stability of chitinolytic enzyme was determined by exposure of the enzyme solution in 50 mM Tris-HCl buffer, pH: 9.0 (optimum pH) to different temperatures (30–110°C) for 3 h. The residual activity was then assayed under standard conditions using chitin-azure as the substrate [14, 15].

2.9. pH Stability. For pH stability assay, the enzyme solution was preincubated for 3 h at room temperature in the buffers of various pH values (pH: 5.0–12.0) and then the residual activity was determined under standard conditions [14, 15].

2.10. Plasmid Curing. All bacteria which produced chitinase were tested with EtBr for plasmid elimination according to Hardy [16].

2.11. Identification of Bacillus sp. Criteria used for classification and identification of *Bacillus* strain HBK-51 were based on morphological [7], physiological and biochemical tests (Table 1) [17], fatty acid analyses (FAME) (Sherlock-MIDI Automated Microbial Identification System version 4.0, MIDI inc., Newark, DE), and 16S RNA/DNA sequence analyses [14, 18, 19].

TABLE 1: Morphological and physiological characteristics of chitinase-producing bacteria strain HBK-51.

Morphological characteristics	
Form	Rod
Gram stain	Positive
Spore	Terminal spore forming
Capsule	Negative
Motility	Positive
Physiological characteristics	
Catalase	Positive
Hydrolysis of starch	Positive
β-haemolysis	Positive
Cimons' Citrate	Positive
Hydrolysis of gelatin	Negative
Hydrolysis of l lecithin	Negative
Indol formation	Negative
VP-Test	Negative
Urease	Negative

TABLE 2: Influence of chemicals on chitinase activity.

Chemical	Con.	Act. (%)
control		100
DMSO	1%	92
	5%	92
EDTA	1 mM	95
	5 mM	93
Ethyl acetimidate	1 mM	90
	5 mM	80
SDS	1 mM	97
	5 mM	93
Phenol Gliaksol	1 mM	92
	5 mM	104
1,10-Phenanthroline	1 mM	90
	5 mM	100
N-Ethylmaleimide	1 mM	106
	5 mM	89
PMSF	1 mM	108
	5 mM	81
Urea	1 mM	106
	5 mM	80
Na-Sulphite	1 mM	98
	5 mM	96
$MgCl_2$	1 mM	100
	5 mM	90
$NaCl_2$	1 mM	97
	5 mM	93
$BaCl_2$	1 mM	101
	5 mM	87
$FeCl_3$	1 mM	81
	5 mM	106
$MnCl_2$	1 mM	118
	5 mM	77
$CuCl_2$	1 mM	127
	5 mM	155
$CoCl_2$	1 mM	115
	5 mM	116
$NiCl_2$	1 mM	132
	5 mM	123
$ZnCl_2$	1 mM	118
	5 mM	106
KCl_2	1 mM	127
	5 mM	144
$CaCl_2$	1 mM	94
	5 mM	106
$HgCl_2$	1 mM	93
	5 mM	89

Con: Concentration, Act: Activity.
The enzyme was incubated with various ions and reagents at room temperature for 30 min., then chitinase activity was assayed under standard conditions. The enzyme activity without any additive was taken as 100%.

2.12. 16S rRNA Gene Sequence Comparison. The 16S rRNA gene was amplified by polymerase chain reaction (PCR) with forward primers:

27F 5′-AGAGTTTGATCMTGGCTCAG-3′ (8–27 position),

530F 5′-GTGCCAGAGCMGCCGCGGTAA-3′ (515–533 position)

and reverse primers:

1525R 5′-AAGGAGGTGWTCCRCC-3′ (1541–1525 position)

1100R 5′-AGGGTTGCGCTCGTTG-3′ (1115–1100 position).

The amplified PCR product was sequenced by the Beckman Coulter-CQ 8800 model sequence analyzer with their methods. The 16S rDNA sequence was aligned with other 16S rDNA bacterial sequence obtained from GenBank by basic local alignment search tool (BLAST) program [20].

2.13. The Influence of Chemicals (Metal Ions, Chelators, and Detergents). The chemicals were used in two different concentrations (Table 2). These chemicals' effect on chitinolytic activity was determined by incubating 0.5 mL enzyme with 1–5 mM EDTA, PMSF, 1,10-Phenanthroline, ethyl-acetimidate, Phenol Gliaksol, N-Ethylmaleimide, Urea, Na_2SO_3, NaCl, $CaCl_2$, $ZnCl_2$, $MgCl_2$, $BaCl_2$, $FeCl_3$, $MnCl_2$, $CuCl_2$, $CoCl_2$, $NiCl_2$, KCl, $HgCl_2$, and 1–5% SDS and DMSO for 30 min at room temperature after which the residual activity was measured with standard assay [21].

2.14. SDS-PAGE Analyses. The enzyme molecular weight was determined with sodium dodecyl sulphate polyacrylamide gel electrophoresis (PAGE) [22]. The gel was prepared using 5% stacking gel and 12% of separating gel (Acr : Bis = 29 : 1). After electrophoresis gel was stained with CBB-R

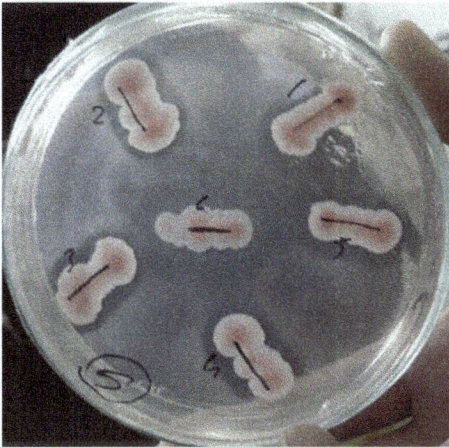

FIGURE 1: Chitinase activity on CHDA plate.

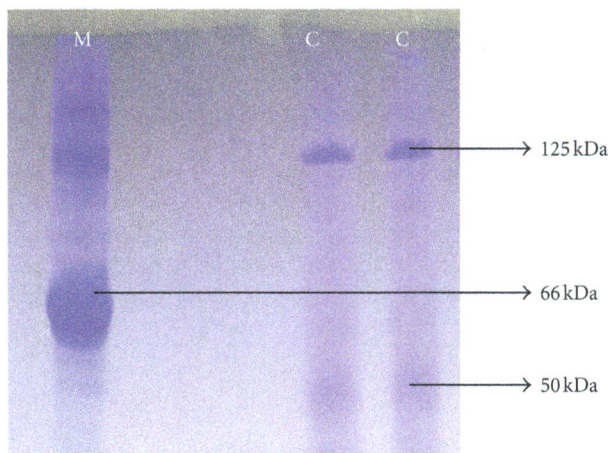

FIGURE 2: Chitinase molecular weight pattern of *Bacillus thuringiensis* subsp. *kurstaki* strain HBK-51. M: Protein Marker (BSA, 66 kDa), C: Chitinase.

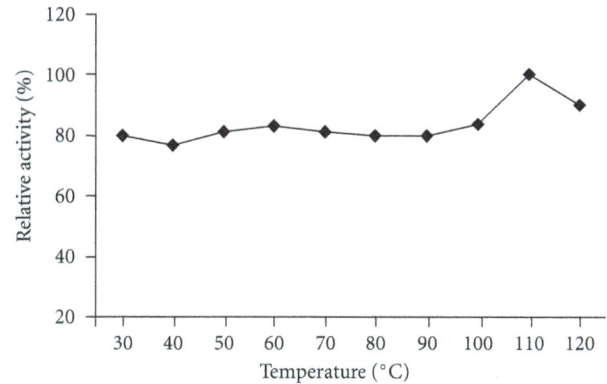

FIGURE 3: Effect of temperature on the chitinase activity.

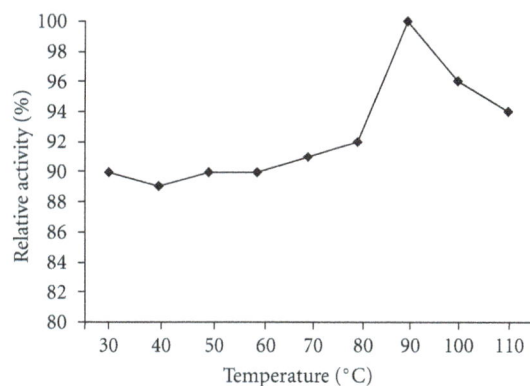

FIGURE 4: Thermal stability of chitinase.

250 then visualized by placing the gel to the Minibus Gel Apparatus, BSA (Sigma) was used as a MW marker.

3. Results and Discussion

Chitinase producing bacteria were isolated from chitin wastes on CHDA (chitinase detection agar, Figure 1) and chitinase was produced in the media containing 1% of colloidal chitin [3]. In total, 550 strains were selected from chitin wastes and tested for chitinolytic activity. After screening the strains for their ability to utilize chitin, 12 of them (2.18%) showed chitinolytic activity. 6 strains were showed the highest hydrolysis zone (4–10 mm) on CHDA agar. They were termed HBK-30, HBK-36, HBK-37, HBK-42, HBK-43, and HBK-51. HBK-51 strain was selected as chitinase producer.

The HBK-51 strain was identified according to morphological and physiological characteristics (are presented in Table 1) [17]. According to Bergey's Manual of Systematic Bacteriology [7], strain HBK-51 was classified as a bacteria belonging to the genus *Bacillus*. Further sequence analysis of the gene encoding 16S rRNA and fatty acid analyses (FAME, data not shown) confirmed the isolate as being *Bacillus* genus and according to BLAST confirmation, it was *Bacillus thuringiensis* subsp. *kurstaki*. The amplified 16S rRNA sequence was submitted to GenBank for possible identification. The result showed 99.00% identity with *Bacillus thuringiensis* subsp. *kurstaki*. The accession number is EU153549. Chitinolytic activity of HBK-51 on CHDA is shown in Figure 1.

According to the plasmid-curing tests, the gene of encoding chitinase of *Bacillus thuringiensis* subsp. *kurstaki* HBK-51 was located on chromosomal.

3.1. pH and Temperature Optima. The pH effect on the chitinase activity was determined using three buffer systems at various pH values. The HBK-51 chitinase was active at broad range of pH (3.0–10.0) but had optimum activity at pH 9.0 (data shown Figure 5), when assayed with chitin azure as a substrate. On the other hand, enzyme was active from 30 to 120°C and exhibited maximum activity at 110°C (Figure 3). The enzyme was stable (after 3 hours incubation period) at 30–120°C and pH 9.0–12.0 and generally, protected original activity approximately 92.4% (Figure 4) and 98% (Figure 6), respectively, at 90°C chitinase was 100% active and at 100–110°C had shown 96% retain activity.

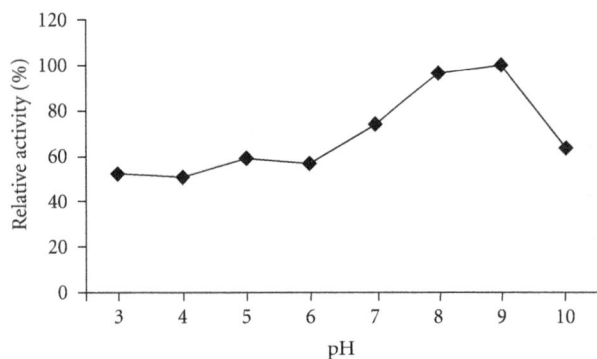

FIGURE 5: Effect of pH on chitinase activity.

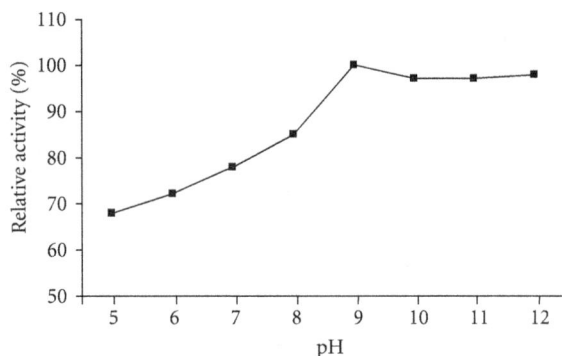

FIGURE 6: pH stability of chitinase.

Therefore, HBK-51 chitinase was called as thermostable at high temperatures. Siwayaprahm et al. [23] reported that for *Bacillus circulans* No. 4.1. recombinant chitinase, the optima pH and temperature were 7.0 and 45°C, respectively, and it was stable in the pH range of 5.0–9.0 and at temperatures up to 50°C. Lee et al. [24] reported that *Bacillus* sp. DAU101 chitinase had optimum activity pH 7.5 and 60°C, Wen et al. [3] reported chitinase activity of *Bacillus* sp. NCTU2 had in the range of 50–60°C at pH 7.0. Kim et al. [14] reported *Serratia* sp. KCK chitinase had broad range of pH (5.0–10.0) with an optimum value 0f 8.0 and 40°C, respectively. Li et al. [25] reported *Bacillus cereus* strain CH2 had optimum activity at pH 7.1 and temperature at 40°C. According to these results, *Bacillus thuringiensis* subsp. *kurstaki* HBK-51 chitinase is a thermotolerant and alkaline enzyme. Guo et al. [26] reported that *Thermomyces lanuginosus* SY2 chitinase exhibited optimum catalytic activity at pH 4.5, 55°C and enzyme was stable at 50°C and its half-life time at 65°C was 25 min. On the other hand, they reported that the thermostable chitinase had major advantages over industrial catalysis for its high activity at high temperature. Thermostable chitinase could be useful for the chitin industry and biotechnological applications.

3.2. Influence of Chemicals and SDS-PAGE Analyses. The effects of various chemicals on enzyme activity were tested (Table 2) in the presence (at 1–5 mM concentrations) of Co^{2+}, Zn^{2+}, Ni^{2+}, K^+, Cu^{2+} enzyme activity was slightly

enhanced by 16%, 18%, 32%, 44%, 56%, respectively. Wen et al. [3] reported that *Bacillus* sp. NCTU2 chitinase activity was enhanced by ≈100% in the presence of 10 mM Ca^{2+}. In the presence of Hg^{2+} and Cu^{2+} (at 10 mM) enzyme activity lost by 95%. Yuli et al. [27] reported that effect of metal ions on chitinase activity was quite different. In the presence of EDTA, SDS, $HgCl_2$, ethyl=acetimidate the enzyme activity was partially inhibited by 7%, 7%, 11%, 20%, respectively, but (Table 2) other chemicals weakly affected enzyme activity. The enzyme molecular weight revealed as 125 and 50 kDa (Figure 2). Similar results were indicated by the other chitinase researchers by SDS-PAGE analyses: *Aeromonas* sp. chitinases was different in the range of 89–120 kDa [28], *Bacillus* sp. chitinase around 25–80 kDa [29], *Streptomyces* strain 68 chitinase, 25, 35, 67 and 200 kDa [30], *Bacillus* sp. NCTU2 chitinase 36.5 kDa [3], *Bacillus cereus* strain CH2 chitinase 15 kDa [25], and *Thermomyces lanuginosus* SY2 chitinase molecular size was estimated to be 48 kDa [26].

Chitinases have a broad spectrum of different industrial applications such as bio-control agents against plant pathogens (fungi) and insects (bio-insecticide friend with environment) and bioconversion of chitin waste to single-cell protein, Et-OH and fertilizer [14].

The thermostable chitinase had major advantages over industrial catalysis for its high activity at high temperature [26]. The properties of thermostable chitinase suggest that it could be useful for the chitin industry and biotechnological applications.

References

[1] J. N. Tabudravu, V. G. H. Eijsink, G. W. Gooday et al., "Psammaplin A, a chitinase inhibitor isolated from the Fijian marine sponge *Aplysinella rhax*," *Bioorganic and Medicinal Chemistry*, vol. 10, no. 4, pp. 1123–1128, 2002.

[2] H. S. Lee, D. S. Han, S. J. Choi et al., "Purification, characterization, and primary structure of a chitinase from *Pseudomonas* sp. YHS-A2," *Applied Microbiology and Biotechnology*, vol. 54, no. 3, pp. 397–405, 2000.

[3] C. M. Wen, C. S. Tseng, C. Y. Cheng, and Y. K. Li, "Purification, characterization and cloning of a chitinase from *Bacillus* sp. NCTU2," *Biotechnology and Applied Biochemistry*, vol. 35, no. 3, pp. 213–219, 2002.

[4] W. T. Chang, Y. C. Chen, and C. L. Jao, "Antifungal activity and enhancement of plant growth by *Bacillus cereus* grown on shellfish chitin wastes," *Bioresource Technology*, vol. 98, no. 6, pp. 1224–1230, 2007.

[5] S. Kudan and R. Pichyangkura, "Purification and characterization of thermostable chitinase from *Bacillus licheniformis* SK-1," *Applied Biochemistry and Biotechnology*, vol. 157, no. 1, pp. 23–35, 2009.

[6] S. A. Shaikh and M. V. Deshpande, "Chitinolytic enzymes: their contribution to basic and applied research," *World Journal of Microbiology and Biotechnology*, vol. 9, no. 4, pp. 468–475, 1993.

[7] J. D. Holt, *Manual of Determinative Bacteriology*, Williams & Wilkins, Baltimore, Md, USA, 1993.

[8] R. M. Cody, N. D. Davis, J. Lin, and D. Shaw, "Screening microorganisms for chitin hydrolysis and production of

ethanol from amino sugars," *Biomass*, vol. 21, no. 4, pp. 285–295, 1990.

[9] J. Frankowski, M. Lorito, F. Scala, R. Schmid, G. Berg, and H. Bahl, "Purification and properties of two chitinolytic enzymes of *Serratia plymuthica* HRO-C48," *Archives of Microbiology*, vol. 176, no. 6, pp. 421–426, 2001.

[10] R. A. K. Srivastava, "Purification and chemical characterization of thermostable amylases produced by *Bacillus stearothermophilus*," *Enzyme and Microbial Technology*, vol. 9, no. 12, pp. 749–754, 1987.

[11] J. Folders, J. Tommassen, L. C. van Loon, and W. Bitter, "Identification of a chitin-binding protein secreted by *Pseudomonas aeruginosa*," *Journal of Bacteriology*, vol. 182, no. 5, pp. 1257–1263, 2000.

[12] M. G. Ramírez, L. I. R. Avelizapa, N. G. R. Avelizapa, and R. C. Camarillo, "Colloidal chitin stained with Remazol Brilliant Blue R, a useful substrate to select chitinolytic microorganisms and to evaluate chitinases," *Journal of Microbiological Methods*, vol. 56, no. 2, pp. 213–219, 2004.

[13] N. J. R. Shih and K. A. McDonald, "Purification and characterization of chitinases from transformed callus suspension cultures of *Trichosanthes kirilowii* Maxim," *Journal of Fermentation and Bioengineering*, vol. 84, no. 1, pp. 28–34, 1997.

[14] H. S. Kim, K. N. Timmis, and P. N. Golyshin, "Characterization of a chitinolytic enzyme from *Serratia* sp. KCK isolated from kimchi juice," *Applied Microbiology and Biotechnology*, vol. 75, no. 6, pp. 1275–1283, 2007.

[15] J. Favaloro, R. Treisman, R. Kamen, L. Grosman, and K. Moldave, *Methods in Enzymology*, Academic Pres, New York, NY, USA, 1980.

[16] K. G. Hardy, *Plasmids*, The Practical Approach Serie, 1993.

[17] J. F. MacFaddin, *Biochemical Tests For Identification of Medical Bacteria*, Williams & Wilkins, Baltimore, Md, USA, 2000.

[18] W. P. Chen and T. T. Kuo, "A simple and rapid method for the preparation of gram-negative bacterial genomic DNA," *Nucleic Acids Research*, vol. 21, no. 9, p. 2260, 1993.

[19] A. A. Denizci, D. Kazan, E. C. A. Abeln, and A. Erarslan, "Newly isolated *Bacillus clausii* GMBAE 42: an alkaline protease producer capable to grow under higly alkaline conditions," *Journal of Applied Microbiology*, vol. 96, no. 2, pp. 320–327, 2004.

[20] S. F. Altschul, W. Gish, W. Miller, E. W. Myers, and D. J. Lipman, "Basic local alignment search tool," *Journal of Molecular Biology*, vol. 215, no. 3, pp. 403–410, 1990.

[21] N. N. Nawani and B. P. Kapadnis, "Production dynamics and characterization of chitinolytic system of *Streptomyces* sp. NK1057, a well equipped chitin degrader," *World Journal of Microbiology and Biotechnology*, vol. 20, no. 5, pp. 487–494, 2004.

[22] D. M. Bollag, S. J. Edelstein, and M. D. Rozycki, *Protein Methods*, Villey-Liss Pres, 1996.

[23] P. Siwayaprahm, M. Audtho, K. Ohmiya, and C. Wiwat, "Purification and characterization of a *Bacillus circulans* No. 4.1 chitinase expressed in Escherichia coli," *World Journal of Microbiology and Biotechnology*, vol. 22, no. 4, pp. 331–335, 2006.

[24] Y. S. Lee, I. H. Park, J. S. Yoo et al., "Cloning, purification, and characterization of chitinase from *Bacillus* sp. DAU101," *Bioresource Technology*, vol. 98, no. 14, pp. 2734–2741, 2007.

[25] J. G. Li, Z. Q. Jiang, L. P. Xu, F. F. Sun, and J. H. Guo, "Characterization of chitinase secreted by *Bacillus cereus* strain CH2 and evaluation of its efficacy against Verticillium wilt of eggplant," *BioControl*, vol. 53, no. 6, pp. 931–944, 2008.

[26] R. F. Guo, B. S. Shi, D. C. Li, W. Ma, and Q. Wei, "Purification and characterization of a novel thermostable chitinase from *Thermomyces lanuginosus* SY2 and cloning of its encoding gene," *Agricultural Sciences in China*, vol. 7, no. 12, pp. 1458–1465, 2008.

[27] P. E. Yuli, M. T. Suhartono, Y. Rukayadi, J. K. Hwang, and Y. R. Pyun, "Characteristics of thermostable chitinase enzymes from the indonesian *Bacillus* sp.13.26," *Enzyme and Microbial Technology*, vol. 35, no. 2-3, pp. 147–153, 2004.

[28] M. Ueda, A. Fujiwara, T. Kawaguchi, and M. Arai, "Purification and some properties of six chitinases from *Aeromonas* sp. No. 10S-24," *Bioscience, Biotechnology, and Biochemistry*, vol. 59, no. 11, pp. 2162–2164, 1995.

[29] K. Sakai, A. Yokota, H. Kurokawa, M. Wakayama, and M. Moriguchi, "Purification and characterization of three thermostable endochitinases of *Bacillus* noble strain MH-1 isolated from chitin containing compost," *Applied and Environmental Microbiology*, vol. 64, no. 9, pp. 3340–3497, 1998.

[30] R. C. Gomes, L. T. A. S. Semêdo, R. M. A. Soares, C. S. Alviano, L. F. Linhares, and R. R. R. Coelho, "Chitinolytic activity of *Actinomycetes* from a cerrado soil and their potential in biocontrol," *Letters in Applied Microbiology*, vol. 30, no. 2, pp. 146–150, 2000.

3

Statistical Analysis of Metal Chelating Activity of *Centella asiatica* and *Erythroxylum cuneatum* Using Response Surface Methodology

R. J. Mohd Salim,[1,2] M. I. Adenan,[1,2] A. Amid,[3] M. H. Jauri,[1] and A. S. Sued[1]

[1] *Natural Product Division, Drug Discovery Centre (DDC), Forest Research Institute Malaysia (FRIM), Jalan Kepong, Selangor Darul Ehsan, 52109 Kepong, Malaysia*

[2] *Malaysian Institute of Pharmaceuticals and Nutraceuticals, Ministry of Science, Technology and Innovation, USM, 10 Persiaran Bukit Jambul, 11900 Bukit Jambul, Malaysia*

[3] *Department of Biotechnology Engineering, International Islamic University Malaysia, Gombak, P.O. Box 10, 50728 Kuala Lumpur, Malaysia*

Correspondence should be addressed to R. J. Mohd Salim; roshanjahn@frim.gov.my

Academic Editor: Triantafyllos Roukas

The purpose of the study is to evaluate the relationship between the extraction parameters and the metal chelating activity of *Centella asiatica* (CA) and *Erythroxylum cuneatum* (EC). The response surface methodology was used to optimize the extraction parameters of methanolic extract of CA and EC with respect to the metal chelating activity. For CA, Run 17 gave optimum chelating activity with IC_{50} = 0.93 mg/mL at an extraction temperature of 25°C, speed of agitation at 200 rpm, ratio of plant material to solvent at 1 g : 45 mL and extraction time at 1.5 hour. As for EC, Run 13 with 60°C, 200 rpm, 1 g : 35 mL and 1 hour had metal chelating activity at IC_{50} = 0.3817 mg/mL. Both optimized extracts were further partitioned using a solvent system to evaluate the fraction responsible for the chelating activity of the plants. The hexane fraction of CA showed potential activity with chelating activity at IC_{50} = 0.090 and the ethyl acetate fraction of EC had IC_{50} = 0.120 mg/mL. The study showed that the response surface methodology helped to reduce the extraction time, temperature and agitation and subsequently improve the chelating activity of the plants in comparison to the conventional method.

1. Introduction

The knowledge and practice of traditional medicine are universal amongst the respected ethnic groups in each country. In Malaysia the benefits of herbal medicine are being conveyed down from one generation to another. Latif et al. [1] state that there are four sources of traditional Malaysian medicine, namely, Malay village medicine (including Orang Asli medicine), Chinese medicine (introduced from China), Indian medicine (introduced from India), and other forms of traditional medicine (including those introduced by the Javanese, Sumatrans, Arabs, Persians, Europeans, etc.).

Centella asiatica (CA) also locally known as pegaga is a crawling plant usually growing wildly in a humid climate around the globe. Its wide medicinal benefits include wound healing, enhancing memory, treating mental weariness [2], anti-inflammatory property [3], anticancer activity [4], antilipid peroxidativity [5], and free radical scavenger [6].

Erythroxylum cuneatum forma cuneatum (Miq.) Kurz (EC) is a genus of tropical flowering plants in the family of Erythroxylaceae [7]. While CAs are being well studied for their various medicinal fortunes *Erythroxylum cuneatum* (EC) on the other hand has a very limited report on its medicinal value. In Terengganu, the leaves are pounded and applied on the forehead of women after miscarriage. In Bunguran, Indonesia leaves are reported to be used in Sajur (vegetable soup) [8]. It is used in Thai traditional medicine for antifever purposes as well as an anti-inflammatory agent [9].

TABLE 1: Parameters to be optimized using response surface methodology for CA and EC.

	CA	EC
Temperature (°C) (X_1)	25, 30, 35	55, 60, 65
Speed (rpm) (X_2)	100, 150, 200	150, 200, 250
Ratio (g : mL) (X_3)	1 : 35, 1 : 40, 1 : 45	1 : 30, 1 : 35, 1 : 40
Time (min) (X_4)	30, 60, 90	30, 60, 90

Neurodegenerative disease (ND) results from the deterioration of neurons which functionalize the intellectual and cognition ability of a human body [10]. Zecca et al. [11] reported that iron may engage in a mechanism involving many neurodegenerative disorders. It was deduced that, as the brain ages, iron accumulates in regions that are affected by Alzheimer and Parkinson diseases, diseases categorized under ND. Thus, it is the interest of the research to study the ability of CA and EC to chelate the metal iron and further optimize the extraction process of the plants with respect to their chelating activity.

The extraction of plant material for example bioactive compounds can be affected by more than one factor such as particle size, extraction solvent, temperature, and time [12]. Response surface methodology is a software tool used to study the interaction that may occur between variable factors [13]. This statistical experimental design is a powerful tool that enables the extraction process conducted effectively by verifying the effects of operational factors and their interactions [14]. The traditional empirical methods only study a single factor at a time and fail to acknowledge the interaction that they have between each other [15].

2. Materials and Methods

2.1. Materials. *Centella asiatica* (CA) was purchased from local market, Pasar Borong Selayang, Selangor, and Erythroxylum cuneatum (EC) was collected from FRIM's compound. Methanol was purchased from Fisher Scientific, ethanol from J. Kollin Corporation, Germany, and hexane, ethyl acetate, and n-butanol from Merck, USA. All chemicals and solvents used were of analytical grade. Iron (II) sulfate heptahydrate ($FeSO_4$) was a product of Aldrich, USA, 4,4′-[3-(2-pyridinyl)-1,2,4-triazine-5,6-diyl]bis also known as ferrozine from Aldrich, USA.

2.2. Methods

2.2.1. Response Surface Methodology (RSM). RSM was used to optimize the conditions for extraction of CA and EC to give the optimum metal chelating activity. A face-centered cube design (FCD) in RSM consisting of 30 experimental runs including six replications at the center point was chosen

to evaluate the combined effect of the independent variables. Three levels were adopted and coded to low, center, and high levels. The experiments were performed in random order to minimize the effects of unexplained variability in the observed responses due to systematic errors [15]. The independent variables were temperature (°C), speed of rotation (rpm), ratio of raw material to solvent (g : mL), and time of extraction (h), while the response is the metal chelating activity reported in $1/IC_{50}$. As the software was meant to display the response at maximum, the inverse IC_{50} ($1/IC_{50}$) was reported in this study so that the IC_{50} will be displayed at its optimum activity.

The total of 30 runs designed by Design Expert by combining the parameters for extraction was shown in Table 1. The figures for each parameter were deduced from preliminary experiment. Each run was performed in triplicate.

2.2.2. Extraction Process. A constant weight of 2 g plants was used for all the 30 runs while adjusting accordingly to the ratio of methanol solvent that was needed in each run as outlined by Design Expert software. The plants were extracted in incubator shaker according to the combination parameters as given by each run. The extracts were then separated from the filtrate, and the methanol solvent was removed using rotary evaporator at 40°C and at a reduced pressure. The extracts from each run were then subjected to the metal chelating activity.

2.2.3. Metal Chelating Activity. The assay was initiated by adding 250 μL of 2.5 mM $FeSO_4$ to 500 μL sample solutions; CA and EC crude extracts were prepared in a series of concentrations. This mixture was vortexed briefly for 10 seconds before adding 250 μL of 6 mM ferrozine. The mixture was vortexed again briefly for 10 seconds and allowed to equilibrate for 10 min at room temperature. The absorbance of the mixture (formation of the ferrous iron-ferrozine complex) was measured at 562 nm [16]. Sample solutions with appropriate dilutions were used as blanks. The ability of extracts to chelate ferrous ion was calculated relative to the control (consisting of iron and ferrozine only) using the following formula [17]:

Chelating effect %

$$= \frac{(\text{Absorbance of control} - \text{Absorbance of sample})}{\text{Absorbance of control}} \quad (1)$$

$$\times 100.$$

2.2.4. Partitioning Process. The crude methanolic extracts were weighed to be 50 g and were suspended in water and then subjected to liquid-liquid partition by adding hexane, ethyl acetate, and n-butanol successively. The residual part that was suspended in water which was the water residue fraction [18] and the hexane, ethyl acetate, and n-butanol fraction were collected and subjected to metal chelating assay as described above.

TABLE 2: Face centered, central composite design setting with the independent variables and their responses in CA.

Run number	X_1	X_2	X_3	X_4	Y $1/\text{IC}_{50}$	IC_{50}
1	30	150	40	60	1.72	0.5814
2	30	150	40	60	1.429	0.6998
3	35	100	35	30	5.2	0.1923
4	30	150	40	60	1.55	0.6452
5	30	150	40	90	1.96	0.5102
6	35	200	35	30	2.4	0.4167
7	30	150	40	60	1.3426	0.7448
8	30	150	35	60	0.84	1.1905
9	30	150	40	30	1.57	0.6369
10	25	100	45	30	2.3	0.4348
11	35	200	45	90	2.703	0.3700
12	35	100	45	30	2.01	0.4975
13	25	100	35	30	4.6	0.2174
14	35	150	40	60	4.56	0.2193
15	30	200	40	60	1.399	0.7148
16	30	150	45	60	0.94	1.0638
17	25	200	45	90	10.753	0.0930
18	35	200	35	90	1.98	0.5051
19	30	150	40	60	1.49	0.6711
20	35	100	35	90	0.84	1.1905
21	30	150	40	60	1.49	0.6711
22	25	200	45	30	3	0.3333
23	35	200	45	30	0.29	3.4483
24	25	150	40	60	7.98	0.1253
25	25	100	45	90	3.49	0.2865
26	25	100	35	90	1.12	0.8929
27	25	200	35	30	4.167	0.2400
28	30	100	45	90	0.098	10.2041
29	35	100	45	90	0.93	1.0753
30	25	200	35	90	4.35	0.2299

TABLE 3: Face centered, central composite design setting with the independent variables and their responses in EC.

Run number	X_1	X_2	X_3	X_4	Y $1/\text{IC}_{50}$	IC_{50}
1	60	200	35	60	2.4	0.4167
2	55	150	40	30	1.3	0.7692
3	55	250	30	90	1.4	0.7143
4	60	150	35	60	2.1	0.4762
5	60	200	30	60	1.5	0.6667
6	65	150	30	90	0.76	1.3158
7	65	150	30	30	1.18	0.8475
8	60	200	40	60	1.59	0.6289
9	60	200	35	60	2.6	0.3846
10	55	250	30	30	0.909	1.1001
11	65	150	40	30	1.1	0.9091
12	55	150	30	90	1.59	0.6289
13	60	200	35	60	2.62	0.3817
14	65	150	40	90	0.7	1.4286
15	60	200	35	60	2.57	0.3891
16	55	250	40	90	1.39	0.7194
17	60	200	35	90	1.57	0.6369
18	65	200	35	60	1.68	0.5952
19	60	200	35	60	2.56	0.3906
20	55	200	35	60	1.89	0.5291
21	60	200	35	30	1.55	0.6452
22	60	250	35	60	2.1	0.4762
23	65	250	30	90	0.833	1.2005
24	65	250	30	30	1.25	0.8000
25	60	200	35	60	2.56	0.3906
26	55	150	30	30	1.28	0.7813
27	65	250	40	30	1.36	0.7353
28	55	150	40	90	1.66	0.6024
29	65	250	40	90	0.906	1.1038
30	55	250	40	30	1.1	0.9091

3. Results and Discussion

3.1. Optimization of Extraction with respect to Metal Chelating Activity. The optimum $1/\text{IC}_{50}$ value for CA (referred to in Table 2) was 10.753 mg/mL (IC_{50} = 0.093 mg/mL) obtained in the combined interaction of the independent parameter at Run 17 with 25°C, 200 rpm, 1 g : 45 mL ratio, and for duration of 1.5 hour.

Table 3 summarized the experimental results for EC. The optimum $1/\text{IC}_{50}$ value of 2.6196 mg/mL (IC_{50} = 0.3817 mg/mL) was obtained in Run 13 with temperature of 60°C, agitation at 200 rpm, and ratio of raw material to solvent 1 g : 35 mL ratio for extraction duration of 1 hour.

3.2. Multiple Regression Analysis. The statistical model was developed by applying multiple regression analysis methods on using the experimental data for the metal chelating activity which is given in (2) for CA and in (3) for EC. The response function (y) measured the $1/\text{IC}_{50}$ value of the metal chelating activity of the crude extracts CA and EC. This value was related to the variables (A, B, C, D) by a second-degree polynomial using (2) and (3) which is displayed in terms of coded factors. The coefficients of the polynomial were represented by a constant term, A, B, C, and D (linear effects), A^2, B^2, C^2, and D^2 (quadratic effects), and AB, AC, AD, BC, BD, and CD (interaction effects). The analysis of variance (ANOVA) tables were generated, and the effect and regression coefficients of individual linear, quadratic, and interaction terms were determined. The significances of all terms in the polynomial were judged statistically by computing the F-value at a probability (P) of 0.001, 0.01, or 0.05. In this case A, B, A^2, B^2, C^2, AB, AC, AD, BC, BD, and CD are significant model terms. On the other hand, values greater than 0.1000 indicate that the model terms are

TABLE 4: Fit statistics for the response of $1/IC_{50}$ value of CA.

Standard deviation	0.14	R-squared	0.9746
Mean	1.49	Adjusted R-squared	0.9509
C.V.	9.40	Predicted R-squared	0.8608
PRESS	1.62	Adequate precision	27.272

TABLE 5: Fit statistics for the response of $1/IC_{50}$ value of EC.

Standard deviation	0.26	R-squared	0.9028
Mean	1.60	Adjusted R-squared	0.8120
C.V.	16.12	Predicted R-squared	0.7243
PRESS	2.83	Adequate precision	9.404

not significant. The regression coefficients were then used to make statistical calculation to generate contour maps from the regression models:

$$
\begin{aligned}
IC_{50} = {} & +1.79 - 1.16 * X_1 + 0.58 * X_2 + 0.051 * X_3 \\
& + 0.14 * X_4 + 4.20 * X_1{}^2 - 1.32 * X_2{}^2 \\
& - 1.18 * X_3{}^2 0.31 * X_4{}^2 - 0.77 * X_1 * X_2 \\
& - 0.61 * X_1 * X_3 - 0.57 * X_1 * X_4 \\
& + 0.43 * X_2 * X_3 + 1.10 * X_2 * X_4 \\
& + 1.15 * X_3 * X_4,
\end{aligned}
\tag{2}
$$

$$
\begin{aligned}
IC_{50} = {} & +2.29 - 0.15 * X_1 - 0.023 * X_2 + 0.022 * X_3 \\
& - 0.012 * X_4 - 0.25 * X_2{}^2 + 0.063 * X_2{}^2 \\
& + 0.022 * X_3{}^2 - 0.48 * X_4{}^2 + 0.10 * X_1 * X_2 \\
& - 0.014 * X_1 * X_3 - 0.20 * X_1 * X_4 \\
& + 0.026 * X_2 * X_3 + 3.750 \exp -003 * X_2 * X_4 \\
& - 0.010 * X_3 * X_4.
\end{aligned}
\tag{3}
$$

3.3. Fit Statistics for the Response. Some characteristics of the constructed model can be explained by details in Table 4 and Table 5. The statistical analysis indicates that the proposed model was adequate, possessing no significant lack of fit and with satisfactory values of the R-squared. The quality of fit of the polynomial model equation was expressed by the coefficient of determination (R^2, adjusted R^2, and adequate precision). R^2 is a measure of the amount of variation around the mean explained by the model and equal to 0.9569 (CA) and 0.9028 (EC). The closer the value of R-squared is to the unity, the better the empirical model fits the actual data. The smaller the value of R-squared is, the less relevant the dependent variables in the model have to explain the behavior variation [18] and [19]. The adjusted-R^2 is adjusted for the number of terms in the model. It decreases as the number of terms in the model increases, if those additional terms do not add value to the model. Adequate precision is a signal-to-noise ratio. It compares the range of the predicted values

at the design points to the average prediction error. Ratios greater than four indicate adequate model discrimination. As for CA it was 21.064 whereas for EC it was 9.404. The standard deviation of 0.66 (CA) and 0.26 (EC) indicates that the model designed was acceptable with a minimum deviation. Coefficient of variation (C.V.) is the standard deviation expressed as a percentage of the mean which is 25.34% (CA) and 16.12 (EC). CV describes the extent to which the data were dispersed. The small values of CV give better reproducibility. In general, a high CV indicates that variation in the mean value is high and does not satisfactorily develop an adequate response model [20].

The predicted residual error sum of squares (PRESS) is a measure of model fitness to each point in the design which gave an amount of 46.29 (CA) and 16.12 (EC).

4. Conclusion

The metal chelating activity of CA and EC was optimized using statistical analysis to improve the chelating activity of the both plants by varying the parameters for the extraction. It shows that the extraction parameters had been optimized (IC_{50} = 0.093 mg/mL at extraction temperature of 25°C, speed of agitation at 200 rpm, ratio of plant material to solvent at 1 g : 45 mL, and extraction time at 1.5 hour). As for EC, Run 13 with extraction temperature at 60°C, speed of agitation at 200 rpm, ratio of plant material to solvent at 1 g : 35 mL, and extraction time at 1 hour had metal chelating activity at IC_{50} = 0.3817 mg/mL.

Acknowledgment

The authors would like to thank FRIM and MOSTI for providing fund for the research.

References

[1] M. Latiff, "Genetic resources of medicinal plants in Malaysia," in *Genetic Resources of Under-Utilised Plants in Malaysia: Proceedings of the National Workshop on Plant Genetic Resources Held in Subang Jaya, Malaysia*, A. H. Zakri, Ed., Malaysian National Committee on Plant Genetic Resources, Kuala Lumpur, Malaysia, 1989.

[2] M. T. Thomas, R. Kurup, A. J. Johnson et al., "Elite genotypes/chemotypes, with high contents of madecassoside and asiaticoside, from sixty accessions of *Centella asiatica* of south India and the Andaman Islands: for cultivation and utility in cosmetic and herbal drug applications," *Industrial Crops and Products*, vol. 32, no. 3, pp. 545–550, 2010.

[3] L. Suguna, P. Sivakumar, and G. Chandrakasan, "Effects of *Centella asiatica* extract on dermal wound healing in rats," *Indian Journal of Experimental Biology*, vol. 34, no. 12, pp. 1208–1211, 1996.

[4] T. D. Babu, G. Kuttan, and J. Padikkala, "Cytotoxic and anti-tumour properties of certain taxa of Umbelliferae with special reference to *Centella asiatica* (L.) Urban," *Journal of Ethnopharmacology*, vol. 48, no. 1, pp. 53–57, 1995.

[5] S. S. Katare and M. S. Ganachari, "Effect of *Centella asiatica* on hypoxia induced convulsions and lithium-pilocarpine induced

status epilepticus and antilipid peroxidation activity," *Indian Journal of Pharmacology*, vol. 33, article 128, 2001.

[6] G. Jayashree, G. K. Muraleedhara, S. Sudarslal, and V. B. Jacob, "Anti-oxidant activity of *Centella asiatica* on lymphoma-bearing mice," *Fitoterapia*, vol. 74, no. 5, pp. 431–434, 2003.

[7] R. J. M. Salim, *Optimisation of extraction of Centella asiatica and Erythroxylum cuneatum and their evaluation as a neuroprotective agent [Master's thesis]*, International Islamic University Malaysia, Selangor, Malaysia, 2010.

[8] A. S. Sued, *Kesan ekstrak Centella asiatica Linnaeus dan Erythroxylum cuneatumForma Cuneatum (Miquel) Kurz ke atas symptom tarikan pada tikus ketagihan morfina dan protein serum mereka [Master's thesis]*, Universiti Kebangsaan Malaysia, Selangor, Malaysia, 2009.

[9] T. Kanchanapoom, A. Sirikatitham, H. Otsuka, and S. Ruchirawat, "Cuneatoside, a new megastigmane diglycoside from *Erythroxylum cuneatum* Blume," *Journal of Asian Natural Products Research*, vol. 8, no. 8, pp. 747–751, 2006.

[10] J. F. Emard, J. P. Thouez, and D. Gauvreau, "Neurodecgnerative diseases and risk factors: a literature review," *Social Science and Medicine*, vol. 40, no. 6, pp. 847–858, 1995.

[11] L. Zecca, M. B. H. Youdim, P. Riederer, J. R. Connor, and R. R. Crichton, "Iron, brain ageing and neurodegenerative disorders," *Nature Reviews Neuroscience*, vol. 5, no. 11, pp. 863–873, 2004.

[12] B. Yang, X. Liu, and Y. Gao, "Extraction optimization of bioactive compounds (crocin, geniposide and total phenolic compounds) from Gardenia (*Gardenia jasminoides* Ellis) fruits with response surface methodology," *Innovative Food Science and Emerging Technologies*, vol. 10, no. 4, pp. 610–615, 2009.

[13] J. H. Kwon, J. M. R. Belanger, and J. R. J. Pare, "Optimization of microwave assisted extraction (MAP) for ginseng components by response surface methodology," *Journal of Agricultural and Food Chemistry*, vol. 51, pp. 1807–1810, 2003.

[14] W. Huang, Z. Li, H. Niu, D. Li, and J. Zhang, "Optimization of operating parameters for supercritical carbon dioxide extraction of lycopene by response surface methodology," *Journal of Food Engineering*, vol. 89, no. 3, pp. 298–302, 2008.

[15] C. S. Ku and S. P. Mun, "Optimization of the extraction of anthocyanin from Bokbunja (*Rubus coreanus* Miq.) marc produced during traditional wine processing and characterization of the extracts," *Bioresource Technology*, vol. 99, no. 17, pp. 8325–8330, 2008.

[16] T. C. P. Dinis, V. M. C. Madeira, and L. M. Almeida, "Action of phenolic derivatives (acetaminophen, salicylate, and 5-aminosalicylate) as inhibitors of membrane lipid peroxidation and as peroxyl radical scavengers," *Archives of Biochemistry and Biophysics*, vol. 315, no. 1, pp. 161–169, 1994.

[17] Y. Wu, S. W. Cui, J. Tang, and X. Gu, "Optimization of extraction process of crude polysaccharides from boat-fruited sterculia seeds by response surface methodology," *Food Chemistry*, vol. 105, no. 4, pp. 1599–1605, 2007.

[18] T. Satake, K. Kamiya, Y. An, T. Oishi, and J. Yamamoto, "The anti-thrombotic active constituents from *Centella asiatica*," *Biological and Pharmaceutical Bulletin*, vol. 30, no. 5, pp. 935–940, 2007.

[19] T. M. Little and F. J. Hills, *Agricultural Experimentation Design and Analysis*, John Wiley, New York, NY, USA, 1978.

[20] W. W. Daniel, *Biostatistics: A Foundation for Analysis in the Health Sciences*, vol. 503, Wiley, New York, NY, USA, 5th edition, 1991.

4

Physiochemical Characterization of Briquettes Made from Different Feedstocks

C. Karunanithy,[1] Y. Wang,[1] K. Muthukumarappan,[1] and S. Pugalendhi[2]

[1] Department of Agricultural and Biosystems Engineering, South Dakota State University, Brookings, SD 57007, USA
[2] Department of Bioenergy, Tamil Nadu Agricultural University, Coimbatore 641003, India

Correspondence should be addressed to C. Karunanithy, chinnadurai.karunani@sdstate.edu

Academic Editor: Jianmin Xing

Densification of biomass can address handling, transportation, and storage problems and also lend itself to an automated loading and unloading of transport vehicles and storage systems. The purpose of this study is to compare the physicochemical properties of briquettes made from different feedstocks. Feedstocks such as corn stover, switchgrass, prairie cord grass, sawdust, pigeon pea grass, and cotton stalk were densified using a briquetting system. Physical characterization includes particle size distribution, geometrical mean diameter (GMD), densities (bulk and true), porosity, and glass transition temperature. The compositional analysis of control and briquettes was also performed. Statistical analyses confirmed the existence of significant differences in these physical properties and chemical composition of control and briquettes. Correlation analysis confirms the contribution of lignin to bulk density and durability. Among the feedstocks tested, cotton stalk had the highest bulk density of 964 kg/m³ which is an elevenfold increase compared to control cotton stalk. Corn stover and pigeon pea grass had the highest (96.6%) and lowest (61%) durability.

1. Introduction

In the last four decades, researchers have been focusing on alternate fuel resources to meet the ever-increasing energy demand and to avoid dependence on crude oil. Biomass appears to be an attractive feedstock because of its renewability, abundance, and positive environmental impacts resulting in no net release of carbon dioxide and very low sulfur content. Biomass is very difficult to handle, transport, store, and utilize in its original form due to factors that can include high moisture content, irregular shape and sizes, and low bulk density. Densification can produce densified products with uniform shape and sizes that can be more easily handled using existing handling and storage equipment and thereby reduce cost associated with transportation, handling, and storage. Tumuluru et al. [1] classified conventional biomass densification processes into baling, pelleting, extrusion, and briquetting, which are carried out using a bailer, pelletizer, screw press, piston press, or roller press. Baling, briquetting, and pelleting are the most common biomass densification methods; pelleting and briquetting are the most common densifications used for solid fuel applications.

In general, biomass/feedstock is a cellular material of high porosity since cells interior consists mainly of large vacuole-filled air in dry conditions [2]. In general, natural binders such as lignin, protein, and starches present in the feedstocks enhance the bonding between particles during densification process. Because of the application of high pressures, particles are brought close together, causing interparticle attraction forces, and the natural binding components in the feedstocks are squeezed out of the cells, which make solid bridges between the particles [3]. Many feedstocks, densification machines, and process variables affect the quality of densified products. Several researchers have reported that feedstock composition such as lignin, hemicellulose, and extractives, types of feedstock, fraction of the same feedstock, feedstock particle size and moisture content, percentage of fines, type of densification machine, die diameter, preheating/steam injection, temperature, and pressure are the major variables that contribute to the quality of densified materials [4–12]. Feedstock composition is one of the major variables; therefore, understanding the compositional changes due to densification can be useful in understanding their compaction behavior [1]. The Literature survey revealed that only Theerarattananoon et al. [13] reported the changes in chemical composition before and after pelleting different feedstocks, none on briquetting. The dimensions of pellet, friction/shear development during pelleting, and briquetting would be different. Hence, this study was undertaken with two objectives: (1) to study the changes in chemical composition of different feedstocks due to briquetting and (2) to validate the relation of different variables that contribute to bulk density and durability.

2. Materials and Methods

2.1. Feedstocks Preparation and Characterization . Switchgrass and prairie cord grass obtained from different local farms were ground in hammer mill (Speedy King, Winona Attrition Mill Co, MN) using an 8 mm sieve for densification and sent to Tamil Nadu Agricultural University (TNAU), Coimbatore, India. Similarly, corn stover, pigeon pea grass, and cotton stalk were collected from experimental field at TNAU, Coimbatore, India. Sawdust was obtained from local sawmill located at Coimbatore, India. The compositional analyses of the feedstocks and briquettes such as total solids, cellulose, hemicellulose, lignin, ash, and extractives content were carried out in triplicate as outlined by Sluiter et al. [14–16] using muffle furnace and HPLC and reported in Table 1.

2.2. Particle Size Analysis . Prior to briquetting, the geometric mean diameter of ground feedstocks was determined using ASAE Standard S319.4 [17] with the help of a Ro-Tap sieve shaker (W. S. Tyler Inc., Mentor, OH, USA) with US sieve numbers 6, 7, 10, 16, 20, 30, 50, 70 100, 140, 200, and 325 (sieve opening sizes: 3.35, 2.80, 2.00, 1.190, 0.841, 0.595, 0.297, 0.210, and 0.149 mm, resp.). For each test, a 100 g sample was placed on a stack of sieves arranged from the largest to the smallest opening. A 10-minute sieve shaking time was used as mentioned in the ASAE Standard S319. The geometric mean diameter (dgw) of the sample and geometric standard deviation of particle diameter (Sgw) were calculated in replicates of three for each feedstock.

2.2.1. Briquetting . The briquetting system consists of 40 hp motor, feed hopper, and die section, and the capacity is 150–200 kg/h. The system had a provision to select 60 or 90 mm die section. For this study, 60 mm die was used. Figure 1 shows the briquetting system along with feedstocks and briquettes. The briquetting machine is the simple horizontal briquetting press. Material handling screw conveyor with 2 hp electric motor coupled with reduction gear and variable pulley with V belt. The shaft moves an eccentric disc through connecting rod where circular motion is connected to linear motion. The eccentric disc is connected to an alloy steel piston which is having to-and-fro movement in stationary cast iron cylinder. The piston carries a hardened and ground alloy steel punch. The hardened ground alloy steel die is held in steel die holder. The raw material, passed into hopper of the machine, is transferred to a chamber where punch pushes the material into the die, forms the cylindrical briquette, and pushes it further into split die and cooling line. Briquettes

were collected and sent through FedEx to South Dakota State University for further analysis.

2.2.2. Density and Porosity. Bulk densities of ground feedstocks and briquettes were measured following the ASAE standard method S269.4 DEC01 [18]. The container used is a 2000 mL glass container. The bulk density was calculated from the mass of feedstocks and briquettes that occupied the container.

The Micromeritics Multivolume Pycnometer and cell (125 cm^3) provided with the equipment was used for the measurement of the true density of the samples. The measurement is based on the pressure difference between a known reference volume and the volume of the sample cell. Helium is used as the gas to fill the reference and sample cells at 19.5 ± 0.2 psi as specified in the instrument manual. True density of the material was measured using equation

$$\text{True density} = \frac{m}{\left\{ V_{\text{cell}} - V_{\text{exp}}/[(P1/P2) - 1] \right\}}, \quad (1)$$

where m is the weight of the sample, V_{cell} is empty volume of the sample cell, V_{exp} is expansion volume, $P1$ is pressure before expansion, and $P2$ is the pressure after expansion.

Porosity is a measure of the void spaces in a material and is a fraction of the volume of voids over the total volume; it generally lies between 0-1. The porosity is calculated by the true density and bulk density measured as explained previously:

$$\text{Porosity} = \left(1 - \frac{\text{Bulk density}}{\text{True density}} \right). \quad (2)$$

2.3. Durability. The durability of the briquettes was determined using a pellet durability tester (model PDT-110, Seedburo Equipment Company, Chicago, IL) following method S269.4 [18]. About 200 g of briquettes were divided into two batches of 100 g each. Each batch was placed in the pellet durability tester for a period of 10 min and operated at 50 rpm. The sample was placed on a no. 4 sieve (4.75 mm) before and after tumbling and measured for the mass retained on the screen. The pellet durability was then calculated using the following equation:

$$\text{Durability} = \left(\frac{M_{\text{at}}}{M_{\text{bt}}} \right), \quad (3)$$

where M_{at} is the mass of the briquettes retained on the screen after tumbling (g), and M_{bt} is the mass of the briquettes retained on the screen before tumbling (g).

2.4. Glass Transition Temperature. The glass transition temperature (T_g) of the feedstocks was evaluated using a differential scanning calorimeter (DSC) (Q series, TM Model Q200, TA Instruments, New Castle, DE). A refrigerated cooling system (RCS40), provided with DSC module, has the ability to control the sample temperature from −40°C to 400°C. About 2.0–2.2 mg feedstock was in T_{zero} aluminum pan and subjected to a heating range of 10 to 150°C with a heating rate of 5°C/min. An empty T_{zero} aluminum pan was considered the reference cell. Universal analyzer software provided by TA instruments (New Castle, DE) was used to analyze T_g from the thermograms, using the half-height integration method [19].

2.5. Statistical Analysis. All physical and chemical properties measurements were made in triplicate, and the data were analyzed with Proc GLM procedure to determine the statistical significance using SAS 9.2 [20] using a type I error (α) of 0.05.

3. Results and Discussion

Briquetting machines can handle larger particles with wide range of moisture content without additional binders, not the pellet mills. Further, friction/shear between the particles and the briquetting machine is much less than that of pelleting/cubing [21]. The standard shape of a fuel pellet is cylindrical, with a diameter of 6 to 8 mm and a length of no more than 38 mm. If the pellets are with more than 25 mm in diameter, they are usually referred to as "briquettes." The dimensions of the pellets found in the literatures are 4–7 mm diameter and 13–23 mm length [22, 23], whereas briquettes can have diameter between 25 and 100 mm with length between 25 and 280 mm [24]. The dimensions, friction/shear, steam injection/preheating, and binder would make much more differences in the resultant compacts, which should be considered to compare the briquettes data presented in this study.

3.1. Particle Size Analysis. Apart from moisture content, particle size distribution and particle size are two important factors that affect the bulk physical properties of feedstocks. Bulk density of ground feedstocks depends on the particle size and particle size distribution. Particle size distribution also reflects on the available surface area. Particle sizes affect the true density of the feedstocks [25] and also influence durability [9]. Particle size analyses of the feedstocks are shown in Figure 2. In general, all the feedstocks had more than 50% of the particle size in the range of 0.297–1.68 mm as evident from the figure. A major fraction of the PCG was shifted towards lower particle size because of their needle-like shape (Figure 1 PCG control). Switchgrass, pigeon pea grass, and cotton stalk had a similar distribution as evident from Figure 2. Though different screen/sieves were used during grinding, similar trend of particle size distribution (normal distribution) was reported for switchgrass [26], olive tree pruning [12], barley, canola, oat, and wheat straws [27]. Colley et al. [26] reported that sieves with aperture sizes of 0.595 and 0.850 mm retained 29.5 and 38.6% switchgrass ground using 3.18 mm screen; in this study, 8 mm screen is used for grinding which explains the difference in particle retention recorded. Sawdust particle distribution was different from Rhén et al. [7] where they dried and milled the sawdust using 4 mm sieve; hence, they could get particles less than 0.5 mm about 44%.

The percentage of fines has influence during densification. In general, fines would result in more durable product, and it comes with grinding cost, which is not desirable.

Table 1: Changes in chemical composition (%) due to briquetting.

	Glucose	Xylose	Lignin	Ash	Extractives
	Control				
CS	36.0^f	15.3^c	22.4^d	10.9^c	11.3^e
SG	31.2^g	19.5^a	24.7^c	5.6^d	18.5^b
PCG	31.5^g	15.5^{bc}	21.4^d	5.6^d	20.3^a
Sawdust	39.1^e	10.5^f	33.6^a	5.3^d	7.5^f
Pigeon pea	50.3^a	10.8^f	24.2^c	3.2^f	6.1^g
Cotton stalk	42.5^d	16.5^b	24.4^c	5.2^d	6.3^g
	Briquettes				
CS	38.4^e	10.1^{gf}	21.9^d	11.9^b	12.9^d
SG	36.0^f	19.0^a	24.8^c	3.7^{ef}	16.5^b
PCG	37.0^{ef}	12.0^e	22.5^d	5.3^d	17.0^b
Sawdust	44.8^c	13.3^d	39.1^b	3.9^e	4.2^h
Pigeon pea	47.3^b	9.2^g	21.0^d	4.2^e	13.9^c
Cotton stalk	38.8^e	14.8^c	22.2^d	14.6^a	6.2^g

Different letters within the same column indicates that means are statistically different ($P < 0.05$).

In general, the finer the feedstock grinds, the higher the quality of compact [9]. Tabil and Sokhansanj [28] considered that particles with sizes below 0.400 mm are fine and highly compressible. Taking this criterion into account, PCG had a maximum fine of 48.3%, followed by cotton stalk (26.7%), and corn stover had the least (13.9%). Olive tree pruning had 18% fines when 4 mm screen was used [12], 14% fines from switchgrass when 3.18 mm screen was used [26], and more than 60% fines from barley, canola, oat, and wheat straws when 1.98 mm screen was used [27]. The differences in fines are mainly due to variation in screen sizes and inherent characteristics of the feedstocks. According to MacBain [29], large particles are fissure points that cause cracks and fractures in compacts. Further, large particles in compact mean inhomogeneous shrinking, which would develop cracks [5]. The cracks on the surface of the briquettes (Figure 1) might be due to larger particles. Several researchers have reported that mixture of different particle sizes would result in better quality due to interparticle bonding with no interparticle space [29, 30].

The order of geometrical mean diameter (GMD) was corn stover (0.833 mm), switchgrass (0.736 mm), sawdust (0.708 mm), pigeon pea grass (0.657 mm), cotton stalk (0.639 mm), and PCG (0.0432 mm), and their geometric standard deviation of particle diameter (Sgw) was 0.422, 0.300, 0.455, 0.341, 0.347, and 0.251 mm, respectively. For switchgrass, Colley et al. [26] recorded a high GMD of 0.867 mm with the geometric standard deviation of 0.357 mm when 3.18 mm screen used. Mani et al. [8] reported a lower GMD of 0.193–0.412 and 0.253–0.456 mm with geometric standard deviation of 0.261–0.447 and 0.255–0.438 mm, respectively, for corn stover and switchgrass ground using 0.8–3.2 mm screen. Similarly, Kaliyan and Morey [21] reported a lower GMD of 0.56–0.66 mm for corn stover and switchgrass when 3 mm screen was used for grinding. Adapa et al. [27] also reported lower GMD in the range of 0.347–0.398 mm for barely, canola, oat, and

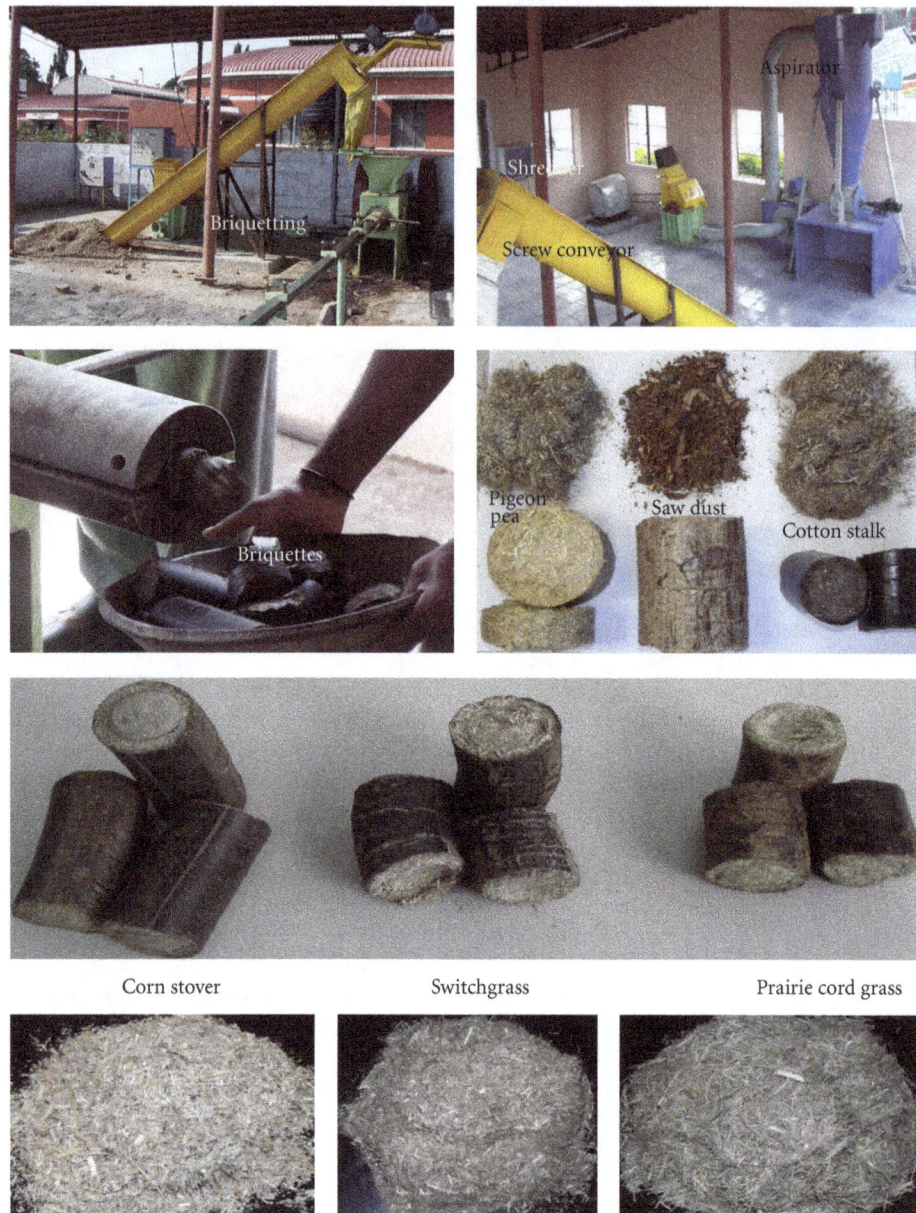

FIGURE 1: Briquetting system along with control and briquettes.

wheat straws. These differences are mainly due to variation in screen sizes used during grinding (0.8–3.2 versus 8 mm). In a recent study, when Adapa et al. [31] used screen size 6.4 mm, GMD of barely, canola, oat, and wheat straws was 0.883, 0.885, 0.935, and 0.997 mm, respectively. Though they used lower screen size (6.4 mm) than this study (8 mm), GMD was higher than the feedstocks used in this study and that might be due to inherent characteristics of the feedstocks.

3.2. Moisture Content. Moisture content has strong influence on density, durability, and storage. Several researchers have recommended a range of moisture content for pelleting or briquetting of different feedstocks. Moisture content (wb) for pelleting pruning of olive residues would be less than 10% wb [12]: about 10% for switchgrass [10], about 8-9% for alfalfa

[32], 6–12% for wood [33], and 5–10% for corn stover [34]. The moisture content of the feedstocks ranged between 6.8 and 10.4% wb, whereas it was 4.9–9.2% wb for briquettes as depicted in Figure 3; the values are well within the range of moisture content reported in the above literatures. The moisture content decrease is due to rise in feedstocks temperature during briquetting. Though PCG had the lowest moisture content, the highest decrease of 28% was observed. Similar observation was reported when Kaliyan and Morey [21] briquetted corn stover and switchgrass with feedstock moisture content in the range of 15–20% wb, the resulted briquettes had an average moisture content in the range of 11–14.5%, which was equivalent to 25–29% decrease in moisture. A minimum change in moisture content due to briquetting for sawdust, among the feedstocks is studied. In

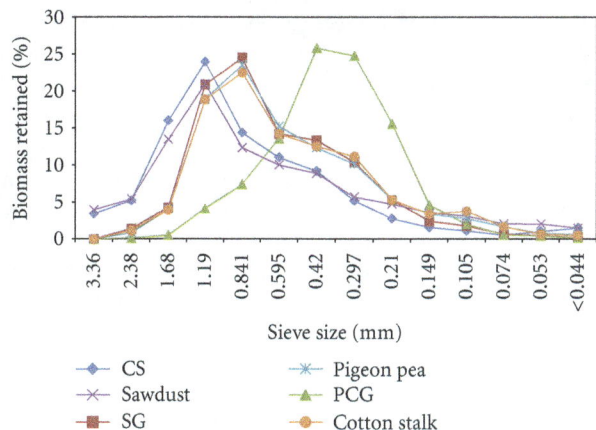

FIGURE 2: Particle size distribution of different feedstocks used in this study.

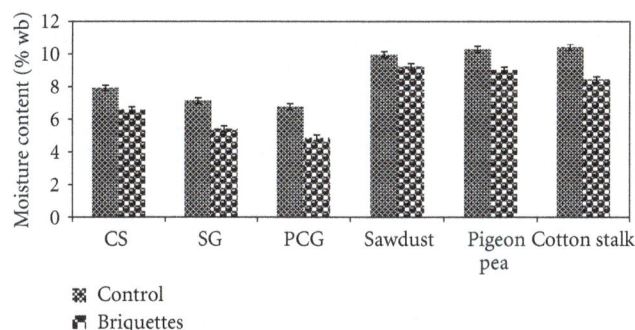

FIGURE 3: Effect of briquetting on moisture content.

general, the compacted/densified biomass would have the moisture content between 7 and 14% [35], and the briquettes produced for this study had moisture content within the range that indicates better density, durability, and storage.

3.3. Chemical Composition.

3.3. Chemical Composition. The composition of feedstock is one of the major variables that contribute to the quality of briquettes/compacts/pellets. The feedstock has low-molecular-weight substances such as organic matter, inorganic matter, and macromolecular substances include cellulose, hemicellulose, and lignin [36]. Understanding the major compositional changes that take place during briquetting can be useful in understanding their compaction behavior [1]. Compositional analysis of the feedstock and briquettes is presented in Table 1. Because of moisture and volatile matters loss due to temperature rise during briquetting, the chemical compositions of briquettes change slightly. Significant change in glucose content was observed irrespective of the feedstocks. Among the feedstocks, PCG had a maximum of 17.5% increase in glucose, whereas cotton stalk had a decrease of 8.7%. For most of the briquettes, xylose content has decreased when compared to their respective feedstocks. Similarly lignin content of the briquettes was less than that of the feedstocks except sawdust. Recently, Theerarattananoon et al. [13] reported similar observation for glucose, xylose, and lignin of pellets produced from wheat straw, corn stover, big bluestem, and sorghum stalk.

Briquettes made from PCG and sawdust have shown an increase in lignin. However, increase in lignin was not significant for PCG and it was in agreement with the findings of Theerarattananoon et al. [13]. The change in ash content is inconsistent across the feedstocks, and similar results were reported for different feedstocks [13]. When compared to other feedstocks, cotton stalk had high volatile matters that reflected on high smoke production as well as brown liquid oozed out during briquetting, thereby more changes in the chemical composition including ash content. According to Kaliyan and Morey [21], ash/mineral content of the feedstocks would show their relative abrasiveness to equipment when there is high friction/shear during densification; the higher the ash content, the higher the abrasion. Ash content of corn stover, pigeon pea grass, and cotton stalk briquettes increased significantly, whereas switchgrass and sawdust briquettes had a significant decrease. Kaliyan and Morey [21] have also reported similar ash/mineral contents of corn stover (11.2%) and switchgrass (5.0%). Ash content of the cotton stalk briquettes increased by 1.8 times.

Extractive is the material present in the feedstock which is soluble in either water or ethanol during extraction and that is not an integral part of the cellular structure [16, 37]. Inorganic material, nonstructural sugars, and nitrogenous material are water soluble, whereas ethanol soluble includes chlorophyll, waxes, or other minor components. Nonstructural component refers to nonchemically bound components of the feedstock that include but are not limited to sucrose, nitrate/nitrites, protein, ash, chlorophyll, and waxes [16]. A mix of long-chain fatty acids, fatty alcohols, sterols, and alkanes are the main constituents of wax [38, 39]. The types of extractives found in the feedstocks are entirely dependent upon the feedstock itself [37]. In general, grasses contain higher amount of extractives than wood, and it can be observed in the Table 1. The change in extractives was not the same for all the feedstocks, and this observation was in agreement with Theerarattananoon et al. [13]. Corn stover and pigeon pea grass briquettes had significantly higher extractives than that of their respective feedstocks. Extractives of corn stover briquettes increased about 14%, which is similar to increase in extractives of corn stover pellet [13]. A maximum increase and decrease in extractives of 130 and 44%, respectively, were recorded for pigeon pea grass and sawdust briquettes. Higher percentages of extractives (waxes, resins, and starches) affect gluability, contribute to the reduction of shrinkage, and would increase the bonding and the overall pellet strength [4, 5].

3.4. Glass Transition Temperature (T_g).

3.4. Glass Transition Temperature (T_g). Lignin could be the deciding factor, and it has strong influence on binding characteristics, thereby the briquette and pellet quality [11, 27]. Lignin content varies depending upon the type of feedstocks [11] and between the fractions of the same feedstock [40]. As noted in Table 1, lignin content of sawdust differed from other feedstocks. According to Back and Salmen [41], lignin and hemicellulose undergo plastic deformation at temperature in their glass transition/softening temperatures. Softening temperature is of high importance, because many properties including elastic modulus would change remarkably when the material passes from a glassy into a rubbery state. The higher the temperature above the T_g, the greater and easier is the flow of these molecules [2]. Corn stover, switchgrass, and PCG had a glass transition temperature of 79.2, 82.5, and 80°C, respectively, and they are very close to each other and it was in agreement with the average glass transition temperature (75°C) of corn stover and switchgrass reported by Kaliyan and Morey [21]. Pigeon pea grass, saw dust, and cotton stalk had a glass transition temperature of 75, 72, and 82°C, respectively. Van Dam et al. [42] reported that lignin has low melting point (~140°C) and thermosetting properties that would help for active binding. The temperatures of the briquettes at the exit were in the range of 130–140°C and confirm that lignin would have crossed its glass transition and melting point.

3.5. Bulk and True Densities. Bulk density plays vital role in transportation and storage efficiency. In addition, bulk density influences the engineering design of transport equipment, storages, and conversion process [43]. Bulk and true densities of the feedstocks and briquettes are depicted in Figure 4. As noted from the figure, bulk density of the feedstocks ranged between 66 to 191 kg/m^3, whereas the briquettes bulk density varied between 285–964 kg/m^3. Among the feedstocks, corn stover had the lowest and sawdust had the highest bulk density. Bulk density of the feedstocks used in this study was well within the range reported for barely and oat straws by Adapa et al. [31]; though they used lower screen sizes (0.8–6.4 mm) during grinding. Mani et al. [8] have reported higher bulk density for corn stover (131–158 kg/m^3) and switchgrass (115–182 kg/m^3). Similarly, Kaliyan and Morey [21] also reported a higher bulk density of 103–160 and 181–220 kg/m^3, respectively, corn stover and switchgrass. Possible reason for their higher bulk density is the use of lower screen sizes (0.8–3.2 and 2.4–4.6 mm) for grinding. Several researchers have reported that densification would result in bulk densities in the range of 450 to 700 kg/m^3 depending upon feedstock and densification conditions [3, 21, 26, 44]. Among the briquettes, PCG and cotton stalk had the lowest and highest bulk density. Irrespective of the feedstocks, briquettes bulk density increased, which is one of the purposes of briquetting. The lowest increase of 1.9 times was observed for sawdust, and the highest increase of 11.3 times was noted for cotton stalk. Depending upon the briquetting machine used, the bulk density of feedstock would increase approximately 10–20 times of its original bulk density [35]. Except bulk density of cotton stalk, briquettes made from other feedstocks did not fall within the expected range, that is, 10–20 times increase. Possible reason might be the type of briquetting machine used, the feedstock properties, and the process conditions employed in this study. However, increase in bulk density of corn stover and switchgrass briquettes were higher than that of Kaliyan et al. [3] reported for corn stover (2.9–3.4 times) and switchgrass (1.6–2.3 times) depending upon the feedstock particle size and temperature, roller speed, and feeder screw speed. The increase in bulk density correlates

well with porosity of the feedstocks as discussed in the next subheading. Though switchgrass pellets [26] had higher bulk density (536–708 kg/m3) than that of this study, but increase in bulk density was only threefold which was lower than that of this study (4.5 times).

True density of the feedstocks varied between 830 and 1376 kg/m^3, whereas it varied between 1340 and 2190 kg/m^3 for briquettes as shown in Figure 4. An increase in true density was in the range of 1.03–2.35 times, whereas Adapa et al. [31] reported decrease in true density of pellets made from different feedstocks. A maximum and minimum increase in true density was observed for corn stover and pigeon pea grass, which is attributed to their structures. Because of lower screen sizes (0.8–3.2 mm) used for grinding, Mani et al. [8] have reported higher true density for corn stover (1170–1399 kg/m^3) and switchgrass (946–1173 kg/m^3). The true density of the feedstocks used in this study was in agreement with barely, canola, oat, and wheat straws [8, 31]. Though Kaliyan and Morey [21] used lower particle sizes of corn stover and switchgrass for briquetting, they reported true density of briquettes in the range of 825–1162 and 417–1065 kg/m^3, respectively, depending on the pressure, preheating, feedstock particle size, and moisture content.

As mentioned earlier, density and durability depends on many feedstocks, machines, and process variables. In order to have a comprehensive understanding of these variables, Table 2 is presented here. In general, bulk density of the briquette/pellet increased 2–13 times depending upon the feedstock, densification equipment, and process conditions. Corn stover briquette had higher true density than the studies listed in the table. True density of switchgrass briquette was lower than that of switchgrass pellet reported by Colley et al. [26] that might be the use of steam for raising the switchgrass grind temperature. Though Kaliyan and Morey [21] used preheating temperature of 25–150°C, their true density of switchgrass pellet was lower than that of the true density obtained in this study. Despite of the fact that Lehtikangas [5] used sawdust with less than 3 mm particle sizes, the true density was lower than that of sawdust true density obtained in the present study. In general, true density of briquettes produced in this study was higher than that of the values listed in Table 2.

3.6. Porosity. Porosity has influence in transportation and storage. Porosity of the feedstocks and briquettes are presented in Figure 5. Sawdust and cotton stalk had the lowest (0.85) and highest (0.93) porosity among the feedstocks studied. Briquettes had lower porosity than that of their respective feedstocks. Among the feedstocks studied, a maximum of 56% decrease in porosity was recorded and that corresponds well with the bulk density. Cotton stalk briquettes had the lowest bulk density of 0.41 indicating less void space and more briquettes, which reflects on high bulk density (964 kg/m^3). Low porosity of the feedstock which indicates that the void space is less and the feedstock within the given volume is more would result in low compressability, whereas high porosity would result in high compressability which was observed for sawdust and cotton stalk. Except

TABLE 2: Effect of briquetting and pelleting conditions of different feedstocks on bulk density, true density, and durability.

Densification	Process conditions	Feedstock conditions	Feedstock bulk density, kg/m³	Product bulk density, kg/m³	True density, kg/m³	Durability	Reference
Briquetting	15.7–31.4 MPa; 80–105°C	Soda weed, 7–13% wb, < 10 mm	172	600–950	NR	NR	[45]
Hydraulic press pelletizer	55.2–552.3 bar, 50–125°C	Switchgrass, 6.5% wb, 10–70 mm	NR	250–720	NR	0.98–0.99	[10]
Sprout Matador 12 press pelletizer		Sawdust, <3 mm	NR	606–641	1228–1234	0.80–0.90	[5]
Modified SPC PP300	0–2.6% steam addition, die temperature 83°C	Reed canary grass, 4 mm, 14–17% wb, precompacted to 269–356 kg/m³	140–160	600–700	NR	0.92–0.98	[46]
Compact pelletizer	95°C, 30–134 MPa	Barley straw, <1.9 mm, 10% wb	261	NR	907–988	NR	[27]
Single pelletizer	95°C, 30–134 MPa	Canola straw, <1.9 mm, 10% wb	273	NR	823–1003	NR	[27]
Single pelletizer	95°C, 30–134 MPa	Oat straw, <1.9 mm, 10% wb	268	NR	849–1011	NR	[27]
Single pelletizer	95°C, 30–134 MPa	Wheat straw, <1.9 mm, 10% wb	269	NR	813–931	NR	[27]
Lab-scale CPM CL-5 pellet mill	Steam at 118°C,	Sun-cured and dehydrated alfalfa, 1.98–3.2 mm, 7–9.3%	NR	NR	1181–1341	0.43–0.92	[22]
Glomera extrusion press	5.7–8.3 MPa, 28.9–49.4°C	Wheat straw, 6–25 mm, 8.1–17.8%wb	NR	NR	1056	0.99	[47]
Glomera extrusion press	3.5–9.0 MPa, 27.8–59.4°C	Flax straw, 6–25 mm, 9.4–19% wb	NR	NR	1069–1260	0.99	[47]
Glomera extrusion press	3.1–9.6 MPa, 36.7–60°C	Sunflower stalk, 6–25 mm, 9–19% wb	NR	NR	940–1620	0.99	[47]
Lab-scale CPM CL5 pellet mill	Steam injection	Switchgrass, <3.18 mm, 6.3–17% wb	169.5	536–708	1410–1430	0.89–0.96	[26]
Pilot-scale roll press briquetting-CS-25 compactor/briquetter	Steam conditioning (25, 75, and 100°C), roll speed 1.5–2.3	Corn stover, 2.4 and 4 mm, 10 and 15% wb	139–160	422–481	NR	0.67–0.90	[3]
Pilot-scale roll press briquetting-CS-25 compactor/briquetter	Steam conditioning (25, 75, and 100°C), roll speed 1.3–2.5	Switchgrass, 2.4 and 4 mm, 10 and 15% wb,	184–220	351–527	NR	0.39–0.70	[3]

TABLE 2: Continued.

Densification	Process conditions	Feedstock conditions	Feedstock bulk density, kg/m³	Product bulk density, kg/m³	True density, kg/m³	Durability	Reference
Pilot-scale conventional ring-die CPM Master model 818806 pelleting machine		Corn stover, 2.4 and 4 mm, 20% wb,	139–160	548–610	NR	0.94–0.95	[3]
Pilot-scale conventional ring-die CPM Master model 818806 pelleting machine		Switchgrass, 4 mm, 20% wb,	184–220	528–570	NR	0.75–0.86	[3]
Laboratory-scale CPM CL-5 pellet mill		Barely straw, 0.8–6.4 mm, 10% wb	96–180	NR	726–1033	0.49–0.98	[31]
Laboratory-scale CPM CL-5 pellet mill		Canola straw, 0.8–6.4 mm, 10% wb	144–247	NR	742–1015	0.22–0.98	[31]
Laboratory-scale CPM CL-5 pellet mill		Oat straw, 0.8–6.4 mm, 10% wb	111–196	NR	771–1002	0.44–0.99	[31]
Laboratory-scale CPM CL-5 pellet mill		Wheat straw, 0.8–6.4 mm, 10% wb	107–203	NR	760–1047	0.52–0.98	[31]
Ring-die pellet mill CPM Master model series 2000		Corn stover, 3.2–6.5 mm, 10% wb	50.9	469–625	529–843	0.96–0.98	[13, 23]
Ring-die pellet mill CPM Master model series 2000		Wheat straw, 3.2–6.5 mm, 10% wb	47.7	496–649	613–852	0.96–0.98	[13, 23]
Ring-die pellet mill CPM Master model series 2000		Big bluestem, 3.2–6.5 mm, 10% wb	46.6	467–648	517–778	0.96–0.98	[13, 23]
Ring-die pellet mill CPM Master model series 2000		Sorghum stalk, 3.2–6.5 mm, 10% wb	59.3	365–479	435–560	0.86–0.94	[13, 23]
Piston cylindercompression/densification apparatus, pressure applied through an Instron universal testing machine	100 and 150 MPa, preheating temperature 25–150°C	Corn stover, 2.4–4.6 mm, 7.3–15% wb	103–160	NR	825–1162	0.50–0.97	[21]
Piston cylindercompression/densification apparatus, pressure applied through an Instron universal testing machine	100 and 150 MPa, preheating temperature 25–150°C	Switchgrass, 2.4–4.6 mm, 9.2–15.1% wb	181–220	NR	417–1065	0–0.68	[21]
Pellet press (SPC 300)	Steam injection 2–6 kg/h	Fresh and 140 days stored sawdust from Scot Pine and Norway Spruce, 7.8–12.5% wb	NR	501–706	NR	0.79–0.99	[48]
Ultrasonic-vibration-assisted pelleting		Switchgrass, 1 mm, 15% wb	NR	415–560	NR	0.39–0.63	[49]

NR: not reported.

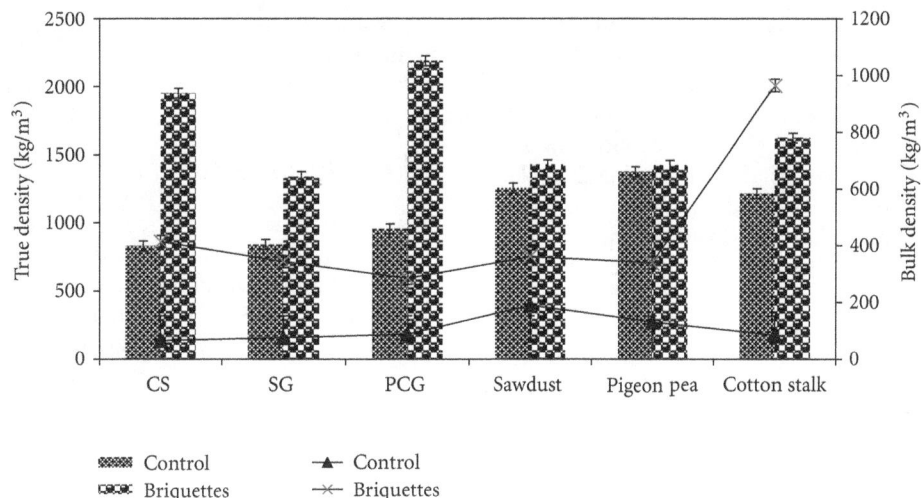

FIGURE 4: Effect of briquetting process on bulk and true densities.

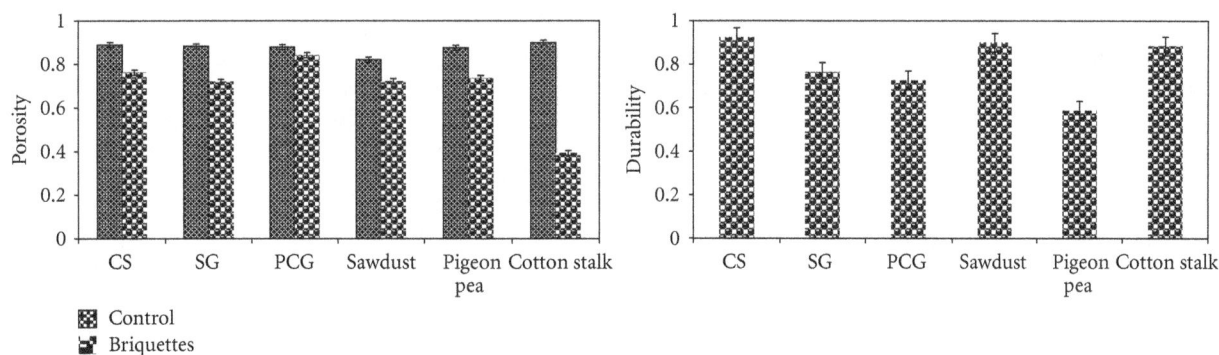

FIGURE 5: Effect of briquetting on porosity and durability.

TABLE 3: Correlation coefficients.

Property	Fines	GMD	Lignin	Extractives	Moisture content
Bulk density	−0.03	−0.02	0.88	−0.49	0.57
Durability	−0.40	0.44	0.33	−0.15	0.03

cotton stalk briquette, briquettes made from other feedstocks had higher porosity than that of switchgrass pellets (0.516–0.626) [26]. This might be due to the differences in size or dimensions of briquettes and pellets.

3.7. Durability. Durability is a measure of the briquettes ability to withstand destructive forces such as compression, impact, and shear during handling and transportation. The production of fines or dust during handling, transport, and storage would create health hazard and inconvenient environment for the workers [50]. There is no limit for the production of fines in place. However, Dobie [51] suggested that fines up to 5% (by weight) would be an acceptable level, and greater than 5% would reduce storage capacity and create problems in flow characteristics. Depending upon the values, researchers have classified the durability into high (>0.8), medium (0.7-0.8), and low (<0.7) [6, 28]. Figure 5 shows

durability of the briquettes that varied between 0.61 (PCG) and 0.97 (CS). High durability might be possible with larger particle size due to mechanical interlocking of relatively long fibers [52]. A noteworthy point is that these briquettes were produced in India and transported to Brookings, SD, USA through Fedex, wherein multiple handling has been involved, in spite of that, these briquettes had high durability. The differences in durability between briquettes might be due to chemical composition including lignin, extractive, cellulose, and hemicellulose, structure, fraction of leaf to stem or stalk, glass transition temperature, and compressibility. According to the above durability classification, corn stover, sawdust, and cotton stalk briquettes come under high, switchgrass and PCG briquettes fall under medium, and pigeon pea grass briquettes under low durability category. Kaliyan and Morey [21] reported a comparable durability of 0.50–0.97 for corn stover briquettes when corn stover grind size of 3 and 4.6 mm with a moisture content of 10–20% (wb) was preheated between 25 and 150°C and applied pressure in the range of 100–150 MPa. However, preheating temperature of 25°C resulted in low durability. Irrespective of switchgrass grind size, moisture content, pretreating temperature, and applied pressure, durability was in the range of 0–0.68 [21] that was lower than the durability of switchgrass briquettes produced in this study. Considering the moisture content of switchgrass, durability of switchgrass briquettes was in agreement with durability of switchgrass pellets [26]. The desirable durability of briquettes depends on the targeted use, that is, high durability for fuel application; low and medium durability would be sufficient for biochemical platform, because more surface area is desirable for enzymatic hydrolysis in biochemical platform. In addition, densification process would disturb the feedstock structure which would facilitate enzymatic hydrolysis.

Overall, the durability of corn stover briquette is comparable to corn stover briquettes and pellets listed in Table 2. Durability of the switchgrass briquette was higher than that of switchgrass briquettes produced using preheating and pressure [21] and ultrasonic vibration-assisted switchgrass pelleting [49]. In general, durability of briquettes used in this study had been either high or comparable to briquettes and pellets made from different feedstocks as listed in the table.

3.8. Correlation Analysis. In general, bulk density depends on chemical composition, particle size distribution, particle shape and size, orientation of particles, true density of individual particles, moisture content, and applied axial pressure [3, 7, 8, 21, 53–55]. Durability depends on the types of feedstock, fraction of different components (leaf, stem), lignin, extractives, particle size/GMD, fines, and moisture content apart from densification machine and process variables [3–12, 21]. Six different feedstocks were used for briquettes production in this study, and it would be appropriate to validate the relationship of different variables with bulk density and durability; accordingly correlation analysis was performed, and the coefficients are presented in Table 3. As noted in table, moisture and lignin contents had strong positive influence, whereas extractives had negative influence on bulk density of the briquettes. In general, lignin is heavier than extractives which would explain their contribution towards bulk density. Extractives and fines showed negative influence; lignin and GMD showed positive influence on durability of the briquettes. When lignin and extractives content exceed a threshold level of 34% in wood samples, pellet durability decreased [56]. Considering this threshold level, the correlation analysis showed a poor relation ($r = 0.08$). Since moisture contents of the briquettes were in a small range (4.9–9.2% wb), the relation with durability might not be prominent. This correlation analysis reveals that there is a need to determine the threshold level of each variable with respect to density and durability.

4. Conclusions

Briquettes were produced from variety of feedstocks to compare their physical and chemical properties. Statistical analyses revealed the existence of significant changes in chemical compositions, differences in density, porosity, and durability. Correlation analyses confirmed the contribution of lignin, extractives, fines, and particle size towards durability. This study confirms that medium size feedstock with low moisture content and lignin content in the range of

21–25% would result in 2–11-fold increase in density with medium and high durability. Cotton stalk briquettes had a bulk density of 964 kg/m^3 with a durability of 0.923.

Acknowledgments

This research was supported by funding from the Agricultural Experiment Station and North Central Sun Grant Center at South Dakota State University through a Grant provided by the US Department of Transportation, Office of the Secretary, Grant no. DTOS59-07-G-00054.

References

[1] J. S. Tumuluru, C. T. Wright, K. L. Kenney, and J. R. Hess, "A technical review on biomass processing: densification, pre-processing, modeling and optimization," Paper #1009401, ASABE, St. Joseph, Mich, USA, 2010.

[2] W. Stelte, J. K. Holm, A. R. Sanadi, S. Barsberg, J. Ahrenfeldt, and U. B. Henriksen, "A study of bonding and failure mechanisms in fuel pellets from different biomass resources," *Biomass and Bioenergy*, vol. 35, no. 2, pp. 910–918, 2011.

[3] N. Kaliyan, R. V. Morey, M. D. White, and A. Doering, "Roll press briquetting and pelleting of corn stover and switchgrass," *Transactions of the ASABE*, vol. 52, no. 2, pp. 543–555, 2009.

[4] P. Y. S. Chen, J. G. Haygreen, and M. A. Graham, "Evaluation of wood/coal pellets made in a laboratory pelletizer," *Forest Products Journal*, vol. 39, no. 7-8, pp. 53–58, 1989.

[5] P. Lehtikangas, "Quality properties of pelletised sawdust, logging residues and bark," *Biomass and Bioenergy*, vol. 20, no. 5, pp. 351–360, 2001.

[6] P. K. Adapa, G. J. Schoenau, L. G. Tabil, S. Sokhansanj, and B. Crerar, "Pelleting of fractionated alfalfa products," ASAE Paper 036069, ASABE, St. Joseph, Mich, USA, 2003.

[7] C. Rhén, R. Gref, M. Sjöström, and I. Wästerlund, "Effects of raw material moisture content, densification pressure and temperature on some properties of Norway spruce pellets," *Fuel Processing Technology*, vol. 87, no. 1, pp. 11–16, 2005.

[8] S. Mani, L. G. Tabil, and S. Sokhansanj, "Specific energy requirement for compacting corn stover," *Bioresource Technology*, vol. 97, no. 12, pp. 1420–1426, 2006.

[9] N. Kaliyan and R. Vance Morey, "Factors affecting strength and durability of densified biomass products," *Biomass and Bioenergy*, vol. 33, no. 3, pp. 337–359, 2009.

[10] P. Gilbert, C. Ryu, V. Sharifi, and J. Swithenbank, "Effect of process parameters on pelletisation of herbaceous crops," *Fuel*, vol. 88, no. 8, pp. 1491–1497, 2009.

[11] M. Alaru, L. Kukk, J. Olt et al., "Lignin content and briquette quality of different fibre hemp plant types and energy sunflower," *Field Crops Research*, 2011.

[12] M. T. Carone, A. Pantaleo, and A. Pellerano, "Influence of process parameters and biomass characteristics on the durability of pellets from the pruning residues of Olea europaea L," *Biomass and Bioenergy*, vol. 35, no. 1, pp. 402–410, 2011.

[13] K. Theerarattananoon, F. Xu, J. Wilson et al., "Effects of the pelleting conditions on chemical composition and sugar yield of corn stover, big bluestem, wheat straw, and sorghum stalk pellets," *Bioprocess and Biosystems Engineering*, vol. 35, no. 4, pp. 615–623, 2012.

[14] A. Sluiter, B. Hames, D. Hyman et al., "Determination of total solids in biomass and total dissolved solids in liquid process

samples," Tech. Rep. NREL/TP-510-42621, National Renewable Energy Laboratory (NREL), Golden, Colo, USA, 2008.

[15] A. Sluiter, B. Hames, R. Ruiz et al., "Determination of structural carbohydrates and lignin in biomass," Tech. Rep. NREL/TP-510-42618, National Renewable Energy Laboratory (NREL), Golden, Colo, USA, 2008.

[16] A. Sluiter, R. Ruiz, C. Scarlata, J. Sluiter, and D. Templeton, "Determination of extractives in biomass," Tech. Rep. NREL/TP-510-42619, National Renewable Energy Laboratory (NREL), Golden, Colo, USA, 2005.

[17] ASABE Standards 319.3, "Method of determining and expressing fineness of feed materials by sieving," in ASABE Standards, vol. 608, American Society of Agricultural and Biological Engineers, St. Joseph, Mich, USA, 2006.

[18] ASABE Standards 269.4. Cubes, "Pellets and crumbles—definitions and methods for determining density, durability and moisture content," in ASABE Standards, American Society of Agricultural and Biological Engineers, St. Joseph, Mich, USA, 2007.

[19] Z. Zhong and X. S. Sun, "Thermal characterization and phase behavior of cornstarch studied by differential scanning calorimetry," Journal of Food Engineering, vol. 69, no. 4, pp. 453–459, 2005.

[20] SAS Institute, User's Guide: Statistics, Version 9.2, Statistical Analysis System, Inc., Cary, NC, USA, 2010.

[21] N. Kaliyan and R. V. Morey, "Densification characteristics of corn stover and switchgrass," Transactions of the ASABE, vol. 52, no. 3, pp. 907–920, 2009.

[22] P. K. Adapa, L. G. Tabil, G. J. Schoenau, and S. Sokhansanj, "Pelleting characteristics of fractionated sun-cured and dehydrated alfalfa grinds," Applied Engineering in Agriculture, vol. 20, no. 6, pp. 813–820, 2004.

[23] K. Theerarattananoon, F. Xu, J. Wilson et al., "Physical properties of pellets made from sorghum stalk, corn stover, wheat straw, and big bluestem," Industrial Crops and Products, vol. 33, no. 2, pp. 325–332, 2011.

[24] W. Pietsch, Size Enlargement by Agglomeration, John Wiley and Sons, New York, NY, USA, 1991.

[25] B. Zhou, K. E. Ileleji, and G. Ejeta, "Physical property relationships of bulk corn stover particles," Transactions of the ASABE, vol. 51, no. 2, pp. 581–590, 2008.

[26] Z. Colley, O. O. Fasina, D. Bransby, and Y. Y. Lee, "Moisture effect on the physical characteristics of switchgrass pellets," Transactions of the ASABE, vol. 49, no. 6, pp. 1845–1851, 2006.

[27] P. K. Adapa, L. Tabil, and G. Schoenau, "Compaction characteristics of barley, canola, oat and wheat straw," Biosystems Engineering, vol. 104, no. 3, pp. 335–344, 2009.

[28] L. Tabil and S. Sokhansanj, "Process conditions affecting the physical quality of alfalfa pellets," Applied Engineering in Agriculture, vol. 12, no. 3, pp. 345–350, 1996.

[29] R. MacBain, Pelleting Animal Feed, American Feed Manufacturing Association, Chicago, Ill, USA, 1966.

[30] P. D. Grover and S. K. Mishra, "Biomass briquetting: technology and practices. Regional wood energy development program in Asia," Field Document 46, Food and Agriculture Organization of the United Nations, Bangkok, Thailand, 1996.

[31] P. K. Adapa, L. Tabil, G. Schoenau, and A. Opoku, "Pelleting characteristics of selected biomass with and without steam explosion pretreatment," International Journal of Agricultural and Biological Engineering, vol. 3, no. 3, pp. 62–79, 2010.

[32] B. Hill and D. A. Pulkinen, "A study of the factors affecting pellet durability and pelleting efficiency in the production

of dehydrated alfalfa pellets," Special Report, Saskatchewan Dehydrators Association, Tisdale, SK, Canada, 1988.

[33] Y. Li and H. Liu, "High-pressure densification of wood residues to form an upgraded fuel," Biomass and Bioenergy, vol. 19, no. 3, pp. 177–186, 2000.

[34] S. Mani, L. G. Tabil, and S. Sokhansanj, "Evaluation of compaction equations applied to four biomass species," Canadian Biosystems Engineering, vol. 46, pp. 355–361, 2004.

[35] V. Panwar, B. Prasad, and K. L. Wasewar, "Biomass residue briquetting and characterization," Journal of Energy Engineering, vol. 137, no. 2, pp. 108–114, 2011.

[36] D. Mohan, C. U. Pittman, and P. H. Steele, "Pyrolysis of wood/biomass for bio-oil: a critical review," Energy and Fuels, vol. 20, no. 3, pp. 848–889, 2006.

[37] D. K. Lee, V. N. Owens, A. Boe, and P. Jeranyama, Composition of Herbaceous Biomass Feedstocks. SGINC1-07, Sun Grant Initiative North Central Sun Grant Center, South Dakota State University, South Dakota, SD, USA, 2007.

[38] R. J. Hamilton, "Analysis of waxes," in Chemistry, Molecular Biology and Functions, Waxes, Ed., The Oily Press, Dundee, UK, 1995.

[39] F. E. I. Deswarte, J. H. Clark, J. J. E. Hardy, and P. M. Rose, "The fractionation of valuable wax products from wheat straw using CO_2," Green Chemistry, vol. 8, no. 1, pp. 39–42, 2006.

[40] W. Jensen, K. E. Kremer, P. Sieril, and V. Vartiovaara, "The chemistry of bark," in The Chemistry of Wood, B. L. Browning, Ed., pp. 587–666, Interscience Publishers, New York, NY, USA, 1963.

[41] E. L. Back and N. L. Salmen, "Glass transitions of wood components hold implications for molding and pulping processes," Tappi, vol. 65, no. 7, pp. 107–110, 1982.

[42] J. E. G. Van Dam, M. J. A. Van Den Oever, W. Teunissen, E. R. P. Keijsers, and A. G. Peralta, "Process for production of high density/high performance binderless boards from whole coconut husk. Part 1: lignin as intrinsic thermosetting binder resin," Industrial Crops and Products, vol. 19, no. 3, pp. 207–216, 2004.

[43] C. R. Woodcock and J. S. Mason, Bulk Solids Handling: An Introduction to the Practice and Technology, Blackie and Son Ltd, Glasgow, Scotland, 1987.

[44] S. Sokhansanj and A. F. Turhollow, "Biomass densification: cubing operations and costs for corn stover," Applied Engineering in Agriculture, vol. 20, no. 4, pp. 495–499, 2004.

[45] H. Yumak, T. Ucar, and N. Seyidbekiroglu, "Briquetting soda weed (Salsola tragus) to be used as a rural fuel source," Biomass and Bioenergy, vol. 34, no. 5, pp. 630–636, 2010.

[46] S. H. Larsson, M. Thyrel, P. Geladi, and T. A. Lestander, "High quality biofuel pellet production from pre-compacted low density raw materials," Bioresource Technology, vol. 99, no. 15, pp. 7176–7182, 2008.

[47] J. A. Lindley and M. Vossoughi, "Physical properties of biomass briquets," Transactions of the American Society of Agricultural Engineers, vol. 32, no. 2, pp. 361–366, 1989.

[48] R. Samuelsson, M. Thyrel, M. Sjöström, and T. A. Lestander, "Effect of biomaterial characteristics on pelletizing properties and biofuel pellet quality," Fuel Processing Technology, vol. 90, no. 9, pp. 1129–1134, 2009.

[49] C. Weilong, P. Zhijian, Z. Pengfei, N. Qin, T. W. Deines, and B. Lin, "Ultrasonic-vibration-assisted pelleting of switchgrass: effects of ultrasonic vibration," Transactions of Tianjin University, vol. 17, no. 5, pp. 313–319, 2011.

[50] J. Vinterbäck, "Pellets 2002: the first world conference on pellets," Biomass and Bioenergy, vol. 27, no. 6, pp. 513–520, 2004.

[51] J. B. Dobie, "Materials-handling systems for hay wafers," *Agricultural Engineering*, vol. 42, pp. 692–697, 1961.

[52] L. Tabil, P. Adapa, and M. Kashaninejad, "Biomass feedstock pre-processing—part 2: densification," in *Biofuel's Engineering Process Technology*, M. A. Dos Santos Bernardes, Ed., chapter 19, pp. 439–464, 2011.

[53] M. Peleg, "Physical characteristics of food powders," in *Physical Properties of Food*, pp. 293–321, AVI Publishing Co., Inc., Westport, Conn, USA, 1983.

[54] W. Lang and S. Sokhansanj, "Bulk volume shrinkage during drying of wheat and canola," *Journal of Food Process Engineering*, vol. 16, no. 4, pp. 305–314, 1993.

[55] S. Sokhansanj and W. Lang, "Prediction of kernel and bulk volume of wheat and canola during adsorption and desorption," *Journal of Agricultural Engineering Research*, vol. 63, no. 2, pp. 129–136, 1996.

[56] J. Bradfield and M. P. Levi, "Effect of species and wood to bark ratio on pelleting of southern woods," *Forest Products Journal*, vol. 34, no. 1, pp. 61–63, 1984.

5

In Vitro Propagation of Muña-Muña (*Clinopodium odorum* (Griseb.) Harley)

María Soledad Diaz, Lorena Palacio, Ana Cristina Figueroa, and Marta Ester Goleniowski

CEPROCOR, Science and Technology Ministry, Arenales 230, X5004APP Córdoba, Argentina

Correspondence should be addressed to Marta Ester Goleniowski, mgoleniowski@gmail.com

Academic Editor: Jochen Kumlehn

A micropropagation protocol was developed which may assist in the safeguarding and augmentation of dwindling natural populations of *Clinopodium odorum* (Griseb.) Harley, a critically and endangered medicinal plant. Factors affecting culture initiation bud sprouting and growth, rooting, and acclimatization were studied, using nodal segments of *in vitro* germinated seedling as primary explants on six media supplemented with different concentrations and combinations of 6-benzylaminopurine (BAP) (0.5–1.5 and 2-Naphthalene acetic acid (NAA) (0.5–1.5). Best results for culture initiation with sustainable multiplication rates (100%) were obtained on WP medium without any growth regulator. WP with the addition of 0.5 : 1 or 0.5 : 1.5) of BAP and NAA promoted a higher elongation; however, the optimum number of nodes were obtained in plantlets grown on 1/2 MS with the addition of 1 : 1.5 of BAP and NAA. Culture of sectioned individual nodes transferred to the media with different rates of BAP and NAA 1/2 MS-9 (1.5 : 1.5), SH-8 (1.5 : 1.0), and 1/2 B5-4 (1.0 : 0.5) media resulted in no proliferated shoots. The *in vitro* plants were successfully acclimatized garden soil and sand (2 : 1) in the greenhouse, with over 90% survival rate. The *in vitro*-grown plants could be transferred to *ex vitro* conditions and the efficacy in supporting *ex vitro* growth was assessed, with a view to develope longer-term strategies for the transfer and reintroduction into natural habitats.

1. Introduction

Clinopodium odorum (Griseb.) Harley is a small deciduous shrub of the family Lamiaceae commonly known as muña-muña. In Argentine, the species is restricted to a very specific niche; plant exploration studies in the region has revealed the occurrence of only small populations that is especially characteristic of Pampa de Achala (Córdoba) distributed at an elevation of 1200 m [1].

The fresh herb is used as a flavoring agent for aliments and an infusion of the aerial parts is utilized as an anti-catarrhal, antispasmodic, stringent, carminative, digestive, diuretic, laxative, stomachic, soporific, vermifuge, menstrual suppression, flatulent, colic, and tonic digestive and antispasmodic and to help in parturition [2, 3].

This plant species have been excessively collected from its habitats and become endangered due to different contributory factors: extensive denudation of the forest floor, caused by cattle grazing and collection of leaf litter, and removal from the wilderness which is highly used in the preparation of liquor companies "Amargos serranos" [4, 5].

Seed is only the means of propagating *C. odorum*, but seeds have reduced probability of germination when adult plants are already growing in the area, thus reducing species dissemination [6].

The use of *in vitro* techniques for rapid and mass propagation offers possibilities for recovery of endangered species thus reducing the risk of extinction. This technique could enable production of large numbers of clonal plants in relatively short-time periods using very little starting material [7].

The establishment of *in vitro* germplasm banks in developing countries has great importance, but these techniques must be associated with other plant genetic resources conservation practices [8, 9]. The *in vitro* conservation techniques allow material exchanges among germplasm banks, and the germplasm keeps its sanitary conditions and viability during the transport [10, 11].

To our knowledge, there are no reports for *in vitro* propagation of *C. odorum*; it is known that developing efficient micropropagation procedures for particular species generally requires detailed studies to define specific composition of mineral salts, plant growth regulators, and organic compounds in the culture medium.

In view of the importance of this species, a holistic approach for the propagation using *in vitro* methods for its conservation have been described in this paper.

2. Materials and Methods

2.1. Plant Material. Seeds of *C. odorum* were collected in 2010 in their natural habitat during the months of February and March when the fruits are ripened and kept at room temperature until the initiation of the experiments. A voucher specimen was deposited in the International Herbarium of the National University of Río Cuarto, Argentine.

2.2. Seeds Germination. For *in vitro* germination different conditions were assessed: (A) control, (B) washing overnight in running tap water, (C) soaking at $1 \, \text{mg} \, \text{L}^{-1}$ of Gibberellic acid (GA_3) during 12 h, (D) soaking at $10 \, \text{mg} \, \text{L}^{-1}$ GA3 during 12 h, (E) soaking at $100 \, \text{mg} \, \text{L}^{-1}$ GA_3 during 12 h, (F) washing overnight in running tap water plus soaking at $1 \, \text{mg} \, \text{L}^{-1}$ GA_3 during 12 h, (G) washing overnight in running tap water plus soaking at $10 \, \text{mg} \, \text{L}^{-1}$ GA_3 during 12 h and (H) washing overnight in running tap water plus soaking at $100 \, \text{mg} \, \text{L}^{-1}$ of GA_3 during 12 h.

Seeds were surface sterilized with a solution of 70% (v/v) ethanol for 2 min and rinsed 3 times with sterile distilled water, followed by 15 min in a solution of sodium hypochlorite (NaOCl) 1.5% (v/v) for 15 min and finally rinsed 3 times with sterile distilled water. These surface sterilized seeds were explanted onto Murashige and Skoog, 1962 (MS) [12], germination culture medium, supplemented with 3% (w/v) sucrose, 0.7% (w/v) agar, and pH 5.8, using 20 seeds/lasks. Culture tubes were incubated at $25 \pm 2°C$ for 15 days in darkness. Germination rates were measured 15 days after starting the culture. The plantlets were kept under a photoperiod of 16 h of cool white fluorescent light ($30 \, \mu E$ m-2S-1) followed by 8 h darkness.

Each treatment (in triplicate) consisted of 20 seeds and germination was monitored weekly up to 10 weeks.

2.3. Culture Initiation. Clonal plants (plants originally derived from seed) were initially propagated on media for 4-5 months to increase stock populations for multiplication rate assessment. Multiplication experiments were carried out using nodal segments with axillaries' buds of *in vitro* germinated seedling aged on half-strength MS major and minor salts medium with $1 \, \text{mg} \, \text{L}^{-1}$ of indole-3-butyric acid (IBA) (initiation medium).

The regeneration potential of nodes with axillaries' buds (0.5–1 cm) obtained from plants grown in the above *in vitro* culture condition was evaluated in terms of frequency (survival percentage), principal shoot length, nodes number, shoot formation, and rooting by culturing on six media:

MS, Schenk and Hildebrandt, 1972 (SH) [13], Lloyd and McCown, 1980 (WP) [14], Gamborg et al., 1968 (B5) [15], and MS and B5 at half-strength salt medium (12 MS and 12 B5), containing 3% (w/v) of sucrose and supplemented with 6-benzylaminopurine (BAP) and 2-naphthalene acetic acid (NAA) at different concentrations ($0.5–1.5 \, \text{mg} \, \text{L}^{-1}$) and combinations. The pH media was adjusted to 5.6 with NaOH or HCl, prior to gelling with 0.7% (w/v) agar-agar, dispensed (10 mL) into culture tubes and sterilized by autoclaving (121°C for 15 min.). In all the experiments, the chemicals used were of analytical grade (Sigma-Aldrich, St. Louis, MO, USA and E. Merck, Darmstadt, Germany).

All cultures were maintained under culture conditions: $25 \pm 2°C$, 45–55% relative humidity, and 16 h photoperiod under cool white fluorescent light ($30 \, \mu E$ m-2S-1). Explants were observed daily for the first week for signs of contamination and thereafter weekly for signs of growth and development.

2.4. Acclimatization. Young plants with well-developed roots were carefully removed from the glass culture vessels; the roots washed with sterile water to eliminate the excess of agar and transferred to plastic cups containing autoclaved garden soil and sand (2 : 1). Each pot was covered with plastic bag for plantlets acclimatization during 15 days, which was punctured the last 5 days to allow air exchange.

Each plantlet was irrigated with distilled water every 2 days for 2 weeks followed by tap water for two other weeks. The potted plantlets were initially maintained under culture room conditions (4 weeks) and later transferred to normal greenhouse conditions.

2.5. Statistical Analysis. All experiments were conducted at least three times. For each treatment, a minimum of 14 and a maximum of 25 explants were used. For statistical analysis, all quantitative data expressed as percentages were first submitted to arcsine transformation and the means corrected for bias before a new conversion of the means and standard error back into percentages [16]. Statistical analysis was performed by ANOVA, and significantly different means were identified using Duncan's test $P \leq 0.05$.

3. Results and Discussion

3.1. Effect of Different Strategies on Seed Germination. Many studies evaluating adequate *in vitro* conditions for propagation of several species have been conducted using *in vitro*-germinated plants, to avoid disinfection of explants, because of extreme sensitivity to common disinfection procedures and subsequent low survival rates [17, 18]. A seed was scored as "germinated" when the root emerged and grew to 1 cm. *C. odorum* seeds were capable of germinating in all the treatments used, although several rates of germination were obtained.

As shown in Table 1, only at 28.1% germination was observed in the control. It is known that in some plant species, seed dormancy is a condition which prevents the

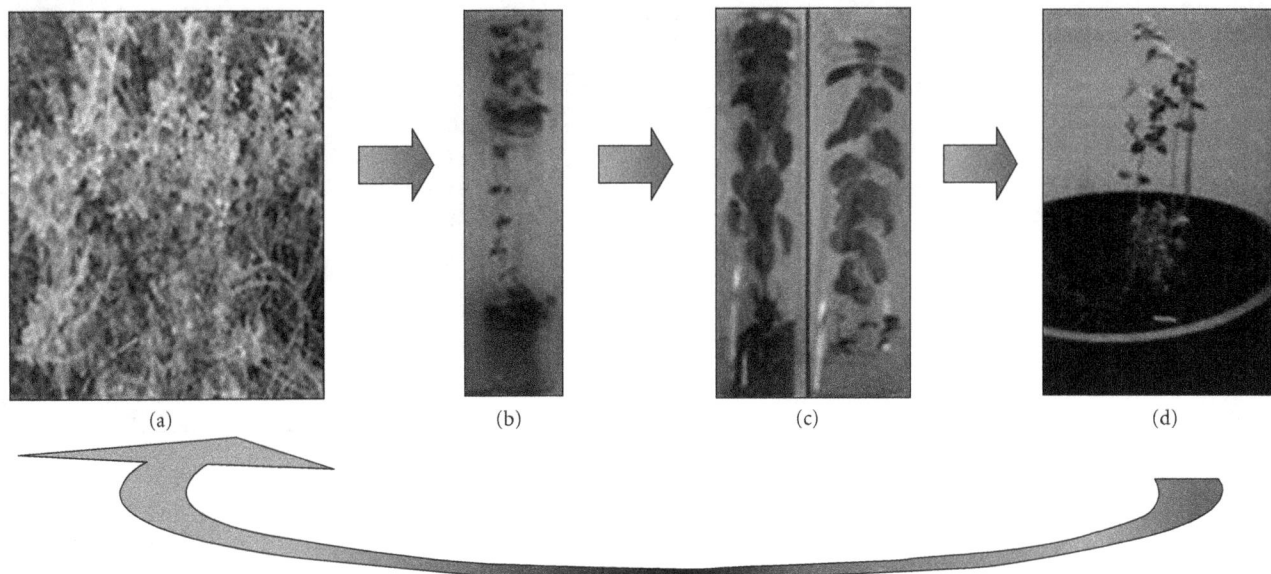

FIGURE 1: *In vitro* plant regeneration from single node of *C. odorum*. (a) General aspect of the wild plant in its habitat, (b) *in vitro* germinated plant after 10 weeks on MS medium, (c) *in vitro* grown plant after 2 months of culture on WP-2 medium, and (d) acclimatized plant, 1 month after potting into a mixture of garden soil and sand (2 : 1) subtracts.

TABLE 1: Effect of pregermination conditions on *in vitro C. odorum* culture, (A) control, (B) washing overnight in running tap water, (C) soaking at $1 \, mg \, L^{-1}$ GA_3 during 12 h, (D) soaking at $1 \, mg \, L^{-1}$ of GA_3 during 12 h, (E) soaking at $100 \, mg \, L^{-1}$ of GA_3 during 12 h, (F) washing overnight in running tap water + soaking at $1 \, mg \, L^{-1}$ GA_3 during 12 h, (G) washing overnight in running tap water + soaking at $10 \, mg \, L^{-1}$ GA_3 during 12 h, and (H) washing overnight in running tap water + soaking at $100 \, mg \, L^{-1}$ GA_3 during 12 h.

Pregermination condition	Percentage of germination
A	28.1 %
B	41.8 %
C	60.2 %
D	37.9 %
E	55.8 %
F	62.2 %
G	42.9 %
H	54.8 %

TABLE 2: Percentage of survival (%) of *C. odorum* explants on different culture media (1/2 MS, SH, WP, 1/2 B5) with the addition of different concentrations and combination of PGRs.

PGRs ($mg \, L^{-1}$)	Culture media			
BAP/NAA	1/2 MS	SH	WP	1/2 B5
0.00	72.22	62.50	100.00	15.78
0.5/0.5	87.50	70.00	84.61	45.00
0.5/1.00	77.77	50.00	93.75	40.00
0.5/1.50	62.50	28.57	57.14	45.00
1.00/0.5	53.84	62.50	38.88	21.05
1.00/1.00	40.00	63.63	53.33	0.00
1.00/1.50	88.88	50.00	83.33	31.57
1.50/0.50	37.50	40.00	61.90	30.43
1.50/1.00	52.63	0.00	15.78	5.26
1.50/1.50	0.00	33.33	33.33	31.57

seed from germinating, even when it is perfectly healthy and all conditions for germination are at an optimum.

Seed dormancy is also a prevalent cause of very slow and erratic germination in the majority of wild plants [19–22].

Therefore, strategies to induce physiological and mechanical seed dormancy rupture were carried out, which caused a positive reaction on the *C. odorum* seed germination response. The washing overnight in running tap water condition was not enough mechanism to prevent total seed dormancy, with the resulting germination rate being only 41.8%. Germination rate is the "speed or velocity" of germination and can be expressed as the time it takes for

a defined percentage of seed to germinate in our case at 15 days.

More than 2-fold improvement in germination was recorded in seeds washing overnight in running tap water plus soaking at $1 \, mg \, L^{-1}$ GA_3 during 12 h, in which a physiological dormancy release increasing germination rates at 62.2% was obtained (Table 1).

This germination value was similar to those obtained with soaking $1 \, mg \, L^{-1}$ GA_3; it suggests that *C. odorum* seeds had among others a requirement to GAs which is known that induce the production of enzymes to digest the endosperm that would otherwise from a mechanical barrier to radicle emergence reported that the use of this plant growth regulator (GA_3) is an improved alternative to release

TABLE 3: Effect of nutrient media and PGRs (mg^{-1}) concentrations on *in vitro C. odorum*. Values followed by the same letter within a column are not significantly different at the $P \leq 0.05$, according ANOVA analysis and by Duncan's test.

Culture media	BAP	NAA	Principal shoot length (cm)	Number of nodes (principal shoot)	Number of axillaries shoots
1/2 MS-0	0.00	0.00	6.06 ± 0.71[a]	9.46 ± 0.91[a]	1.00 ± 0.35[a]
1/2 MS-1	0.50	0.50	8.91 ± 0.68[a]	11.14 ± 0.88[a]	0.79 ± 0.34[a]
1/2 MS-2	0.50	1.00	9.81 ± 0.68[a]	11.93 ± 0.88[a]	0.79 ± 0.34[a]
1/2 MS-3	0.50	1.50	7.13 ± 0.81[a]	7.89 ± 1.09[a]	0.44 ± 0.43[a]
1/2 MS-4	1.00	0.50	3.91 ± 0.96[a]	5.14 ± 1.24[a]	1.00 ± 0.48[a]
1/2 MS-5	1.00	1.00	4.69 ± 0.90[a]	7.89 ± 1.09[a]	1.22 ± 0.43[a]
1/2 MS-6	1.00	1.50	12.14 ± 0.64[b]	16.38 ± 0.82[b]	1.75 ± 0.32[a]
1/2 MS-7	1.50	0.50	2.70 ± 1.04[a]	4.17 ± 1.34[a]	1.50 ± 0.52[a]
1/2 MS-8	1.50	1.00	5.22 ± 0.81[a]	7.50 ± 1.04[a]	3.60 ± 0.40[a]
1/2 MS-9	1.50	1.50	—	—	—
SH-0	0.00	0.00	7.46 ± 1.14[a]	6.60 ± 1.47[a]	1.80 ± 0.57[a]
SH-1	0.50	0.50	8.89 ± 0.96[a]	10.14 ± 1.24[b]	1.29 ± 0.48[a]
SH-2	0.50	1.00	8.60 ± 1.47[a]	10.00 ± 1.89[a]	1.67 ± 0.74[a]
SH-3	0.50	1.50	4.15 ± 1.80[a]	4.50 ± 1.32[a]	1.00 ± 0.90[a]
SH-4	1.00	0.50	3.56 ± 1.14[a]	4.40 ± 1.47[a]	0.80 ± 0.57[a]
SH-5	1.00	1.00	6.84 ± 0.96[a]	7.71 ± 1.24[a]	0.71 ± 0.48[a]
SH-6	1.00	1.50	5.85 ± 1.04[a]	8.33 ± 1.34[a]	1.17 ± 0.52[a]
SH-7	1.50	0.50	4.38 ± 1.27[a]	6.25 ± 1.64[a]	1.03 ± 0.48[a]
SH-8	1.50	1.00	—	—	—
SH-9	1.50	1.50	6.20 ± 1.47[a]	8.33 ± 1.89[a]	4.67 ± 0.74[a]
WP-0	0.00	0.00	7.74 ± 0.68[a]	7.71 ± 0.88[a]	0.93 ± 0.34[a]
WP-1	0.50	0.50	11.24 ± 0.77[a]	12.91 ± 0.99[a]	1.27 ± 0.39[a]
WP-2	0.50	1.00	12.67 ± 0.66[b]	13.07 ± 0.85[a]	1.53 ± 0.33[a]
WP-3	0.50	1.50	12.61 ± 0.74[b]	14.42 ± 0.95[a]	1.17 ± 0.37[a]
WP-4	1.00	0.50	9.39 ± 0.96[a]	9.71 ± 1.24[a]	0.29 ± 0.42[a]
WP-5	1.00	1.00	10.19 ± 0.90[a]	10.38 ± 1.16[a]	1.25 ± 0.45[a]
WP-6	1.00	1.50	9.14 ± 0.81[a]	10.50 ± 1.04[a]	0.80 ± 0.40[a]
WP-7	1.50	0.50	5.74 ± 0.71[a]	8.85 ± 0.91[a]	0.69 ± 0.35[a]
WP-8	1.50	1.00	2.57 ± 1.47[a]	6.33 ± 1.89[a]	0.33 ± 0.74[a]
WP-9	1.50	1.50	2.55 ± 1.27[a]	6.25 ± 1.64[a]	0.75 ± 0.64[a]
1/2 B5-0	0.00	0.00	2.37 ± 1.44[a]	4.33 ± 1.77[a]	0.67 ± 1.00[a]
1/2 B5-1	0.50	0.50	3.70 ± 0.83[a]	4.33 ± 1.02[a]	5.33 ± 0.58[a]
1/2 B5-2	0.50	1.00	6.21 ± 0.88[a]	6.63 ± 1.08[a]	6.63 ± 0.61[a]
1/2 B5-3	0.50	1.50	4.93 ± 0.83[a]	6.78 ± 1.02[a]	7.00 ± 0.58[b]
1/2 B5-4	1.00	0.50	4.72 ± 1.25[a]	7.00 ± 1.53[a]	6.75 ± 0.87[a]
1/2 B5-5	1.00	1.00	—	—	—
1/2 B5-6	1.00	1.50	6.12 ± 1.02[a]	6.83 ± 1.25[a]	3.00 ± 0.71[a]
1/2 B5-7	1.50	0.50	2.39 ± 0.94[a]	3.71 ± 1.16[a]	4.14 ± 0.65[a]
1/2 B5-8	1.50	1.00	1.90 ± 2.49[a]	5.00 ± 3.06[b]	—
1/2 B5-9	1.50	1.50	4.03 ± 1.02[a]	4.33 ± 1.25[a]	3.00 ± 0.71[a]

dormancy seed; this induction mechanism resulted to be more effective on other species [19, 22, 23].

3.2. Plant Tissue Culture. Successful plant tissue culture depends on the choice of nutrient medium. The explants of most plant species can be grown on completely defined media and there is a small number of standard culture media that are widely used with inorganic supplements. No single medium can be used for all types of plants and organs, so the composition of the culture medium for each plant material

has to be worked out [24]. Best results for culture initiation with sustainable multiplication rates (100%) were obtained on WP medium without any growth regulator.

Significant differences were observed among the media for shoot elongation (Table 2). The maximum shoot length was significantly higher on hormone-free media WP and SH media resulting 7.74 ± 0.68 and 7.46 ± 1.14 cm, respectively (Table 3).

A scarce elongation (2.37 ± 1.44 cm) was observed when the explants were placed on 12 B5. The plants produced on 12

MS showed significant differences for the number of nodes per plant among other treatments resulting with a value of 9.46 ± 0.91 at $P \leq 0.05$.

The media 12 MS-9, SH-8, and 12 B5-4 were not effective to induce growth and caused necrosis, thus resulting in death of nodal segments.

The steps in plant tissue culture protocols can impose a series of PGRs; in many plants, a balance between two groups of PGRs; auxins and cytokinins, constitutes the most conspicuous group of plant hormones regulating cell division and elongation and determining morphogenesis. An adequate assessment of the suitability requirement of PGRs depends upon the type of plant tissues or explants that are used for culture.

The use of culture media supplemented with BAP and NAA has been associated with *in vitro* morphogenetic events and it is known that cytokinins have the ability to induce the shoot formation.

Mc Cown and Sellmer, 1987 [25], reported that the effect of growth regulators can be strongly modified by the medium on which the culture is grown. When the experiments were conducted to determine how different ratios of plant growth regulators would support microplant growth, significant difference with an excellent growth response (length of principal shoot and node number) corresponded to plantlets grown on 12 MS-6 (12.14 ± 0.64 cm) and WP-2 (12.67 ± 0.66 cm) and WP-3 (12.61 ± 0.74 cm). 12 B5-3 affected significantly shoot proliferation as it recorded the highest number of new shoots/explants, 7.0 ± 0.58 after 8 weeks in culture (Table 3).

A rooting rate of 90% was attained when the plantlets were placed in inductive media or included in hormone-free, suggesting NAA incorporated to the media was not necessary.

Plantlets were acclimatized successfully when transferred to trays containing A mix substrate of plastic cups containing autoclaved garden soil and sand ($2 : 1$). Rooted plantlets were transferred into pots and the survival rate was the 85% after 1 month (Figure 1(d)).

4. Conclusion

In conclusion, an efficient protocol for micropropagation of an important medicinal plant *C. odorum* was developed by testing various concentrations of growth regulators and nutrition conditions. The success of plant tissue culture as a mean of plant propagation is greatly influenced by nature of the culture medium used; they are grown *in vitro* on artificial media, which supplies the nutrients necessary for growth.

Different nutrient media hormone-free tested to find out the suitable starting nutrient medium supplemented with 3% (w/v) sucrose, for 8 weeks for *in vitro* nodal cuttings establishment, showed that an efficient survival percentage was obtained on plantlets grown on free-hormone WP medium. The free-hormone medium contains the proportion of inorganic nutrients to satisfy the nutritional as well as the physiological needs of this species *in vitro* culture. However, the optimum growth depended on the addition of

PGRs. The positive response to these additions indicated a requirement of the explants for a morphogenetic induction as was observed when the explants grown on 12 B5-3, this medium appear to be adequate for new shoots formation.

Our investigation revealed that no auxin supplementation was necessary for rooting differentiation (100%) in proliferation of regenerated shoots.

The *in vitro*-grown plants could be transferred to ex vitro conditions, with a view to develop longer-term strategies for the transfer and reintroduction of micropropagated *C. odorum* plants into natural habitat [26, 27]. The results will make the conservation and propagation of the species much easier.

Abbreviations

BAP: 6-Benzylaminopurine
B5:   Gamborg medium
GA3:   Gibberellic acid
IBA: Indole-3-butyric acid
NAA: 2-Naphthalene acetic acid
PGRs: Plant growth regulators
MS: Murashige and Skoog medium
SH: Schenk and Hildebrandt medium
WP: Woody plant medium (Lloyd and McCown) medium
12MS: Half-strength Murashige and Skoog macronutrients
12B5:   Half-strength Gamborg macronutrients.

Acknowledgment

The authors thank the Consejo Nacional de Investigaciones Científicas y Tecnicas (CONICET) Project grant PIP 114-200801-00408.

References

[1] G. E. Barboza, J. J. Cantero, C. Núñez, A. Pacciaroni, and L. Ariza Espinar, *Medicinal Plants: Review and a phytochemical and ethnopharmacological Screening of the Native Argentine Flora*, vol. 34, Kurtziana, 2009.

[2] G. B. Mahady, "Medicinal plants for the prevention and treatment of bacterial infections," *Current Pharmaceutical Design*, vol. 11, no. 19, pp. 2405–2427, 2005.

[3] R. M. Harley and A. G. Paucar, "List of species of tropical American *Clinopodium* (Labiatae), with new combinations," *Kew Bulletin*, vol. 55, no. 4, pp. 917–927, 2000.

[4] M. E. Goleniowski, G. A. Bongiovanni, L. Palacio, C. O. Nuñez, and J. J. Cantero, "Medicinal plants from the "Sierra de Comechingones", Argentina," *Journal of Ethnopharmacology*, vol. 107, no. 3, pp. 324–341, 2006.

[5] G. J. Martínez, A. M. Planchuelo, E. Fuentes, and M. Ojeda, "A numeric index to establish conservation priorities for medicinal plants in the Paravachasca Valley, Córdoba, Argentina," *Biodiversity and Conservation*, vol. 15, no. 8, pp. 2457–2475, 2006.

[6] J. J. Cantero and C. A. Bianco, "Las plantas vasculares del suroeste de la provincia de Córdoba. Catálogo preliminar de

las especies," *Revista Universidad Nacional De Río Cuarto*, vol. 6, pp. 65–75, 1986.

[7] A. Rubluo, V. Chávez, A. P. Martínez, and O. Martínez-Vázquez, "Strategies for the recovery of endangered orchids and cacti through in-vitro culture," *Biological Conservation*, vol. 63, no. 2, pp. 163–169, 1993.

[8] G. M. Alves and M. P. Guerra, "Micropropagation for mass propagation and conservation of Vrieseafri burgensis var. paludosa from microbuds," *Journal of the Bromeliad Society*, vol. 515, pp. 202–212, 2001.

[9] H. K. Badola, H. K. Badola, and B. K. Pradhan, "Chemical stimulation of seed germination in ex situ produced seeds in *Swertia chirayita*, a critically endangered medicinal herb," *Research Journal of Seed Science*, vol. 3, no. 3, pp. 139–149, 2010.

[10] F. Engelmann, "*In vitro* conservation methods," in *Biotechnology and Plant Genetic Resources*, J. A. Callow, B. V. Ford-Lloyd, and H. J. Newbury, Eds., pp. 119–161, CAB International, Oxon, England, 1997.

[11] A. Rech Filho, L. L. Dal Vesco, R. O. Nodari, R. W. Lischka, C. V. Müller, and M. P. Guerra, "Tissue culture for the conservation and mass propagation of Vriesea reitzii Leme and Costa, a bromeliad threatened of extinction from the Brazilian Atlantic Forest," *Biodiversity and Conservation*, vol. 14, no. 8, pp. 1799–1808, 2005.

[12] T. Murashige and F. Skoog, "A revised medium for rapid growth and bio assays with tobacco tissue cultures," *Physiology Plant*, vol. 15, pp. 473–497, 1962.

[13] R. U. Schenk and A. Hildebrandt, "Medium and techniques for induction and growth of Monocotyledonous and dicotyledonous plant cell cultures," *Canadian Journal of Botany*, vol. 350, pp. 199–204, 1972.

[14] G. Lloyd and B. McCown, "Commercially-feasibible micropropagation of mountain laurel, *Kalmia latifolia*, by use of shoot-tip culture," *Proceedings International Plant Propagators Society*, vol. 30, pp. 421–427, 1980.

[15] O. L. Gamborg, R. A. Miller, and K. Ojima, "Nutrient requirements of suspension cultures of soybean root cells," *Experimental Cell Research*, vol. 50, no. 1, pp. 151–158, 1968.

[16] J. H. Zar, *Biostatistical Analysis*, Prentice-Hall, Englewood Clifffs, NJ, USA, 1996.

[17] E. Pérez-Molphe-Balch and C. A. Dávila-Figueroa, "*In vitro* propagation of *Pelecyphora aselliformis* Ehrenberg and *P. strobiliformis* Werdermann (Cactaceae)," *In vitro Cellular and Developmental Biology-Plant*, vol. 38, no. 1, pp. 73–78, 2002.

[18] M. D. S. Santos-Díaz, R. Méndez-Ontiveros, A. Arredondo-Gómez, and M. D. L. Santos-Díaz, "*In vitro* organogenesis of *Pelecyphora aselliformis* Erhenberg (Cactaceae)," *In vitro Cellular and Developmental Biology-Plant*, vol. 39, no. 5, pp. 480–484, 2003.

[19] P. Grappin, D. Bouinot, B. Sotta, E. Miginiac, and M. Jullien, "Control of seed dormancy in *Nicotiana plumbaginifolia*: post-imbibition abscisic acid synthesis imposes dormancy maintenance," *Planta*, vol. 210, no. 2, pp. 279–285, 2000.

[20] G. Metzger, "Plant hormone interactions during seed dormancy release and germination," *Seed Science Research*, vol. 15, no. 4, pp. 281–307, 2005.

[21] W. E. Finch-Savage and G. Leubner-Metzger, "Seed dormancy and the control of germination," *New Phytologist*, vol. 171, no. 3, pp. 501–523, 2006.

[22] J. V. Jacobsen, D. W. Pearce, A. T. Poole, R. P. Pharis, and L. N. Mander, "Abscisic acid, phaseic acid and gibberellin contents associated with dormancy and germination in barley," *Physiologia Plantarum*, vol. 115, no. 3, pp. 428–441, 2002.

[23] B. Kucera, M. A. Cohn, and G. Leubner-Metzger, "Plant hormone interactions during seed dormancy release and germination," *Seed Science Research*, vol. 15, no. 4, pp. 281–307, 2005.

[24] J. Wu, X. Zhang, Y. Nie, S. Jin, and S. Liang, "Factors affecting somatic embryogenesis and plant regeneration from a range of recalcitrant genotypes of Chinese cottons (*Gossypium hirsutum* L.)," *In vitro Cellular and Developmental Biology - Plant*, vol. 40, no. 4, pp. 371–375, 2004.

[25] B. H. Mc Cown and J. C. Sellmer, "General media and vessels suitable for woody plant culture," in *Cell and Tissue Culture in Forestry*, J. M. Bonga and D. J. Durzan, Eds., vol. 1 of *General Principles and Biotechnology*, pp. 4–16, Martinus Nijhoff Publishers, Dordrecht, The Netherlands, 1987.

[26] M. Lambardi and A. De Carlo, "Application of tissue culture to the germplasm conservation of temperate broad-leaf trees," in *Micropropagation of Woody Trees and Fruits*, S. M. Jain and K. I. Ishii, Eds., pp. 241–248, Kluwer Academic, Dordrecht, The Netherlands, 2003.

[27] L. Palacio, M. C. Baeza, J. J. Cantero, R. Cusidó, and M. E. Goleniowski, "*In vitro* propagation of "Jarilla" (*Larrea divaricata* Cav.) and Secondary Metabolite Production," *Biological and Pharmaceutical Bulletin*, vol. 31, no. 12, pp. 2321–2325, 2008.

6

Antioxidant and Hepatoprotective Properties of Tofu (*Curdle Soymilk*) against Acetaminophen-Induced Liver Damage in Rats

Ndatsu Yakubu,[1] **Ganiyu Oboh,**[2] **and Amuzat Aliyu Olalekan**[1]

[1] *Department of Biochemistry, Ibrahim Badamasi Babangida University, Niger State, Lapai, Nigeria*
[2] *Department of Biochemistry, Federal University of Technology, Ondo State, Akure, Nigeria*

Correspondence should be addressed to Ndatsu Yakubu; ndatsuyakubu2011@hotmail.com

Academic Editor: Gabriel A. Monteiro

The antioxidant and hepatoprotective properties of tofu using acetaminophen to induce liver damage in albino rats were evaluated. Tofus were prepared using calcium chloride, alum, and steep water as coagulants. The polyphenols of tofu were extracted and their antioxidant properties were determined. The weight gain and feed intake of the rats were measured. The analysis of serum alanine aminotransferase (ALT), alkaline phosphatase (ALP), aspartate aminotransferase (AST), and lactate dehydrogenase (LDH) activities and the concentrations of albumin, total protein, cholesterol, and bilirubin were analyzed. The result reveals that the antioxidant property of both soluble and bound polyphenolic extracts was significantly higher in all tofus, but the steep water coagulated tofu was recorded higher. Rats fed with various tofus and acetaminophen had their serum ALP, ALT, AST, and LDH activities; total cholesterol; and bilirubin levels significantly ($P < 0.05$) reduced, and total protein and albumin concentrations increased when compared with basal diet and acetaminophen administered group. Therefore, all tofus curdled with various coagulants could be used to prevent liver damage caused by oxidative stress.

1. Introduction

Reactive oxygen species (ROS) have been implicated in more than 100 diseases [1]. Foods (tubers, grains, fruits, and vegetables) provide a wide variety of ROS-scavenging antioxidants such as phytochemical and antioxidant vitamins [2, 3]. The increased consumption of fruits and vegetables, containing high levels of phytochemicals, has been recommended to prevent or reduce oxidative stress in the human body [2–4]. The natural antioxidant defense mechanisms can be insufficient and hence dietary intake of antioxidant components is important and recommended [5].

Liver disease is a worldwide problem. Conventional drugs used in the treatment of liver diseases are sometimes inadequate and can have serious adverse effects [6]. Soybeans are inexpensive and serve as high quality protein source. Soymilk and tofu consumption is increasing in Nigeria due to animal diseases such as mad cow disease, global shortage of animal protein, strong demand for healthy (cholesterol-free and low in saturated fat) and religious halal food, and economic reasons [7]. The greater acceptance of soy foods by the general population is due to increased recognition of the health benefits of soy foods, especially by those who want to reduce their consumption of animal products [8]. Tofu, also known as soybean curd, is a soft cheese-like food made by curdling fresh hot soymilk with a coagulant [9]. Traditionally, in Nigeria, it is produced by curdling fresh hot soymilk with $CaCl_2$, $MgSO_4$ alum, or steep water (effluence from pap produced from maize) [10–12]. Tofu is rich in proteins; low in saturated fats; higher in polyunsaturated fatty acids; cholesterol-free; and a good source of β-vitamin, minerals, isoflavones, and antioxidants (carotenoids, vitamins C and E, phenolic and thiol (SH) compounds, and essential amino acids) [8, 12]. It has been demonstrated that phenolic compounds are effective antioxidants, due to the formation of stable phenoxyl radical [13], and that flavonoids have potent antioxidant activities by scavenging hydroxyl radicals, superoxide anions, and lipid peroxy radicals [14]. Soybean products reduce the risk of heart diseases by lowering levels of oxidized cholesterol, which is taken up more rapidly by coronary artery walls to form dangerous plaques [15]. Previous research has shown that soy consumption reduces cholesterol in general while also decreasing the amount of bad cholesterol (low-density lipoprotein, LDL) in the body and maintaining the amount of good cholesterol (LDL) [16, 17].

Acetaminophen is a commonly used and safe analgesic drug, which is known to cause centribular necrosis upon overdose [18]. Its toxicity accounts for many emergency hospital admission and continues to be associated with high mortality. The hepatotoxic effect of acute paracetamol overdose is well known and has been extensively reviewed [19]. However, tofu (soybean product) contains compounds that are valuable antioxidants and protecting molecules, which can trap or destroy free radicals and subsequently protect us from damage due to oxidative stress [9]. In view of this, people have started to take an interest in tofu consumption due to its good nutritional and health benefit to human. Therefore, this research is designed to evaluate its polyphenol distribution and the antioxidant properties in tofu produced using different coagulants, and to compare its hepatoprotective properties on acetaminophen-induced toxicity in rat's liver.

2. Materials and Methods

The soybeans (*Glycine max*), TGX923-1E variety, were obtained directly from Ibrahim Badamasi Babangida University, Lapai experimental farm in Niger State, Nigeria. The alum and calcium salt ($CaCl_2$) were industrial grade, while the steep water was collected from domestically processed pap. The water used in the analysis was glass distilled. The weaning albino rats used were of the same litter origin obtained from the rat colony of the Department of Biochemistry, University of Ilorin, Nigeria.

2.1. Tofu Preparation. Soybeans (2.0 kg) were soaked in water (6 litres) at $20–40°C$ for 9 hours. The soaked beans were drained, weighed, and ground with grinder; tap water was added at a ratio of 6 : 1 with raw bean and then filtered to separate soy cake from soymilk. The soymilk was subsequently heated to $98°C$ and maintained for 1 minute before delivering to the mix tank. When cooled to $87°C$, 1 litre of soymilk was mixed at 420 rpm with each of the coagulants (50 mL). The mixed solutions were held for 5 seconds and then filled onto tofu trays and allowed to coagulate for 10 minutes. The bean curd was pressed after which the tofu weight was recorded. Tofu was stored in water at $4°C$ overnight prior to analysis.

2.2. Animals. Twenty-five of 3-week old strain albino rats with an average weight of 50.2 g were used in this study. They were obtained from the animal house of the College of Medicine, University of Ilorin, Kwara State, Nigeria. The animals were housed in metabolic cage in the laboratory under ambient temperature and 12-hour light and dark periodicities. They were fed with commercial rat pellets (Neimeth Livestock Feeds Ltd., Ikeja) and water *ad libitum* and allowed to acclimatize for 2 weeks. Animal experiments were conducted in accordance with the internationally accepted principle for laboratory animal use and care [20].

2.3. Extraction of Free Soluble Polyphenols. Free soluble phenols were extracted using the method modified by Nwanna and Oboh [21]. About 200 g of each tofu was homogenized in 80% acetone (1 : 2 w/v) separately using chilled Waring blender for 5 minutes. Thereafter, the homogenates were filtered through Whatman number 2 filter paper on a Buchner funnel under vacuum. The residues were kept for extractions of bound phenols. The filtrates were evaporated using a rotary evaporator under vacuum at $45°C$ until 90% of the filtrates had been evaporated. The extracts were frozen at $-40°C$.

2.4. Extraction of Bound Polyphenols. These were extracted using the method modified by Nwanna and Oboh [21]. The bound phenolic contents were extracted from the residue from the free soluble polyphenol extracts. The drained residues from soluble free extraction were hydrolyzed directly with 20 mL of 4 M NaOH at room temperature for 1 hour with shaking. The mixture was acidified to PH 2 with conc. HCl and extracted six times with ethyl acetate. The ethyl acetate fractions were evaporated at $45°C$ under vacuum to dryness.

2.5. Antioxidant Activity

2.5.1. Total Phenol Content. These were determined by the modified method of Nwanna and Oboh [21]. About 0.5 mL of each extract was dissolved in 20 mL of 70% acetone with equal volume of water; 0.5 mL Folin-Cioalteus reagent and 2.5 mL of sodium carbonate were subsequently added and the absorbance was measured after 40 minutes at 725 nm, using tannic acid as standard.

2.5.2. Reducing Property. These were determined by assessing the ability of the tofu extracts to reduce $FeCl_3$ solution using modified method of Nwanna and Oboh [21]. About 2.5 mL of each extract was dissolved in 20 mL of methanol and mixed with 2.5 mL of 200 M sodium phosphate buffer (PH 6.6) and 2.5 mL of 100% potassium ferricyanide. The mixtures were incubated at $50°C$ for 20 minutes; thereafter 2.5 mL, 10%, trichloroacetic acid was added and subsequently centrifuged at 650 rpm for 10 minutes; 5 mL of the supernatant was mixed with equal volume of water and 1 mL of 0.1% ferric chloride. The absorbances were then measured at 700 nm, and a higher absorbance indicates a higher reducing power.

2.5.3. Free Radical Scavenging Ability. The free radical scavenging ability of the soluble free and bound extracts against DPPH (1, 1-diphenyl-2-picrylhydrazyl) free radical was also evaluated [22]. About 1 mL of the extract was dissolved in 20 mL methanol and mixed with 1 mL of 0.4 M methanolic solution containing 1, 1-diphenyl-2-picrylhydrazyl (DPPH) radicals. The mixtures were left in the dark for 30 minutes before measuring the absorbance at 516 nm.

2.6. Experimental Design. Animals were weighed and randomly assigned into five groups, namely,

normal control group ($n = 5$) was placed on a basal diet of 20 g/day only;

negative control group (n = 5) was placed on 20 g/day and 100 mg/100 g/day of basal diet and acetaminophen orally administered, respectively;

treated group 1 (n = 5) was placed on steep water tofu (20 g) and acetaminophen (100 mg/100 g/day) orally administered, respectively;

treated group 2 (n = 5) was placed on alum tofu (20 g) and acetaminophen (100 mg/100 g/day) orally administered, respectively;

treated group 3 (n = 5) was placed on calcium tofu (20 g/day) and acetaminophen (100 mg/100 g/day) orally administered, respectively.

The experimental period was 14 days. During the experiment, the weights of the rats were measured two times in a week during the 2 weeks of the experiment.

2.7. Measurements. The mean average weight of the rats was determined at the beginning of the experiment at every 2 days. The weight of the rats was determined using weighing scale (OHAUS MODEL Cs 5000, CAPACITY 500 × 2 g). This was done by placing a container on the scale, with the balance adjusted to zero, after which the rats in each group were placed into container and the measurement taken [23]. Total feed consumed in percentage was calculated using the following formula:

total feed consumed (%) = 100 (feed supplied (20 g) − leftover (g)/feed supplied (20 g).

total weight gain (TWG) in percentage was calculated as

total weight gain (%) = 100 × (final weight − initial weight)/initial weight [24].

On the 15th day, the rats were killed by cervical dislocation, the bloods collected were centrifuged, and the supernatants were collected for the assessment of liver function. In addition, the livers were harvested for histopathological analysis.

2.7.1. Assessment of Liver Function. The serum was used for the assay of marker enzymes (aspartate aminotransferase (AST), alanine aminotransferase (ALT), alkaline phosphatase (ALP), lactate dehydrogenase (LDH), and bilirubin) concentrations using Roche Modular Autoanalyzer. Total protein, albumin, sugar, and total cholesterol were analyzed by using standard randox laboratory kits.

2.8. Statistical Analysis. The results were reported as mean ± SEM (standard error of mean). One-way analysis of variance and least significant difference (LSD) test were used to evaluate the significant difference. Probability levels of less than 0.05 were considered significant.

3. Results

Table 1 shows the result of antioxidants properties of the tofu studied. The results revealed that the steep water coagulated tofu significantly recorded high total phenols (15.0 ± 2.24%), followed by the alum-coagulated tofu (13.0 ± 2.08), and the calcium salt coagulated tofu appeared significantly higher in

free radical scavenging activities of the free soluble phenol (65.5 ± 4.67%). The alum coagulated tofu (47.3 ± 3.97%) and the steep water had the least (38.2 ± 3.57%). Conversely, steep water coagulated tofu significantly recorded higher free radical scavenging activity of the bond phenol (69.1 ± 4.80%), followed by alum coagulated tofu (56.5 ± 3.36%), and calcium salt coagulated tofu had the lowest (27.3 ± 3.02%) free radical scavenging activity of the bound phenol. Also, the reducing power of free soluble and bound phenol of steep water and calcium salt coagulated tofu were not significantly different (0.4 ± 0.20), while alum coagulated tofu had the reducing power of free soluble phenol (0.3 ± 0.17) to be lower than that of bound phenol (0.5 ± 0.30).

The result of total feed intake and total weight gain are presented in Table 2 and revealed that there was decrease in total feed intake (141.8 ± 5.31%) and total weight gain (9.11 ± 1.35%) in rats fed with basal diet and given acetaminophen when compared with those rats fed with basal diet without acetaminophen (total feed intake, 443.8 ± 9.40%; total weight gain, 14.5 ± 1.70%). However, there was a significant increase (P < 0.05) in total feed intake (44.2 ± 2.10–73.5 ± 3.80%) of rats fed with tofu curdled with various coagulants and given acetaminophen orally compared to those rats fed with basal diet and given acetaminophen. In addition, the total weight gain of rats fed with tofu curdled with various coagulants plus acetaminophen decreased significantly (−5.1 ± 1.01–2.1 ± 0.60%) when compared with those fed with acetaminophen (9.1 ± 1.35%).

The results of liver function tests of the rats studied are shown in Table 3 and revealed that serum AST, ALT, ALP, and LDH levels of rats treated with basal diet and given acetaminophen orally were quite higher than that of the group treated with basal diet without acetaminophen. In contrast, the rats treated with various tofus and acetaminophen orally had significantly lower levels of AST, ALT, ALP, and LDH when compared with the group treated with basal diet and acetaminophen orally (Table 3).

The results of serum chemistry of rats studied are presented in Table 4 and revealed that bilirubin (0.153 ± 0.12 mg/dL), cholesterol (243.45 ± 7.00 mg/dL), and glucose (162.68 ± 5.70 mg/dL) levels significantly (P < 0.05) increased in rats treated with basal diet without acetaminophen intubation when compared with those treated with basal diet and acetaminophen intubation. Nevertheless, the content of total protein (3.45 ± 0.80 mg/dL) and albumin (1.65 ± 0.57 mg/dL), respectively, significantly decreased (P < 0.05). In addition, rats fed with various coagulated tofu and acetaminophen showed significant (P < 0.05) reduction in the level of bilirubin, cholesterol, and glucose when compared with rats fed with basal diet and acetaminophen intubation. In contrast, the level of total protein and albumin of rats fed with various tofus and acetaminophen were significantly increased (P < 0.05) when compared with those fed with basal diet and acetaminophen.

4. Discussions

The antioxidant properties of polyphenolic extractions of tofu produced using different coagulants and the effects of these

TABLE 1: Antioxidant properties of tofu studied.

Sample (tofu)	TP	FSFP (%)	FSBP	RPFP	RPBP
SWT	15.0 ± 2.24^a	38.2 ± 3.57^c	69.1 ± 4.80^a	0.4 ± 0.20^a	0.4 ± 0.20^b
ALT	13.0 ± 2.08^b	47.3 ± 3.97^b	56.5 ± 3.36^b	0.3 ± 0.17^b	0.5 ± 0.30^a
CST	11.0 ± 1.91^c	65.5 ± 4.67^a	27.3 ± 3.02^c	0.4 ± 0.20^a	0.4 ± 0.20^b

Values with the same superscript letter(s) along the same column are not significant different ($P \leq 0.05$). SWT: steep water coagulated tofu, ALT: alum coagulated tofu; CST: calcium salt coagulated tofu' TP: total phenol; FSFP: free radical scavenging ability of free soluble phenols; FSBP: free radical scavenging ability of bound phenols; RPFP: reducing power of free soluble phenols; RPBP: reducing power of bound phenols.

TABLE 2: Total feed intake and weight gain of albino rats fed with commercial diet and various coagulated tofu (g/rats/days).

Sample	Total food taken (%)	Total weight gained (%)
BDA	141.8 ± 5.31^b	9.11 ± 1.35^b
BDWA	443.8 ± 9.40^a	14.5 ± 1.70^a
CSTA	64.7 ± 3.60^d	-3.2 ± 0.80^d
SWTA	44.2 ± 2.10^e	-2.1 ± 0.60^e
ALTA	73.5 ± 3.80^c	-5.1 ± 1.01^c

Values represent mean of triplicate reading. Means with the same superscript letter(s) within the same column are not significantly different ($P \leq 0.05$). BDA: basal diet with acetaminophen orally administered; BDWA: basal diet without acetaminophen orally administered; ALTA: alum coagulated tofu with acetaminophen; SWTA: tofu coagulated with effluent from pap with acetaminophen; CSTA: calcium coagulated tofu with acetaminophen orally administered.

tofus on acetaminophen-induced hepatotoxicity in albino rats were evaluated. Polyphenols, particularly the flavonoids, are among the most potent plant antioxidants. Polyphenol can form complexes with reactive metals such as iron, zinc, and copper reducing their absorption [21]. It seems to reduce nutrients absorption, but excess levels of such elements (metals cations) in the body can promote the generation of free radicals and contribute to the oxidative damage of cell membranes and cellular DNA [25]. In addition, polyphenols also function as potent free radical scavengers within the body, where they can neutralize free radicals before they can cause cellular damage [26]. The results of polyphenolic antioxidant properties of various tofus studied as indicated in Table 1, revealed that steep water coagulated tofu had a significantly higher ($P < 0.05$) total phenol, followed by alum coagulated tofu, and calcium salt tofu had the least value of total phenol. These values generally compared well with what Shokunbi et al. [9] reported for some varieties of commercial mushrooms (0.01 g/100 g). The results are also in line with the report of Oboh and Rocha [27] on some hot pepper (*Capsicum annum*, Tepin, and *Capsicum pubescens*). In contrast, free radical scavenging activity of the free phenols of various types of tofu analyzed had a significantly higher ($P < 0.05$) calcium coagulated tofu, followed by alum coagulated tofu, and steep water coagulated tofu, which is the least (Table 1). However, the reverse was the case for the bound phenols where steep water coagulated tofu was recorded high, followed by alum coagulated tofu, and calcium-coagulated tofu had the least value of bound phenol (Table 1).

These values are higher than what Oboh and Akindahunsi [28] reported on some commonly consumed green leafy vegetables in Nigeria. Oxidative stress is the state of imbalance between the level of antioxidant defense system and the production of the oxygen-derived species. The increased level of oxygen and oxygen-derived species causes oxidative stress. The finding suggests that various tofus produced using different coagulants are good source of antioxidants and could be used to reduce biological oxidative stress and prevent cellular damage by scavenging free radical activity of the oxidative stress. Table 2 shows the results of total feed intake and total weight gain of rats studied. The result reveals that there was a decrease in total feed intact and total weight gain of the rats fed with basal diet and acetaminophen when compared with those rats fed with basal diet without acetaminophen (Table 2). This drastic reduction in feed intake and total weight gain could be because of overdose of paracetamol intake that generated oxidative stress in rats. The free radicals produced in rats might cause certain abnormality in food consumed and weight gain. It was also observed that rats fed with alum-coagulated tofu consumed the highest amount of tofu in the duration of the experiment, followed by those fed with calcium chloride coagulated tofu, while those fed with tofu coagulated with steep water consumed the lowest amount of tofu (Table 2). The total weight gain of rats fed with tofu produced using different coagulants was significantly higher ($P < 0.05$) in alum-coagulated tofu, and tofu coagulated with steep water significantly had the least weight gain (Table 2). The higher consumption of alum and calcium chloride coagulated tofu might be because of good taste and odour as indicated in the tofu, and the negative value of weight gain recorded by the rats fed with steep water tofu could be because of very low feed intake. While, the low intake of the tofu coagulated with steep water could be because of the unpleasant odor imparted by the steep water to the tofu. The wide variation for tofu consumed by the various rats could be because of the difference in the taste, nutritional quality, and acceptability of the various coagulated tofu [16, 29]. As shown in Table 3, overdosing rats with 100 mg/mL/day acetaminophen alone for 14 days caused a significant increase ($P < 0.05$) in the serum marker enzymes (AST, ALT, ALP, and LDH) (Table 3). Increase in AST levels signified liver damage. This finding suggested that the mega doses of acetaminophen administered induce the production of free radicals, which cause damage to the hepatocytes of rats.

This result correlates with the finding of Eriksson et al., [30] that toxicity with acetaminophen occurs when too much of it is taken. The elevations of serum liver enzymes indicate liver damage. This correlates with the report of Sai et al., [31] that a significant increase in serum AST,

TABLE 3: Serum enzyme marker level of rats studied.

Sample (rats)	ALP (mg/dL)	AST (mg/dL)	ALT (mg/dL)	LDH (mg/dL)
BDA	122.0 ± 4.93^a	101.0 ± 4.49^a	31.0 ± 2.49^a	156.00 ± 5.59^a
BDWA	90.0 ± 4.24^c	77.0 ± 3.92^c	26.0 ± 2.28^b	78.40 ± 3.96^c
SWTA	110.0 ± 4.70^b	82.0 ± 4.05^b	26.0 ± 2.28^b	85.52 ± 4.14^c
ALTA	120.0 ± 4.90^{ab}	94.0 ± 4.34^b	27.0 ± 2.32^b	90.00 ± 4.24^b
CSTA	113.0 ± 4.75^b	80.0 ± 4.0^c	30.0 ± 2.45^{ab}	90.23 ± 4.25^b

Means with the same superscript letter(s) along the same column are not significantly different ($P \leq 0.05$). BDA: basal diet with acetaminophen orally administered; BDWA: basal diet without acetaminophen orally administered; SWTA: steep water coagulated tofu with acetaminophen orally administered; ALTA: alum coagulated tofu with acetaminophen orally administered; CSTA: calcium chloride coagulated tofu.

TABLE 4: Serum chemistry level of rats studied.

Sample (rats)	BIL	CHL (Mg/dL)	GLU	TPN	ALB
BDA	0.153 ± 0.12^a	243.45 ± 7.00^a	162.68 ± 5.70^a	3.45 ± 0.83^d	1.65 ± 0.57^d
BDWA	0.125 ± 0.14^b	95.20 ± 4.36^d	87.34 ± 4.18^c	7.75 ± 1.24^a	3.93 ± 0.89^a
SWTA	0.118 ± 0.04^c	145.32 ± 5.40^c	93.50 ± 4.32^b	7.05 ± 1.19^b	3.00 ± 0.77^b
ALTA	0.122 ± 0.12^c	157.50 ± 5.61^b	95.00 ± 4.36^b	6.45 ± 1.14^c	2.60 ± 0.72^c
CSTA	0.128 ± 0.16^b	148.43 ± 5.45^c	93.76 ± 4.33^b	6.74 ± 1.16^c	2.91 ± 0.76^c

Means with the same superscript letter(s) along the same column are not significantly different ($P \leq 0.05$). BDA: basal diet with acetaminophen orally administered; BDWA: basal diet without acetaminophen orally administered; SWTA: steep water coagulated tofu with acetaminophen orally administered; ALTA: alum coagulated tofu with acetaminophen orally administered; CSTA: calcium chloride coagulated tofu; BIL: bilirubin; CHL: cholesterol; GLU: glucose; TPN: total protein; ALB: albumin.

ALT, and ALP levels suggests liver damage. Under normal therapeutic dose of acetaminophen, the excessive metabolites produced by the cytochrome P-450 system can be reduced by glutathione. However, if acetaminophen is overdosed, the glutathione stores will be depleted and the excessive metabolites will react with the liver macromolecules and cause hepatic cell death. The hepatic cellular enzyme ALP in serum will therefore increase. In addition, the hepatic malondialdehyde level will increase invariably, hence resulting in the generation of free radicals in the body [21]. This suggests that acetaminophen hepatotoxicity appears to be critically dependent on the depletion of cellular glutathione, and a relatively high reduction in the intracellular level of reduced glutathione leads to a situation of oxidative stress [32, 33]. Conversely, a simultaneous administration of albino rats with 100 mg/mL/acetaminophen alongside various tofus produced caused a significant decrease ($P \leq 0.05$) in serum AST, ALT, and ALP (Table 3). Decrease in serum marker enzyme levels (liver enzymes) indicates the ability of various tofus to protect the hepatocytes from oxidative damage caused by overdosing rats with acetaminophen. This is an indication that antioxidant mechanism may be involved in the protection of the liver cell by the various tofus from acetaminophen-induced oxidative stress. It was suggested that soy products contain high antioxidant properties, which serve as an extracellular neutralizer of free radicals [34].

In addition, soy food and vegetables had been reported to be rich in many phenols such as flavonoid [35]. Flavonoids have antioxidants capacity that is much stronger than those of vitamins C and E, reportedly used to prevent free radical production [35]. In contrast, the serum LDH level of rats fed with basal diet and acetaminophen was increased significantly ($P < 0.05$) when compared with those fed

with basal diet without acetaminophen (Table 3). This is in line with the findings of Ravikumar et al., [36] that the serum LDH level was elevated in rat administered CCl_4. The increased level of LDH is an indication of abnormality in liver functioning, which may be due to the formation of highly reactive free radicals caused by acetaminophen overdose. The hepatotoxicity of acetaminophen may directly affect the polyunsaturated fatty acids and alter the liver microsomal membranes in the rats. In a reverse case, the LDH levels of rats fed with tofu and acetaminophen were significantly decreased ($P < 0.05$) when compared with that of rats in group BDA (Table 3), which is a sign of an improvement by the various tofus over the damage done to the liver by acetaminophen. The result of serum total cholesterol level as indicated in Table 4 revealed that there was a significant increase in total cholesterol of rats fed with basal diet and acetaminophen orally when compared to those fed with basal diet without acetaminophen orally administered (Table 4).

The elevation in serum total cholesterol level could be attributed to the ability of acetaminophen to induce the production of free radicals, which results in hypercholesterolemia and the atherosclerosis. This correlates with the findings of Oboh [15] that an increase in the serum levels of cholesterol and LDL is associated with hypercholesterolemia and atherosclerosis, respectively. It is also supported by the findings of Ravikumar et al., [36] that intoxication of rat with CCl_4 elevated total cholesterol levels, which suggests the inhibition of bile acid synthesis and leads to increased level of cholesterol. However, supplementation of tofu curdled with various coagulants and acetaminophen orally administered causes a significant decrease ($P > 0.05$) in serum total cholesterol, compared to those fed with basal diet with acetaminophen orally administered (Table 4). It indicated that tofus produced using three coagulated agents are capable of preventing acetaminophen inducing oxidative stress and inhibiting bile acids synthesis.

This could be attributed to high antioxidant potential of soybeans, which serve as an extracellular neutralizer of free radicals [37]. This finding is in line with the study of Oboh [15] that the antihypercholesterolemic effect of soy protein was found to decrease the plasma concentrations of LDL as well as the ratio of plasma LDL to high density lipoprotein (HDL). It was reported that flavonoids could protect membrane lipids from oxidation, and a major source of flavonoids is vegetables, fruits, and soybeans [35]. The significant ($P <$ 0.05) reduction in total protein and albumin levels in rats fed with basal diet with acetaminophen intubation, compared to those fed with basal diet without acetaminophen orally administered (Table 4), indicates cellular damage produced. The damage produced might be due to the functional failure of endoplasmic reticulum, which leads to decrease in protein synthesis and accumulation of triglycerides [36]. Increased serum bilirubin level in rats fed with basal diet with acetaminophen intubation (Table 4) could be looked upon as a compensatory/retaliatory phenomenon in response to cellular peroxidative changes, which cause damage to the biliary gland. This is because bilirubin functions *in vivo* as a powerful antioxidant, antimutagen, and an endogenous tissue protector [37]. Reduction of bilirubin and elevation of total protein and albumin levels in rats treated with various coagulated tofus and acetaminophen orally administered (Table 4) were most effective and stabilized the biliary cell function and endoplasmic reticulum leading to bile acid and protein synthesis [38]. This indicates hepatoprotection. The administration of acetaminophen alone may adversely interfere with protein metabolism probably by inhibiting the synthesis of proteins such as albumin in the liver.

Simultaneous administration of acetaminophen to rats with supplementation of various tofus produced reversed these changes, maybe by increasing protein synthesis. Stimulation has been advanced as a contributory hepatoprotective mechanism, which accelerates regeneration of cells [39]. LDH activities all point to the fact that tofu has hepatoprotective potential against hepatotoxin caused by acetaminophen. It could be possible that a probable mechanism of hepatoprotection of various tofus against acetaminophen-induced damage is the antioxidant activity. The antioxidant activity of the various tofus supplemented may be attributed to the presence of phenolics and flavonoids [40]. Therefore, lower level of serum enzyme markers, total cholesterol, and bilirubin observed in rats on tofu coagulated with various coagulants and acetaminophen orally administered suggested liver damage repair, but steep water coagulated tofu looked more promising in terms of liver repair than other forms of tofu diet. It can be concluded from the present finding that tofus curdled with various coagulants could be efficiently used to prevent liver damage caused by high doses of acetaminophen orally administered after successful clinical trials. In addition, the intake of acetaminophen in excess should be discouraged in homes because of its destructive effects on the liver. Tofu consumption should be encouraged in diets as it can be used as a functional food to prevent liver damage due to its antioxidant properties, and as such, tofu could be recommended for clinical trial.

Conflict of Interests

The authors declare that they have no competing interests apart from the possible ones already acknowledged below.

Acknowledgments

The authors would like to acknowledge the Department of Biochemistry, Federal University of Technology, Akure, Ondo State, Nigeria, for providing facilities for this work. One of the authors, Mr. N. Yakubu, is grateful to the Niger State College of Agriculture, Mokwa, Niger State, Nigeria, for financial sponsorship that enabled him to participate in this study.

References

[1] Y. Ali, O. Munir, and B. Vahit, "The antioxidant activity of leaves of *Cydonia vulgaris*," *Turkish Journal of Medical Sciences*, vol. 31, pp. 23–27, 2001.

[2] J. Sun, Y. F. Chu, X. Wu, and R. H. Liu, "Antioxidant and antiproliferative activities of common fruits," *Journal of Agricultural and Food Chemistry*, vol. 50, no. 25, pp. 7449–7454, 2002.

[3] R. H. Liu, "Health benefits of fruit and vegetables are from additive and synergistic combinations of phytochemicals," *American Journal of Clinical Nutrition*, vol. 78, no. 3, pp. 517S–520S, 2003.

[4] Y. F. Chu, J. Sun, X. Wu, and R. H. Liu, "Antioxidant and antiproliferative activities of common vegetables," *Journal of Agricultural and Food Chemistry*, vol. 50, no. 23, pp. 6910–6916, 2002.

[5] P. D. Duh, "Antioxidant activity of Burdock, its scavenging effect on free radical and active oxygen," *Journal of the American Oil Chemists' Society*, vol. 75, no. 4, pp. 455–461, 1998.

[6] T. Patel, D. Shirode, S. Pal Roy, S. Kumar, and S. Ramachandra Setty, "Evaluation of antioxidant and hepatoprotective effects of 70% ethanolic bark extract of *Albizzia lebbeck* in rats," *International Journal of Research in Pharmaceutical Sciences*, vol. 1, no. 3, pp. 270–276, 2010.

[7] M. A. Asgar, A. Fazilah, N. Huda, R. Bhat, and A. A. Karim, "Nonmeat protein alternatives as meat extenders and meat analogs," *Comprehensive Reviews in Food Science and Food Safety*, vol. 9, no. 5, pp. 513–529, 2010.

[8] V. Poysa and L. Woodrow, "Stability of soybean seed composition and its effect on soymilk and tofu yield and quality," *Food Research International*, vol. 35, no. 4, pp. 337–345, 2002.

[9] O. S. Shokunbi, O. O. Babajide, D. O. Otaigbe, and G. O. Tayo, "Effect of coagulants on the yield nutrient and antinutrient composition of Tofu," *Archives of Applied Science Research*, vol. 3, no. 3, pp. 522–527, 2011.

[10] K. Descheemaeker and I. Debruyne, *Clinical Evidence, Dietetic Applications*, Garant, 2001.

[11] P. A. Murphy and L. A. Wilson, "Soybean protein composition and tofu quality," *Food Technology*, vol. 51, no. 3, pp. 86–88, 1997.

[12] G. Paganga, N. Miller, and C. A. Rice-Evans, "The polyphenolic content of fruit and vegetables and their antioxidant activities. What does a serving constitute?" *Free Radical Research*, vol. 30, no. 2, pp. 153–162, 1999.

[13] S. W. Qader, M. A. Abdulla, L. S. Chua, N. Najim, M. M. Zain, and S. Hamdan, "Antioxidant, total phenolic content and cytotoxicity evaluation of selected Malaysian plants," *Molecules*, vol. 16, no. 4, pp. 3433–3443, 2011.

[14] A. L. Miller, "Antioxidant flavonoids: structure, function and clinical usage," *Alternative Medicine Review*, vol. 1, no. 2, pp. 103–111, 1996.

[15] G. Oboh, "Coagulants modulate the hypocholesterolemic effect of tofu (coagulated soymilk)," *African Journal of Biotechnology*, vol. 5, no. 3, pp. 290–294, 2006.

[16] F. M. Sacks, A. Lichtenstein, L. Van Horn, W. Harris, P. Kris-Etherton, and M. Winston, "Soy protein, isoflavones, and cardiovascular health: an American Heart Association Science Advisory for professionals from the Nutrition Committee," *Circulation*, vol. 113, no. 7, pp. 1034–1044, 2006.

[17] J. Parma, H. R.Shama, and R. Verma, "Effect of source and coagulants on the Physicochemical and organoleptic evaluation of soy tofu," *Journal of Dairying Foods & Home Sciences* , vol. 26, no. 2, pp. 69–74, 2007.

[18] D. G. Davidson and W. N. Eastham, "Acute liver necrosis following overdose of paracetamol," *British Medical Journal*, vol. 5512, pp. 497–499, 1966.

[19] L. P. James, P. R. Mayeux, and J. A. Hinson, "Acetaminophen-induced hepatotoxicity," *Drug Metabolism and Disposition*, vol. 31, no. 12, pp. 1499–1506, 2003.

[20] NIH Publication no. 85-23, "Respect for life," National Institute of Environmental Health Sciences-NIEHS, 1985, http://www .niehs.nih.gov/.

[21] E. E. Nwanna and G. Oboh, "Antioxidant and hepatoprotective properties of polyphenol extracts from Telfairia occidentalis (Fluted Pumpkin) leaves on acetaminophen induced liver damage," *Pakistan Journal of Biological Sciences*, vol. 10, no. 16, pp. 2682–2687, 2007.

[22] F. Ursini, M. Maiorino, P. Morazzoni, A. Roveri, and G. Pifferi, "A novel antioxidant flavonoid (IdB 1031) affecting molecular mechanisms of cellular activation," *Free Radical Biology and Medicine*, vol. 16, no. 5, pp. 547–553, 1994.

[23] O. A. Semeon, "Haematological Characteristics of *Clarias gariepinus* (Buchell, 1822) Juveniles Fed on Poultry Hatchery Waste," *American-Eurasian Journal of Toxicological Sciences*, vol. 2, no. 4, pp. 190–195, 2010.

[24] H. Öbek, S. Uğraş, I. Bayram et al., "Hepatoprotective effect of *Foeniculum vulgare* essential oil: a carbon-tetrachloride induced liver fibrosis model in rats," *Scandinavian Journal of Laboratory Animal Science*, vol. 31, no. 1, pp. 201–211, 2004.

[25] O. Y. Okafor, O. L. Erukainure, J. A. Ajiboye, R. O. Adejobi, F. O. Owolabi, and S. B. Kosoko, "Pineapple peel extract modulates lipid peroxidation, catalase activity and hepatic biomarker levels in blood plasma of alcohol-induced oxidative stressed rats," *Asian Pacific Journal of Tropical Medicine*, vol. 1, no. 1, pp. 11–14, 2011.

[26] N. Ara and H. Nur, "In vitro antioxidant methanolic leaves and flowers extracts of *Lippia alba*," *Research Journal of Medical Sciences*, vol. 4, no. 1, pp. 107–110, 2009.

[27] G. Oboh and J. B. T. Rocha, "Polyphenols in red pepper [*Capsicum annuum var. aviculare* (Tepin)] and their protective effect on some pro-oxidants induced lipid peroxidation in brain and liver," *European Food Research and Technology*, vol. 225, no. 2, pp. 239–247, 2007.

[28] G. Oboh and A. A. Akindahunsi, "Change in the ascorbic acid, total phenol and antioxidant activity of sun-dried commonly consumed green leafy vegetables in Nigeria," *Nutrition and Health*, vol. 18, no. 1, pp. 29–36, 2004.

[29] K. G. Aning, A. G. Ologun, A. Onifade, J. A. Alokan, A. I. Adekola, and V. A. Aletor, "Effect of replacing dried brewer's grain with 'sorghum rootlets' on growth, nutrient utilisation and some blood constituents in the rat," *Animal Feed Science and Technology*, vol. 71, no. 1-2, pp. 185–190, 1998.

[30] L. S. Eriksson, U. Broome, M. Kalin, and M. Lindholm, "Hepatotoxicity due to repeated intake of low doses of paracetamol," *Journal of Internal Medicine*, vol. 231, no. 5, pp. 567–570, 1992.

[31] K. Sai, A. Takagi, T. Umemura, and Y. Kurokawa, "Toxicology," *Journal of Environmental Pathology*, vol. 11, pp. 139–143, 1992.

[32] S. L. Arnaiz, S. Llesuy, J. C. Cutrin, and A. Boveris, "Oxidative stress by acute acetaminophen administration in mouse liver," *Free Radical Biology and Medicine*, vol. 19, no. 3, pp. 303–310, 1995.

[33] M. A. Tirmenstein and S. D. Nelson, "Acetaminophen-induced oxidation of protein thiols. Contribution of impaired thiol-metabolizing enzymes and the breakdown of adenine nucleotides," *Journal of Biological Chemistry*, vol. 265, no. 6, pp. 3059–3065, 1990.

[34] T. Anderson and A. J. Theron, "Antioxidant and tissue protective function of ascorbic acid," *World Review of Nutritional and Diabetes*, vol. 62, pp. 37–38, 1990.

[35] M. Alía, C. Horcajo, L. Bravo, and L. Goya, "Effect of grape antioxidant dietary fiber on the total antioxidant capacity and the activity of liver antioxidant enzymes in rats," *Nutrition Research*, vol. 23, no. 9, pp. 1251–1267, 2003.

[36] S. Ravikumar, M. Gnanadesigan, J. Seshserebiah, and S. Jacob Inbanseon, "Hepatoprotective effect of an Indian salt marsh herb Suaeda monoica Forsk. Ex. Gmel against concanavalin: an induced toxicity in rats," *Life Sciences and Medicine Research*, vol. 2010, p. LSMR-2, 2010.

[37] K. Pratibha, U. Anand, and R. Agarwal, "Serum adenosine deaminase, $5'$ nucleotidase and malondialdehyde in acute infective hepatitis," *Indian Journal of Clinical Biochemistry*, vol. 19, no. 2, pp. 128–131, 2004.

[38] S. V. Sureshkumar and S. H. Mishra, "Hepatoprotective activity of extracts from *Pergularia daemia* Forsk. against carbon tetrachloride induced toxicity in rats," *Pharmacognosy Magazine*, vol. 3, no. 11, pp. 187–191, 2007.

[39] D. Awang, "Milk Thistle," *Canadian Pharmacists Journal*, vol. 23, pp. 749–754, 1993.

[40] O. M. Iniaghe, S. O. Malomo, J. O. Adebayo, and R. O. Arise, "Evaluation of the antioxidant and hepatoprotective properties of the methanolic extract of *Acalypha racemosa* leaf in carbon tetrachloride-treated rats," *African Journal of Biotechnology*, vol. 7, no. 11, pp. 1716–1720, 2008.

7

Involvement of the Ligninolytic System of White-Rot and Litter-Decomposing Fungi in the Degradation of Polycyclic Aromatic Hydrocarbons

Natalia N. Pozdnyakova

Institute of Biochemistry and Physiology of Plants and Microorganisms, Russian Academy of Sciences, 13 Prospekt Entuziastov, Saratov 410049, Russia

Correspondence should be addressed to Natalia N. Pozdnyakova, nataliapozdnyakova@yahoo.com

Academic Editor: Susana Rodríguez-Couto

Polycyclic aromatic hydrocarbons (PAHs) are natural and anthropogenic aromatic hydrocarbons with two or more fused benzene rings. Because of their ubiquitous occurrence, recalcitrance, bioaccumulation potential and carcinogenic activity, PAHs are a significant environmental concern. Ligninolytic fungi, such as *Phanerochaete chrysosporium*, *Bjerkandera adusta*, and *Pleurotus ostreatus*, have the capacity of PAH degradation. The enzymes involved in the degradation of PAHs are ligninolytic and include lignin peroxidase, versatile peroxidase, Mn-peroxidase, and laccase. This paper summarizes the data available on PAH degradation by fungi belonging to different ecophysiological groups (white-rot and litter-decomposing fungi) under submerged cultivation and during mycoremediation of PAH-contaminated soils. The role of the ligninolytic enzymes of these fungi in PAH degradation is discussed.

1. Introduction

The use of fossil fuels for energy and raw material in the past century has led to widespread environmental pollution. Among these pollutants are polycyclic aromatic hydrocarbons (PAHs), which are considered a potential health risk because of their possible carcinogenic and mutagenic activities [1]. PAHs consist of benzene analogs having two or more aromatic rings in various alignments (Figure 1). Most of the low-molecular-weight PAHs (up to three aromatic rings) are very toxic [2], and most of the high-molecular-weight PAHs (four and more aromatic rings) are highly mutagenic, teratogenic, and carcinogenic for humans and animals [3]. PAHs are compounds of great environmental significance and are considered by the Environmental Protection Agency (USA) and other national institutions to be of toxicological relevance [http://www.defra.gov.uk/Environment/consult/airqual01/11.htm].

The general and scientific interest in the fate of PAHs in the environment and their microbial degradation, especially of higher-molecular-weight PAHs consisting of more than four rings, is based on their carcinogenic and mutagenic properties. Many reviews are available on different aspects of PAH degradation [4–6]. Several fungi are known to have the property of degradation PAHs. The degradation of these compounds by ligninolytic fungi, including white-rot and litter-decomposing fungi, has been intensively studied. They produce extracellular enzymes with very low substrate specificity, making them suitable for degradation of lignin and different low- and higher-molecular-weight aromatic compounds [6].

Investigations into the microbial bioconversion of PAHs has shown that wood- and litter-decay fungi are efficient degraders of these organopollutants [7–11]. They can mineralize PAHs with four and more condensed aromatic rings, in contrast to bacteria and soil fungi. They also can metabolize both individual PAHs and their complex mixtures, such as creosote [12–23]. The toxicity of PAHs underlies the use of creosote, a PAH mixture, as a fungicidal wood preservative. However, many fungi, including white-rot basidiomycetes

Involvement of the Ligninolytic System of White-Rot and Litter-Decomposing Fungi in the Degradation of
Polycyclic Aromatic Hydrocarbons

47

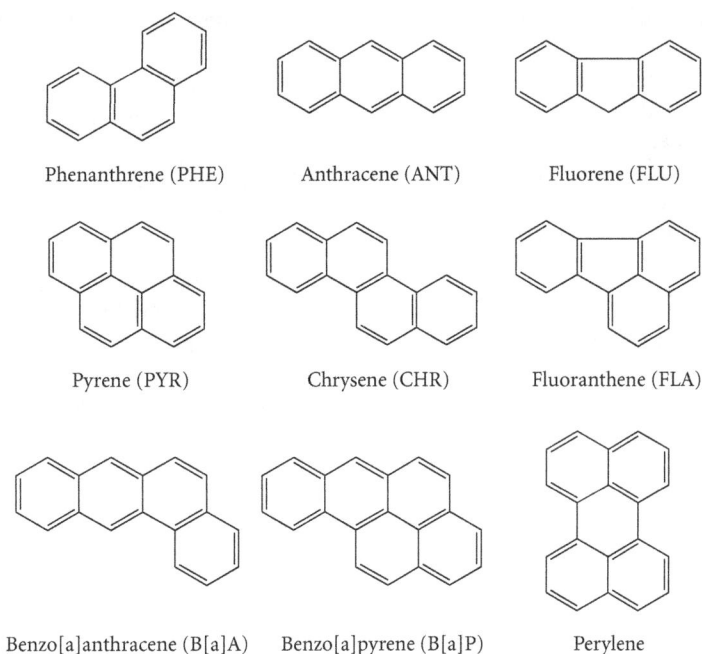

FIGURE 1: Chemical formulas of some 3-, 4- and 5-ring PAHs [http://www.chemport.ru/].

and litter-decomposing fungi, tolerate this treatment and grow on creosote-treated wood. These fungi have also been shown to deplete and detoxify PAHs in contaminated soil [24–31].

The first studies on the potential of ligninolytic fungi for use in PAH biodegradation can be attributed to 1985, when Bumpus et al. [32] reported that the white-rot basidiomycete *Phanerochaete chrysosporium* partially degraded benzo[a]pyrene to carbon dioxide. Later, PAH degradation has been reported, among others, for the genera *Phanerochaete* [7, 13], *Trametes* [8], *Bjerkandera* [33], *Coriolus* [34], *Nematoloma* [18], *Irpex* [35], and *Pleurotus* [14–16, 21–23, 28, 30, 33, 36].

Different authors have associated the ability of these fungi to degrade PAHs with the extracellular ligninolytic system, which includes lignin peroxidase (LiP), Mn-peroxidase (MnP), versatile peroxidase (VP), and laccase (LAC) [8, 17, 37, 38]. Fungal laccases and peroxidases have been suggested to play a key role in lignin degradation and to enable their producers—the wood- and litter-decomposing fungi—to detoxify xenobiotic compounds by partial degradation or complete mineralization [35]. The catalytic action of these enzymes generates more polar and water-soluble metabolites, such as quinones, which are more susceptible to further degradation by indigenous bacteria present in soils and sediments [24]. Knowledge of all the metabolites formed from fungal metabolism is a key requirement to validate soil bioremediation. In soil, quinonic metabolites serve as substrates for microbial populations and are mineralized to carbon dioxide. They also may undergo polymerization and become part of the humus pool [27].

Several enzymatic mechanisms of PAH degradation by these fungi have been discussed: (a) LiP and MnP directly catalyze one-electron oxidation of PAHs with an ionization potential (IP) of 7.55 eV to produce PAH quinones [39–41], which can be further metabolized via ring fission [42]; (b) LAC catalyzes one-electron oxidation of PAHs, for example anthracene (ANT) and benzo[a]pyrene (B[a]P), whose efficiency is enhanced in the presence of mediators such as 1-hydroxybenzotriazole (HBT) or 2,2′-azinobis(3-ethylbenzthiazoline-6-sulfonic acid (ABTS) [43, 44]; (c) some PAH compounds containing up to six rings were shown to be degradable via MnP-dependent lipid peroxidation reactions both *in vitro* and *in vivo* [13, 25]; (d) intracellular cytochrome P-450 monooxygenase activity followed by epoxide hydrolase-catalyzed hydroxylation of 3-, 4-, and 5-ring PAHs are believed to initially metabolize PAH molecules, including phenanthrene (PHE), having an IP of 8.03 eV [14–16, 36, 45, 46].

In organisms that use the cytochrome P-450 system, the *trans*-dihydrodiol product cannot be used as an energy source, although further metabolism may occur. For example, B[a]P is oxidized by the cytochrome P-450 monooxygenase system; among other products, benzo[a]pyrene dihydrodiol epoxide is formed [46]. However, in white-rot fungi, such as *Phanerochaete chrysosporium* and *Trametes versicolor*, there occurs mineralization of PAHs, indicating the complete breakdown of PAHs [19]. At present time, substantial and conclusive evidence exists that ligninolytic enzymes are involved in PAH degradation by these fungi.

Two possible roles of ligninolytic system have been discussed up to now: (a) LiP, MnP, and LAC were found to have a pivotal role in the degradation of PAHs, catalyzing the first attack of molecule [10, 44]; (b) cytochrome P-450 monooxygenase could be responsible for this initial step [14–16, 46], and the ligninolytic mechanism was supposed to be involved in later steps of metabolism leading to CO_2 evolution [16].

This paper summarizes the data available on PAH degradation by fungi belonging to different ecophysiological groups (white-rot and litter-decomposing fungi) under submerged cultivation and during mycoremediation of PAH-contaminated soils. The possible functions of ligninolytic enzymes of these fungi in PAH degradation are discussed.

2. PAH Degradation, the Key Products, and the Time Course of Ligninolytic Enzyme Production

2.1. Submerged Cultivation Conditions. Various publications have shown the ability of white-rot and litter-decomposing fungi to degrade different PAHs (Table 1), including phenanthrene (PHE) (by *T. versicolor*, *Kuehneromyces mutabilis* [9], *P. chrysosporium* [45], and *Pleurotus ostreatus* [14–16]), anthracene (ANT) (by *Bjerkandera* sp. BOS55 [47], *P. ostreatus* [14], *P. chrysosporium*, *T. versicolor*, *B. adusta* [12], and *I. lacteus* [48]), fluorene (FLU) (by *P. ostreatus* [14]), pyrene (PYR) (by *Bjerkandera* sp. BOS55 [47], *P. ostreatus* [14, 49], *Irpex lacteus* [48, 49], *T. versicolor* [49], and *P. chrysosporium* [49]), fluoranthene (FLA) (by *I. lacteus* [48], *T. versicolor*, and *Kuehneromyces mutabilis* [9]), chrysene (CHR) (by *P. ostreatus* [21]), benzo[a]pyrene (B[a]P) (by *Bjerkandera* sp. BOS55 [47], *P. chrysosporium*, *T. versicolor*, and *B. adusta* [12]), benz[a]anthracene (B[a]A) (by *I. lacteus* [50]), and dibenzothiophene (by *P. ostreatus* [14]).

In various studies, quinones have been identified as major products in the degradation of PAHs by fungi (Table 1) [12, 42, 51]. For example, PYR was metabolized by *P. ostreatus* predominantly to pyrene-4,5-dihydrodiol, ANT to antracene-1,2-dihydrodiol and 9,10-anthraquinone, FLU to 9-fluorenol and 9-fluorenone, and dibenzothiophene to corresponding sulfoxide and sulfone [14].

Many white-rot fungi can mineralize PAHs to CO_2 as well. The PHE mineralization has been shown for *C. versicolor*, *I. lacteus* [52], *T. versicolor*, and *Kuehneromyces mutabilis* [9]. *C. versicolor* and *I. lacteus* also demonstrated high mineralization rates for 4-ring PYR [52]. The mineralization of PYR by *T. versicolor* and *Kuehneromyces mutabilis* was also found [9]. *P. ostreatus*, in addition to PYR [16, 49], mineralized B[a]P, ANT, and FLU but did not mineralize FLA [16].

2.1.1. Phenanthrene. The most studied is the metabolism of PHE by *P. chrysosporium* [45, 53] and *P. ostreatus* [14–16]. In an early work Sutherland et al. [45] studied PHE degradation by *P. chrysosporium*; phenanthrene-*trans*-9,10-dihydrodiol, phenanthrene-*trans*-3,4-dihydrodiol, 9-phenanthrol, 3-phenanthrol, 4-phenanthrol, and the conjugate 9-phenanthryl-*D*-glucopyranoside were identified as metabolites. Since LiP was not detected in the culture medium, the authors suggested the involvement of monooxygenase and epoxide hydrolase activity in the initial oxidation and hydration of PHE by this fungus [45]. Dhawale et al. [53] showed that homokaryotic isolates of

P. chrysosporium caused the disappearance of PHE when they were grown in low- as well as high-nitrogen media. Moreover, LiP and MnP activities were not detected in any of the cultures incubated in the presence of PHE. Additionally, mineralization of PHE was observed even under nonligninolytic conditions. The authors suggested that LiP and MnP are not essential for the degradation of PHE by *P. chrysosporium* [53].

Later, Song [52] found that the fungus *P. chrysosporium* oxidized PHE and phenanthrene-9,10-quinone at their C-9 and C-10 positions to give a ring-fission product, 2,2′-diphenic acid. 2,2′-diphenic acid formation from PHE was somewhat greater in low-nitrogen (ligninolytic) cultures than in high-nitrogen (nonligninolytic) cultures. The oxidation of phenanthrene-9,10-quinone to 2,2′-diphenic acid was unaffected by the level of nitrogen added, and it was significantly faster than the cleavage of PHE to 2,2′-diphenic acid. Phenanthrene-*trans*-9,10-dihydrodiol, previously shown to be the principal PHE metabolite in nonligninolytic *P. chrysosporium* cultures, was not formed in the ligninolytic cultures of this fungus. The authors suggested that PHE degradation by *P. chrysosporium* proceeds in the order PHE → phenanthrene-9,10-quinone → 2,2′-diphenic acid, involves both ligninolytic and nonligninolytic enzymes, and is not initiated by the classic microsomal cytochrome P-450. The extracellular LiP of *P. chrysosporium* was not able to oxidize PHE *in vitro*, and therefore, is also unlikely to catalyze the first step of PHE degradation *in vivo* [54].

Later, the important roles of both cytochrome P-450 and MnP in PHE metabolism by *P. chrysosporium* were found [55]. Ning et al. [55] showed that the microsomal P-450 degraded PHE with a NADPH-dependent activity. One of the major detectable metabolites of PHE in the ligninolytic cultures and microsomal fractions was identified as phenanthrene-*trans*-9,10-dihydrodiol. Piperonyl butoxide, a P-450 inhibitor that had no effect on MnP activity, significantly inhibited PHE degradation and *trans*-9,10-dihydrodiol formation in both intact cultures and microsomal fractions. Furthermore, PHE was also efficiently degraded by the extracellular fraction with high MnP activity. The authors suggested the involvement of both cytochrome P-450 and MnP in PHE metabolism by *P. chrysosporium* [55].

The other much studied white-rot fungus, *Pleurotus ostreatus*, when grown in basidiomycetes-rich medium, metabolized 94% of PHE added; 3% was mineralized to CO_2. Approximately 52% of PHE was metabolized to *trans*-9,10-dihydroxy-9,10-dihydrophenanthrene (phenanthrene-*trans*-9,10-dihydrodiol), 2,2′-diphenic acid, and unidentified metabolites. $^{18}O_2$-labeling experiments indicated that one atom of oxygen was incorporated into the phenanthrene-*trans*-9,10-dihydrodiol. Significantly less phenanthrene-*trans*-9,10-dihydrodiol was observed in incubations with cytochrome P-450 inhibitors. The experiments with cytochrome P-450 inhibitors and $^{18}O_2$ labeling and the formation of phenanthrene-*trans*-9R,10R-dihydrodiol

Involvement of the Ligninolytic System of White-Rot and Litter-Decomposing Fungi in the Degradation of
Polycyclic Aromatic Hydrocarbons

49

TABLE 1: PAH degradation under submerged cultivation conditions.

PAH	Fungus	Metabolites	Enzymes	References
PHE	P. chrysosporium (WRF)	PHE-*trans*-9,10-dihydrodiol; PHE-*trans*-3,4-dihydrodiol; 9-phenantrol, 3-phenanthrol; 4-phenanthrol; 9-phenanthryl-*D*-glucopyranoside	Monooxygenase; epoxide hydrolase	[45]
		CO_2	ND	[53]
		PHE-9,10-quinone; 2,2′-diphenic acid	ND	[52]
		PHE-*trans*-9,10-dihydrodiol	Cytochrome P-450; MnP	[55]
	P. sordida (WRF)	ND	MnP	[56]
	P. ostreatus (WRF)	PHE-*trans*-9,10-dihydrodiol	Cytochrome P-450; epoxide hydrolase	[13]
	T. versicolor (WRF)	ND	ND	[20]
	Agrocybe sp. (WRF)	ND	ND	[57]
	Ganoderma lucidum (WRF)	ND	LAC	[58]
ANT	Agrocybe sp. (WRF)	9,10-anthraquinone	ND	[57]
	B. adusta (WRF)	9,10-anthraquinone	ND	[59]
	P. ostreatus (WRF)	9,10-anthraquinone	MnP; LAC	[43, 44, 59, 60]
	I. lacteus (WRF)	9,10-anthraquinone	ND	[48]
	Trametes versicolor	9,10-anthraquinone	ND	[12, 60]
	Coriolopsis polyzona (WRF)	9,10-anthraquinone	ND	[60]
	P. chrysosporium (WRF)	9,10-anthraquinone; phthalic acid; CO_2	LiP; MnP	[42, 60]
	Stropharia coronilla (LDF)	ND	MnP	[10]
	T. trogii (WRF)	9,10-anthraquinone	LAC	[61]
FLU	Agrocybe sp. (WRF)	9-fluorenol; 9-fluorenone	ND	[57]
	B. adusta (WRF)	ND	ND	[59]
	P. ostreatus (WRF)	ND	ND	[59]
PYR	I. lacteus (WRF)	Quinonic metabolites	MnP; Mn-inhibited peroxidase	[52, 62]
	P. ostreatus (WRF)	PYR-4,5-dihydrodiol	LAC	[22]
		PYR-4,5-dihydrodiol; phthalic acid	LAC, VP	[22]
	Ganoderma lucidum (WRF)	ND	LAC	[58]
FLA	P. sordida (WRF)	ND	MnP	[56]
B[a]A	P. laevis (WRF)	Quinone metabolite	ND	[7]
	P. chrysosporium (WRF)	Quinone metabolite	ND	[7]
	I. lacteus (WRF)	B[a]A-7,12-dione; phthalic acid, 1,2-naphthalenedicarboxylic acid; 2-hydroxymethyl benzoic acid; mono- and di-methyl esters of phthalic acid; 1-tetralone; 1,4-naphthalenedione; 1,4-naphthalenediol; 1,2,3,4-tetrahydro-1-hydroxynaphthalene	ND	[50]
B[a]P	P. chrysosporium	Quinone metabolite; CO_2	LiP; MnP	[11]
	Bjerkandera sp. (WRF)	Quinone metabolite	ND	[11]
	P. ostreatus (WRF)	Quinone metabolite	ND	[11]
	Stropharia coronilla (LDF)	Quinone metabolite; CO_2	LiP; MnP	[11]
	Stropharia rugosoannulata (LDF)	CO_2	MnP	[10]

as the predominant metabolite suggested that *P. ostreatus* initially oxidizes PHE stereoselectively by a cytochrome P-450 monooxygenase and that this is followed by epoxide hydrolase-catalyzed hydration reactions [13].

Another white-rot fungus, *T. versicolor*, removed about 46% and 65% of PHE added in shaken and static cultures. PHE degradation was maximal at pH 6, and the optimal temperature was 30°C. Although the PHE removal percentage was highest (76.7%) at 10 mg/L of PHE, the transformation rate was maximal (0.82 mg/h) at 100 mg/L of PHE in the fungal culture [20]. *Agrocybe* sp. CU-43, a white-rot fungus isolated from Thailand, metabolized about 99% of 100 ppm PHE [57].

2.1.2. Anthracene. The degradation of another 3-ring PAH, ANT, was found in most of the studied fungi. For example, *Agrocybe* sp. CU-43 removed about 92% of ANT [57]; *B. adusta*, 38%; *P. ostreatus*, about 60% of these compounds [59]. All the tested white-rot fungi oxidized ANT to anthraquinone. The appearance of anthraquinone, coinciding with ANT degradation, is common to white-rot fungi as the first step [60]. Field et al. [12] concluded that anthraquinone behaves like a dead-end metabolite in certain white-rot fungi, including strains of the genera *Bjerkandera* and *Phanerochaete*. In liquid culture containing ANT, *I. lacteus* accumulated the degradation product anthraquinone [48]. However, four *Trametes* strains removed ANT without significant accumulation of the quinone; the ability of these fungi to metabolize anthraquinone was confirmed as well [12]. Anthraquinone did not accumulate in *P. ostreatus*, *Coriolopsis polyzona*, or *T. versicolor*, indicating that it was degraded further. *P. ostreatus* and *C. polyzona* failed to degrade anthraquinone in the absence of ANT [60]. Hammel et al. [42] showed also that the oxidation of ANT by *P. chrysosporium* to anthraquinone was rapid and that both compounds were significantly mineralized. Both compounds were cleaved by the fungus to give the same ring-fission metabolite, phthalic acid, and phthalate production from anthraquinone was shown to occur only under ligninolytic culture conditions. The results suggest that the major pathway for ANT degradation proceeds in the order ANT → 9,1-anthraquinone → phthalate → CO_2 and that it is probably mediated by enzymes of ligninolytic metabolism [42].

2.1.3. Fluorene. The degradation of FLU by *Agrocybe* sp. CU-43 [57], *B. adusta*, and *P. ostreatus* [59] was also found. Two of the metabolites from FLU degradation by *Agrocybe* sp. CU-43 were identified as 9-fluorenol and 9-fluorenone, the less toxic intermediates of FLU. However, 9-fluorenol is not an end product for the degradation [57].

2.1.4. Pyrene. *I. lacteus* metabolized most of the added 4-ring PAH PYR; almost 50% of PYR was converted to polar metabolites and was recovered from an aqueous phase of the culture [52]. The accumulation of pyrene-4,5-dihydrodiol during PYR degradation in Kirk's medium by *P. ostreatus* D1 was found also [22]. However, in basidiomycetes-rich medium, the accumulation of pyrene-4,5-dihydrodiol by this fungus was not found, and PYR degradation was complete, with a PHE-derivative and phthalic acid being formed as intermediates. Phthalic acid, in turn, can be involved in basal metabolism [22].

2.1.5. Chrysene. The same data were obtained for CHR bioconversion by the fungus *P. ostreatus* D1 [21]. In this case, the dependence of the completeness of CHR degradation on the cultivation conditions was found as well [21]. In Kirk's medium, accumulation of the quinone metabolite was found; in basidiomycetes-rich medium, no quinone was accumulated and CHR degradation was complete, with phthalic acid being formed as an intermediate [21].

2.1.6. Benzo[a]Anthracene. Studies with B[a]A and its 7,12-dione indicated that only small amounts of quinone products were ever present in *Phanerochaete laevis* cultures and that quinone intermediates of PAH metabolism were degraded faster and more extensively by *P. laevis* than by *P. chrysosporium* [7]. *I. lacteus* incubated in a nutrient liquid medium degraded more than 70% of the initially applied B[a]A. At the first step of metabolization, B[a]A was transformed via a typical pathway of ligninolytic fungi to benz[a]anthracene-7,12-dione. The product was further transformed *via* at least two routes, one being similar to the ANT metabolic pathway of *I. lacteus*. Benz[a]anthracene-7,12-dione was degraded to 1,2-naphthalenedicarboxylic and phthalic acids, which was followed by production of 2-hydroxymethyl benzoic acid or monomethyl and dimethyl esters of phthalic acid. Another degradation product of benz[a]anthracene-7,12-dione was identified as 1-tetralone. Its transformation *via* 1,4-naphthalenedione, 1,4-naphthalene-diol and 1,2,3,4-tetrahydro-1-hydroxynaphthalene again resulted in phthalic acid. None of the intermediates were identified as dead-end metabolites [50].

2.1.7. Benzo[a]Pyrene. First studies with B[a]P were conducted with *P. chrysosporium* and demonstrated B[a]P oxidation to quinones and partial mineralization of [^{14}C]BaP. More recently, other fungi of these ecophysiological groups, such as *Bjerkandera* sp. BOS55 and *P. ostreatus*, were found to degrade B[a]P. The litter-decomposing basidiomycete *Stropharia coronilla*, which preferably colonizes grasslands, was found to be capable of metabolizing and mineralizing B[a]P in liquid culture [11]. In all cases, the degradation of B[a]P has been attributed to the activity of ligninolytic enzymes [11].

Finally, the nature of the transformation products formed during pollutant degradation differs among white-rot species (Table 1). This has been best demonstrated for ANT degradation. Significant accumulations of 9,10-anthraquinone were detected concomitant with depletion of ANT from liquid cultures of LAC-free species belonging to several genera (*Bjerkandera*, *Phanerochaete*, and *Ramaria*), with none or very little detected in LAC-producing species (*Trametes*, *Pleurotus*, and *Daedaleopsis*). LiP, MnP, VP, and LAC are all known to be produced by white-rot fungi,

Involvement of the Ligninolytic System of White-Rot and Litter-Decomposing Fungi in the Degradation of
Polycyclic Aromatic Hydrocarbons

51

although the specific enzyme complements of different species are highly variable. Undoubtedly, all three levels of variability are consequences of both differences in the enzymology of the various white-rot species and differences in growth and enzyme production responses of various fungi to different culture media. Typical ligninolytic enzyme patterns of the white-rot fungi in liquid cultures are the following: *P. chrysosporium*, LiP and MnP; *P. ostreatus*, MnP, VP, and LAC; *T. versicolor*, LiP, MnP, and LAC. Each of these enzyme classes has been implicated in pollutant degradation by these fungi [7].

Summarizing the data presented above it should be noted that involvement of intracellular cytochrome P-450 in PAH degradation clearly showed only for PHE degradation [13, 45, 54, 55]. In many other cases, the degradation of PAHs has been attributed to the activity of ligninolytic enzymes [11, 21, 22, 42, 50]. The transport of PAHs inside fungal cell can be limited by solubility of these compounds. It is known that the solubility of PAHs in aqueous solutions is very low (0.003–1.3 mg/L) [4], and PHE is one from more soluble (1.3 mg/L). It can be suggested that the relatively well-soluble PHE can penetrate inside fungal cell where it is available to intracellular cytochrome P-450. However the less-soluble compounds cannot penetrate to fungal cell and should be metabolized firstly by extracellular enzymes.

2.1.8. Ligninolytic Enzymes Production during PAH Degradation.

Several studies have shown that diverse white-rot fungi are capable of PAH mineralization and that mineralization rates correlate with the production of ligninolytic enzymes [9, 12, 37]. It was found that PAH conversion was correlated with the production of MnP and LAC [63]. These enzymes have been repeatedly implicated in biodegradation of PAHs, including PHE [17, 25, 43, 44]. In *P. ostreatus* cultures, these enzymes were present at least during the first 3 weeks of the experiment, and thus their involvement in the removal of PAHs was possible, including the production of anthraquinone from ANT [43, 44]. In the case of *P. chrysosporium* and *C. polyzona*, ANT degradation to anthraquinone started earlier than the production of LiP but during the same time slot as the secretion of MnP [60].

Clear indications have been found that extracellular MnP is involved in the conversion process, since Mn^{2+} supplementation considerably stimulated both enzyme production and degradation of PAHs [10]. For example, Steffen et al. [10] showed that the litter-decomposing *Stropharia rugosoannulata* was the most efficient degrader, removing or transforming B[a]P almost completely, and about 95% of ANT and 85% of PYR, in cultures supplemented with 200 μM Mn^{2+}. In contrast, less than 40, 18, and 50% of B[a]P, ANT, and PYR, respectively, were degraded in the absence of supplemental Mn^{2+}. In the case of *Stropharia coronilla*, the presence of Mn^{2+} led to a 20 fold increase in ANT conversion. The effect of manganese could be attributed to the stimulation of MnP. The Mn^{2+}-supplemented cultures degraded about 6% of ^{14}C-labeled B[a]P to $^{14}CO_2$, whereas only 0.7% was mineralized in the absence of Mn^{2+} [10].

Mn^{2+} stimulated considerably both the conversion and the mineralization of B[a]P by the litter-decomposing basidiomycete *Stropharia coronilla*; the fungus metabolized and mineralized about four and twelve times, respectively, more of the B[a]P in the presence of supplemental Mn^{2+} than in the basal medium. This stimulating effect could be attributed to MnP, whose activity increased after the addition of Mn^{2+} [11].

Collins and Dobson [8] reported high MnP activities and substantial oxidation of PHE and FLU when liquid cultures of the white-rot fungus *T. versicolor* were supplemented with Mn^{2+}. Despite the apparent reliance of the strain primarily on MnP, liquid cultures of *P. laevis* were capable of extensive transformation of ANT, PHE, B[a]A, and B[a]P. No LiP was found in the culture medium [7].

Two strains of *P. sordida* degraded a significantly greater amount of PHE and FLA than *P. chrysosporium*. The production of MnP, the only extracellular ligninolytic enzyme detected during the cultivation, was evaluated [56]. The fungus *I. lacteus* was shown to be an efficient degrader of PAHs possessing 3–6 aromatic rings. The strain produced mainly MnP in pollutant-free media. However, after contamination with PAHs (especially PYR), the values increased, and significant activity of Mn-independent peroxidase appeared in the complex medium [62].

LAC activity was also implicated in the degradation of PAHs by white-rot fungi [44]. A high and relatively stable activity of LAC was observed during degradation of ANT by *T. trogii* [61]. Hovewer, Han et al. [20] showed that LAC production by *T. versicolor* 951022 was not enhanced by addition of PHE. The addition of $CuSO_4$, citric acid, gallic acid, tartaric acid, veratryl alcohol (VA), guaiacol, and ABTS enhanced the degradation of PHE and PYR and increased laccase activities in submerged culture of *Ganoderma lucidum* [58].

The effect of PAHs on the time course of LAC production by *P. ostreatus* D1 under conditions of submerged cultivation in Kirk's medium was also studied. It was shown that PHE, FLA, PYR, and CHR actively induced this enzyme, whereas FLU and ANT had a smaller effect [64].

Pozdnyakova et al. [23] showed that the activities of two ligninolytic enzymes, LAC and versatile peroxidase (VP), of *P. ostreatus* D1 were induced by the PAHs, their derivatives, and their degradation products under conditions of submerged cultivation in basidiomycetes rich medium. LAC was produced mostly in the first 7–10 days, whereas the production of VP began after 5–7 days of cultivation. The difference in the production time for these enzymes suggests that LAC can be involved in the first stages of PAH degradation and that VP can be necessary for oxidation of some degradation products. That was the first report on VP induction by PAHs and their derivatives [23].

Furthermore, LAC activity was revealed on the mycelial surface of *P. ostreatus* D1 cultivated in the presence of PYR and CHR [21, 22].

2.1.9. Bioavailability of PAHs.

As was mentioned above, the solubility of PAHs in aqueous solutions is very low.

Bioavailability of PAHs may be the limiting factor for fungal and microbial attack [4]. In experimental conditions the addition of a detergent increases the solubility of PAHs and allows repeatable determination of the substrate and products. For increasing the substrate availability to ligninolytic enzymes and cells, nonionic surfactants such as Tween-20 and Tween-80 are usually used [47]. For example, various surfactants could increase the rate of ANT, PYR, and B[a]P oxidation by *Bjerkandera* sp. BOS55 by two-to-fivefold. The stimulating effect of surfactants was found to be solely due to the increased bioavailability of PAHs, indicating that the oxidation of PAHs by the extracellular ligninolytic enzymes is limited by low compound bioavailability [47].

At the present time, there are few reports on emulsifying compound production during PAH degradation by ligninolytic fungi. Song [49] showed that when *P. chrysosporium* was grown shaken, some foam could be seen, possibly because of production of a biosurfactant. The author suggested that this surfactant is responsible for the solubilization of pyrene in aqueous medium [49].

The production of an emulsifying agent during the degradation of phthalic, 2,2'-diphenic, and α-hydroxy-β-naphthoic acids, PHE, ANT, FLU, PYR, FLA, and CHR by the white-rot fungus *P. ostreatus* was found as well [65]. Emulsifying activity was inversely dependent on the water solubility of the compounds used. Maximal emulsifying activities were found in the presence of CHR (48.4%) and α-hydroxy-β-naphthoic acids (52.2%). The obtained data suggest that *P. ostreatus* D1 can produce some emulsifying agent as a response to the presence of PAHs and some products of their degradation and that the emulsifying agent can promote solubilization of these compounds [65]. Three different possible functions of the found emulsifying agent can be proposed: (a) this agent can be essential for increasing the solubility of hydrophobic compounds, (b) it could be involved in the oxidation of hydrophobic compounds catalyzed by ligninolytic enzymes [65], and (c) it can have a positive effect on the production of extracellular ligninolytic enzymes in agitated culture similar to that was described by Jager et al. [66].

The positive effect of an emulsifying agent on the production of extracellular ligninolytic enzymes in agitated culture was described by Jager et al. [66]. Those authors showed an increase in the production of LiP by *P. chrysosporium* in the presence of some detergents (Tween-80, Tween-20, and 3-[(3-colamidopropyl)dimethylammonio]1-propanesulfonate) [66]. The biosurfactant rhamnolipid increased the activities of LiP and MnP in *P. chrysosporium* and the activities of LiP and LAC in *Penicillium simplicissimum* [67].

The involvement of some surfactants in PAH oxidation by ligninolytic enzymes has been reported as well. For example, Böhmer et al. [68] showed PHE oxidation by the LAC/HBT pair in the presence of synthetic detergent (Tween-80) containing unsaturated lipids. They assumed that two coupling reactions take place: lipid peroxidation and PHE oxidation [68]. In line with the results of Böhmer et al., the data of Pozdnyakova et al. [69] showed that the *P. ostreatus* laccase/ABTS pair was able to oxidize PHE to a limited extent

and that this reaction was greatly enhanced by Tween-80 [69]. Moen and Hammel [25] found a similar effect in a system containing *P. chrysosporium* MnP and Tween-80, and they showed coupling of the lipid peroxidation and PHE oxidation reactions [25].

Summarizing the data presented above it should be noted that the main studies of PAH degradation are carried out under different conditions (different medium and pH ligninolytic and nonligninolytic conditions) and with fungi, producing different sets of ligninolytic enzymes (e.g., *P. chrysosporium*: LiP and MnP, *T. versicolor*: LiP, MnP, LAC; and *P. ostreatus*: MnP, VP, LAC). It complicates the generalization and discussion of the data obtained by different authors. In my opinion, the same conditions of different fungi and correspondingly the different conditions of the same fungus can clarify the mechanisms of PAH degradation and the involvement of different group of enzymes. Now similar dependence of both bioconversion of PYR and CHR and ligninolytic enzyme production on the cultivation conditions was found for *P. ostreatus* D1 only [21, 22]. Under LAC production conditions, transformation of these PAHs occurred with accumulation of the quinone metabolites. Under both LAC and versatile peroxidase (VP) production conditions, CHR and PYR degradation occurred, with the stages leading to phthalic acid formation and its further utilization.

The second problem of PAH degradation which should be solved is bioavailability of these compounds. The study of natural emulsifying agent producing by ligninolytic fungi during xenobiotics degradation can promote the explanation of this problem.

Finally two groups of enzymes should be studied: cytochrome P-450 and cell-associated/intracellular forms of ligninolytic enzymes. The presence of many cytochrome P-450-related genes in white-rot fungi and some recent data on cytochrome P-450-catalyzed oxidation of PHE by *P. chrysosporium* [55] suggest the involvement of these enzymes in PAH degradation. However, our resent data [21, 22] suggest that the initial attack on the PAH molecule may be catalyzed by cell-associated enzymes (at least by LAC), because some time is required for the extracellular enzymes to appear in the culture medium at concentrations sufficient for substrate degradation.

2.2. Mycoremediation of PAH- and Oil-Contaminated Soils. Mycoremediation is a process by which fungi degrade or transform hazardous organic contaminants to less toxic compounds [70]. White-rot and litter-decomposing fungi are potential candidates for the treatment of contaminated soils because of their high capability to degrade a wide range of xenobiotics not only in liquid culture [10, 11, 71–76] but also in contaminated soil (Table 2) [72, 77–82]. Attempts have, therefore, been made to apply these fungi to the bioremediation of soils contaminated with compounds not sufficiently degradable by soil microorganisms [28].

For example, *L. lacteus* and *P. ostreatus* degraded three- and four-ring unsubstituted PAHs, including FLU, PHE, ANT, FLA, PYR, CHR, and B[a]A, in two contaminated

Involvement of the Ligninolytic System of White-Rot and Litter-Decomposing Fungi in the Degradation of
Polycyclic Aromatic Hydrocarbons

53

TABLE 2: PAH degradation under mycoremediation.

Fungus	PAH	Metabolites	Enzymes	References
Agrocybe aegerita	PYR	CO_2	ND	[83]
Anthracophyllum discolor	PHE; ANT; FLA; PYR; B[a]P	9,10-anthraquinone; phthalic acid; 4-hydroxy-9-fluorenone; 9-fluorenone; 4,5-dihydropyrene; CO_2 from PHE	MnP	[84]
Bjerkandera sp.	FLA; PYR; CHR	ND	ND	[82]
I. lacteus	FLU; PHE; ANT; FLA; PYR; CHR; B[a]A; B[a]P; benzo[k]fluoranthene	ND	LiP; MnP; LAC	[48, 80, 85]
Kuehneromyces mutabilis	PYR	CO_2	ND	[83]
P. chrysosporium	FLU; CHR; ANT; PHE; PYR	9-fluorenone	MnP	[26, 86]
Pleurotus sp. Florida	PYR; CHR; B[a]A; B[a]P; benzo[b]fluoranthene; benzo[k]fluoranthene; dibenzo[a,h]anthracene; benzo[ghi]perylene	CO_2 from PYR; CHR; B[a]A; B[a]P	ND	[87]
P. ostreatus	FLU; PHE; ANT; FLA; PYR; CHR; B[a]A; B[a]P; benzo[b]fluoranthene; benzo[k]fluoranthene; dibenzo[a,h]anthracene; benzo[ghi]perylene	9,10-anthraquinone	MnP; LAC	[30, 80, 88, 89]
Stropharia rugosoannulata	B[a]A; B[a]P; dibenzo[a,h]anthracene	ND	ND	[90]
Stropharia coronilla	B[a]A; B[a]P; dibenzo [a,h]anthracene	ND	ND	[90]
T. versicolor	ANT; PHE; PYR	ND	MnP; LAC	[89]
Consortium: T. versicolor B. adusta B. fumosa	PYR	ND	LAC; Mn-independent peroxidase	[91]

industrial soils [80]. The biodegradation in soil by *P. chrysosporium* was approximately 80–85% of FLU, 52% of fluorenone, and 94% of 1,4-naphthoquinone [86]. It was found that the native soil microflora can be prompted into full mineralization of PAHs in some contaminated soils and that this mineralization can be enhanced when supplemented with the white-rot fungus *P. chrysosporium* [24]. The degradation of 4-ring PAHs including dibenzothiophene, fluoranthene, pyrene, and chrysene by *Bjerkandera* sp. BOS55 was shown too [82].

The degradation of PAHs in contaminated soil can also be attributed to the activity of ligninolytic enzymes. For example, periods of high *mnp* transcript levels and extractable MnP enzyme activity coincided with maximal rates of depletion of FLU and corresponding accumulation of 9-fluorenone, and CHR disappearance in soil cultures, supporting the hypothesis that PAHs are oxidized in soil via MnP-dependent mechanisms and that these reactions play a role in soil bioremediation by these fungi [26].

The removal efficiency of three-, four-, and five-ring PAHs in contaminated soil bioaugmented with *Anthracophyllum discolor* was investigated, and the production of lignin-degrading enzymes and PAH mineralization in the soil were also determined. A high removal capability for PHE (62%), ANT (73%), FLA (54%), PYR (60%), and B[a]P (75%) was observed in autoclaved soil inoculated with *A. discolor* in the absence of indigenous microorganisms, and it was associated with the production of MnP. The metabolites found in PAH degradation were anthraquinone, phthalic acid, 4-hydroxy-9-fluorenone, 9-fluorenone, and 4,5-dihydropyrene. *A. discolor* was able to mineralize 9% of PHE [84].

The effect of the indigenous soil microflora on growth, extracellular enzyme production, and PAH degradation efficiency in soil by the white-rot fungi *T. versicolor* and *I. lacteus* was investigated too. Both fungi were able to colonize soil. LAC was produced in *T. versicolor* cultures in the presence or absence of bacteria, but live bacteria decreased the LAC levels in soil by about 1/5. MnP was not detected in *T. versicolor* cultures. The amounts of MnP and LAC in *I. lacteus* cultures were not affected by the presence of bacteria. The rates of PAH removal by *T. versicolor* in sterile soil were 1.5, 5.8, and 1.8 fold for 2-3-ring, 4-ring, and 5-6-ring PAHs, compared to *I. lacteus*, respectively. *I. lacteus* showed a low efficiency of removal of PYR, B[a]A, and benzo[k]fluoranthene, compared to *T. versicolor*, whereas CHR and benzo[b]fluoranthene were degraded by neither fungus. Weak fungal/bacterial synergistic effects were observed in the case of removal of acenapthylene, B[a]P, dibenzo[a,h]anthracene and benzo[g,h,i]perylene by *I. lacteus* and acenaphthylene by *T. versicolor* [85].

Initially, the potential of *P. ostreatus* for mycoremediation of PAH-contaminated soil was evaluated in a model system. For example, the degradation of eight unlabeled highly condensed PAHs and the mineralization of three ^{14}C-labeled PAHs by the white-rot fungus *Pleurotus* sp. Florida were investigated in sterile sea sand. PAH-loaded sea sand was mixed into a straw substrate and incubated. The disappearance of the unlabeled four-to-six-ring PAHs, including PYR, B[a]A, CHR, benzo[b]fluoranthene, benzo[k]fluoranthene, B[a]P, dibenzo[a,h]anthracene, and benzo[ghi]perylene, was determined. *Pleurotus* sp. Florida mineralized 53% of [^{14}C]PYR, 25% of [^{14}C]B[a]A, and 39% of [^{14}C]B[a]P to [14]CO$_2$ in the presence of eight unlabeled PAHs (50 lg applied) within 15 weeks [87].

Later, Zebulun et al. [88] showed that after 90 days of incubation of *P. ostreatus* in soil, the level of contamination decreased, and fungal treatment affected the rate of degradation of all levels of ANT contamination. The inoculated soil showed more degradation of ANT (76–89%) than did the control soil. The release of ligninolytic peroxidases and LAC by *P. ostreatus* was associated with the observed ANT degradation [91].

The effect of salinity of the soil on PAH degradation and ligninolytic enzymes production was studied by Valentín et al. [81]. The minimal effect of salinity on ligninolytic activities of *I. lacteus* and *Lentinus tigrinus* was shown, while in *B. adusta,* activity was inhibited by salinity level of 32% [81].

A consortium of three basidiomycetes, including *T. versicolor, B. adusta,* and *B. fumosa,* grown on straw was found able to efficiently colonize soil and remove about 56 out of 100 mg/kg of soil dry weight of PYR in 28 days; in the mean time, the germination index increased, indicating a reduction in phytotoxicity. Enzymatic assays showed that LAC and manganese-independent peroxidase activity could have played a role in the degradation process [90].

The effects of concomitant pollutants, such as heavy metals, on PAH degradation in soil was evaluated by Baldrian et al. [30]. It was shown that *P. ostreatus* was able to degrade B[a]A, CHR, benzo[b]fluoranthene, benzo[k]fluoranthene, B[a]P, dibenzo[a,h]anthracene, and benzo[ghi]perylene in nonsterile soil both in the presence and in the absence of

cadmium and mercury. During 15 weeks of incubation, recovery of individual compounds was 16 to 69% in soil without additional metals. While soil microflora contributed mostly to degradation of PYR (82%) and B[a]A (41%), the fungus enhanced the disappearance of less soluble PAHs containing five or six aromatic rings. Although the heavy metals in the soil affected the activity of ligninolytic enzymes produced by the fungus (LAC and MnP), no decrease in PAH degradation was found in soil containing Cd or Hg at 10 to 100 ppm. In the presence of cadmium at 500 ppm in soil, degradation of PAHs by the soil microflora was not affected, whereas the contribution of the fungus was negligible, probably owing to the absence of MnP activity. In the presence of Hg at 50 to 100 ppm or Cd at 100 to 500 ppm, the extent of soil colonization by the fungus was limited [30].

With the focus on alternative microbes for soil bioremediation, eight species of litter-decomposing basidiomycetous fungi were selected for a bioremediation experiment with an artificially contaminated soil. Up to 70%, 86%, and 84% of B[a]A, B[a]P, and dibenzo[a,h]anthracene, respectively, were removed in the presence of fungi, while the indigenous microorganisms converted merely up to 29%, 26%, and 43% of these compounds in 30 days. The low-molecular-mass PAHs studied were easily degraded by the soil microbes, and only ANT degradation was enhanced by the fungi as well. The agaric basidiomycetes *Stropharia rugosoannulata* and *Stropharia coronilla* were the most efficient PAH degraders among the litter-decomposing species used [83].

Similar to what is observed liquid medium, the mineralization of PAHs during mycoremediation of contaminated soil occurs also. For example, the mineralization of ^{14}C-PYR in sterilized and nonsterile soil was investigated by using the wood-decaying fungi *Kuehneromyces mutabilis* and *Agrocybe aegerita*. In sterile soil, 5.1% and 1.5% of PYR was mineralized to [14]CO$_2$ by *K. mutabilis* and *A. aegerita*, respectively. During soil inoculation with the fungi, the mineralization was higher (47.7% for *K. mutabilis* and 38.5% for *A. aegerita*). In comparison with the indigenous microflora, *K. mutabilis* enhanced PYR elimination up to 42% [92].

Compared to wood, soil and litter are a more complex and heterogeneous environment, which may hamper the detection and estimation of enzyme activities. LAC activity reflects the course of degradation of organic substances, and thus it varies with time. Being the most abundant ligninolytic enzyme in soil, LAC also reflects the presence of fungal mycelia and participates in the transformation of lignin in forest litter. It is also generally presumed that LAC is able to react with soil humic substances that can be directly formed from lignin [89].

Novotný et al. [93] compared the abilities of *P. chrysosporium, P. ostreatus,* and *T. versicolor* to degrade PAHs and produce ligninolytic enzymes in soil. They found that colonization of sterilized soilby straw-grown inocula and degradation of ANT, PHE, and PYR were greatest with *P. ostreatus*. The production of MnP and LAC in soil was

Involvement of the Ligninolytic System of White-Rot and Litter-Decomposing Fungi in the Degradation of
Polycyclic Aromatic Hydrocarbons

55

similar for *P. ostreatus* and *T. versicolor* but was extremely low for *P. chrysosporium*. *I. lacteus* efficiently colonized sterile and nonsterile soil by mycelium growing from a wheat straw inoculum. Good colonization of nonsterile gasworks soil contaminated with PAHs and heavy metals was also observed. *I. lacteus* efficiently removed three- and four-ring PAHs, including ANT, FLA, and PYR, from artificially spiked soil. LiP and LAC, but not MnP, were also detected when the fungus colonized the soil [48].

The relationship between ligninolytic activity and PAH degradation by *P. ostreatus* was demonstrated by Eggen [29]. In *P. ostreatus*, LAC and MnP were found, and their involvement in the removal of PAHs was possible, including the production of anthraquinone from ANT [93]. A similar correlation also was reported for the expression of MnP and the removal of FLU and CHR by soil cultures of *P. chrysosporium* [26].

The production of ligninolytic enzymes during mycore-mediation of old oil-contaminated soil by 12 strains of white-rot fungi was studied by Pozdnyakova et al. [31]. LAC activity peaked during the first 2 weeks after sunflower-seed hulls colonized by the fungi had been added to the soil. Thereafter, LAC activity decreased and reached a minimal value, which was maintained in the course of the experiment. Enzyme activities in sterile and nonsterile soil were similar, but LAC activity tended to be higher in sterile soil. The fungi that were good colonizers of contaminated soil were good producers of LAC under the conditions used. The most active producers of LAC under these conditions were *Agaricus* sp. and *P. ostreatus* strains. Only *Pleurotus* and *Agaricus* strains produced peroxidase under these conditions. In either case, peroxidase activity was low and, as in the case of LAC activity, was similar in sterile and nonsterile soil. It should be noted that the peroxidase activities of the *P. ostreatus* strains detected in the presence of Mn^{2+} exceeded those without Mn^{2+} by about 20% only. The peroxidase activities of the *Agaricus* sp. strains could be detected only in the presence of Mn^{2+} [31]. The obtained results demonstrate different production of ligninolytic enzymes with respect to growth yields of various white-rot fungi growing in soil. The growth rates, the mycelium densities, the production of ligninolytic enzymes, and the degradation of old oil-contamination decreased in the order *Agaricus* sp. > *P. ostreatus* > *L. edodes* > *Coriolus* sp. The strains of the white-rot fungus *P. ostreatus* and the litter-decomposing fungus *Agaricus* sp. were the most active producers of ligninolytic enzymes and the most active degraders of old oil-contamination in soil [31].

In summary, it should be noted that white-rot and litter-decomposing fungi can actively degrade PAHs as in a model soil systems as in the conditions of the real soil (Table 2), for example, in soils with high salinity and in soils containing concomitant contamination such as heavy metals. The detection of the metabolites, which are similar to submerged cultures (e.g., anthraquinone, 9-fluorenone, 4,5-dihydropyrene, and other), and production of the same ligninolytic enzymes suggest the using the common mechanisms of PAH degradation. The decrease of the toxicity of the soil

during mycoremediation makes this method perspective for use.

3. Enzymology of PAH Degradation

White-rot fungi produce four major groups of enzymes for the degradation of lignin: lignin peroxidase (known as ligninase in early publications; LiP; EC 1.11.1.14), manganese-dependent peroxidase (manganese peroxidase, MnP; EC 1.11.1.13), versatile peroxidase (VP; EC 1.11.1.16), and laccase (LAC; EC 1.10.3.2) [106].

It is generally known that ligninolytic enzymes, that is, LiP, MnP, and LAC, are also involved in the degradation of a wide range of organopollutants, including PAHs [33, 39]. For example, Vyas et al. [60] proposed that *in vitro* degradation of ANT by a crude enzyme preparation of white-rot fungi attests to the involvement of ligninolytic enzymes in the oxidation of ANT to anthraquinone. Sanglard et al. [107] showed that LiP is responsible for the initial steps in B[a]P metabolism by *P. chrysosporium*. Crude extracellular peroxidases from *P. laevis* transformed ANT, PHE, B[a]A, and B[a]P either in MnP-Mn^{2+} reactions or in MnP-based lipid peroxidation systems. No transformation of B[a]A or PHE was observed. In contrast, MnP-dependent lipid peroxidation reactions supported transformation of all four PAHs [7]. Experiments with purified cell-free enzyme extracts have confirmed the role of ligninolytic enzymes in the attack on PAHs. Extracellular preparations of LiP from *P. chrysosporium* were among the first to be shown as capable of PAH oxidation [39, 51].

The exact role of individual ligninolytic enzymes in degradation is still unclear. One possible way of elucidating the participation in degradation is by performing degradation with isolated enzymes together with identification of degradation products (Table 3) [35]. Since LiP, MnP, and LAC appear to be the predominant ligninolytic enzymes produced during PAH metabolism, most research has focused on these enzymes [33].

3.1. Lignin Peroxidase. Lignin peroxidases [LiP; EC 1.11.1.14, 1,2-bis(3,4-dimethoxyphenyl)propane-1,3-diol: hydrogen peroxide oxidoreductase] catalyze the H_2O_2-dependent oxidative depolymerization of lignin. The overall reaction is represented by 1,2-bis(3,4-dimethoxyphenyl)propane-1,3-diol + H_2O_2 ↔ 3,4-dimethoxybenzaldehyde+1-(3,4-dimethoxyphenyl)ethane-1,2-diol + H_2O. LiP is relatively nonspecific to its substrates and has been known to oxidize phenolic aromatic substrates and a variety of nonphenolic lignin model compounds as well as a range of organic compounds. LiPs have the unique ability to catalyze oxidative cleavage of carbon-carbon bonds and ether bonds (C–O–C) in nonphenolic aromatic substrates of high redox potential.

The enzyme activity of LiP is conveniently measured by the oxidation of veratryl alcohol (VA), the favored LiP substrate, to veratraldehyde by the increase in absorbance at 310 nm [106].

Table 3: PAH oxidation by pure enzymes.

Enzyme	Fungus	PAH	Products	References
LiP	P. chrysosporium	B[a]P	B[a]P-1,6-quinone; B[a]P-3,6-quinone; B[a]P-6,12-quinone	[1, 51]
		ANT	9,10-anthraquinone	[1, 40, 41]
		PYR	PYR-1,6-dione; PYR-1,8-dione	[1, 39, 41]
		FLA	ND	[41]
		1-methylanthracene	1-methylanthraquinone	[41]
		2-methylanthracene	2-methylanthraquinone	[1, 41]
		9-methylanthracene	9-anthraquinone; 9-methyleneanthranone; 9-methanol-9,10-dihydroanthracene	[41]
		Acenaphthene	1-acenaphthenone; 1-acenaphthenol	[1, 41]
		Dibenzothiophene	dibenzothiophene sulfoxide	[41]
MnP	Anthracophyllum discolor	PYR; ANT; FLA; PHE	ND	[94]
	I. lacteus	PHE; ANT; FLA; PYR	9,10-anthraquinone	[35]
		ANT	anthrone; 9,10-anthraquinone; 2-(2'-hydroxybenzoyl)-benzoic acid; phthalic acid	[25, 40, 42, 95]
	P. chrysosporium	FLU	9-fluorenone	[17]
		PHE	PHE-9,10-quinone; 2,2'-diphenic acid	[25]
		dibenzothiophene	4-methoxybenzoic acid	[95]
	Nematoloma frowardii (Phlebia sp.)	PHE; ANT; PYR; FLA; CHR; B[a]A; B[a]P; benzo[b]fluoranthene	CO_2 from PHE; ANT; PYR; B[a]A; B[a]P	[37, 96]
	Stropharia coronilla	ANT; B[a]P	9,10-anthraquinone; CO_2; B[a]P-1,6-quinone	[10, 11]
VP	B. adusta	ANT; PYP; B[a]P	9,10-anthraquinone	[97]
	B. fumosa	ANT; PHE; FLU; PYR; CHR; FLA	9,10-anthraquinone; 9-fluorenone	Pozdnyakova et al., unpublished data
	P. ostreatus D1	ANT; PHE; FLU; PYR; CHR; FLA	9,10-anthraquinone; 9-fluorenone	
LAC	C. hirsutus	ANT; PHE; PYR; FLA; B[a]P; B[a]P; ANT; PHE; FLU; 9-methylanthracene	ND	[34]
	Coriolopsis gallica	2-methylanthracene; Acenaphthene; carbazole; N-ethylcarbazole; Dibenzothiophene	9-fluorenone; dibenzothiophene sulfone	[19, 98]
	Ganoderma lucidum	ANT; FLU; B[a]A; B[a]P; Acenaphthene; Acenaphthylene	ND	[99]
	P. ostreatus	ANT; PHE; FLU; PYR; FLA; perylene	9,10-anthraquinone; 9-fluorenone	[69, 100]
	Pycnoporus cinnabarinus	B[a]P	B[a]P-1,6-quinone; B[a]P-3,6-quinone; B[a]P-6,12-quinone	[101]
	T. versicolor	Acenaphthene; PHE; ANT; Acenaphthylene, B[a]P; ANT; FLA; PYR; B[a]A; CHR; perylene; benzo[b]fluoranthene; benzo[k]fluoranthene; FLU	1,2-acenaphthenedione 1,8-naphthalic acid anhydride; 9,10-anthraquinone; PHE-9,10-quinone, 2,2'-diphenic acid; B[a]P-1,6-quinone; B[a]P-3,6-quinone; B[a]P-6,12-quinone	[20, 44, 68, 102–105]

Involvement of the Ligninolytic System of White-Rot and Litter-Decomposing Fungi in the Degradation of
Polycyclic Aromatic Hydrocarbons

57

As the first of these enzymes, purified LiP from *P. chrysosporium* was shown to attack B[a]P via one-electron abstractions, leading to unstable B[a]P radicals that undergo further spontaneous reactions to hydroxylated metabolites and several B[a]P quinones [51]. Three products of B[a]P oxidation by LiP of *P. chrysosporium*, namely, benzo[a]pyrene-1,6-, 3,6-, and 6,12-quinones, were found. Simultaneously with the appearance of oxidation products, LiP was inactivated. As all peroxidases, LiP is inactivated by the presence of hydrogen peroxide [108]; the enzyme could be stabilized by the addition of VA to the reaction mixture. Addition of VA to the reaction mixture increased the oxidation rate by about 15 times, and the enzyme retained most of its activity during the B[a]P oxidation [51].

The majority of studies of LiP-catalyzed oxidation of PAHs have been done with LiP from *P. chrysosporium* (Table 3). For example, the oxidation of another PAH, ANT, by LiP of *P. chrysosporium* was described [39, 40, 51]. Anthraquinone was identified as the main product of ANT oxidation by LiP from *P. chrysosporium* [40]. Hammel et al. [39] showed that LiP of *P. chrysosporium* catalyzes the oxidation of certain PAHs with ionization potentials (IP) of <7.55 eV. This result demonstrates that the H_2O_2-oxidized states of LiP are more oxidizing than the analogous states of classical peroxidases. Experiments with PYR as the substrate showed that pyrene-1,6-dione and pyrene-1,8-dione are the major oxidation products. Gas chromatography/mass spectrometry analysis of LiP-catalyzed PYR oxidation done in the presence of H_2O_2 showed that the quinone oxygens come from water. The resulting quinones were not substrates for LiP. The one-electron oxidative mechanism of LiP applies not only to lignin and lignin-related substructures but also to certain polycyclic aromatic and heteroaromatic pollutants. The oxidation of PYR by whole cultures of *P. chrysosporium* also resulted in these quinones. The authors suggested that LiP is thus likely to catalyze the first step in the mineralization of these compounds by whole cultures of *P. chrysosporium* [39].

Vazquez-Duhalt et al. [41] used LiP from *P. chrysosporium* to study the oxidation of ANT, 1-, 2-, and 9-methylanthracenes, acenaphthene, FLA, PYR, carbazole, and dibenzothiophene. Of the compounds studied, LiP was able to oxidize those with IP of <8 eV. The highest specific activity toward PAHs was found in systems with pHs between 3.5 and 4.0. The reaction products contain hydroxyl and keto groups. The product of ANT oxidation was thought to be 9,10-anthraquinone. The products of LiP oxidation of 1- and 2-methylanthracene were 1- and 2-methylanthraquinone, respectively.

Three products were detected by these authors [41] from the oxidation of 9-methylanthracene: 9,10-anthraquinone, 9-methyleneanthranone, and 9-methanol-9,10-dihydroanthracene. In this case, carbon-carbon bond cleavage, yielding anthraquinone from 9-methylanthracene, was detected. The mass spectra of the two products from acenaphthene matched those of 1-acenaphthenone and 1-acenaphthenol. The product of dibenzothiophene oxidation by LiP was its sulfoxide, as determined by comparing the GC-mass spectrometry analysis with a sample of authentic dibenzothiophene sulfoxide. The UV spectrum of the product of PYR oxidation most closely matched that of 1,8-pyrenedione. Although FLA and carbazole were oxidized, their products were not identified [41].

Torres et al. [1] tested LiP, cytochrome *c*, and hemoglobin for oxidation of PAHs in the presence of hydrogen peroxide and showed that LiP oxidized ANT, 2-methylanthracene, 9-hexylanthracene, PYR, acenaphthene, and B[a]P; the unreacted compounds included CHR, PHE, naphthalene, triphenylene, biphenyl, and dibenzofuran. The oxidation of the aromatic compounds by LiP correlated with their IPs; only those compounds that had IPs of <8 eV were transformed. It was found that the reaction products from all three hemoproteins (LiP, cytochrome *c*, and hemoglobin) were mainly quinones, suggesting the same oxidation mechanism for the three biocatalysts. The product from ANT was anthraquinone, and the product from 2-methylanthracene was 2-methylanthraquinone. The oxidation products for PYR and B[a]P were pyrenedione and benzo[a]pyrenedione, respectively. The mass spectra of the two products from acenaphthene oxidation with LiP matched 1-acenaphthenone and 1-acenaphthenol [1].

3.2. Mn-Peroxidase. Manganese peroxidase [MnP; EC 1.11.1.13] is a heme-containing glycoprotein that requires hydrogen peroxide (H_2O_2) as an oxidant. MnP oxidizes Mn^{2+} to Mn^{3+}, which then oxidizes phenolic rings to phenoxy radicals, leading finally to the decomposition of compounds. Owing to its high reactivity, Mn^{3+} has to be stabilized via chelation by dicarboxylic acids, such as malonate or lactate. In addition to phenolic structures, the MnP system has been reported to catalyze cleavage of nonphenolic lignin model compounds [18, 106].

Because LiP is not produced by all white-rot fungi, more recent studies have focused on MnP, which is wide spread among the basidiomycetes and is found not only in wood-decay fungi but also in litter-decomposing fungi (Table 3). PAH degradation by MnP was first described in *P. chrysosporium* as a lipid peroxidation-dependent process [25]. Later, it was demonstrated that PAH degradation by some MnPs also occurs directly. PAH degradation experiments showed that MnP from *I. lacteus* was able to efficiently degrade three- and four-ring PAHs, including PHE, ANT, FLA, and PYR; PHE and FLA have IPs higher than 7.8 eV. The major degradation products of ANT were identified [35]. MnP produced by *Anthracophyllum discolor*, a Chilean white-rot fungus, degraded PYR (>86%), and ANT (>65%) alone or in mixture, and, to a lesser extent, it degraded FLA (<15.2%) and PHE (<8.6%) [94].

MnP-catalyzed oxidation of PAHs resulted in corresponding quinones. For example, MnP from *P. chrysosporium* oxidized ANT to anthraquinone [40]. During the degradation of ANT by MnP, the formation of anthrone was also detected, which was an expected intermediate, and it was followed by the appearance of 9,10-anthraquinone. Anthraquinone has earlier been described as the common oxidation product in *in vitro* reactions of peroxidases.

Further oxidation resulted in the formation of phthalic acid, as it was observed in ligninolytic cultures of *P. chrysosporium* [42]. The characteristic ring-cleavage product 2-(2′-hydroxybenzoyl)-benzoic acid indicates that MnP is even able to cleave the aromatic ring of a PAH molecule [25]. I found only a single report suggesting that MnP does not oxidize ANT in the presence of Mn^{2+} [41].

The high hydrophobicity of PAHs greatly hampers their degradation in liquid media. The enzymatic action of MnP in media containing a miscible organic solvent, acetone (36% v/v), was evaluated as a feasible system for the degradation of three PAHs: ANT, dibenzothiophene, and PYR. The order of degradability, in terms of degradation rates, was as follows: ANT > dibenzothiophene > PYR. ANT was degraded to phthalic acid. A ring cleavage product of the oxidation of dibenzothiophene, 4-methoxybenzoic acid, was also detected [95].

The first description of direct enzymatic mineralization of PAHs by MnPs indicates their important role in the oxidation of PAHs by wood-decaying fungi was showed by Sack et al. [37]. Degradation of PAHs by crude MnP preparation of *Nematoloma frowardii* (later been shown to be a *Phlebia* sp. [109]) was demonstrated for a mixture of eight different compounds (PHE, ANT, PYR, FLA, CHR, B[a]A, B[a]P, and benzo[b]fluoranthene) and for five individual PAHs, including PHE, ANT, PYR, FLA, and B[a]A. The oxidation of PAHs was enhanced by the addition of glutathione, a mediator substance, which is able to form reactive thiyl radicals. Glutathione-mediated MnP was capable of mineralizing ^{14}C-PYR (7.3%), ^{14}C-ANT (4.7%), ^{14}C-B[a]P (4.0%), ^{14}C-B[a]A (2.9%), and ^{14}C-PHE (2.5%) [37]. The mineralization of B[a]P and ANT by MnP from the litter-decomposing fungus *Stropharia coronilla* was also shown by Steffen et al. [11]. The oxidation of B[a]P leads to B[a]P-1,6-quinone as a temporary intermediate [11].

The stimulatory effect of glutathione was demonstrated by Günther et al. [96], who reported the degradation of 30% of ANT and 12% of PYR by MnP from *N. frowardii* (later *Phlebia* sp. [109]). The addition of mediating agents, such as reduced glutathione, increased the oxidative strength of MnP; as a result ANT was completely reduced and 60% of PYR was degraded after 24 h. Crude MnP from the agaric white-rot fungus *N. frowardii* (*Phlebia* sp.) oxidized *in vitro* several ^{14}C-PAHs, including PYR and B[a]P, leading to the formation of significant amounts of ^{14}CO$_2$ ("enzymatic combustion"); mineralization increased 3- to 10-fold when reduced glutathione (GSH) was present in the reaction solution. The GSH effect was attributed to the transient formation of particularly reactive thiyl radicals. However, it is rather unlikely that fungi secrete "valuable" substances such as GSH under natural conditions into their microhabitat [96].

Therefore, alternative redox mediators, enhancing the oxidative strength of the MnP system, have been sought. Very promising compounds acting as such mediators were found among the unsaturated fatty acids (e.g., oleic and linoleic acids) and their derivatives (e.g., Tween-80). These substances have been shown to act similarly to GSH and were detected in liquid and solid fungal cultures. In the presence of Tween-80, MnP was able to oxidize FLU, a PAH that cannot be directly oxidized by chelated Mn^{3+}because of its high IP (8.2 eV) as well as a complex PAH mixture (creosote). Tween-80 enabled *Stropharia coronilla* MnP to convert a large amount of B[a]P into polar fragments, and BaP-1,6-quinone was detected as a transient metabolite, which was further broken down to unknown products [11]. *Stropharia coronilla* MnP oxidized the individual PAHs in a mixture of 16 different compounds according to their IP and the presence of Tween-80. Only B[a]P and ANT were oxidized by the simple MnP system, but the initiation of lipid peroxidation *via* unsaturated fatty acid components of Tween-80 resulted in a substantial decrease in the content of all other PAHs. In addition, it was reported that poorly bioavailable PAHs, such as the six-ring compound benzo[g,h,i]perylene, are also subject to MnP attack. Steffen et al. proposed that MnP is the key enzyme in the degradation of B[a]P and other PAHs by litter-decomposing basidiomycetes [10, 11].

It has previously been mentioned that PAH degradation by MnP was first described in *P. chrysosporium*as a lipid peroxidation-dependent process [25]. Compounds with up to six rings are degraded *in vitro* during MnP-dependent lipid peroxidation reactions, and these same compounds are depleted from liquid cultures of *P. chrysosporium* [13].

The oxidation of FLU, a PAH that is not a substrate for fungal LiP, was studied *in vitro* with *P. chrysosporium* extracellular enzymes. Oxidation of FLU to 9-fluorenone was obtained in a system that contained Mn^{2+}, an unsaturated fatty acid, and either crude *P. chrysosporium* peroxidases or purified recombinant MnP. The oxidation of FLU was inhibited by the free-radical scavenger butylated hydroxytoluene but not by the LiP inhibitor $NaVO_3$. Mn^{3+}-malonic acid complexes could not oxidize FLU. Maximal formation of 9-fluorenone in this system required an unsaturated fatty acid, Mn^{2+}, and crude MnP. These results indicate that FLU oxidation *in vitro* was a consequence of lipid peroxidation mediated by *P. chrysosporium* MnP [17].

Crude and purified MnP from *Stropharia coronilla* oxidized B[a]P efficiently, a process that was enhanced by the surfactant Tween-80. A clear indication was found that benzo[a]pyrene-1,6-quinone was formed as a transient metabolite, which disappeared over the further course of the reaction. Treatment of a mixture of 16 PAHs with MnP resulted in concentration decreases of 10 to 100% for the individual compounds, and again, a stimulating effect of Tween-80 was observed. Probably owing to their lower IP, poorly bioavailable, high-molecular-mass PAHs such as B[a]P, benzo[g,h,i]perylene, and indeno[1,2,3-c,d]pyrene were converted to larger extents than low-molecular-mass ones (e.g., PHE and FLU) [11].

The MnP of *P. chrysosporium* supported Mn^{2+}-dependent, H_2O_2-independent lipid peroxidation, as shown by two findings: (a) linolenic acid was peroxidized to give products that reacted with thiobarbituric acid and (b) linoleic acid was peroxidized to give hexanal. MnP also supported the slow oxidation of PHE to 2,2′-diphenic acid in a reaction that required Mn^{2+}, oxygen, and unsaturated lipids. It was shown that fungal peroxidases are associated with the plasma membrane, where they might initiate the peroxidation of the

Involvement of the Ligninolytic System of White-Rot and Litter-Decomposing Fungi in the Degradation of
Polycyclic Aromatic Hydrocarbons

59

membrane lipids. Extracellular membranes are frequently attached to the hyphae of white-rot fungi, and peroxidases are found on these structures. The chemical composition of these extracellular membranes indicated that they contain lipids. An analysis of the extractable lipids in *P. chrysosporium* mycelium showed that they contain unsaturated fatty acids. Lipid peroxidation is frequently initiated by transition metal ions that react with endogenous lipid hydroperoxides, either oxidizing them to peroxy radicals or reducing them to alkoxy radicals. In the case of *P. chrysosporium,* possible initiators include Mn^{2+}, Mn^{3+}, and the peroxidase heme. The oxygen-centered radicals produced during lipid peroxidation are known to trigger xenobiotic cooxidations. PHE is most likely to be oxidized by the latter route, which would yield phenanthrene-9,10-quinone as an intermediate, which subsequently undergoes facile 9,10-cleavage to give 2,2′-diphenic acid [25].

3.3. Versatile Peroxidase.

3.3. Versatile Peroxidase. It has been reported that some MnPs isolated from the fungi *B. adusta, Bjerkandera* sp. strain BOS55, *Bjerkandera* sp. (B33/3), *B. fumosa, P. eryngii, P. ostreatus,* and *P. pulmonarius* exhibit activities on aromatic substrates similar to that of LiP. This group of enzymes, known as versatile peroxidases (VPs), is not only specific for Mn^{2+}, as is MnP, but also oxidizes phenolic and nonphenolic substrates that are typical for LiP, including veratryl alcohol, methoxybenzenes, and lignin model compounds, in the absence of manganese [106].

There have been only a few reports on VP production during PAH degradation by white-rot fungi [21, 22]. Now I find only single report (Table 3) on the oxidation of PAHs by VPs [97]. Wang et al. studied the PAH oxidation by a purified MnLiP hybrid isoenzyme (in fact it is VP) from *B. adusta* in the presence and absence of manganous ions. The substrates were PAHs with ionization potentials of 7.43 eV or lower, including ANT, its methyl derivatives, PYR and B[a]P. The PAH metabolites were identified as the corresponding quinones [97].

The oxidation of 3- and 4-rings PAHs by VPs from *B. fumosa* and *P. ostreatus* D1 to the corresponding quinones was found by Pozdnyakova et al. too (unpublished data).

3.4. Laccase.

3.4. Laccase. Laccase (LAC; EC 1.10.3.2, benzenediol: oxygen oxidoreductase) belongs to a group of polyphenol oxidases containing copper atoms in the catalytic center and usually called multicopper oxidases. LAC catalyzes the reduction of oxygen to water accompanied by the oxidation of a substrate, typically a *p*-dihydroxy phenol or another phenolic compound. LACs have overlapping substrate specificity, which can be extended to non-phenolic aromatic compounds with the use of redox mediators. In the presence of some synthetic and natural mediators, LAC can oxidize such compounds as veratryl and benzyl alcohols, nonphenolic groups of the lignin polymer, and lignin model substances [89].

There have been many reports on PAH oxidation by purified fungal laccases (Table 3). Most such studies have been made with LACs of *T. versicolor, C. hirsutus, P. ostreatus,* and *Coriolopsis gallica.* For example, *T. versicolor*

LAC, in combination with HBT, was able to oxidize two PAHs, acenaphthene and acenaphthylene; ABTS did not significantly influence the oxidation rate. LAC alone oxidized about 35% of the acenaphthene and only 3% of acenaphthylene. The main products detected after incubation were 1,2-acenaphthenedione and 1,8-naphthalic acid anhydride [102]. The purified LAC of *T. versicolor* did not transform PHE. The addition of a redox mediator, ABTS or HBT, to the reaction mixture increased the oxidation of PHE by LAC about 40% and 30%, respectively [20]. The *in vitro* oxidation of ANT and B[a]P, which have IPs of ≤7.45 eV, is catalyzed by LAC from *T. versicolor.* Oxidation of ANT was enhanced in the presence of ABTS, whereas ABTS was essential for the oxidation of B[a]P. Anthraquinone was identified as the major end product of ANT oxidation [44].

The oxidation of five PAHs, including ANT, B[a]P, FLA, PHE, and PYR, was catalyzed by LAC from *C. hirsutus* in the presence of the redox mediators ABTS and HBT. In the ABTS-mediated system, B[α]P was the most rapidly oxidized substrate, with ANT being the most rapidly oxidized in the HBT-mediated system. There was no clear relationship between the IP and the oxidation of the substrates. The degree of oxidation, by LAC of *C. hirsutus,* for the PAHs tested ranged from 10.9 to 97.2%. FLA and PYR were readily oxidized by *C. hirsutus* LAC in the presence of all the redox mediators used, ranging from 37.9 to 92.7%. PYR, one of the least oxidizable PAHs, was still oxidized by about 40% in the presence of all the mediators. From this, it was concluded that the PAH-oxidizing abilities of LAC are different, depending on the fungal species from which it was obtained [34].

The biotransformation of B[a]P by purified LAC of *Pycnoporus cinnabarinus* was shown, with the reaction requiring the presence of the exogenous mediator ABTS. Most of the substrate (95%) was converted within 24 hours. The enzyme oxidized the substrate mainly to benzo[a]pyrene-1,6-, 3,6- and 6,12-quinones at a 2/1/1 ratio [101].

LAC of the white-rot fungus *Ganoderma lucidum* degraded ANT completely without a redox mediator and also degraded B[a]P, FLU, acenapthene, acenaphthylene, and B[a]A up to 100.0, 98.6, 95.4, 90.1, and 85.3%, respectively, when the mediator was present [99].

Majcherczyk et al. [103] demonstrated that LAC of *T. versicolor* was able to oxidize *in vitro* most of the 14 PAHs tested. Acenaphthylene was removed by 37%, followed by ANT and B[a]P, which were oxidized by 18 and 19%, respectively. Lower but significant oxidation of about 10% was found for eight additional PAHs: acenaphthene, FLA, PYR, B[a]A, CHR, benzo[b]fluoranthene, benzo[k]fluoranthene, and perylene. Naphthalene, FLU, and PHE were recovered unchanged after incubation for 72 h with laccase. Addition of HBT to the reaction mixture increased the oxidation of PAHs: acenaphthylene, acenaphthene, FLU, ANT, B[a]P, and perylene were almost completely removed from the reaction mixture. Oxidation of PYR and B[a]A increased from 8 and 6% without a mediator to 48 and 53% in the presence of HBT. PAH-quinones as oxidation products were formed from all PAHs to different extents. Some of the PAHs were polymerized in the LAC/mediator system to products of average molecular weight (MW) of approximately 1,500 Da [103].

The effect of different mediators on LAC oxidation was studied by Pickard et al. [19]. The following seven PAHs were oxidized by LAC of *Coriolopsis gallica* UAMH 8260: B[a]P, 9-methylanthracene, 2-methylanthracene, ANT, biphenylene, acenaphthene, and PHE. 9-Methylanthracene was the substrate that was the most rapidly oxidized. There was no clear relationship between the ionization potential of the substrate and the first-order rate constant for substrate loss *in vitro* in the presence of ABTS. The effects of mediating substrates were examined further by using ANT as a substrate. HBT supported approximately one-half ANT oxidation rate that ABTS supported, but HBT plus ABTS increased the oxidation rate nine-fold, compared with the oxidation rate in the presence of ABTS. A synergistic effect of the two mediators was found [19].

LAC of the white-rot fungus *P. ostreatus* D1 was able to degrade ANT (91%), PHE (72%), FLU (53.5%), PYR (65.5%), FLA (69.7%), and perylene (73%) only in the presence of a synthetic mediator. The degradation of PHE in the presence of detergents varied from 49 to 72%, whereas in the absence of any detergent, it reached 10%. Investigating the effect of various mediators on PAH degradation showed that ABTS was a better mediator of ANT oxidation and that HBT was a better mediator of FLU oxidation. PYR and ANT were degraded more rapidly in a mixture than separately. The degradation yield depends on the structure of the PAH molecule, type of the organic solvent, the presence and type of a detergent, enzyme concentration, and duration of the reaction. It does not correlate with the IP values, solubility, and recalcitrance of the studied PAHs. Apparently, it is necessary to take into account all these factors for studies of the catalytic mechanism responsible for LAC-catalyzed oxidation of PAHs [69].

LAC from *Coriolopsis gallica* oxidized not only FLU (75%) but also its polycyclic heterocyclic analogs such as carbazole (100% loss), *N*-ethylcarbazole (100%), and dibenzothiophene (60%) in the presence of HBT and ABTS as free radical mediators. Susceptibility to LAC oxidation appears related to the ionization potential of the substrate: the compounds with an IP of above 8.52, namely, dibenzofuran (IP = 8.77) and benzothiophene (IP = 8.73) were not attacked. Carbazole (IP = 7.68) was the most sensitive to oxidation, with >99% being transformed after 1 h. 9-fluorenone was identified as the product of FLU (IP = 8.52) oxidation, and dibenzothiophene sulfone as product of dibenzothiophene (IP = 8.44) was found [98].

PHE was efficiently oxidized by LAC in the presence of both HBT and unsaturated lipids, with 73% of the initially added PHE being degraded. The system was also able to peroxidize linoleic acid to its corresponding hydroperoxides, suggesting the involvement of lipid peroxidation in LAC-catalyzed PHE oxidation. Lipid peroxidation by LAC required HBT and did not depend on Mn^{2+} or H_2O_2, suggesting that the chemical reactions involved differ from those previously reported for MnP. LAC efficiently oxidized PHE in the presence of HBT and Tween-80. Two major products were formed 2,2'-diphenic acid and phenanthrene-9,10-quinone. In contrast to HBT, neither ABTS nor 3-hydroxyanthranilic acid stimulated PHE oxidation under the experimental conditions used. This finding was not further investigated, but it seems likely that HBT forms more reactive intermediates than ABTS and 3-hydroxyanthranilic acid do when oxidized by LAC. PHE was poorly degraded when Tween-80 was replaced by Tween-20. Tween-80 contains unsaturated fatty acid esters, whereas Tween-20 does not. The results show that the LAC/HBT couple was able to oxidize PHE to a limited extent, and that this reaction was greatly enhanced by unsaturated lipids [68].

The white-rot fungi secrete a large number of low-molecular-weight aromatic compounds, some of which are phenol derivatives and potential LAC substrates. The oxidation of PAHs was studied by Cañas et al. [104] in systems consisting of LAC from *T. versicolor* and compounds known as mediator compounds. The enzymatic oxidation of acenaphthene, acenaphthylene, ANT, and FLU was mediated by various LAC substrates (phenols and aromatic amines) or compounds produced and secreted by white-rot fungi. The best natural mediators, such as phenol, aniline, 4-hydroxybenzoic acid, and 4-hydroxybenzyl alcohol were as efficient as the previously described synthetic compounds ABTS and HBT. The oxidation efficiency increased proportionally with the redox potentials of the phenolic mediators up to a maximum value of 0.9 V and decreased thereafter with redox potentials exceeding this value.

Natural compounds such as methionine, cysteine, and reduced glutathione, containing sulfhydryl groups, were also active as mediator compounds [105]. Efficient transformation of several PAHs was obtained by using a fungal LAC in the presence of phenolic compounds related to those formed in nature during the turnover of lignin and humus. The effect of these natural mediators, namely, vanillin, acetovanillone, acetosyringone, syringaldehyde, 2,4,6-trimethylphenol, *p*-coumaric acid, ferulic acid, and sinapic acid, was compared with that of synthetic mediators such as ABTS and HBT. ANT was significantly degraded by LAC in the absence of mediators, whereas B[a]P and PYR were weakly transformed (less than 15% after 24 h). Vanillin, acetovanillone, 2,4,6-trimethylphenol, and, above all, *p*-coumaric acid strongly promoted the removal of PAHs by LAC. 9,10-Anthraquinone was the main product detected from ANT oxidation by all the LAC mediator systems. The yield of anthraquinone formed was directly correlated with the amount of *p*-coumaric acid used. This compound resulted in a better LAC mediator than ABTS and close similarity to HBT, attaining 95% removal of ANT and B[a]P and about 50% of PYR within 24 h. Benzo[a]pyrene-1,6-, 3,6-, and 6,12-quinones were produced during B[a]P oxidation with LAC and *p*-coumaric acid, HBT, or ABTS as mediators, although use of the latter mediator gave further oxidation products that were not produced by the two other systems [105].

During solid-state fermentation of a natural lignin-containing substrate, the white-rot fungi produce what is known as a yellow form of LAC. The active center of this enzyme is modified by lignin-degradation products. As a result of this modification, LAC gains the ability to catalyze

Involvement of the Ligninolytic System of White-Rot and Litter-Decomposing Fungi in the Degradation of
Polycyclic Aromatic Hydrocarbons

61

the oxidation of nonphenolic compounds without addition of mediators [100]. The catalytic activity of the yellow LAC from *P. ostreatus* D1 toward PAHs, their derivatives, and anthracene-like synthetic dyes, was investigated. Yellow LAC did not catalyze the oxidation of the two-ring PAH naphthalene, but the naphthalene derivatives α- and β-naphthols, α-nitroso-β-naphthol, α-hydroxy-β-naphthoic acid, and α-naphthylamine were all good LAC substrates. Yellow LAC degraded all the PAHs containing from three to five rings, with the following efficiencies: 91% for ANT, 40% for PYR, 95% for FLU, 47% for FLA, 82% for PHE, and 100% for perylene. These efficiencies were higher than that observed for a blue LAC from the same fungus in both absence and presence of the typical synthetic mediators ABTS and HBT under the same experimental conditions. Yellow LAC oxidized a model mixture of PAHs and all the synthetic dyes. The same product of ANT oxidation and various unidentified products of FLU oxidation were observed in solutions of various solvents [100].

Comparison of the catalytic properties of ligninolytic enzymes suggests that the role of LiP in PAH degradation limited by narrow range of compounds according to their IP values. In my opinion, MnP (perhaps VP too) and LAC are the most important enzymes in PAH oxidation. Their catalytic possibilities are extended the following factors (a) the presence of some natural and synthetic compounds (e.g., gluthatione for MnP and LAC; ABTS for LAC); (b) the modification of the active centre of LAC during cultivation of a fungi on lignin-containing natural substrates as it occurs in case of yellow LACs; (c) the coupling of PAH oxidation and lipid peroxidation (MnP and LAC). Thus, MnP and LAC probably plays a more important role than simply that of the initial oxidation and production of quinones.

4. Conclusions

In summary, according to the presented data the following conclusions can be made.

All the studied white-rot and litter-decomposing fungi can metabolize and mineralize PAHs both in liquid medium and in soil. The initial products of degradation in the case of all the studied PAHs by all the studied fungi are the corresponding quinones, regardless of the cultivation conditions.

The accumulation or the subsequent metabolization of the quinonic products can be determined by (a) the species of the white-rot or litter-decomposing fungus and the composition of their ligninolytic enzyme system; (b) the cultivation conditions and the produced enzyme complex under these conditions.

Large size of molecule and the poor solubility can determine the nature of enzyme system involved in their first attack. This system should be extracellular and nonspecific. The ligninolytic enzyme system of white-rot and litter-decomposing fungi is sufficient for such requirements. All the studied ligninolytic enzymes, LiP, MnP, VP, and LAC, oxidized PAHs to the corresponding quinones. MnPs not only oxidize PAHs but can cleave the aromatic ring and

mineralize the resulting products to CO_2. The efficiency of PAH oxidation by ligninolytic enzymes can be improved by different natural factors, such as natural and synthetic mediators, and by coupling PAH oxidation and lipid peroxidation.

In addition, the poor solubility of these compounds can be surmount by emulsifying compounds production during PAH degradation. The production of an emulsifying agent suggests three possibilities: (a) this agent can be essential for increasing the solubility of hydrophobic compounds, (b) it can have a positive effect on the production of extracellular ligninolytic enzymes in agitated culture, and (c) it could be involved in the oxidation of hydrophobic compounds catalyzed by ligninolytic enzymes.

Summarizing the data of different authors, I propose that the ligninolytic enzyme system plays a key role in the initial step of PAH degradation by white-rot and litter-decomposing fungi. The resulting metabolites are more soluble and can be taken inside the cell, where different intracellular enzymes (e.g., cytochrome P-450) can act. The main contradiction is that PAH degradation occurs before extracellular enzyme production. This contradiction can be solved if one takes into account the presence of a mycelia surface-bound LAC pool, which may be involved in the initial stages of PAH degradation.

In my opinion, more attention should be given to studies on the fate of PAH metabolites outside and inside of fungal cell. The intracellular enzymes probably involved in the utilization of these metabolites should be studied carefully. In this connection, the study of the intracellular and/or cell surface-bound pool of ligninolytic enzymes seems very important.

Abbreviations

P:	Phanerochaete
C:	Coriolus
T:	Trametes
P:	Pleurotus
I:	Irpex
LiP:	Lignin peroxidase
MnP:	Mn-peroxidase
VP:	Versatile peroxidase
LAC:	Laccase
HBT:	1-hydroxybenzotriazole
ABTS:	2,2′-azinobis(3-ethylbenzthiazoline-6-sulfonic acid)
VA:	3,4-dimethoxybenzyl alcohol, veratryl alcohol
PAHs:	Polycyclic aromatic hydrocarbons
ANT:	Anthracene
PHE:	Phenanthrene
FLU:	Fluorene
PYR:	Pyrene
FLA:	Fluoranthene
CHR:	Chrysene
B[a]P:	Benzo[a]pyrene
B[a]A:	Benzo[a]anthracene
IP:	Ionization potential

WRF: White-rot fungi
LDF: Litter-decomposing fungi.

Acknowledgment

The author acknowledges Dr. Dmitry Tychinin (IBPPM RAS) for his assistance in preparation of the English text of this paper.

References

[1] E. Torres, R. Tinoco, and R. Vazquez-Duhalt, "Biocatalytic oxidation of polycyclic aromatic hydrocarbons in media containing organic solvents," *Water Science and Technology*, vol. 36, no. 10, pp. 37–44, 1997.

[2] M. Moore, D. Livingstone, and J. Widdows, "Hydrocarbons in marine mollusks: biological effects and ecological consequences," in *Metabolism of Polycyclic Aromatic Hydrocarbons in the Aquatic Environment*, U. Varanishi, Ed., pp. 291–328, CRC Press, Boca Raton, Fla, USA, 1989.

[3] R. Pahlman and O. Pelkonen, "Mutagenicity studies of different polycyclic aromatic hydrocarbons: the significance of enzymatic factors and molecular structure," *Carcinogenesis*, vol. 8, no. 6, pp. 773–778, 1987.

[4] C. E. Cerniglia, "Biodegradation of polycyclic aromatic hydrocarbons," *Current Opinion in Biotechnology*, vol. 4, no. 3, pp. 331–338, 1993.

[5] A. L. Juhasz and R. Naidu, "Bioremediation of high molecular weight polycyclic aromatic hydrocarbons: a review of the microbial degradation of benzo[a]pyrene," *International Biodeterioration and Biodegradation*, vol. 45, no. 1-2, pp. 57–88, 2000.

[6] A. K. Haritash and C. P. Kaushik, "Biodegradation aspects of Polycyclic Aromatic Hydrocarbons (PAHs): a review," *Journal of Hazardous Materials*, vol. 169, no. 1-3, pp. 1–15, 2009.

[7] B. W. Bogan and R. T. Lamar, "Polycyclic aromatic hydrocarbon-degrading capabilities of *Phanerochaete laevis* HHB-1625 and its extracellular ligninolytic enzymes," *Applied and Environmental Microbiology*, vol. 62, no. 5, pp. 1597–1603, 1996.

[8] P. J. Collins and A. D. W. Dobson, "Oxidation of fluorene and phenanthrene by Mn^{2+}-dependent peroxidase activity in whole cultures of *Trametes (Coriolus) versicolor*," *Biotechnology Letters*, vol. 18, no. 7, pp. 801–804, 1996.

[9] U. Sack, T. M. Heinze, J. Deck et al., "Comparison of phenanthrene and pyrene degradation by different wood-decaying fungi," *Applied and Environmental Microbiology*, vol. 63, no. 10, pp. 3919–3925, 1997.

[10] K. Steffen, A. Hatakka, and M. Hofrichter, "Removal and mineralization of polycyclic aromatic hydrocarbons by litter-decomposing basidiomycetous fungi," *Applied Microbiology and Biotechnology*, vol. 60, no. 1-2, pp. 212–217, 2003.

[11] K. T. Steffen, A. Hatakka, and M. Hofrichter, "Degradation of benzo[a]pyrene by the litter-decomposing basidiomycete *Stropharia coronilla*: role of manganese peroxidase," *Applied and Environmental Microbiology*, vol. 69, no. 7, pp. 3957–3964, 2003.

[12] J. A. Field, E. De Jong, G. F. Costa, and J. A. M. De Bont, "Biodegradation of polycyclic aromatic hydrocarbons by new isolates of white rot fungi," *Applied and Environmental Microbiology*, vol. 58, no. 7, pp. 2219–2226, 1992.

[13] B. W. Bogan and R. T. Lamar, "One-electron oxidation in the degradation of creosote polycyclic aromatic hydrocarbons by *Phanerochaete chrysosporium*," *Applied and Environmental Microbiology*, vol. 61, no. 7, pp. 2631–2635, 1995.

[14] L. Bezalel, Y. Hadar, P. P. Fu, J. P. Freeman, and C. E. Cerniglia, "Initial oxidation products in the metabolism of pyrene, anthracene, fluorene, and dibenzothiophene by the white rot fungus *Pleurotus ostreatus*," *Applied and Environmental Microbiology*, vol. 62, no. 7, pp. 2554–2559, 1996.

[15] L. Bezalel, Y. Hadar, P. P. Fu, J. P. Freeman, and C. E. FCerniglia, "Metabolism of phenanthrene by the white rot fungus *Pleurotus ostreatus*," *Applied and Environmental Microbiology*, vol. 62, no. 7, pp. 2547–2553, 1996.

[16] L. Bezalel, Y. Hadar, and C. E. Cerniglia, "Mineralization of polycyclic aromatic hydrocarbons by the white rot fungus *Pleurotus ostreatus*," *Applied and Environmental Microbiology*, vol. 62, no. 1, pp. 292–295, 1996.

[17] B. W. Bogan, R. T. Lamar, and K. E. Hammel, "Fluorene oxidation in vivo by *Phanerochaete chrysosporium* and in vitro during manganese peroxidase-dependent lipid peroxidation," *Applied and Environmental Microbiology*, vol. 62, no. 5, pp. 1788–1792, 1996.

[18] M. Hofrichter, K. Scheibner, I. Schneegaß, and W. Fritsche, "Enzymatic combustion of aromatic and aliphatic compounds by manganese peroxidase from *Nematoloma frowardii*," *Applied and Environmental Microbiology*, vol. 64, no. 2, pp. 399–404, 1998.

[19] M. A. Pickard, R. Roman, R. Tinoco, and R. Vazquez-Duhalt, "Polycyclic aromatic hydrocarbon metabolism by white rot fungi and oxidation by *Coriolopsis gallica* UAMH 8260 laccase," *Applied and Environmental Microbiology*, vol. 65, no. 9, pp. 3805–3809, 1999.

[20] M. J. Han, H. T. Choi, and H. G. Song, "Degradation of phenanthrene by *Trametes versicolor* and its laccase," *Journal of Microbiology*, vol. 42, no. 2, pp. 94–98, 2004.

[21] S. V. Nikiforova, N. N. Pozdnyakova, O. E. Makarov, M. P. Chernyshova, and O. V. Turkovskaya, "Chrysene bioconversion by the white rot fungus *Pleurotus ostreatus* D1," *Microbiology*, vol. 79, no. 4, pp. 456–460, 2010.

[22] N. N. Pozdnyakova, S. V. Nikiforova, O. E. Makarov, M. P. Chernyshova, K. E. Pankin, and O. V. Turkovskaya, "Influence of cultivation conditions on pyrene degradation by the fungus *Pleurotus ostreatus* D1," *World Journal of Microbiology and Biotechnology*, vol. 26, no. 2, pp. 205–211, 2010.

[23] N. N. Pozdnyakova, S. V. Nikiforova, and O. V. Turkovskaya, "Influence of PAHs on ligninolytic enzymes of the fungus *Pleurotus ostreatus* D1," *Central European Journal of Biology*, vol. 5, no. 1, pp. 83–94, 2010.

[24] T. S. Brodkorb and R. L. Legge, "Enhanced biodegradation of phenanthrene in oil tar-contaminated soils supplemented with *Phanerochaete chrysosporium*," *Applied and Environmental Microbiology*, vol. 58, no. 9, pp. 3117–3121, 1992.

[25] M. A. Moen and K. E. Hammel, "Lipid peroxidation by the manganese peroxidase of *Phanerochaete chrysosporium* is the basis for phenanthrene oxidation by the intact fungus," *Applied and Environmental Microbiology*, vol. 60, no. 6, pp. 1956–1961, 1994.

[26] B. W. Bogan, B. Schoenike, R. T. Lamar, and D. Cullen, "Manganese peroxidase mRNA and enzyme activity levels during bioremediation of polycyclic aromatic hydrocarbon-contaminated soil with *Phanerochaete chrysosporium*," *Applied and Environmental Microbiology*, vol. 62, no. 7, pp. 2381–2386, 1996.

Involvement of the Ligninolytic System of White-Rot and Litter-Decomposing Fungi in the Degradation of
Polycyclic Aromatic Hydrocarbons

63

[27] R. May, P. Schröder, and H. Sandermann, "Ex-situ process for treating PAH-contaminated soil with *Phanerochaete chrysosporium*," *Environmental Science and Technology*, vol. 31, no. 9, pp. 2626–2633, 1997.

[28] E. Lang, F. Nerud, and F. Zadrazil, "Production of ligninolytic enzymes by *Pleurotus* sp. and *Dichomitus squalens* in soil and lignocellulose substrate as influenced by soil microorganisms," *FEMS Microbiology Letters*, vol. 167, no. 2, pp. 239–244, 1998.

[29] T. Eggen, "Application of fungal substrate from commercial mushroom production—*Pleurotus ostreatus*—for bioremediation of creosote contaminated soil," *International Biodeterioration and Biodegradation*, vol. 44, no. 2-3, pp. 117–126, 1999.

[30] P. Baldrian, C. In Der Wiesche, J. Gabriel, F. Nerud, and F. Zadražil, "Influence of cadmium and mercury on activities of ligninolytic enzymes and degradation of polycyclic aromatic hydrocarbons by *Pleurotus ostreatus* in soil," *Applied and Environmental Microbiology*, vol. 66, no. 6, pp. 2471–2478, 2000.

[31] N. Pozdnyakova, E. Dubrovskaya, O. Makarov, V. Nikitina, and O. Turkovskaya, "Production of ligninolytic enzymes by white-rot fungi during bioremediation of oil-contaminated soil," in *Soil Enzymology, Soil Biology*, G. Shukla and A. Varma, Eds., vol. 22, pp. 363–377, Springer, Berlin, Germany, 2011.

[32] J. A. Bumpus, M. Tien, D. Wright, and S. D. Aust, "Oxidation of persistent environmental pollutants by a white rot fungus," *Science*, vol. 228, no. 4706, pp. 1434–1436, 1985.

[33] S. B. Pointing, "Feasibility of bioremediation by white-rot fungi," *Applied Microbiology and Biotechnology*, vol. 57, no. 1-2, pp. 20–33, 2001.

[34] S. J. Cho, S. J. Park, J. S. Lim, Y. H. Rhee, and K. S. Shin, "Oxidation of polycyclic aromatic hydrocarbons by laccase of *Coriolus hirsutus*," *Biotechnology Letters*, vol. 24, no. 16, pp. 1337–1340, 2002.

[35] P. Baborová, M. Möder, P. Baldrian, K. Cajthamlová, and T. Cajthaml, "Purification of a new manganese peroxidase of the white-rot fungus *Irpex lacteus*, and degradation of polycyclic aromatic hydrocarbons by the enzyme," *Research in Microbiology*, vol. 157, no. 3, pp. 248–253, 2006.

[36] L. Bezalel, Y. Hadar, and C. E. Cerniglia, "Enzymatic mechanisms involved in phenanthrene degradation by the white rot fungus *Pleurotus ostreatus*," *Applied and Environmental Microbiology*, vol. 63, no. 7, pp. 2495–2501, 1997.

[37] U. Sack, M. Hofrichter, and W. Fritsche, "Degradation of polycyclic aromatic hydrocarbons by manganese peroxidase of *Nematoloma frowardii*," *FEMS Microbiology Letters*, vol. 152, no. 2, pp. 227–234, 1997.

[38] M. I. S. Kim, E. J. Huh, H. K. Kim, and K. W. Moon, "Degradation of polycyclic aromatic hydrocarbons by selected white-rot fungi and the influence of lignin peroxidase," *Journal of Microbiology and Biotechnology*, vol. 8, no. 2, pp. 129–133, 1998.

[39] K. E. Hammel, B. Kalyanaraman, and T. K. Kirk, "Oxidation of polycyclic aromatic hydrocarbons and dibenzo[p]dioxins by *Phanerochaete chrysosporium* ligninase," *Journal of Biological Chemistry*, vol. 261, no. 36, pp. 16948–16952, 1986.

[40] J. A. Field, R. H. Vledder, J. G. Van Zelst, and W. H. Rulkens, "The tolerance of lignin peroxidase and manganese-dependent peroxidase to miscible solvents and the *in vitro* oxidation of anthracene in solvent: water mixtures," *Enzyme and Microbial Technology*, vol. 18, no. 4, pp. 300–308, 1996.

[41] R. Vazquez-Duhalt, D. W. S. Westlake, and P. M. Fedorak, "Lignin peroxidase oxidation of aromatic compounds in systems containing organic solvents," *Applied and Environmental Microbiology*, vol. 60, no. 2, pp. 459–466, 1994.

[42] K. E. Hammel, B. Green, and Wen Zhi Gai, "Ring fission of anthracene by a eukaryote," *Proceedings of the National Academy of Sciences of the United States of America*, vol. 88, no. 23, pp. 10605–10608, 1991.

[43] C. Johannes, A. Majcherczyk, and A. Hüttermann, "Degradation of anthracene by laccase of *Trametes versicolor* in the presence of different mediator compounds," *Applied Microbiology and Biotechnology*, vol. 46, no. 3, pp. 313–317, 1996.

[44] P. J. Collins, M. J. J. Kotterman, J. A. Field, and A. D. W. Dobson, "Oxidation of anthracene and benzo[a]pyrene by laccases from *Trametes versicolor*," *Applied and Environmental Microbiology*, vol. 62, no. 12, pp. 4563–4567, 1996.

[45] J. B. Sutherland, A. L. Selby, J. P. Freeman, F. E. Evans, and C. E. Cerniglia, "Metabolism of phenanthrene by *Phanerochaete chrysosporium*," *Applied and Environmental Microbiology*, vol. 57, no. 11, pp. 3310–3316, 1991.

[46] S. Masaphy, D. Levanon, Y. Henis, K. Venkateswarlu, and S. L. Kelly, "Evidence for cytochrome P-450 and P-450-mediated benzol[a]pyrene hydroxylation in the white rot fungus *Phanerochaete chrysosporium*," *FEMS Microbiology Letters*, vol. 135, no. 1, pp. 51–55, 1996.

[47] M. J. J. Kotterman, H.-J. Rietberg, A. Hage, and J. A. Field, "Polycyclic aromatic hydrocarbon oxidation by the white-rot fungus *Bjerkandera* sp. strain BOS55 in the presence of nonionic surfactants," *Biotechnology and Bioengineering*, vol. 57, no. 2, pp. 220–227, 1998.

[48] C. Novotný, P. Erbanová, T. Cajthaml, N. Rothschild, C. Dosoretz, and V. Sasek, "*Irpex lacteus*, a white rot fungus applicable to water and soil bioremediation," *Applied Microbiology and Biotechnology*, vol. 54, no. 6, pp. 850–853, 2000.

[49] H. G. Song, "Comparison of pyrene biodegradation by white rot fungi," *World Journal of Microbiology and Biotechnology*, vol. 15, no. 6, pp. 669–672, 1999.

[50] T. Cajthaml, P. Erbanová, V. Sasek, and M. Moeder, "Breakdown products on metabolic pathway of degradation of benz[a]anthracene by a ligninolytic fungus," *Chemosphere*, vol. 64, no. 4, pp. 560–564, 2006.

[51] S. D. Haemmerli, M. S. A. Leisola, D. Sanglard, and A. Fiechter, "Oxidation of benzo[a]pyrene by extracellular ligninases of *Phanerochaete chrysosporium*. Veratryl alcohol and stability of ligninase," *Journal of Biological Chemistry*, vol. 261, no. 15, pp. 6900–6903, 1986.

[52] H. G. Song, "Biodegradation of aromatic hydrocarbons by several white-rot fungi," *Journal of Microbiology*, vol. 35, no. 1, pp. 66–71, 1997.

[53] S. W. Dhawale, S. S. Dhawale, and D. Dean-Ross, "Degradation of phenanthrene by *Phanerochaete chrysosporium* occurs under ligninolytic as well as nonligninolytic conditions," *Applied and Environmental Microbiology*, vol. 58, no. 9, pp. 3000–3006, 1992.

[54] K. E. Hammel, W. Z. Gai, B. Green, and M. A. Moen, "Oxidative degradation of phenanthrene by the ligninolytic fungus *Phanerochaete chrysosporium*," *Applied and Environmental Microbiology*, vol. 58, no. 6, pp. 1832–1838, 1992.

[55] D. Ning, H. Wang, C. Ding, and H. Lu, "Novel evidence of cytochrome P450-catalyzed oxidation of phenanthrene in *Phanerochaete chrysosporium* under ligninolytic conditions," *Biodegradation*, vol. 21, no. 6, pp. 889–901, 2010.

[56] H. Lee, Y. S. Choi, M. J. Kim et al., "Degradation ability of oligocyclic aromates by *Phanerochaete sordida* selected via screening of white-rot fungi," *Folia Microbiologica*, vol. 55, no. 5, pp. 447–453, 2010.

[57] K. Chupungars, P. Rerngsamran, and S. Thaniyavarn, "Polycyclic aromatic hydrocarbons degradation by *Agrocybe* sp. CU-43 and its fluorene transformation," *International Biodeterioration and Biodegradation*, vol. 63, no. 1, pp. 93–99, 2009.

[58] W. T. E. Ting, S. Y. Yuan, S. D. Wu, and B. V. Chang, "Biodegradation of phenanthrene and pyrene by *Ganoderma lucidum*," *International Biodeterioration and Biodegradation*, vol. 65, no. 1, pp. 238–242, 2011.

[59] A. Schützendübel, A. Majcherczyk, C. Johannes, and A. Hüttermann, "Degradation of fluorene, anthracene, phenanthrene, fluoranthene, and pyrene lacks connection to the production of extracellular enzymes by *Pleurotus ostreatus* and *Bjerkandera adusta*," *International Biodeterioration and Biodegradation*, vol. 43, no. 3, pp. 93–100, 1999.

[60] B. R. M. Vyas, S. Bakowski, V. Sasek, and M. Matucha, "Degradation of anthracene by selected white rot fungi," *FEMS Microbiology Ecology*, vol. 14, no. 1, pp. 65–70, 1994.

[61] L. Levin, A. Viale, and A. Forchiassin, "Degradation of organic pollutants by the white rot basidiomycete *Trametes trogii*," *International Biodeterioration and Biodegradation*, vol. 52, no. 1, pp. 1–5, 2003.

[62] T. Cajthaml, P. Erbanová, A. Kollmann, C. Novotný, V. Sasek, and C. Mougin, "Degradation of PAHs by ligninolytic enzymes of *Irpex lacteus*," *Folia Microbiologica*, vol. 53, no. 4, pp. 289–294, 2008.

[63] G. Gramss, B. Kirsche, K. D. Voigt, T. Günther, and W. Fritsche, "Conversion rates of five polycyclic aromatic hydrocarbons in liquid cultures of fifty-eight fungi and the concomitant production of oxidative enzymes," *Mycological Research*, vol. 103, no. 8, pp. 1009–1018, 1999.

[64] N. N. Pozdnyakova, S. V. Nikiforova, O. E. Makarov, and O. V. Turkovskaya, "Effect of polycyclic aromatic hydrocarbons on laccase production by white rot fungus *Pleurotus ostreatus* D1," *Applied Biochemistry and Microbiology*, vol. 47, no. 5, pp. 543–548, 2011.

[65] S. V. Nikiforova, N. N. Pozdnyakova, and O. V. Turkovskaya, "Emulsifying agent production during PAHs degradation by the white rot fungus *Pleurotus ostreatus* D1," *Current Microbiology*, vol. 58, no. 6, pp. 554–558, 2009.

[66] A. Jager, S. Croan, and T. K. Kirk, "Production of ligninases and degradation of lignin in agitated submerged cultures of *Phanerochaete chrysosporium*," *Applied and Environmental Microbiology*, vol. 50, no. 5, pp. 1274–1278, 1985.

[67] R. Y. Wang, J. X. Liu, H. L. Huang, Z. Yu, X. M. Xu, and G. M. Zeng, "Effect of rhamnolipid on the enzyme production of two species of lignin-degrading fungi," *Journal of Hunan University Natural Sciences*, vol. 35, no. 10, pp. 70–74, 2008.

[68] S. Böhmer, K. Messner, and E. Srebotnik, "Oxidation of phenanthrene by a fungal laccase in the presence of 1-hydroxybenzotriazole and unsaturated lipids," *Biochemical and Biophysical Research Communications*, vol. 244, no. 1, pp. 233–238, 1998.

[69] N. N. Pozdnyakova, J. Rodakiewicz-Nowak, O. V. Turkovskaya, and J. Haber, "Oxidative degradation of polyaromatic hydrocarbons catalyzed by blue laccase from *Pleurotus ostreatus* D1 in the presence of synthetic mediators," *Enzyme and Microbial Technology*, vol. 39, no. 6, pp. 1242–1249, 2006.

[70] V. Sasek, T. Cajthaml, and M. Bhatt, "Use of fungal technology in soil remediation: a case study," *Water, Air, and Soil Pollution*, vol. 3, no. 3, pp. 5–14, 2003.

[71] C. Cripps, J. A. Bumpus, and S. D. Aust, "Biodegradation of azo and heterocyclic dyes by *Phanerochaete chrysosporium*," *Applied and Environmental Microbiology*, vol. 56, no. 4, pp. 1114–1118, 1990.

[72] P. Morgan, S. T. Lewis, and R. J. Watkinson, "Comparison of abilities of white-rot fungi to mineralize selected xenobiotic compounds," *Applied Microbiology and Biotechnology*, vol. 34, no. 5, pp. 693–696, 1991.

[73] D. P. Barr and S. D. Aust, "Mechanisms white rot fungi use to degrade pollutants," *Environmental Science and Technology*, vol. 28, no. 2, pp. 78A–87A, 1994.

[74] C. A. Reddy, "The potential for white-rot fungi in the treatment of pollutants," *Current Opinion in Biotechnology*, vol. 6, no. 3, pp. 320–328, 1995.

[75] J. M. Bollag, H. L. Chu, M. A. Rao, and L. Gianfreda, "Enzymatic oxidative transformation of chlorophenol mixtures," *Journal of Environmental Quality*, vol. 32, no. 1, pp. 63–69, 2003.

[76] H. Hou, J. Zhou, J. Wang, C. Du, and B. Yan, "Enhancement of laccase production by *Pleurotus ostreatus* and its use for the decolorization of anthraquinone dye," *Process Biochemistry*, vol. 39, no. 11, pp. 1415–1419, 2004.

[77] R. T. Lamar and D. M. Dietrich, "*In situ* depletion of pentachlorophenol from contaminated soil by *Phanerochaete* spp.," *Applied and Environmental Microbiology*, vol. 56, no. 10, pp. 3093–3100, 1990.

[78] A. Khadrani, F. Seigle-Murandi, R. Steiman, and T. Vroumsia, "Degradation of three phenylurea herbicides (chlortoluron, isoproturon and diuron) by micromycetes isolated from soil," *Chemosphere*, vol. 38, no. 13, pp. 3041–3050, 1999.

[79] A. Kubátová, P. Erbanová, I. Eichlerová, L. Homolka, F. Nerud, and V. Sasek, "PCB congener selective biodegradation by the white rot fungus *Pleurotus ostreatus* in contaminated soil," *Chemosphere*, vol. 43, no. 2, pp. 207–215, 2001.

[80] M. Bhatt, T. Cajthaml, and V. Sasek, "Mycoremediation of PAH-contaminated soil," *Folia Microbiologica*, vol. 47, no. 3, pp. 255–258, 2002.

[81] L. Valentín, G. Feijoo, M. T. Moreira, and J. M. Lema, "Biodegradation of polycyclic aromatic hydrocarbons in forest and salt marsh soils by white-rot fungi," *International Biodeterioration and Biodegradation*, vol. 58, no. 1, pp. 15–21, 2006.

[82] L. Valentín, T. A. Lu-Chau, C. López, G. Feijoo, M. T. Moreira, and J. M. Lema, "Biodegradation of dibenzothiophene, fluoranthene, pyrene and chrysene in a soil slurry reactor by the white-rot fungus *Bjerkandera* sp. BOS55," *Process Biochemistry*, vol. 42, no. 4, pp. 641–648, 2007.

[83] K. T. Steffen, S. Schubert, M. Tuomela, A. Hatakka, and M. Hofrichter, "Enhancement of bioconversion of high-molecular mass polycyclic aromatic hydrocarbons in contaminated non-sterile soil by litter-decomposing fungi," *Biodegradation*, vol. 18, no. 3, pp. 359–369, 2007.

[84] F. Acevedo, L. Pizzul, M. D. P. Castillo, R. Cuevas, and M. C. Diez, "Degradation of polycyclic aromatic hydrocarbons by the Chilean white-rot fungus *Anthracophyllum discolor*," *Journal of Hazardous Materials*, vol. 185, no. 1, pp. 212–219, 2011.

[85] E. Borràs, G. Caminal, M. Sarrà, and C. Novotny, "Effect of soil bacteria on the ability of polycyclic aromatic hydrocarbons (PAHs) removal by *Trametes versicolor* and *Irpex lacteus* from contaminated soil," *Soil Biology and Biochemistry*, vol. 42, no. 12, pp. 2087–2093, 2010.

Involvement of the Ligninolytic System of White-Rot and Litter-Decomposing Fungi in the Degradation of
Polycyclic Aromatic Hydrocarbons

65

[86] E. J. George and R. D. Neufeld, "Degradation of fluorene in soil by fungus *Phanerochaete chrysosporium*," *Biotechnology and Bioengineering*, vol. 33, no. 10, pp. 1306–1310, 1989.

[87] M. Wolter, F. Zadrazil, R. Martens, and M. Bahadir, "Degradation of eight highly condensed polycyclic aromatic hydrocarbons by *Pleurotus* sp. Florida in solid wheat straw substrate," *Applied Microbiology and Biotechnology*, vol. 48, no. 3, pp. 398–404, 1997.

[88] H. O. Zebulun, O. S. Isikhuemhen, and H. Inyang, "Decontamination of anthracene-polluted soil through white rot fungus-induced biodegradation," *Environmentalist*, vol. 31, no. 1, pp. 11–19, 2011.

[89] P. Baldrian, "Fungal laccases-occurrence and properties," *FEMS Microbiology Reviews*, vol. 30, no. 2, pp. 215–242, 2006.

[90] A. Anastasi, T. Coppola, V. Prigione, and G. C. Varese, "Pyrene degradation and detoxification in soil by a consortium of basidiomycetes isolated from compost: role of laccases and peroxidases," *Journal of Hazardous Materials*, vol. 165, no. 1-3, pp. 1229–1233, 2009.

[91] E. Rodríguez, O. Nuero, F. Guillén, A. T. Martínez, and M. J. Martínez, "Degradation of phenolic and non-phenolic aromatic pollutants by four *Pleurotus* species: the role of laccase and versatile peroxidase," *Soil Biology and Biochemistry*, vol. 36, no. 6, pp. 909–916, 2004.

[92] U. Sack and W. Fritsche, "Enhancement of pyrene mineralization in soil by wood-decaying fungi," *FEMS Microbiology Ecology*, vol. 22, no. 1, pp. 77–83, 1997.

[93] C. Novotný, P. Erbanová, V. Šašek et al., "Extracellular oxidative enzyme production and PAH removal in soil by exploratory mycelium of white rot fungi," *Biodegradation*, vol. 10, no. 3, pp. 159–168, 1999.

[94] F. Acevedo, L. Pizzul, M. Castillo et al., "Degradation of polycyclic aromatic hydrocarbons by free and nanoclay-immobilized manganese peroxidase from *Anthracophyllum discolor*," *Chemosphere*, vol. 80, no. 3, pp. 271–278, 2010.

[95] G. Eibes, T. Cajthaml, M. T. Moreira, G. Feijoo, and J. M. Lema, "Enzymatic degradation of anthracene, dibenzothiophene and pyrene by manganese peroxidase in media containing acetone," *Chemosphere*, vol. 64, no. 3, pp. 408–414, 2006.

[96] T. Günther, U. Sack, M. Hofrichter, and M. Lätz, "Oxidation of PAH and PAH-derivatives by fungal and plant oxidoreductases," *Journal of Basic Microbiology*, vol. 38, no. 2, pp. 113–122, 1998.

[97] Y. Wang, R. Vasquez-Duhalt, and M. A. Pickard, "Manganese-lignin peroxidase hybrid from *Bjerkandera adusta* oxidizes polycyclic aromatic hydrocarbons more actively in the absence of manganese," *Canadian Journal of Microbiology*, vol. 49, no. 11, pp. 675–682, 2003.

[98] D. C. Bressler, P. M. Fedorak, and M. A. Pickard, "Oxidation of carbazole, *N*-ethylcarbazole, fluorene, and dibenzothiophene by the laccase of *Coriolopsis gallica*," *Biotechnology Letters*, vol. 22, no. 14, pp. 1119–1125, 2000.

[99] H. Punnapayak, S. Prasongsuk, K. Messner, K. Danmek, and P. Lotrakul, "Polycyclic aromatic hydrocarbons (PAHs) degradation by laccase from a tropical white rot fungus *Ganoderma lucidum*," *African Journal of Biotechnology*, vol. 8, no. 21, pp. 5897–5900, 2009.

[100] N. N. Pozdnyakova, J. Rodakiewicz-Nowak, O. V. Turkovskaya, and J. Haber, "Oxidative degradation of polyaromatic hydrocarbons and their derivatives catalyzed directly by the yellow laccase from *Pleurotus ostreatus* D1," *Journal of Molecular Catalysis B*, vol. 41, no. 1-2, pp. 8–15, 2006.

[101] R. Rama, C. Mougin, F. D. Boyer, A. Kollmann, C. Malosse, and J. C. Sigoillot, "Biotransformation of benzo[a]pyrene in bench scale reactor using laccase of *Pycnoporus cinnabarinus*," *Biotechnology Letters*, vol. 20, no. 12, pp. 1101–1104, 1998.

[102] C. Johannes, A. Majcherczyk, and A. Hüttermann, "Oxidation of acenaphthene and acenaphthylene by laccase of *Trametes versicolor* in a laccase-mediator system," *Journal of Biotechnology*, vol. 61, no. 2, pp. 151–156, 1998.

[103] A. Majcherczyk, C. Johannes, and A. Hüttermann, "Oxidation of polycyclic aromatic hydrocarbons (PAH) by laccase of *Trametes versicolor*," *Enzyme and Microbial Technology*, vol. 22, no. 5, pp. 335–341, 1998.

[104] A. I. Cañas, M. Alcalde, F. Plou, M. J. Martínez, A. Martínez, and S. Camarero, "Transformation of polycyclic aromatic hydrocarbons by laccase is strongly enhanced by phenolic compounds present in soil," *Environmental Science and Technology*, vol. 41, no. 8, pp. 2964–2971, 2007.

[105] C. Johannes and A. Majcherczyk, "Natural mediators in the oxidation of polycyclic aromatic hydrocarbons by laccase mediator systems," *Applied and Environmental Microbiology*, vol. 66, no. 2, pp. 524–528, 2000.

[106] D. W. S. Wong, "Structure and action mechanism of ligninolytic enzymes," *Applied Biochemistry and Biotechnology*, vol. 157, no. 2, pp. 174–209, 2009.

[107] D. Sanglard, M. S. A. Leisola, and A. Fiechter, "Role of extracellular ligninases in biodegradation of benzo(a)pyrene by *Phanerochaete chrysosporium*," *Enzyme and Microbial Technology*, vol. 8, no. 4, pp. 209–212, 1986.

[108] B. Valderrama, M. Ayala, and R. Vazquez-Duhalt, "Suicide inactivation of peroxidases and the challenge of engineering more robust enzymes," *Chemistry and Biology*, vol. 9, no. 5, pp. 555–565, 2002.

[109] K. S. Hildén, R. Bortfeldt, M. Hofrichter, A. Hatakka, and T. K. Lundell, "Molecular characterization of the basidiomycete isolate *Nematoloma frowardii* b19 and its manganese peroxidase places the fungus in the corticioid genus *Phlebia*," *Microbiology*, vol. 154, no. 8, pp. 2371–2379, 2008.

8

DNA Damage Protecting Activity and Free Radical Scavenging Activity of Anthocyanins from Red Sorghum (*Sorghum bicolor*) Bran

P. Suganya Devi,[1] M. Saravana Kumar,[1] and S. Mohan Das[2]

[1] *P.G. Department of Biotechnology, Dr. Mahalingam Centre for Research and Development, N.G.M.College, Pollachi 642001, India*
[2] *Kaamadhenu Arts and Science College, Sathyamangalam 638503, India*

Correspondence should be addressed to P. Suganya Devi, suganyabiotech@yahoo.co.in

Academic Editor: Sneh Singla-Pareek

There is increasing interest in natural food colorants like carotenoids and anthocyanins with functional properties. Red sorghum bran is known as a rich source for anthocyanins. The anthocyanin contents extracted from red sorghum bran were evaluated by biochemical analysis. Among the three solvent system used, the acidified methanol extract showed a highest anthocyanin content (4.7 mg/g of sorghum bran) followed by methanol (1.95 mg/g) and acetone (1 mg/g). Similarly, the highest total flavonoids (143 mg/g) and total phenolic contents (0.93 mg/g) were obtained in acidified methanol extracts than methanol and acetone extracts. To study the health benefits of anthocyanin from red sorghum bran, the total antioxidant activity was evaluated by biochemical and molecular methods. The highest antioxidant activity was observed in acidified methanol extracts of anthocyanin in dose-dependent manner. The antioxidant activity of the red sorghum bran was directly related to the total anthocyanin found in red sorghum bran.

1. Introduction

Anthocyanins are becoming increasingly important not only as food colorants but also as antioxidants. Anthocyanins are reported to have some therapeutic benefits including vasoprotective and anti-inflammatory properties [1] and anticancer [2] as well as hypoglycemic effects [3]. There is a rising demand for natural sources of food colorants with nutraceutical benefits [4], and alternative sources of natural anthocyanins are becoming more important. Special features of the sorghum crop are very important in the world's human diet, with over 300 million people dependent on it [5]. Sorghum is one of the major staple foods in Africa, Middle East, and Asia. It is drought resistant and is, therefore, an extremely important commodity that provides necessary food and feed for millions of people living in semiarid environment worldwide. Sorghums have high levels of anthocyanins and other phenols concentrated in their brans [6, 7]. These sorghum bran fractions are potentially useful

ingredients in various functional food applications and were shown to produce desirable attributes (e.g., attractive natural colour) without adversely affecting other sensory properties of foods. These ingredients are bound to play a crucial role in food applications as a diversified functional food base. However, to ensure their economic potential, the sorghum bran anthocyanins must be extracted in an efficient manner in which their original form is preserved as much as possible.

Oxidation is essential to many living organisms for the production of energy necessary for biological processes. Oxygen-centered free radicals, also known as reactive oxygen species (ROS), including superoxide, hydrogen peroxide, hydroxyl (HO–), peroxyl (ROO–), and alkoxyl (RO–), are produced *in vivo* during oxidation [8]. ROS are not only strongly associated with lipid peroxidation, leading to food deterioration, but are also involved in development of a variety of diseases, including cellular aging, mutagenesis, carcinogenesis, coronary heart disease, diabetes, and neurodegeneration [9, 10]. Although almost all organisms possess antioxidant defense and repair systems to protect against oxidative damage, these systems are insufficient to prevent the damage entirely [11, 12].

Antioxidative properties of anthocyanins arise from their high reactivity as hydrogen or electron donors, from the ability of the polyphenol-derived radicals to stabilize and delocalize the unpaired electron, and from their ability to chelate transition metal ions (termination of the Fenton reaction) [13]. Thus, anthocyanins may play a role as antioxidants.

The objective of the current research is to evaluate the antioxidant properties of anthocyanins from red sorghum bran.

2. Experimental Procedures

2.1. Samples.
The brans of *Sorghum bicolor* (L.) red sorghum were collected from farmer's field in Tamil Nadu, India, and was stored at $-20°C$ in the dark. All samples were decorticated using a PRL dehuller to obtain bran. Bran yield was 15% for each sample. Bran samples were ground through a UDY mill (1 mm mesh) before extraction and analysis. All analyses were conducted in triplicate.

2.2. Sample Extraction.
The anthocyanin from red sorghum bran was extracted by the method followed by Joseph et al. [14]. Two solvents, methanol and acidified methanol, were used for extraction procedure. For 0.5 g sample, 10 mL of solvent was added in 50 mL centrifuge tubes, and the samples were kept in an orbit shaker (Neolab) for 2 hrs at low speed. Samples were then stored at $-20°C$ for over night in the dark to allow for maximum diffusion of phenolics from the cellular matrix. Samples were then kept at room temperature and centrifuged at $7,000 \times g$ for 10 min, taken for analysis. Residues were rinsed with two additional 10 mL volumes of solvent for 5 min and centrifuged at $7,000 \times g$ for 10 min, and taken for analysis. The three aliquots were mixed and stored at $-20°C$ in the dark until further biochemical analysis.

2.3. Analytical Procedures

2.3.1. Flavanoid Confirmation Test.
The flavanoid confirmation test was determined by the method of Harbone 1998.

2.3.2. Stability at Variable pH.
The anthocyanin stability was tested by treating 1 mL of sample with 1 mL of pH 1.0 and 4.5 solutions. The color change was observed [15].

2.3.3. Determination of Total Phenolic Content.
Total phenolic contents of anthocyanin samples were determined by the Folin-Ciocalteu's method [16]. Briefly, aliquots of 0.1 g powder of anthocyanin samples were dissolved in 1 mL of deionised water. This solution (0.1 mL) was mixed with 2.8 mL of deionized water, 2 mL of 2% sodium carbonate (Na_2CO_3), and 0.1 mL of 50% Folin-Ciocalteu's reagent. After incubation at room temperature for 30 min, the absorbance of the reaction mixture was measured at 750 nm against a deionized water blank using a spectrophotometer

(Geneysis 5). Gallic acid was chosen as a standard to get a seven-point standard curve (0–200 mg/L). The levels of total phenolic contents in sorghum bran were determined using the standards curve. The data obtained from sorghum bran were expressed as milligram of gallic acid equivalents (GAEs)/g powder and converted to milligram gallic acid equivalents (GAEs)/100 g of sorghum bran.

2.3.4. Determination of Total Flavonoid Content.
The total flavonoid content was quantified by using aluminum chloride colorimetric method described by Chang et al. [17, 18]. In brief, aliquots of 0.1 g of samples were dissolved in 1 mL of deionized water. This solution (0.5 mL) was mixed with 1.5 mL of 95% alcohol, 0.1 mL of 10% aluminium chloride hexahydrate ($AlCl_3$), 0.1 mL of 1 M potassium acetate (CH_3COOK), and 2.8 mL of deionised water. After incubation at room temperature for 40 min, the reaction mixture absorbance was analysed using spectrophotometer (Geneysis 5) at 415 nm. The deionized water was used as a blank. Quercetin was chosen as a standard to get a seven-point standard curve (0–50 mg/L). The levels of total flavonoid contents in sorghum bran were determined in triplicate, respectively. The data obtained from sorghum bran were expressed as milligram of quercetin equivalents (GAE)/g powder. The data were expressed as milligram equivalents (GAE)/g powder. Finally, the data were converted to milligram quercetin equivalents (GAE)/100 g of sorghum bran.

2.3.5. Determination of Total Anthocyanin.
The total amount of anthocyanin content was determined by using pH differential method. A spectrophotometer was used for the spectral measurements at 210 nm and 750 nm [19]. The absorbance of the samples (A) was calculated as follows:

$$
\text{anthocyanin pigment content (mg/liter)} = \frac{A \times MW \times DF \times 1000}{\varepsilon \times 1}, \tag{1}
$$

where A = (Absorbance λ vis-max-A750) at pH 1.0–(Absorbance λ vis-max-A750) at pH 4.5 molecular weight of anthocyanin (cyd-3-glu) = 449, extraction coefficient (ε) = 29,600, and DF = diluted factor.

2.4. Antioxidant Assays

2.4.1. Scavenging Activity of DPPH Radical.
Scavenging activity of anthocyanins against DPPH radicals was assessed according to the method of Larrauri et al. [20] with some modifications. Briefly, 0.1 mL of various concentrations of sorghum anthocyanins (1, 10, 50, 100 μg/mL) was mixed with 2.9 mL of 0.1 mM DPPH-methanol solution. Mixed solutions were incubated for 30 min at 25°C in dark; the decrease in the absorbance at 517 nm was measured. Methanol was used as control instead of antioxidant solution while blanks contained methanol instead of DPPH solution. In the experiment, L-ascorbic acid and BHT were used as

standards. The inhibition of DPPH radicals by the samples was calculated according to the following equation:

DPPH-scavenging activity (%)

$$= \left[1 - \frac{(\text{absorbance of the sample} - \text{absorbance of blank})}{\text{absorbance of the control}} \right] \times 100.$$

(2)

2.4.2. Hydroxyl Radical Scavenging Activity. The hydroxyl radical scavenging activity was determined based on the protocol described by Singh et al. [21]. of various concentrations of sorghum anthocyanins (1, 10, 50, 100 μg/mL) was taken in different test tubes. 1.0 mL of iron-EDTA solution (0.1% ferrous ammonium sulfate and 0.26% EDTA), 0.5 mL of DMSO (0.85% v/v in 0.1 M phosphate buffer, pH 7.4) were added to these tubes, and the reaction was initiated by adding 0.5 mL of 0.22% ascorbic acid. Test tubes were capped tightly and heated on a water bath at 80–90°C for 15 min. The reaction was terminated by the addition of 1 mL of ice-cold TCA (17.5%w/v). 3 mL of the Nash reagent (75 g of ammonium acetate, 3 mL of glacial acetic acid, and 2 mL of acetyl acetone were mixed and raised to 1 L with distilled water) was added to all of the tubes and left at room temperature for 15 min for the color development. The intensity of the yellow color formed was measured spectrophotometrically at 412 nm against the reagent blank. L-ascorbic acid and butylhydroxytoluene (BHT) were used as standards. The percentage of hydroxyl radical scavenging activity was calculated by using the formula:

$$\% \text{ of hydroxyl radical scavenging activity}$$
$$= 1 - \frac{\text{absorbance of sample}}{\text{absorbance of blank}} \times 100.$$

(3)

2.4.3. Determination of Reducing Power. The reducing power was determined based on the method of Oyaizu [22]. A 0.25 mL aliquot of various concentrations of sorghum anthocyanins (1, 10, 50, 100 μg/mL) was mixed with 2.5 mL of 200 mM sodium phosphate buffer (pH 6.6) and 2.5 mL of 1% potassium ferricyanide. The mixture was then incubated at 50°C for 20 min. After incubation, 2.5 mL of 10% trichloroacetic acid (w/v) were added; the mixture was centrifuged at 650 ×g for 10 min. About 5 mL aliquot of the upper layer was mixed with 5 mL of distilled water and 1 mL of 0.1% ferric chloride, and the absorbance of the mixture was measured at 700 nm. L-ascorbic acid and butylhydroxytoluene (BHT) were used as standards. A higher absorbance indicated a higher reducing power.

2.4.4. Determination of Superoxide Radical-Scavenging Activity. Superoxide radicals were generated by the modified protocol described by Siddhurajuna et al. [23] all the solutions were prepared using 0.05 M phosphate buffer (pH 7.8). The photoinduced reactions were performed in aluminium foil-lined box with two 30 W fluorescent lamps. The distance between the reaction solution and the lamp was adjusted until the intensity of illumination reached about 4000 lux.

A 30 μL aliquot of various concentrations of sorghum anthocyanins (1, 10, 50, 100 μg/mL) was mixed with 3 mL of reaction buffer solution (1.3 mm riboflavin, 13 mM methionine, 63 μM nitro blue tetrazolium and 100 μM EDTA, pH 7.8). The reaction solution was illuminated for 15 min at 25°C. The reaction mixture, without sample, was used as a control. L-ascorbic acid and butylhydroxytoluene (BHT) were used as standards. The scavenging activity was calculated as follows:

$$\text{scavenging activity (\%)}$$
$$= \left(1 - \frac{\text{absorbance of the sample}}{\text{absorbance}} \right) \times 100.$$

(4)

2.4.5. Metal Chelating Activity. The chelation of ferrous ions by the extract was estimated by the method of Dinis et al. [24] with slight modification and compared it with EDTA, BHT, and that of ascorbic acid. The chelation test initially includes the addition of ferrous chloride. The antioxidants present in the samples chelate the ferrous ions from the ferrous chloride. The remaining ferrous combine with ferrozine to form ferrous-ferrozine complex. The intensity of the ferrous-ferrozine complex formation depends on the chelating capacity of the sample, and the colour formation was measured at 562 nm (Shimadzu UV-Vis 2450). Different concentrations of standard and sorghum anthocyanins (1, 10, 50, 100 μg/mL) were added to a solution of 100 μL FeCl$_2$ (1 mM). The reaction was initiated by the addition of 250 μL ferrozine (1 mM). The mixture was finally quantified to 1.3 mL with methanol, shaken vigorously and left at room temperature for 10 min., after the mixture had reached equilibrium, the absorbance of the solution was measured spectrophotometrically at 562 nm. All the test and analysis were done in duplicates, and average values were taken. L-ascorbic acid and butylhydroxytoluene (BHT) were used as positive control. The percentage inhibition of ferrous ferrozine complex formation was calculated using the formula: $\% = 1 - \text{As/Ac} \times 100$, where "Ac" is the absorbance of the control and "As" is the absorbance of the sample.

2.4.6. Estimation of Anti-FeCl$_2$-H$_2$O$_2$-Stimulated Linoleic Acid Peroxidation. The effect of anti-FeCl$_2$-H$_2$O$_2$-stimulated linoleic acid peroxidation was determined by the method as described by Duh [25]. In brief, 0.2 mL of various concentration sorghum anthocyanins (1, 10, 50, 100 μg/mL) were added to a solution of 0.1 M linoleic acid (0.2 mL), 2.0 mM FeCl$_2$. H$_2$O (0.2 mL), and 0.2 M phosphate buffer (pH 7.4, 5 mL). The reaction mixture was incubated at 370°C for 24 h. After incubation, 0.2 mL of BHA (20 mg/mL), 1 mL of thiobarbituric acid (TBA)(1%), and 1 mL of trichloro acetic acid (TCA)(10%) were added to the mixture, which was heated for 30 min in a boiling water bath. After cooling, 5 mL of chloroform was added, and the mixture was centrifuged at 1000 ×g to give a supernatant. Absorbance of the supernatant was measured using spectrophotometer at 532 nm.

2.4.7. Determination of the Inhibitory Effect on Deoxyribose Degradation (Molecular Antioxidant Analysis). Inhibitory

TABLE 1: Phenolic composition, flavonoid and anthocyanin content of red sorghum bran in three different solvent extracts.

Red sorghum bran	Total phenols (mg gallic acid equiv/g)	Total flavonoids (mg queractin equiv/g)	Anthocyanins (mg cyanidin 3-glucoside equiv/g) (mg/g)
Acetone extract	0.14 ± 0.03	65 ± 0.19	1.00 ± 0.11
Methanol extract	0.58 ± 0.04	111 ± 0.20	1.95 ± 0.13
Acidified methanol extract	0.93 ± 0.06	143 ± 0.23	4.7 ± 0.20

Values are mean ($n = 3$) $\pm$ SD ($n = 3$, $P < 0.05$).

effect of the anthocyanins on deoxyribose degradation was determined by measuring the reaction activity between either antioxidants or hydroxyl radicals (referred to as non-site-specific scavenging assay) or antioxidants and iron ions (referred to as site-specific scavenging assay) described by Lee et al. [26]. For non-site-specific scavenging assay, a 0.1 mL aliquot of different concentration of anthocyanin was mixed with 1 mL of reaction buffer ($100 \, \mu M$ $FeCl_3$, $104 \, \mu M$ EDTA, 1.5 mM H_2O_2, 2.5 mM deoxyribose, and $100 \, \mu M$ L-ascorbic acid, pH 7.4) and incubated for 1 h at 37°C. A 1 mL aliquot of 0.5% 2-thiobarbituric acid in 0.025 M NaOH and 1 mL of 2.8% trichloroacetic acid were added to the mixture, and it was heated for 30 min at 80°C. The mixture was cooled on ice, and the absorbance was measured at 532 nm. Site-specific scavenging activity, which represented the ability of anthocyanins to chelate iron ions and interfere with hydroxyl radical generation, was measured using the same reaction buffer without EDTA. Percent inhibition of deoxyribose degradation was calculated as (1 − absorbance of sample/absorbance of control) × 100. Control—without sample.

2.4.8. DNA Nicking Assay. DNA nicking assay was performed using genomic DNA by the method of Lee et al. [26]. A mixture of $10 \, \mu L$ of sorghum anthocyanins ($1 \, \mu g/mL$) and genomic DNA was incubated for 10 min at room temperature followed by the addition of $10 \, \mu L$ of Fenton's reagent (30 mM H_2O_2, $50 \, \mu M$ ascorbic acid, and $80 \, \mu M$ $FeCl_3$). The final volume of the mixture was made up to $20 \, \mu L$ and incubated for 30 min at 37°C. The DNA was analysed on 1% agarose gel using ethidium bromide staining.

3. Results and Discussion

3.1. Anthocyanin Extraction. The total anthocyanin was extracted by 70% aqueous acetone, methanol, and acidified methanol as solvent system. Acidified methanol resulted in significantly higher values for the total anthocyanins than methanol and aqueous acetone (Table 1). In our results the anthocyanins extracted by aqueous acetone was much lower than methanol and acidified methanol. However, Lu and Foo [27] observed significant anthocyanin interaction with aqueous acetone to form pyranoanthocyanidins which significantly lowered quantities of detectable anthocyanins. This reaction was significantly affected by time of anthocyanin-acetone interaction and temperature. Our results did not observe such anthocyanin—solvent reactions

in methanol and acidified methanol— and subsequently acidified methanol was proposed as a better solvent than methanol and aqueous acetone.

3.2. Flavonoid Confirmation Test. In the presence of $FeCl_3$, the methanol and acidified methanol extracts showed brown color which confirmed the presence of Flavanoids [28]. In the presence of $AlCl_3$ dark color was observed in methanol and acidified methanol extracts.

3.3. Stability at Variable pH. The sample appears red color at pH 1 and color disappeared at pH 4.5. The results were found to be same in the extracts of methanol and acidified methanol. Giusti and Wrolstad [29] reported that the anthocyanins are stable in low pH.

3.4. Total Phenolic Content. A wide variation in total phenolic content (TPC) was observed in three different solvent extracts from red sorghum bran (Table 1). The highest TPC was observed in acidified methanol extracts, 0.93 mg/g bran, while in methanol 0.58 mg/g bran and in aqueous acetone 0.14 mg/g bran. Awika et al. [30] reported that the highest concentrations of phenols were present in sorghum bran when acidified methanol was used as a solvent. It was also reported that the white sorghum bran fractions had phenol levels much lower than those measured in the black and brown sorghum brans. The pigmented sorghum varieties are a superior source of these beneficial compounds. Duodu et al. [31] reported that the chemical composition of any extract is dependent on the solvent used for extraction. Their results showed that acidified methanol extracts can preserve wider range of compounds, mainly phenolic compounds than aqueous extracts of sorghum bran.

3.5. Evaluation of Total Flavonoids. The anthocyanins are the major class of flavonoids in sorghum. The flavonoid content ranges from 65 to 142 mg/g bran in red sorghum. (Figure 1). Total flavonoid content in acidified methanol was 143 mg/g bran extract, while, for other extracts, total flavonoid content range was 65–111 mg/g bran of aqueous acetone and methanol extracts. Wharton et al. [32] reported that red pericarp sorghum has flavonol compounds, such as luteoforol and apiforol, which are produced from flavanones (i.e., naringenin and eriodictyol) and may be precursors of anthocyanidins in sorghum. A positive correlation between total phenols and flavanols (Table 1) suggests that total phenols are contributed mostly by flavanols in red pericarp

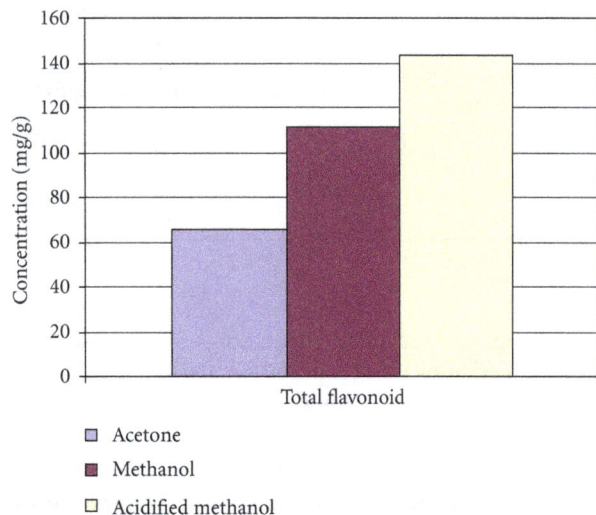

FIGURE 1: Flavonoid content of red sorghum bran in three different solvent extracts.

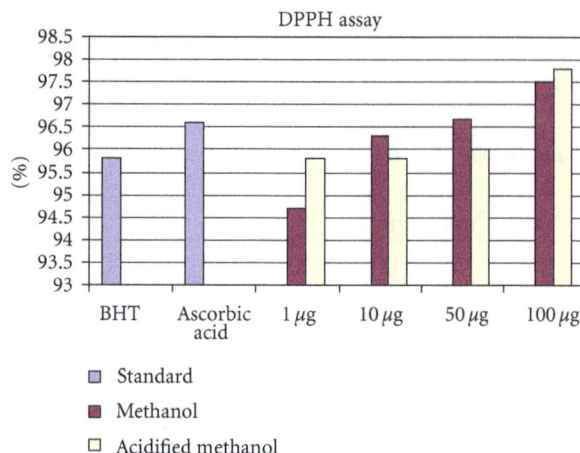

FIGURE 2: Concentration-dependent free radical scavenging activity of methanol and methanol extracts of anthocyanin from red sorghum bran.

sorghum. Similar results were reported by Linda Dykes et al. [33].

3.6. Anthocyanin Content.

Acidified methanol resulted in significantly higher values for the mono- and total anthocyanins than methanol and aqueous acetone (Table 1). The total anthocyanins extracted by acidified methanol extracts were on average 59% higher than aqueous acetone extracts and 28% higher than methanol extracts. Several authors reported that aqueous acetone was better than various alcoholic solvents for extraction of fruit procyanidins, anthocyanins, and other phenols [34, 35]. However, since acidified methanol preservers better the extracted anthocyanins in their original form, it can be the solvent of choice for quantification and analysis of anthocyanins.

3.7. Antioxidant Analysis

3.7.1. DPPH Radical Scavenging Activity. Free radical scavenging is one of the known mechanisms by which antioxidants inhibit lipid peroxidation [8, 13]. The DPPH radical scavenging activity has been extensively used for screening antioxidants from fruits and vegetable juices or extract [36, 37]. DPPH radical scavenging activity of red sorghum bran anthocyanin, ascorbic acid, and BHT was shown in Table 2. The anthocyanin significantly inhibited the activity of DPPH radical in a dose-dependant manner. Anthocyanin had the highest scavenging activity followed by ascorbic acid and BHT. Antioxidant activities of extracted samples from sorghum bran using acidified methanol and methanol were compared (Figure 2). Samples extracted in acidified methanol had significantly higher antioxidant activity than those extracted in methanol. This implies that the acidified methanol is a more powerful solvent than methanol in extracting red sorghum antioxidants. At 1 μg/mL, the scavenging effects were 94.7% and 95.8% for anthocyanins

extracted from methanol and acidified methanol, respectively, while almost complete inhibition of DPPH radical activity was observed when the anthocyanins were used at 100 μg/mL. It appears that red sorghum bran anthocyanins have a strong donating capacity and can efficiently scavenge DPPH radicals. Einbond et al. [38] reported that the DPPH radical scavenging activity of Surinam cherry, Jamaica cherry, and salal, Juboticaba, were due to the presence of large amount of anthocyanins. Awika et al. reported that 3-deoxy anthocyanins found in black sorghum had antioxidant activities that were similar to the anthocyanins found in the fruits and vegetables.

3.7.2. Hydroxyl Radical Scavenging Activity. The sorghum anthocyanins exhibited the highest activity of 99% at 100 μg/mL where as 73% inhibition was noted at 1 μg/mL, respectively. Table 1 shows hydroxyl radical scavenging activity of anthocyanin extracts which increases with increasing concentration. The hydroxyl radical is an extremely reactive free radical formed in biological system and has been implicated as a highly damaging species in free radical pathology capable of damaging almost every molecule found in living cells. This species is considered to be one of the quick initiators of the lipid peroxidation process, abstracting hydrogen atoms from unsaturated fatty acids [39]. The acidified methanol extracts exhibited a highest activity of 99% at 100 μg/mL than methanol extracts of 98.8% at 100 μg/mL. This is similar to observations of several others who have reported a dose-dependent activity in sesame coat, pomegranate peel and seeds and grape pomace [17, 18, 21, 40]. The ability of ethanol extracts of leafy vegetables to quench hydroxyl radicals seems to be directly related to prevent ion in propagation of the process of lipid peroxidation. Hence, the extract seems to be a good scavenger of active oxygen species, thus reducing rate of chain reaction. A high positive correlation was observed between the polyphenol content and the hydroxyl radical scavenging activity. Shyamala et al. [41] reported that the

TABLE 2: Concentration dependent free radical scavenging activity of methanol and methanol extracts of anthocyanin from red sorghum bran.

Antioxidant assays	Standards (%)		Methanol extract (%)				Acidified methanol extract (%)			
	BHT	Ascorbic acid	1 μg	10 μg	50 μg	100 μg	1 μg	10 μg	50 μg	100 μg
DPPH assay	95.8	96.6	94.7	96.3	96.7	97.5	95.8	95.8	96	97.8
Superoxide radical scavenging assay	60.4	69.9	7.6	29	36.5	42.5	11.2	37.5	45	53.8
Hydroxyl radical scavenging assay	99.8	69.9	30.4	84.5	98.5	98.8	73	89.7	98.6	99
Reducing power (Absorbance)	0.138	0.577	0.024	0.184	0.552	0.840	0.014	0.037	0.445	0.984
Metal chelating	79	83	46	74.4	80.3	85.1	48.3	75.2	83.7	85.2
Hydrogen peroxide	94.6	—	40.6	69.5	85.3	90.2	47.9	88.6	90.4	91.3
Deoxyribose degradation site-specific assay	99.1	48.4	85	89.6	91.3	91.8	72	88.9	91.7	91.8
Deoxyribose degradation non-site-specific assay	80.2	33.9	74.8	79.8	81.9	86.6	59.8	79	81.4	82
Anti FeCl$_3$–H$_2$O$_2$ stimulated Linoleic acid	77.2	87.8	82.2	83.5	85.5	86.2	73.1	81	84.3	92

leafy vegetables have a proton radical scavenging action, which is an important mechanism of antioxidants.

3.7.3. Reducing Power.

It has been reported that reducing power is associated with antioxidant activity and may serve as significant reflection on the antioxidant activity [17, 18, 42]. As shown in the Table 2, anthocyanins from red sorghum bran exhibited a higher reducing power than BHT and ascorbic acid, suggesting that it has strong electron-donating capacity. The reducing power of red sorghum bran anthocyanins at 100 μg/mL ascorbic acid and BHT were 0.984, 0.577, and 0.138, respectively. Furthermore, a linear relationship was observed between concentration and reducing power of sorghum anthocyanins. The results indicated that the antioxidant activities of samples extracted in acidified methanol had significantly higher reducing power than those extracted in methanol. Earlier authors [25, 43, 44] have observed a direct correlation between antioxidant activities of certain plant extracts. The reducing properties are generally associated with the presence of reductones [25], which have been shown to exert antioxidant action by breaking the free radical chain by donating a hydrogen atom [45]. Reductones are also reported to react with certain precursors of peroxide, thus preventing peroxide formation.

3.7.4. Superoxide Anion Scavenging Activity.

The relative scavenging effects of red sorghum bran anthocyanin on superoxide radical are shown in Table 2. The anthocyanin extracts exhibited 71.92% scavenging acitivity at 100 μg per mL. Figure 3 shows the result of superoxide radical anion scavenging acitivity of anthocyanin extract, BHT, and ascorbic acid. Superoxide anion radicals are produced by a number of cellular reactions, including various enzyme system such as lipoxygenase, peroxidase, NADPH oxidase, and xanthine oxidase. Superoxide anion plays an important role in plant tissue and also involved in formations of other cell-damaging free radicals [8]. In the present study superoxide radical was generated by illuminating a solution containing riboflavin. Although superoxide is a relatively weak oxidant, it decomposes to form stronger reactive oxidative species, such as singlet oxygen and hydroxyl

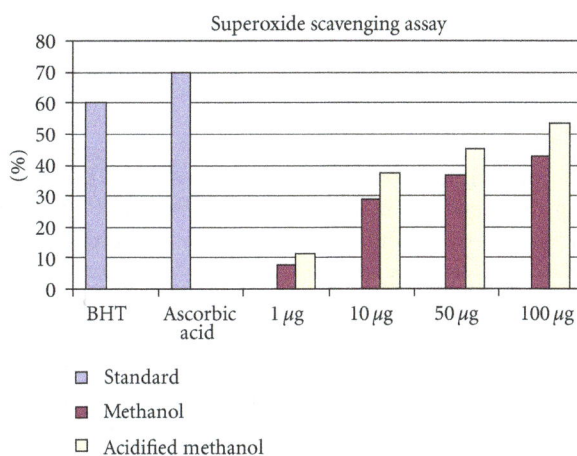

FIGURE 3: Concentration-dependent free radical scavenging activity of methanol and methanol extracts of anthocyanin from red sorghum bran.

radicals, which initiate peroxidation of lipids [46]. In the present study, anthocyanin extracted from red sorghum bran effectively scavenged superoxide in a concentration-dependent manner. Further, superoxides are also known to indirectly initiate lipid peroxidation as a result of hydrogen peroxide formation, creating precursors of hydroxyl radicals [47]. These results clearly suggested that the antioxidant activity of anthocyanin is also related to its ability to scavenge superoxides.

3.7.5. Metal Chelating Activity.

The sorghum anthocyanins exhibited the highest activity of 85.2% at 100 μg/mL whereas 48.3% inhibition was noted at 1 μg/mL, respectively. The percentage inhibition values ranged from 46% to 85.2% (Table 2). The acidified methanol extracts had the highest chelating capacity than methanol, BHT, and ascorbic acid. The samples had better chelating capacity than standards based on percentage inhibiction values in terms of μg/mL. Correlation was found between iron-chelating capacity and phenolic content.

The ability of antioxidant to form insoluble metal complexes with ferrous ion or to generate steric hindrance

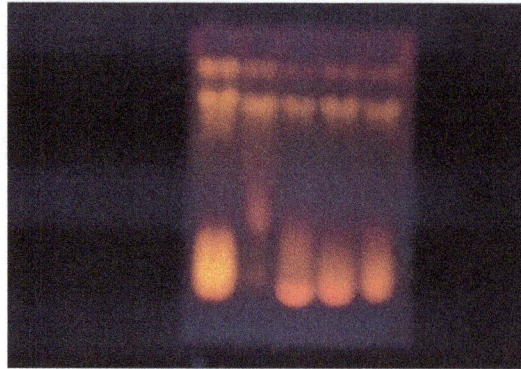

Lane 1 : Control genomic DNA
Lane 2 : Fenton's reagent + genomic DNA
Lane 3 : Fenton's reagent + genomic DNA + sorghum anthocyanin (1 μg/mL)
Lane 4 : Fenton's reagent + genomic DNA + quercetin (100 μg/mL)
Lane 5 : Fenton's reagent + genomic DNA + sorghum anthocyanin (10 μg/mL)

FIGURE 4: DNA nicking assay.

that prevent interation between metal and lipid is evaluated using the metal ion chelating capacity assay [48]. Acitivity is measured by monitoring the decrease in absorbance of the red Fe^{2+} ferrozine complex as antioxidants compete with ferrozine in chelating ferrous ion [49].

3.7.6. Anti-Ferric Chloride Hydrogen Peroxide System. The effect of anthocyanin extracts of red sorghum bran on the formation of malonaldehyde (MDA) from linoleic acid is show in Table 2. As the concentration of the antioxidant extracts increased, the formation of MDA decreased. A dose-dependent MDA inhibition in linoleic acid oxidation was evident. Methanol extracts showed a higher inhibition of MDA ranging from 73% to 92% at 100 μg/mL than BHT and ascorbic acid. Iron salts are thought to react with H_2O_2 called the Fenton reaction, to make hydrogen radicals which bring about peroxide reaction of lipids [25].

3.7.7. Hydrogen Peroxide Scavenging Activity. Hydrogen peroxide scavenging activities of the anthocyanin extracts, BHT, and ascorbic acid were measured at 230 nm. Hydrogen peroxide scavenging activities of the anthocyanin extracted by using acidified methanol were 47.9% at 1 μg/mL, 88.6% at 10 μg/mL, 90.4% at 50 μg/mL, and 91.3% at 100 μL/mL, respectively. Similarly hydrogen peroxide scavenging activities extracted using methanol were 40.6% at 1 μg/mL, 69.5% at 10 μg/mL, 85.3% at 50 μg/mL and 90.2% at 100 μL/mL. Table 2 shows, BHT (94.6%) had higher hydrogen peroxide scavenging activity than anthocyanin extracts. Park et al. [50] reported that the anthocyanin extracted from Black rice (Heugjinjubyeo) had good hydrogen peroxide scavenging activities in bran.

3.7.8. Inhibitory Effects of Deoxyribose Degradation (Molecular Antioxidant Analysis) . Hydroxyl radical can be formed by the Fenton reaction in the presence of reduced transition metals such as Fe^{2+}, and H_2O_2, which is known to be

the most reactive of all the reduced forms of dioxygen and is thought to initiate cell damage in vivo [12, 51]. To determine whether sorghum anthocyanins reduce hydroxyl radical generation by chelating metal ions or by directly scavenging hydroxyl radicals, the effects of the anthocyanins or hydroxyl radical generated by Fe^{3+} ions were analyzed by determining the degree of deoxyribose degradation. Table 2 shows the concentration-dependent inhibition of hydroxyl radical induced deoxy ribose degradation by anthocyanins in both the site-specific and non-site-specific assays using the same concentration, relatively greater antioxidant activity was observed in the site-specific assay than non-site-specific assay, implying that the anthocyanins inhibited deoxyribose degradation mainly by chelating metal ions rather than by scavenging hydroxyl radical directly. In this system, anthocyanin extracts exhibited a stronger concentration-dependent inhibition of deoxyribose oxidation. Smith et al. [52] earlier reported that molecules that can inhibit deoxyribose degradation are those that can chelate iron ions and render them in active or poorly active in a Fenton reaction. In the present study, in site specific system, we have demonstrated the iron chelating ability of the anthocyanin extracts. It is likely that the chelating effect of red sorghum bran anthocyanins on metal ions may be responsible for the inhibition of deoxyribose oxidation. Iron, a transition metal, is capable of generating free radicals from peroxides by the Fenton reaction and is implicated in many diseases [53]. Fe^{2+} has also been shown to produce oxyradicals, and lipid peroxidation and reduction of Fe^{2+} concentration in the Fenton reaction would protect against oxidative damage. Similar results were reported for extracts of Opuntia ficus-indica var. saboten [26] and Hypericum perforatum L. [54].

3.7.9. DNA Nicking Assay. Hydroxyl radicals generated by the Fenton reaction are known to cause oxidatively induced breaks in DNA strands to yield its fragmented forms. The free radical scavenging effect of sorghum anthocyanins were studied on Genomic DNA damage. The anthocyanins showed significant reduction in the formation of nicked DNA and increased native form of DNA. Quercetin effectively protected DNA strand scission from *tert*-butyl hydroperoxide [55]. In biological systems metal binding can occur on DNA leading to partial site-specificity hydroxyl radical formation. Anthocyanins are potential protecting agents against the lethal effects of oxidative stress and offer protection of DNA by chelating redox-active transition metal ions. Mas et al. [56] suggested that anthocyanins have the ability to stabilize DNA triple-helical complex. So far no reports are available on protecting DNA damage of red sorghum anthocyanins (Figure 4).

4. Conclusion

From the results it can be concluded that red sorghum bran has high amount of anthocyanins content on its bran. Reactive oxygen species plays a crucial role in a wide range of common diseases and age-related degenerative conditions including cardiovascular diseases, inflammatory conditions, and neurodegenerative diseases such as Alzheimer's disease, mutations and cancer [57]. So antioxidant capacity is widely used as a parameter to characterize food or medicinal plant, and their bioactive components. In this study, the antioxidant activity of the anthocyanins extracted from sorghum bran was evaluated, and it showed very strong antioxidant activity.

Thus, these results suggest that anthocyanin extracted from red sorghum bran can be used as antioxidant material and food addictives.

References

[1] T. Tsuda, F. Horio, and T. Osawa, "Cyanidin 3-O-β-D-glucoside suppresses nitric oxide production during a zymosan treatment in rats," *Journal of Nutritional Science and Vitaminology*, vol. 48, no. 4, pp. 305–310, 2002.

[2] C. Zhao, M. Giusti, M. Malik, M. Moyer, and B. Magnuson, "Effects of commercial anthocyanin—rich extracts on colonic cancer and non tumorigenic colonic cell growth," *Journal of Agricultural and Food Chemistry*, vol. 52, pp. 6122–6128, 2004.

[3] T. Tsuda, F. Horio, K. Uchida, H. Aoki, and T. Osawa, "Dietary cyanidin 3-O-β-D-glucoside-rich purple corn color prevents obesity and ameliorates hyperglycemia in mice," *Journal of Nutrition*, vol. 133, no. 7, pp. 2125–2130, 2003.

[4] W. Boyd, "Natural colors as functional ingredients in healthy foods," *Cereal Foods World*, vol. 45, no. 5, pp. 221–222, 2000.

[5] R. Bukantis, "Energy inputs in sorghum production," in *Handbook of Energy Utilization in Agriculture*, D. Pimentel, Ed., pp. 103–108, CRC Press, Boca Raton, Fla, USA, 1980.

[6] S. M. Awika, *Sorghum phenols as antioxidants*, M.S. thesis, Texas A&M University: College Station, Tex, USA, 2000.

[7] F. Gous, *Tannins and phenols inblack sorghum*, Ph.D. Dissertation, Texas A&M University: College Station, Tex, USA, 1989.

[8] O. Blokhina, E. Virolainen, and K. V. Fagerstedt, "Antioxidants, oxidative damage and oxygen deprivation stress: a review," *Annals of Botany*, vol. 91, pp. 179–194, 2003.

[9] B. Halliwell and J. M. C. Gutteridge, *Free radicals in biology and medicine*, Oxford University Press, New York, NY, USA, 3rd edition, 1999.

[10] J. Moskovitz, K. A. Yim, and P. B. Chock, "Free radicals and disease," *Archives of Biochemistry and Biophysics*, vol. 397, no. 2, pp. 354–359, 2002.

[11] M. G. Simic, "Mechanisms of inhibition of free-radical processes in mutagenesis and carcinogenesis," *Mutation Research*, vol. 202, no. 2, pp. 377–386, 1988.

[12] Đ. Gülçin, "Antioxidant and antiradical activities of L-carnitine," *Life Sciences*, vol. 78, no. 8, pp. 803–811, 2006.

[13] C. Rice-Evans, N. J. Miller, and G. Paganga, "Antioxidant properties of phenolic compounds," *Trends in Plant Science*, vol. 2, no. 4, pp. 152–159, 1997.

[14] M. A. Joseph, W. R. Lloyd, and D. W. Ralph, "Anthocaynins from black sorghum and their antioxidant properties," *Food Chemistry*, vol. 90, pp. 293–301, 2004.

[15] D. Strack, E. Busch, and E. Klein, "Anthocyanin patterns in european orchids and their taxonomic and phylogenetic relevance," *Phytochemistry*, vol. 28, no. 8, pp. 2127–2139, 1989.

[16] A. Meda, C. E. Lamien, M. Romito, J. Millogo, and O. G. Nacoulma, "Determination of the total phenolic, flavonoid and proline contents in Burkina Fasan honey, as well as their radical scavenging activity," *Food Chemistry*, vol. 91, no. 3, pp. 571–577, 2005.

[17] L. W. Chang, W. J. Yen, S. C. Huang, and P. D. Duh, "Antioxidant activity of sesame coat," *Food Chemistry*, vol. 78, no. 3, pp. 347–354, 2002.

[18] C. Chang, M. H. Yang, H. M. Wen, and J. C. Chern, "Estimation of total flavonoid content in propolis by two complementary colometric methods," *Journal of Food and Drug Analysis*, vol. 10, no. 3, pp. 178–182, 2002.

[19] T. Fuleki and F. J. Francis, "Determination of total anthocyanin and degradation index for cranberry juice," *Food Science*, vol. 33, pp. 78–83, 1968.

[20] J. A. Larrauri, C. Sánchez-Moreno, and F. Saura-Calixto, "Effect of temperature on the free radical scavenging capacity of extracts from red and white grape pomace peels," *Journal of Agricultural and Food Chemistry*, vol. 46, no. 7, pp. 2694–2697, 1998.

[21] R. P. Singh, C. K. N. Murthy, and G. K. Jayaprakash, "Studies on the antioxidant acitivity of pomegranate (Pubica granatum) Peel and seed extract using in vitro models," *Journal of Agricultural and Food Chemistry*, vol. 50, pp. 81–86, 2002.

[22] M. Oyaizu, "Studies on product of browning rection prepared from glucose amine," *Japanese Journal of Nutrition*, vol. 44, pp. 307–315, 1986.

[23] O. Siddhurajua, P. S. Mohanb, and K. Beckera, "Studies on the antioxidant activity of Indian Laburnum (*Cassia fistula* L,): a preliminary assessment of crude extracts from stem bark, leaves, flowers and fruit pulp," *Food Chemistry*, vol. 79, pp. 61–67, 2002.

[24] T. C. P. Dinis, V. M. C. Madeira, and L. M. Almeida, "Action of phenolic derivatives (acetaminophen, salicylate, and 5-aminosalicylate) as inhibitors of membrane lipid peroxidation and as peroxyl radical scavengers," *Archives of Biochemistry and Biophysics*, vol. 315, no. 1, pp. 161–169, 1994.

[25] P. D. Duh, "Antioxidant activity of burdock (*Arctium lappa* Linne): its scavenging effect on free radical and active oxygen," *Journal of the American Oil Chemists Society*, vol. 75, no. 4, pp. 455–461, 1998.

[26] J. C. Lee, H. R. Kim, J. Kim, and Y. S. Jang, "Antioxidant property of an ethanol extract of the stem of *Opuntia ficus*-indica var. saboten," *Journal of Agricultural and Food Chemistry*, vol. 50, no. 22, pp. 6490–6496, 2002.

[27] Y. Lu and L. Y. Foo, "Unusual anthocyanin reaction with acetone leading to pyranoanthocyanin formation," *Tetrahedron Letters*, vol. 42, no. 7, pp. 1371–1373, 2001.

[28] S. Sellappan and C. C. Akoh, "Flavonoids and antioxidant capacity of Georgia-grown Vidalia onions," *Journal of Agricultural and Food Chemistry*, vol. 50, no. 19, pp. 5338–5342, 2002.

[29] M. M. Giusti and R. E. Wrolstad, "Acylated anthocyanins from edible sources and their applications in food systems," *Biochemical Engineering Journal*, vol. 14, no. 3, pp. 217–225, 2003.

[30] M. Awika, C. M. McDonough, and L. W. Rooney, "Decorticating sorghum to concentrate healthy phytochemicals," *Journal of Agricultural and Food Chemistry*, vol. 53, no. 16, pp. 6230–6234, 2005.

[31] K. G. Duodu and P. S. Belton, "Chemical composition and antioxidant effects from sorghum flour and bran," 2006.

[32] P. S. Wharton and R. H. Nicholson, "Temporal synthesis and radiolabelling of the sorghum 3-deoxyanthocyanidin phytoalexins and the anthocyanin, cyanidin 3-dimalonyl glucoside," *New Phytologist*, vol. 145, no. 3, pp. 457–469, 2000.

[33] L. Dykes and L. W. Rooney, "Sorghum and millet phenols and antioxidants," *Journal of Cereal Science*, vol. 44, no. 3, pp. 236–251, 2006.

[34] C. Garcia-Viguera, P. Zafrilla, and F. A. Tomás-Barberán, "The use of acetone as an extraction solvent for anthocyanins from strawberry fruit," *Phytochemical Analysis*, vol. 9, no. 6, pp. 274–277, 1998.

[35] S. Kallithraka, C. Garcia-Viguera, P. Bridle, and J. Bakker, "Survey of solvents for the extraction of grape seed phenolics," *Phytochemical Analysis*, vol. 6, no. 5, pp. 265–267, 1995.

[36] K. Robards, P. D. Prenzler, G. Tucker, P. Swatsitang, and W. Glover, "Phenolic compounds and their role in oxidative processes in fruits," *Food Chemistry*, vol. 66, no. 4, pp. 401–436, 1999.

[37] C. Sánchez-Moreno, "Methods used to evaluate the free radical scavenging acitivity in foods and biological systems," *Food Science and Technology International*, vol. 8, no. 3, pp. 121–137, 2002.

[38] L. S. Einbond, K. A. Reynertson, X. D. Luo, M. J. Basile, and E. J. Kennelly, "Anthocyanin antioxidants from edible fruits," *Food Chemistry*, vol. 84, no. 1, pp. 23–28, 2004.

[39] H. Kappus, "Lipid peroxidation; mechanism and biological relevance," in *Free Radicals and Food Additives*, O. Asuoma and B. Halliwell, Eds., pp. 59–74, Taylor and Francis, New York, NY, USA, 1991.

[40] C. K. N. Murthy, R. P. Singh, and G. K. Jayaprakasha, "Antioxidant acitivities of grape (*Visits vinegera*) Pomace extract," *Journal to Agricultural and Food Chemistry*, vol. 50, pp. 5905–5914, 2002.

[41] B. N. Shyamala, S. Gupta, A. J. Lakshmi, and J. Prakash, "Leafy vegetable extracts-antioxidant activity and effect on storage stability of heated oils," *Innovative Food Science and Emerging Technologies*, vol. 6, no. 2, pp. 239–245, 2005.

[42] G. C. Yen and P. D. Duh, "Antioxidative properties of methanolic extracts from peanut hulls," *Journal of the American Oil Chemists' Society*, vol. 70, no. 4, pp. 383–386, 1993.

[43] X. Pin-Der-Duh, X. Pin-Chan-Du, and X. Gow-Chin Yen, "Action of methanolic extract of mung bean hulls as inhibitors of lipid peroxidation and non-lipid oxidative damage," *Food and Chemical Toxicology*, vol. 37, no. 11, pp. 1055–1061, 1999.

[44] M. Tanaka, C. W. Kuie, X. Nagashima, and T. Taguchi, "Applications of antioxidative maillard reaction products from histidine and glucose to Saradine products," *Nippon Gakkaishi*, vol. 54, pp. 1409–1414, 1988.

[45] L. A. Gordon, *Utilization of sorghum brans and barley flour in bread*, M.S. thesis, Texas A and M University: College Station, Texas, Tex, USA, 2001.

[46] M. K. Dahl and T. Richardson, "Photogeneration of Superoxide anion in serum of bovine milk and in model system containing ribotharin and aminocids," *Journal of Dairy Science*, vol. 61, pp. 400–407, 1978.

[47] A. S. Meyer and A. Isaksen, "Application of enzymes as food antioxidants," *Trends in Food Science and Technology*, vol. 6, no. 9, pp. 300–304, 1995.

[48] C. L. Hsu, W. Chen, Y. M. Weng, and Y. Tseng, "Chemical composition, physical properties, and antioxidant activities of Yam flours as affected by different drying methods," *Food Chemistry*, vol. 83, no. 1, pp. 85–92, 2003.

[49] M. Elmastaş, I. Gülçin, Ö. Işildak, Ö. I. Küfrevioğlu, K. Ibaoğlu, and H. Y. Aboul-Enein, "Radical scavenging activity and antioxidant capacity of bay leaf extracts," *Journal of the Iranian Chemical Society*, vol. 3, no. 3, pp. 258–266, 2006.

[50] Y. S. Park, S. J. Kim, and H. I. Chang, "Isolation of Anthocyanin from Black Rice (Heugjinjibyeu) and Screening of its antioxidant activities," *Korean Journal of Microbiology and Biotechnology*, vol. 36, pp. 55–60, 2008.

[51] E. Rollet-Labelle, M.-J. Grange, C. Elbim, C. Marquetty, M.-A. Gougerot-Pocidalo, and C. Pasquier, "Hydroxyl radical as a potential intracellular mediator of polymorphonuclear neutrophil apoptosis," *Free Radical Biology and Medicine*, vol. 24, no. 4, pp. 563–572, 1998.

[52] C. Smith, B. Halliwell, and O. I. Aruoma, "Protection by albumin against the pro-oxidant actions of phenolic dietary components," *Food and Chemical Toxicology*, vol. 30, no. 6, pp. 483–489, 1992.

[53] B. Halliwell and J. M. C. Gutteridge, "Role of free radicals and catalytic metal ions in human disease: an overview," *Methods in Enzymology*, vol. 186, pp. 1–85, 1990.

[54] Y. Zhou, Y. Lu, and D. Wei, "Antioxidant activity of a flavanoid rich extract of *Hypericum perforatum L. invitro*," *Journal of Agricultural and Food Chemistry*, vol. 52, pp. 5032–5039, 2004.

[55] P. Sestili, A. Guidarelli, M. Dacha, and O. Cantoni, "Quercetin prevents DNA single strand breakage and cytotoxicity caused by tert-butylhydroperoxide: free radical scavenging versus iron chelating mechanism," *Free Radical Biology and Medicine*, vol. 25, no. 2, pp. 196–200, 1998.

[56] T. Mas, J. Susperregui, B. Berké et al., "DNA triplex stabilization property of natural anthocyanins," *Phytochemistry*, vol. 53, no. 6, pp. 679–687, 2000.

[57] S. J. Han, S. N. Ryu, and S. S. Kang, "A new 2-arylbenzofuran with antioxidant activity from the black colored rice (*Oryza sativa L.*) bran," *Chemical and Pharmaceutical Bulletin*, vol. 52, no. 11, pp. 1365–1366, 2004.

9

Structural Variations of Human Glucokinase Glu256Lys in MODY2 Condition Using Molecular Dynamics Study

Nanda Kumar Yellapu,[1] **Kalpana Kandlapalli,**[2] **Koteswara Rao Valasani,**[3] **P. V. G. K. Sarma,**[4] **and Bhaskar Matcha**[1]

[1] *Division of Animal Biotechnology, Department of Zoology, Sri Venkateswara University, Tirupati, Andhra Pradesh 517502, India*
[2] *Department of Biochemistry, Sri Venkateswara Institute of Medical Sciences, Tirupati, Andhra Pradesh 517507, India*
[3] *Department of Pharmacology and Toxicology, University of Kansas, Lawrence, KS 66047, USA*
[4] *Department of Biotechnology, Sri Venkateswara Institute of Medical Sciences, Tirupati, Andhra Pradesh 517507, India*

Correspondence should be addressed to Bhaskar Matcha; matchabhaskar2010@gmail.com

Academic Editor: Yau Hung Chen

Glucokinase (GK) is the predominant hexokinase that acts as glucose sensor and catalyses the formation of Glucose-6-phosphate. The mutations in GK gene influence the affinity for glucose and lead to altered glucose levels in blood causing maturity onset diabetes of the young type 2 (MODY2) condition, which is one of the prominent reasons of type 2 diabetic condition. In view of the importance of mutated GK resulting in hyperglycemic condition, in the present study, molecular dynamics simulations were carried out in intact and 256 E-K mutated GK structures and their energy values and conformational variations were correlated. Energy variations were observed in mutated GK (3500 Kcal/mol) structure with respect to intact GK (5000 Kcal/mol), and it showed increased γ-turns, decreased β-turns, and more helix-helix interactions that affected substrate binding region where its volume increased from 1089.152 Å^2 to 1246.353 Å^2. Molecular docking study revealed variation in docking scores (intact $= -12.199$ and mutated $= -8.383$) and binding mode of glucose in the active site of mutated GK where the involvement of A53, S54, K56, K256, D262 and Q286 has resulted in poor glucose binding which probably explains the loss of catalytic activity and the consequent prevailing of high glucose levels in MODY2 condition.

1. Introduction

Type 2 diabetic condition is the increase in blood glucose levels and is due to many reasons; one of the most important factor being MODY2 condition, which is characterized at an early age and is an autosomal dominant inherited disorder [1]. Glucokinase (GK) is one of the potential candidate genes for type 2 diabetes acting through elevated fasting plasma glucose. It is a glucose sensing enzyme that catalyses the formation of glucose-6-phosphate from glucose by utilizing one molecule of ATP and that determines the threshold for glucose-stimulated insulin secretion in islets and controls gluconeogenesis and glycogen synthesis in hepatocytes. It can regulate the insulin secretion and integration of hepatic intermediatory metabolism [2]. GK gene is 52.15 kilo bases (kb) in length and is present on Chromosome 7 p13 with 12 exons and produces a transcript of 2.7 kb. A number of reports suggest that the existence of mutations in the coding region of GK is associated with MODY2 [3–11]. The mutated structures show variation in the affinity for binding with glucose, which may affect the kinetics of GK [12, 13]. In order to assess the mutations in GK affecting the catalysis process, *in silico* mutagenic studies will help in revealing the effect of structural and functional variations with respect to mutations in the enzyme such that the same can be exploited to explain the MODY2 condition in type 2 diabetic patients. Molecular dynamics simulation techniques can be applied to study the behavior of both intact and mutated GK structures at any specified conditions, which can be used to investigate its specific molecular interaction in the system [14–16]. The dynamic simulations can explain the interaction and charge distribution of GK using density functional theory calculations in both intact and mutated structures [17]. This technique can also explain the impact of environmental conditions such as solvation and temperature on the GK conformations and energy changes which are of fundamental importance to describe the function and activity. The impact of every mutation on GK conformation can be clearly studied within a very less time. The biochemical function of any protein is defined by its 3D structures, and under physiological conditions, the 3D structures of protein are defined by its component residues among which each residue is having its specific impact on the conformation of the protein. These residues have a primary effect on the rate of protein folding, noncovalent interactions, and kinetic stability. Any mutations in the protein will reflect the variations in the biochemical function of the protein [18]. Determining such key residues would greatly enhance to understand the stability and reactivity of GK under normal and MODY2 condition [19]. Mutations that disrupt overall structure and dynamics can often have drastic functional consequences. The knowledge of structure and function relationship combined with the number of solved structures with no biochemical annotations has motivated the development of computational tools for the prediction of molecular function using sequence and structural information [20]. The identification and analysis of such residues will give an important insight into the structure-function correlations.

Hence, the present study is aimed to identify the impact of an active site mutation 256 E-K and its influenced regions, which will give a better idea on the activity of both intact and mutated GK. There was a survey by Bell et al. in 1996, indicating the natural occurrence of 256 E-K mutation first time in a population with MODY2 condition, and even they reported the altered activity of GK under mutated condition [21]. Molnes et al. reported in their site-directed mutagenic study that replacement of Glu with Lys/Ala at the 256th position resulted in enzyme forms that did not bind with α-D-glucose at a concentration of 200 mM and was essentially catalytically inactive [22]. Gidh-Jain et al. induced this mutation in human β-Cell GK by *in vitro* site directed mutagenesis and expressed in *Escherichia coli*, and they observed changes in enzyme activity including a decrease in V_{max} and/or increase in K_m for glucose [12]. We analyzed the impact of this active site mutation on the conformational fluctuations of GK and most interestingly into active site variations through molecular dynamics and docking. We observed variations in both the affinity and the binding mode of glucose in the active site along with energy fluctuations that eventually results in the loss of catalytic activity. Our study is strongly supported by the functional analysis done by the previous researchers explained previously.

2. Materials and Methods

All the molecular dynamics simulations and molecular docking studies were carried out in molecular operating environment software tool (MOE 2011.10. Chemical Computing Group Inc.).

2.1. Preparation of Intact Glucokinase Structure. The X-ray crystallographic structure of GK (PDB ID: 3F9M) at resolution of 1.5 Å was retrieved from Protein Data Bank (http://www.rcsb.org/pdb/home/home.do), which is a huge repository of three-dimensional structures of macromolecules [23]. The water molecules and heteroatoms were removed, polar hydrogens were added, and the structure was protonated. Energy minimization was carried out in MMFF94x force filed at root mean square gradient of 0.05.

2.2. Preparation of Mutated Glucokinase Structure. The MODY2 mutation at the 256th position that was reported in GK entry (ID: P35557) of UniProt database [24] and also in previous studies [12, 21, 22] was introduced where Glutamate was replaced with Lysine residue into the energy minimized intact GK structure, and again energy minimization was carried out with the previously explained conditions.

2.3. Molecular Dynamics Studies of Energy-Minimized Intact and Mutated GK Structures. The energy minimized conformations of both intact and mutated GK structures were subjected to molecular dynamics simulations individually in the same force field. The NPT (number of particles, pressure, and temperature) statistical ensemble in which the simulations generate stable conformations was specified, and both temperature and pressure were held fixed. The algorithm Nose-Poincare-Anderson (NPA) was specified to solve the equations of motion during simulations. This method is the most the accurate and sensitive, and, it generates true ensemble trajectories. The initial temperature was set to 30 K and increased to a run time temperature of 300 K, and pressure was set to 101 kPa. The heat time was set at 0 picoseconds (ps), the total run time of simulations was carried out for 10 nanoseconds (ns) and the final cool time was set to 0 ps. The constraints were applied on light bonds, and a time step of 0.002 ps was used to discretize the equations of motion. The position, velocity, and acceleration of the trajectories were saved for each 0.5 ps. The energy values of each conformation were plotted as graphs to observe the energy variations among intact and mutated GK.

2.4. PDBsum Analysis. PDBsum is a web-based database mainly providing the pictorial summaries of the 3D structures of proteins and their detailed structural analysis [25, 26]. The simulated structures obtained at the end of simulation period were submitted to PDBsum to identify the conformational variations that aroused due to introduction of mutation with respect to intact GK structure. The pictorial representation of mutated structure was correlated with intact structure, and conformational variations were identified.

2.5. Structural Alignment. The structural alignment task was carried out by PyMol software tool using align command [27]. The mutated structure was superimposed with intact structure to get a clear insight about the conformational fluctuations, especially in substrate binding regions. The active site residues, that is, T168, K169, N204, D205, N231, E256, and E290, were identified from PDBsum ligand interaction page of GK entry (http://www.ebi.ac.uk/thornton-srv/databases/cgi-bin/pdbsum/GetPage.pl?pdbcode=3f9m&template=ligands.html&l=1.1). The surface volumes of substrate binding cavities were measured to find out the volume differences.

2.6. Binding Mode Analysis. A comparative molecular docking analysis was carried out to know the binding mode of glucose in the active site, with both intact and mutated structures using MOE dock tool to obtain a population of possible conformations and orientations for glucose at the binding site. Glucose three-dimensional structure was constructed and optimized in MOE working environment. Initially, the simulated and stabilized trajectory of intact GK structure obtained at the end of the simulations was loaded into MOE. The binding site was defined with the residues T168, K169, N204, D205, N231, E256, and E290, and glucose was specified as ligand. Molecular docking was carried out into the specified binding site using triangle matcher docking placement methodology where the poses are generated by aligning ligand triplets of atoms on triplets of alpha spheres of receptor in a systemic way. A dock database was generated containing 30 docked conformations of the receptor and ligand. Londong dG scoring methodology was applied that estimates the free binding energy of the ligand from a given pose and ranks the docked conformations. The total docked conformations were subjected to refinement in the same force field and rescored using the same scoring function. Duplicates were removed from the final list of docked conformations. After docking process, the conformation with the lowest docking score was chosen for further study and analysis.

The same procedure was also carried out separately for the mutated GK docking process, but among the active site residues specified previously there is Lysine residue at the 256th position, and the remaining residues are same.

2.7. Molecular Dynamics Studies of Receptor-Ligand Complexes. The docking complexes of both intact and mutated GK-glucose complexes were subjected to molecular dynamics simulations for 10 ns individually with the same parameters specified previously for GK simulations alone. The energy values of both complexes were plotted as graphs at the end of the simulations to observe the variation. The conformations of ligand and its interaction with active site residues during simulations were analyzed at each 500 ps for both intact and mutated GK-glucose complexes.

3. Results

The stabilized trajectories of intact and mutated GK structures obtained at the end of simulations were observed for their energy variations. The intact GK structure with an initial energy of 525.966 Kcal/mol was stabilized around 5000 Kcal/mol while, the mutated GK structure with an initial energy of 365.061 Kcal/mol was stabilized around 3500 Kcal/mol in a 10 ns of simulation (Figure 1). This energy variation is the result of the substitution of E with K at the 256th position, and this mutation showed its effect not

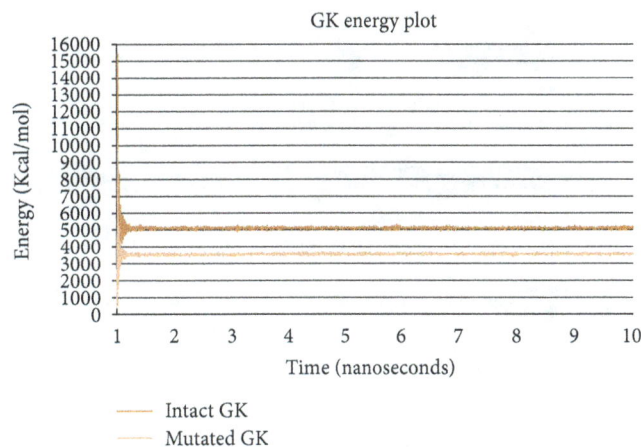

FIGURE 1: GK energy plot showing the energy transitions of intact and mutated GK structures during molecular dynamics simulations for a period of 10 ns. Intact GK conformation is stabilized around the energy levels of 5000 Kcal/mol and mutated GK around 3500 Kcal/mol.

TABLE 1: PDBsum analysis showing the variations in secondary structural conformations of intact and mutated GK structures.

Secondary conformation[a]	Intact GK[b]	Mutated GK[c]
Sheets	3	3
Beta alpha beta unit	1	1
Beta hairpins	5	5
Beta hairpins	5	4
Strands	13	13
Helices	20	22
Helix-helix interactions	24	40
β turns	34	31
γ turn	3	13

[a] Type of secondary conformation.
[b] Number of respective secondary conformations observed in intact GK.
[c] Number of respective secondary conformations observed in mutated GK.

only on the energy of the GK but also on the secondary structure conformation. The mutated GK structure showed increased γ turns, decreased β turns and more helix-helix interactions compared to intact GK structure as revealed from PDBsum analysis, indicating that 256 E-K, that is, acidic to basic amino acid replacement has profound effect on the GK conformation (Figure 2, Table 1).

The superimposition of substrate binding site of mutated GK with intact GK showed distinct changes which is correlated with their molecular surface area. The intact GK substrate binding site showed a surface area of 1089.152 Å^2 where glucose binds and fits into the cavity, and it was changed to 1246.353 Å^2 in the mutated structure (Figure 3).

Further, molecular docking analysis revealed that glucose is binding with the intact GK active site forming hydrogen bonds with P153, L165, K169, E256, Q287, and E290 residues while in mutated GK showed hydrogen bonds with S54, N166, K256 and D262 residues. The docking scores −12.199

Intact GK structure (left):

```
H1              H3                        ββ ββ β    A
ENLYFQGMKKEKVEQILAEFQLQEEDLKKVMRRMQKEMDRGLRLETHEEASVKMLPTYVR
4    10   15    20   25   30   35   40   45   50   55   60

β          B          B            B            H4   H5
STPEGSEVGDFLSLDLGGTNFRVMLVKVGE QWSVKTKHQMYSIPEDAMTGTAEMLFDY
64   70   75   80   85   90   98   105  110  115  120  125

      ββ       B      γ  C   C C   β      β    β      H6
ISECISDFLDKHQMKHKKLPLGFTFSFPVRHEDIDKGILLNWIKGFKASGAEGNNVVGLL
126 130  135  140  145  150  155  160  165  170  175  180  185

      B        H7       ββ β  A     ββ  A     H8      γβ
RDAIKRRGDFEMDVVAMVNDTVATMISCYYEDHQCEVGMIVGTGCNACYMEEMQNVELVE
186 190  195  200  205  210  215  220  225  230  235  240  245

A      H9   β β    H10 H11        β β β  H12 H13
GDEGRMCVNTEWGAFGDSGELDEFLLEYDRLVDESSANPGQQLYEKLIGGKYMGELVRLV
246 250  255  260  265  270  275  280  285  290  295  300  305

H14          βββ  β βγ   H15    H16      H17
LLRLVDENLLFHGEASEQLRTRGAFETRFVSQVESDTGDRKQIYNILSTLGLRPSTTDCD
306 310  315  320  325  330  335  340  345  350  355  360  365

                        β   A     H18     H19
IVRRACESVSTRAAHMCSAGLAGVINRMRESRSEDVMRITVGVDGSVYKLHPSFKERFHA
366 370  375  380  385  390  395  400  405  410  415  420  425

      β    A    β   H20
SVRRLTPSCEITFIESEEGSRGAALVSAVACK
426 430  435  440  445  450  455
```

256 E-K mutated GK structure (right):

```
H1            γ  H3                      H4    β β    A
ENLYFQGMKKEKVEQILAEFQLQEEDLKKVMRRMQKEMDRGLRLETHEEASVKMLPTYVR
4    10   15    20   25   30   35   40   45   50   55   60

β          B        β B        ββ B         H5     H6
STPEGSEVGDFLSLDLGGTNFRVMLVKVGEQWSVKTKHQMYSIPEDAMTGTAEMLFDYIS
64   70   75   80   85   90   98   105  110  115  120  125

   β      B     γ  C      C      β      β  ββ β       H7
ECISECISDFLDKHQMKHKKLPLGFTFSFPVRHEDIDKGILLNWIKGFKASGAEGNNVVGLLRD
128  135  140  145  150  155  160  165  170  175  180  185

  γγ    B       H8      βββ β  γ A     ββ  A      H9   γβ
AIKRRGDFEMDVVAMVNDTVATMISCYYEDHQCEVGMIVGTGCNACYMEEMQNVELVEGD
188  195  200  205  210  215  220  225  230  235  240  245

A      H10   ββ    H11 H12       β β β  H13  H14
EGRMCVNTKWGAFGDSGELDEFLLEYDRLVDESSANPGQQLYEKLIGGKYMGELVRLVLL
248  255  260  265  270  275  280  285  290  295  300  305

H15  γ γ H16   β γ β γ  H17    β    H18       H19
RLVDENLLFHGEASEQLRTRGAFETRFVSQVESDTGDRKQIYNILSTLGLRPSTTDCDIV
308  315  320  325  330  335  340  345  350  355  360  365

                     γ β   A       H20     H21
RRACESVSTRAAHMCSAGLAGVINRMRESRSEDVMRITVGVDGSVYKLHPSFKERFHASV
368  375  380  385  390  395  400  405  410  415  420  425

   β   A    β    H22
RRLTPSCEITFIESEEGSRGAALVSAVACK
428  435  440  445  450  455
```

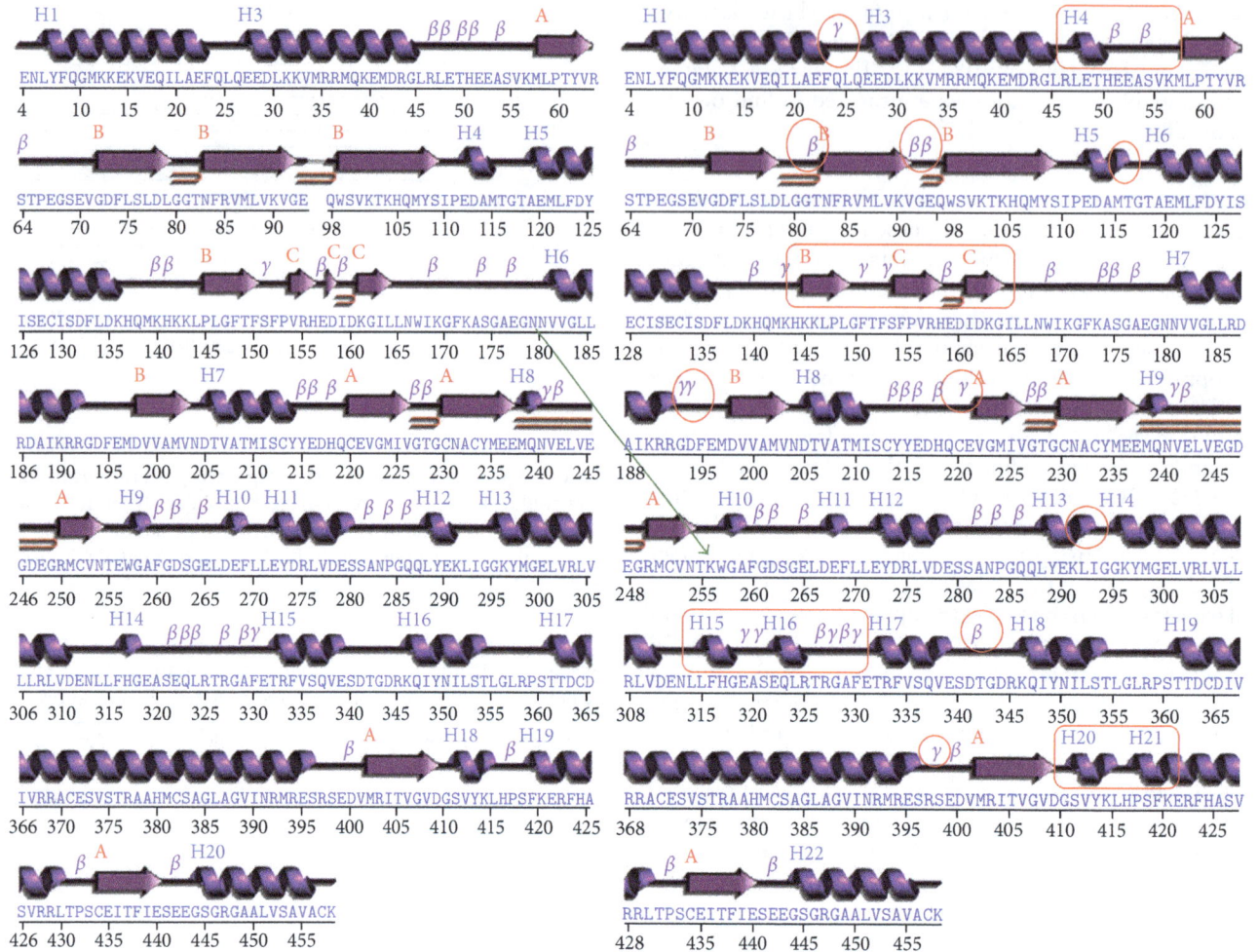

FIGURE 2: PDBsum analysis of intact GK structure (left) and 256 E-K mutated GK structure (right). The changes in the secondary structure conformations of mutated structure are shown in red-colored circles. These changes are due to mutation at position 256 where Glutamate is replaced with Lysine residue (indicated with green arrow).

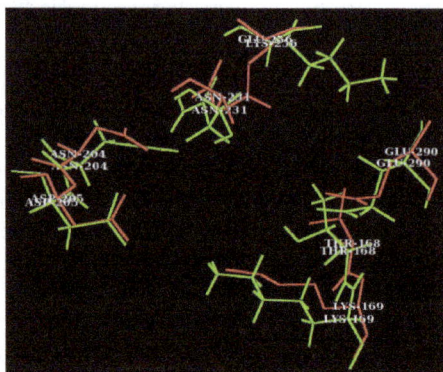

FIGURE 3: Superimposition of substrate binding regions of intact (red) and 256 E-K mutated (green) GK structures. The distance between the superimposed residues explains the variation in volume and surface area of substrate binding region, which in turn influences the binding affinity with glucose.

and −8.383 of intact GK and mutated GK, respectively, showed that the affinity of binding of glucose decreased in mutated GK (Figure 4). Here, the mutated residue lysine at position 256 is found to be interacting with glucose molecule forming two hydrogen bonds. There is a drastic variation in the binding mode of glucose with intact GK active site where it was found to be sitting in the cavity and showed no interaction with the solvent, whereas in the mutated GK active site, the glucose molecule was found to be on the surface of the cavity and was interacting with the solvent. These variations in the glucose interaction were due to the mutation generated in the GK molecule (Table 2).

The comparative molecular dynamics simulations results of the docking complexes of both intact and mutated GK showed variations in energy transitions and conformations during simulation period. The intact GK docking complex showed stability around energy levels of 5000 Kcal/mol which is equal to the energy transitions of intact GK simulations, and no energy fluctuations were observed even after docking process, while mutated GK docking complex showed variations in energy levels of 8600 Kcal/mol; however, the mutated GK alone showed energy levels around 3500 Kcal/mol (Figure 5). These results clearly indicated

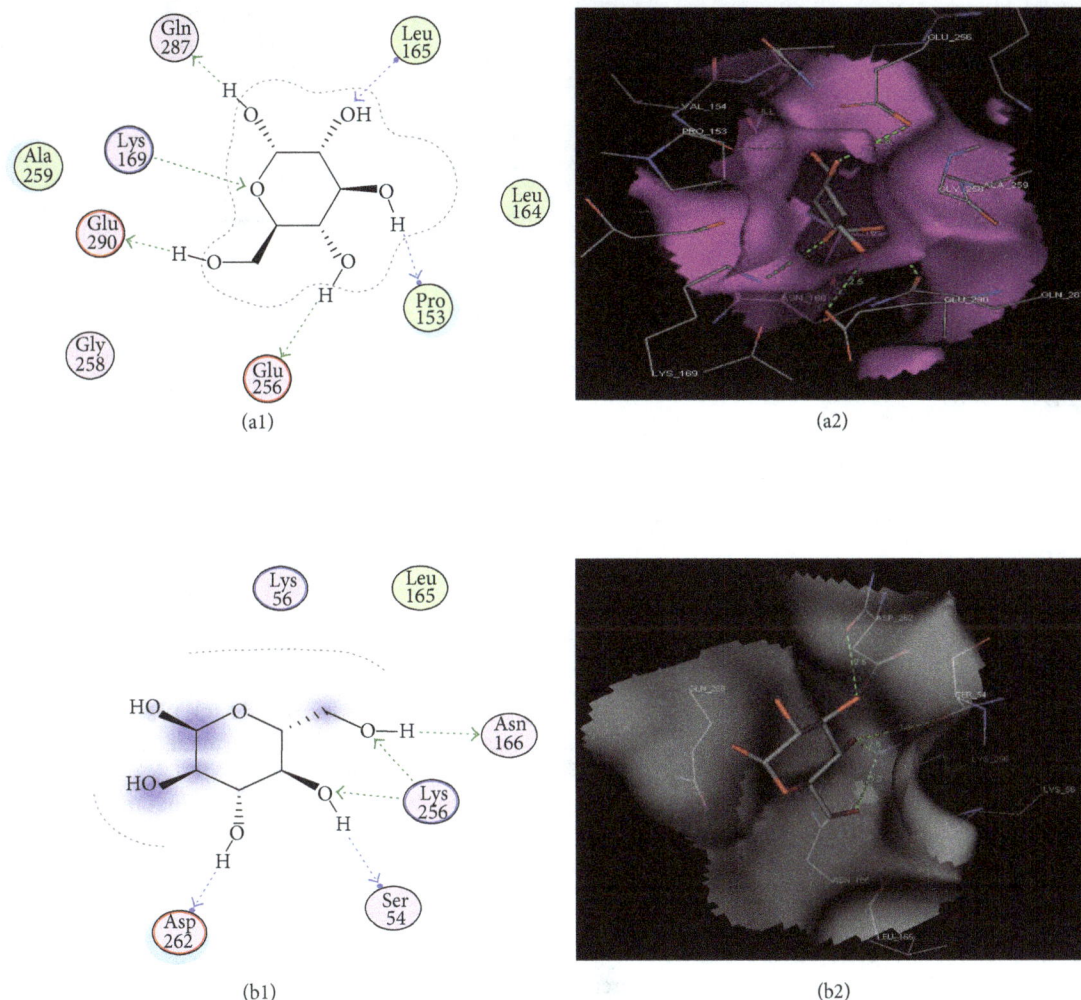

FIGURE 4: Binding mode of glucose with intact and mutated GK active sites after molecular docking. (a1) Two-dimensional linear representation of the glucose interaction with intact GK active site residues showing 6 hydrogen bonds. (a2) Three-dimensional graphical representation of glucose interaction found to be sit in the active site cavity with hydrogen bond interactions. (b1) Two-dimensional linear representation of glucose interaction with mutated GK active site residues showing 5 hydrogen bonds. The blue-colored shade represents the solvent exposure area of glucose molecule. (b2) Three-dimensional graphical representation of glucose interaction found to be on the surface of active site cavity with limited hydrogen bond interactions.

that energy levels were the same in intact GK when it is docked with glucose, while extensive variation in energy levels with mutated GK is due to the change in the acidic to basic amino acid which probably prevented the release of H$^+$ ions in the phosphorylation reaction. Further, the conformational analysis at every 500 ps for both intact and mutated complexes, the binding orientations of glucose, and its interaction with the specific active site residues at specific time period of simulations explain the binding affinity variations of glucose to the active site (Table 3) (see Supplementary information in Tables S1 and S2 in the Supplementary Material available online at http://dx.doi.org/10.1155/2013/264793).

Majority of the conformations of intact GK complex showed the major contribution by K169 to bind with glucose followed by L165, N166, and Q256. A very less frequency of interaction was observed with P153, Q287, and E290.

Mutated GK docking complex conformations revealed that only N166 and Q287 were found to be interacting commonly as the intact GK. The new residues such as A53, S54, K56, K256, D262, and Q286 that are in the surrounding area of the active site came into interaction with glucose among which the major contribution was made by D262 followed by Q286, and a very less frequency of interaction was made by S54, K56, and K256 residues. Interaction of glucose with these residues in mutated GK making it come out from the binding site cavity and showing interaction with solvent. This may be a responsible factor along with drastic energy variations bringing instability in GK-glucose complex which may result in poor binding of glucose and may also result in the disassociation of the complex. Such a mutation is observed in MODY2 condition, which, therefore, explains the loss of catalytic activity resulting in high glucose condition in type 2 diabetes.

TABLE 2: Molecular docking of glucose into the active site cavity of intact and mutated GK. Docking score shown in the second column indicates the binding affinity of glucose to the active site. The lower is the score, the higher will be the stability of the complex. The interacting active site residues of GK that are involved in formation of hydrogen bonds with glucose are shown in the fourth column, and the respective hydrogen bond lengths are indicated in Angstroms in the last column.

GK structure	Docking score	No. H-bonds	Interacting residue of GK	H-bond length (Å)
Intact	−12.199	6	P 153	1.49
			E 256	2.04
			Q 287	1.45
			E 290	1.58
			L 165	2.63
			K 169	2.95
Mutated	−8.383	5	S 54	2.45
			N 166	1.54
			D 262	1.69
			K 256	2.46
			K 256	3.00

4. Discussion

Natural mutations in GK gene result in poor affinity towards glucose resulting in high blood glucose levels, which is one of the condition in type 2 diabetes and these mutations are explained as MODY2 mutations. Basically, the mutations are observed throughout the gene so far. Increased type 2 diabetic population all over the world with different MODY2 mutations in GK gene showing altered affinity towards glucose could be fatal in such patients. In order to elucidate the probable occurrence of such mutations and their impact on GK catalysis, in the present study, we concentrated on an active site MODY2 mutation 256 E-K and carried out comparative molecular dynamics simulations and molecular docking studies. For this purpose, the intact and mutated GK structures were simulated and submitted to PDBsum for the conformational analysis and observed extensive conformational variations not only in the active site but also throughout the mutated GK structure. The active site variations were correlated with its molecular surface area, which in turn explains decreased glucose binding in the mutated structure. This variation of glucose binding affects the catalytic properties of GK. This mutation is not only affecting the conformation of the structure but also results in extremely variable energy levels.

Thus, this kind of variations in both energies and conformations clearly explains not only the decreased affinity for glucose but also increased blood glucose levels in the patients affected with MODY2 mutation. Zhang et al. also demonstrated this kind of study where they explained the importance of K169 residue in the GK catalytic mechanism with the help of molecular dynamics simulations, and they even verified their prediction by experimental mutagenesis and enzymatic analysis to provide a strong evidence for the

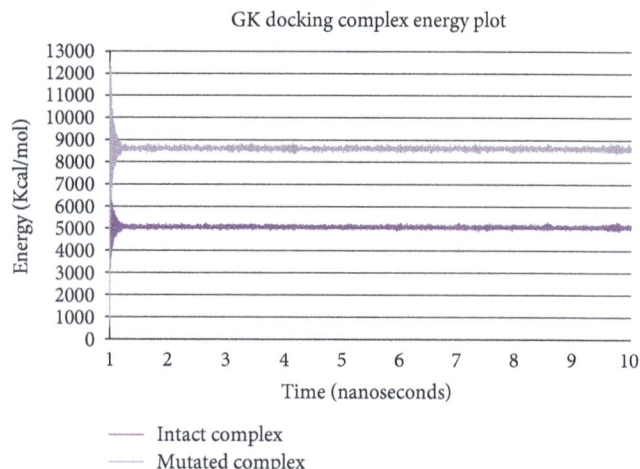

FIGURE 5: Energy transition plot of intact and mutated GK docking complexes during molecular dynamics simulations for a period of 10 ns. Intact GK docking complex is stabilized around the energy levels of 5000 Kcal/mol and mutated GK docking complex around 8800 Kcal/mol.

pathogenic mechanism of MODY2 condition [16]. In the same way, this study can provide the evidence for altered catalytic mechanism of each MODY2 mutated GK. Ramirez et al. also studied in the same manner to identify the mutation inducing variations in the active site of Haemoglobin I from *Lucina pectinata*, and they analyzed the ligand binding kinetics that plays major role in the stabilization process of binding site [28].

Figure 2 can explain clear comparative pictorial variations in the mutated GK secondary structural conformation where two new α helices were formed, three β turns were lost, and ten new γ turns were generated. To observe the impact of this mutation on the substrate binding site, the simulated structures of intact and mutated GK were superimposed, and the change in the cavity volume was clearly observed providing the reason for positional fluctuations of glucose. Figure 4 shows the interaction of glucose with the substrate binding sites of intact and mutated GK structures where the positional changes are clearly observed. This was strengthened by molecular docking analysis where we observed the variation in docking scores and binding mode of glucose among intact and mutated GK structures. Comparatively, the lowest docking score was observed with intact GK which explains the stronger affinity of glucose to the active site than in mutated one.

The molecular dynamics simulations of intact and mutated GK-glucose docking complexes revealed the energy transition variations where the intact GK showed no significant variation even after docking, but mutated GK showed higher energy levels after docking process. Such higher energy levels result in less affinity between enzyme and substrate and may also cause the dissociation of complex, thereby the rate of reaction will be reduced. The intact and mutated docking conformations are showing three common interacting residues, that is, L165, N166, and Q287 (Table 3) indicating the importance of these residues in the substrate

TABLE 3: Interaction of glucose with active site of intact and mutated GK and energy transitions of GK-glucose complexes during molecular dynamics simulations for a period of 10 ns.

Simulation[a] period (ps)	No. H-bonds[b]		Interacting residues of GK active site[c]		Energy of the complex[d] (Kcal/mol)	
	Intact	Mutated	Intact	Mutated	Intact	Mutated
0	6	5	**P153**, *L165*, **K169**, **E256**, **E290**, *Q287*	*S54*, *D262*, *N166*, **K256**, **K256**	527.75	366.85
500	5	8	**P153**, *L165*, *N166*, **K169**, *Q287*	*A53*, **K56**, *N166*, **Q286**, **Q286**, **D262**, **D262**, **D262**	5135.83	8536.07
1000	5	6	**P153**, *L165*, *N166*, **K169**, *Q287*	*A53*, *N166*, **K256**, **D262**, **D262**, **Q286**	5054.68	8600.38
1500	6	8	**P153**, *L165*, *N166*, **K169**, *Q287*, **E290**	*A53*, *L165*, **K256**, **K256**, **D262**, **D262**, **D262**, **Q286**	5066.78	8553.16
2000	6	4	*L165*, *N166*, **K169**, **E256**, *Q287*, *Q287*	*A53*, **D262**, **D262**, **Q286**	5112.49	8625.42
2500	8	4	*L165*, *L165*, *N166*, *N166*, **K169**, **K169**, **E256**, *Q287*	*A53*, **K256**, **D262**, **D262**	5094.22	8536.71
3000	6	4	*L165*, *N166*, **K169**, **K169**, **E256**, **E290**	*A53*, **D262**, **D262**, **Q286**	5087.11	8534.96
3500	8	6	*L165*, *L165*, *N166*, *N166*, **K169**, **K169**, **K169**, **E256**	*A53*, *S54*, **D262**, **D262**, **D262**, **Q286**	5154.06	8633.15
4000	8	7	*L165*, *L165*, *N166*, *N166*, **K169**, **K169**, **K169**, **E256**	*A53*, **D262**, **D262**, **D262**, **Q286**, **Q286**, *Q287*	5088.23	8608.42
4500	8	7	*L165*, *N166*, *N166*, **K169**, **K169**, **K169**, **E256**, *Q287*	*A53*, *S54*, **D262**, **D262**, **D262**, **Q286**, *Q287*	5014.87	8579.74
5000	8	5	*L165*, *L165*, *N166*, *N166*, **K169**, **K169**, **K169**, **E256**	*A53*, **D262**, **D262**, **Q286**, **Q286**	5078.62	8521.10
5500	6	5	*L165*, *N166*, **K169**, **K169**, **K169**, **E256**	*A53*, **D262**, **D262**, **D262**, **Q286**	5087.78	8548.05
6000	8	5	*L165*, *L165*, *N166*, *N166*, **K169**, **K169**, **K169**, **E256**	*A53*, **D262**, **D262**, **D262**, **Q286**	5116.11	8566.65
6500	8	7	*L165*, *L165*, *N166*, *N166*, **K169**, **K169**, **K169**, **E256**	*A53*, *S54*, **D262**, **D262**, **D262**, **Q286**, **Q286**	5059.85	8542.55
7000	7	6	*L165*, *N166*, *N166*, **K169**, **K169**, **K169**, **E256**	*A53*, **D262**, **D262**, **D262**, **Q286**, **Q286**	5030.70	8446.74
7500	8	6	*L165*, *L165*, *N166*, *N166*, **K169**, **K169**, **K169**, **E256**	*A53*, *S54*, **D262**, **D262**, **D262**, **Q286**	5063.48	8597.48
8000	4	7	*N166*, *N166*, **K169**, **E256**	*A53*, *S54*, **D262**, **D262**, **D262**, **Q286**, *Q287*	5106.54	8658.68
8500	8	5	*L165*, *L165*, *N166*, *N166*, **K169**, **K169**, **E256**, *Q287*	*A53*, **D262**, **D262**, **D262**, **Q286**	5104.12	8548.79
9000	6	7	*L165*, *N166*, **K169**, **K169**, **K169**, **E256**	*A53*, *S54*, **D262**, **D262**, **D262**, **Q286**, *Q287*	5127.97	8518.22
9500	7	4	*L165*, *L165*, *N166*, **K169**, **K169**, **K169**, **E256**	*A53*, **D262**, **D262**, **Q286**	5105.28	8574.21
10000	8	6	*L165*, *L165*, *N166*, *N166*, **K169**, **K169**, **K169**, **E256**	*A53*, *S54*, **D262**, **D262**, **D262**, **Q286**	5160.90	8522.43

[a] Duration of simulation period where the respective conformation was analyzed.

[b] Number of hydrogen bonds formed between the glucose and active site residues of intact and mutated GK.

[c] Interacting residues of intact and mutated GK during simulations in a specified conformation. The residues in bold are active site residues that are interacting with glucose specifically from intact GK, the residues in italic are found to be interacting with glucose in both intact and mutated GK, and the residues in bold italic are found to be interacting with glucose in mutated GK only.

[d] Energies of the docking complexes of intact and mutated GK at specified simulation periods.

binding mechanism and in the positional shift of glucose molecule. The remaining residues P153, K169, E256, and E290 that were found to be interacting with glucose in the intact GK active site lost their interaction because of conformational variations due to mutation in the active site where the other new residues A53, S54, K56, K256, D262, and Q286 came into interaction. Because of this, there is drastic variation in the conformation of active site resulting in poor binding of glucose and which eventually resulted in loss of catalytic activity. The significance of K169 residue in the catalytic activity of GK was already experimentally proved [16], so loss of interaction of such key residues of catalysis in the mutated GK could affect the catalytic mechanism of glucose phosphorylation in the active site. This may be explained with the variation seen in docking scores where the mutated GK showed higher docking score than the intact GK that cleared the reduced affinity for glucose.

These variations in mutated structure probably affect the binding affinity of glucose and catalytic activity of GK that will finally affect the phosphorylation and utilization of glucose and in turn results in the hyperglycemic condition. Such variations are characteristic features observed in MODY2. Thus, this study clearly explains the reasons for the increased blood glucose levels due to altered catalytic activities of GK in MODY2 condition.

5. Conclusion

The conformational fluctuations that aroused in the structure of GK are due to the mutation, which may alter its affinity for binding with glucose. This study had best explained the conformational variations of mutated GK structure, in both functional and nonfunctional regions. Finally, it provided a strong reason for the affinity changes in terms of both energy and docking score. Further, the 256 E-K mutation has profound effect on the conformational variation of active site resulting in poor binding of glucose and loss of catalytic activity.

Conflict of Interests

The author, N. K. Yellapu has received the INSPIRE fellowship from DST, Government of India, as monthly stipend for his living expenses and not for funding support of the work. The author, K. R. Valasani, has relationship with Kansas University and has the license policy to use the commercial software MOE from Chemical Computing Groups. This research work has been carried out on the agreement of all the authors, and the paper is submitted after the concurrence of all of them.

Acknowledgments

This work was supported by INSPIRE Division, Department of Science and Technology (DST), Government of India, New Delhi. The authors would like to acknowledge them gratefully for providing DST INSPIRE fellowship for supporting doctoral studies.

References

[1] A. T. Hattersley, R. C. Turner, M. A. Permutt et al., "Linkage of type 2 diabetes to the glucokinase gene," *The Lancet*, vol. 339, no. 8805, pp. 1307–1310, 1992.

[2] L. Agius, "Targeting hepatic glucokinase in type 2 diabetes: weighing the benefits and risks," *Diabetes*, vol. 58, no. 1, pp. 18–20, 2009.

[3] M. Stoffel, P. Froguel, J. Takeda et al., "Human glucokinase gene: Isolation, characterization, and identification of two missense mutations linked to early-onset non-insulin-dependent (type 2) diabetes mellitus," *Proceedings of the National Academy of Sciences of the United States of America*, vol. 89, no. 16, pp. 7698–7702, 1992.

[4] M. Stoffel, P. Patel, Y. M. D. Lo et al., "Missense glucokinase mutation in maturity-onset diabetes of the young and mutation screening in late-onset diabetes," *Nature Genetics*, vol. 2, no. 2, pp. 153–156, 1992.

[5] H. Sakura, K. Eto, H. Kadowaki et al., "Structure of the human glucokinase gene and identification of a missense mutation in a Japanese patient with early-onset non-insulin-dependent diabetes mellitus," *Journal of Clinical Endocrinology and Metabolism*, vol. 75, no. 6, pp. 1571–1573, 1992.

[6] J. Hager, H. Blanche, F. Sun et al., "Six mutations in the glucokinase gene identified in MODY by using a nonradioactive sensitive screening technique," *Diabetes*, vol. 43, no. 5, pp. 730–733, 1994.

[7] B. Guazzini, D. Gaffi, D. Mainieri et al., "Three novel missense mutations in the glucokinase gene (G80S; E221K; G227C) in Italian subjects with maturity-onset diabetes of the young (MODY). Mutations in brief no. 162. Online," *Human Mutation*, vol. 12, no. 2, article 136, 1998.

[8] A. T. Hattersley, F. Beards, E. Ballantyne, M. Appleton, R. Harvey, and S. Ellard, "Mutations in the glucokinase gene of the fetus result in reduced birth weight," *Nature Genetics*, vol. 19, no. 3, pp. 268–270, 1998.

[9] M. C. Y. Ng, B. N. Cockburn, T. H. Lindner et al., "Molecular genetics of diabetes mellitus in chinese subjects: Identification of mutations in glucokinase and hepatocyte nuclear factor-1α genes in patients with early-onset type 2 diabetes mellitus/MODY," *Diabetic Medicine*, vol. 16, no. 11, pp. 956–963, 1999.

[10] J. H. Nam, H. C. Lee, Y. H. Kim et al., "Identification of glucokinase mutation in subjects with post-renal transplantation diabetes mellitus," *Diabetes Research and Clinical Practice*, vol. 50, no. 3, pp. 169–176, 2000.

[11] P. R. Njølstad, O. Søvik, A. Cuesta-Muñoz et al., "Neonatal diabetes mellitus due to complete glucokinase deficiency," *The New England Journal of Medicine*, vol. 344, no. 21, pp. 1588–1592, 2001.

[12] M. Gidh-Jain, J. Takeda, L. Z. Xu et al., "Glucokinase mutations associated with non-insulin-dependent (type 2) diabetes mellitus have decreased enzymatic activity: implications for structure/function relationships," *Proceedings of the National Academy of Sciences of the United States of America*, vol. 90, no. 5, pp. 1932–1936, 1993.

[13] M. Stoffel, K. L. Bell, C. L. Blackburn et al., "Identification of glucokinase mutations in subjects with gestational diabetes mellitus," *Diabetes*, vol. 42, no. 6, pp. 937–940, 1993.

[14] F. Merino and V. Guixé, "Specificity evolution of the ADP-dependent sugar kinase family—in silico studies of the glucokinase/phosphofructokinase bifunctional enzyme from *Methanocaldococcus jannaschii*," *FEBS Journal*, vol. 275, no. 16, pp. 4033–4044, 2008.

[15] C. A. F. de Oliveira, M. Zissen, J. Mongon, and J. A. Mccammon, "Molecular dynamics simulations of metalloproteinases types 2 and 3 reveal differences in the dynamic behavior of the S1' binding pocket," *Current Pharmaceutical Design*, vol. 13, no. 34, pp. 3471–3475, 2007.

[16] J. Zhang, C. Li, T. Shi, K. Chen, X. Shen, and H. Jiang, "Lys169 of human glucokinase is a determinant for glucose phosphorylation: implication for the atomic mechanism of glucokinase catalysis," *PLoS ONE*, vol. 4, no. 7, Article ID e6304, 2009.

[17] S. Nagarajan, J. Rajadas, and E. J. P. Malar, "Density functional theory analysis and spectral studies on amyloid peptide Aβ(28-35) and its mutants A30G and A30I," *Journal of Structural Biology*, vol. 170, no. 3, pp. 439–450, 2010.

[18] J. Takeda, M. Gidh-Jain, L. Z. Xu et al., "Structure/function studies of human β-cell glucokinase. Enzymatic properties of a sequence polymorphism, mutations associated with diabetes, and other site-directed mutants," *The Journal of Biological Chemistry*, vol. 268, no. 20, pp. 15200–15204, 1993.

[19] Z. Dosztányi, C. Magyar, G. E. Tusnády, M. Cserzo, A. Fiser, and I. Simon, "Servers for sequence-structure relationship analysis and prediction," *Nucleic Acids Research*, vol. 31, no. 13, pp. 3359–3363, 2003.

[20] M. I. Sadowski and D. T. Jones, "The sequence-structure relationship and protein function prediction," *Current Opinion in Structural Biology*, vol. 19, no. 3, pp. 357–362, 2009.

[21] G. I. Bell, S. J. Pilkis, I. T. Weber, and K. S. Polonsky, "Glucokinase mutations, insulin secretion, and diabetes mellitus," *Annual Review of Physiology*, vol. 58, pp. 171–186, 1996.

[22] J. Molnes, L. Bjørkhaug, O. Søvik, P. R. Njølstad, and T. Flatmark, "Catalytic activation of human glucokinase by substrate binding—residue contacts involved in the binding of D-glucose to the super-open form and conformational transitions," *FEBS Journal*, vol. 275, no. 10, pp. 2467–2481, 2008.

[23] H. M. Berman, J. Westbrook, Z. Feng et al., "The protein data bank," *Nucleic Acids Research*, vol. 28, no. 1, pp. 235–242, 2000.

[24] C. H. Wu, R. Apweiler, A. Bairoch et al., "The universal protein resource (UniProt): an expanding universe of protein information," *Nucleic Acids Research*, vol. 34, pp. D187–D191, 2006.

[25] R. A. Laskowski, "PDBsum: summaries and analyses of PDB structures," *Nucleic Acids Research*, vol. 29, no. 1, pp. 221–222, 2001.

[26] R. A. Laskowski, E. G. Hutchinson, A. D. Michie, A. C. Wallace, M. L. Jones, and J. M. Thornton, "PDBsum: a web-based database of summaries and analyses of all PDB structures," *Trends in Biochemical Sciences*, vol. 22, no. 12, pp. 488–490, 1997.

[27] D. Seeliger and B. L. de Groot, "Ligand docking and binding site analysis with PyMOL and Autodock/Vina," *Journal of Computer-Aided Molecular Design*, vol. 24, no. 5, pp. 417–422, 2010.

[28] E. Ramirez, A. Cruz, D. Rodriguez et al., "Effects of active site mutations in haemoglobin i from Lucina pectinata: a molecular dynamic study," *Molecular Simulation*, vol. 34, no. 7, pp. 715–725, 2008.

10

Bioprocessing of "Hair Waste" by *Paecilomyces lilacinus* as a Source of a Bleach-Stable, Alkaline, and Thermostable Keratinase with Potential Application as a Laundry Detergent Additive: Characterization and Wash Performance Analysis

Ivana A. Cavello, Roque A. Hours, and Sebastián F. Cavalitto

Research and Development Center for Industrial Fermentations (CINDEFI) (UNLP, CONICET La Plata), Calle 47 y 115, B1900ASH La Plata, Argentina

Correspondence should be addressed to Sebastián F. Cavalitto, cavali@biotec.org.ar

Academic Editor: Michael Hust

Paecilomyces lilacinus (Thom) Samson LPS 876, a locally isolated fungal strain, was grown on minimal mineral medium containing "hair waste," a residue from the hair-saving unhairing process, and produced a protease with keratinolytic activity. This enzyme was biochemically characterized. The optimum reaction conditions, determined with a response surface methodology, were 60°C and pH 6.0. It was remarkably stable in a wide range of pHs and temperatures. Addition of Ca^{2+}, Mg^{2+}, or sorbitol was found to be effective in increasing thermal stability of the protease. PMSF and Hg^{2+} inhibited the proteolytic activity indicating the presence of a thiol-dependent serine protease. It showed high stability toward surfactants, bleaching agents, and solvents. It was also compatible with commercial detergents (7 mg/mL) such as Ariel, Skip, Drive, and Ace, retaining more than 70% of its proteolytic activity in all detergents after 1 h of incubation at 40°C. Wash performance analysis revealed that this protease could effectively remove blood stains. From these properties, this enzyme may be considered as a potential candidate for future use in biotechnological processes, as well as in the formulation of laundry detergents.

1. Introduction

Microbial proteases are the most widely exploited industrial enzymes with major application in detergent formulations [1, 2]. These enzymes are being widely used in detergent industry since their introduction in 1914 as detergent additive. Over the past 30 years, the importance of proteases in detergents has changed from being the minor additives to being the key ingredients. The main areas where use of proteases has expanded are household laundry, automatic dishwashers, and industrial and institutional cleaning. In laundry detergents, protein stains such as grass, blood, food, and human swear, are removed through proteolysis. The performance of proteases is influenced by several factors such as pH of detergent, ionic strength, wash temperature, detergent composition, bleach systems, and mechanical

handling. Thus, the key challenge for the use of enzymes in detergents is their stability. Various attempts have been made to enhance stability of alkaline proteases by site-directed mutagenesis [3] and protein engineering. "Subtilisin Carlsberg" has been protein engineered to obtain a bleach-stable, alkaline protease by molecular modification [4], but still, there is always a need for newer thermostable alkaline proteases which can withstand bleaching agents present in detergent. Among these different proteases, keratinases constitute a group of enzymes capable of disrupting the highly stable keratin structure consisting of disulfide, hydrogen, and hydrophobic bonds in the form of α-helices and β-sheets [5].

Argentine's economy has traditionally been based on agriculture and related industries. Livestock (cattle, sheep, and poultry) and grains have long been the bulwark of its wealth; its cattle herds are among the world's finest. There are more than 50 million of livestock which generate large amounts of waste including insoluble keratin-containing animal material such as feather, hair, wool, nails claws, hooves, horns, and beaks.

Although hair-saving unhairing processes reduce the organic load from beamhouse liquid effluent, a new solid residue called "hair waste" is generated, its appropriate disposal being then necessary. In a hair saving unhairing process almost 10% (wet basis) in weight of each salted bovine hide become hair waste. Because of tanning industry in Argentina processes, on average, more than 100 ton of salted hide per day, more than 10 ton of solid waste are generated per day, generating an environmental problem of considerable magnitud [6]. Since it is a protein waste, it deserves special attention in order to be utilized for practical purposes.

The fungal biotransformation of the "hair waste" implies considering it as a raw material instead of the present idea of disposability. Thus, hair waste would be the substrate, on which the fungus would act, giving rise to a (partially) hydrolyzed protein with different potential uses (i.e., as animal feed, fertilizer, etc.). In addition, the fungus would produce a proteolytic (keratinolytic) extract of biotechnological interest with a variety of potential applications (cosmetics, textiles, detergent industries, etc.). This paper deals with a particularly case of the second aspect above mentioned.

A series of studies dealing with the bioconversion of keratin waste resulted in the discovery of a novel keratinase activity in a culture supernatant of a fungal strain (*Paecilomyces lilacinus* (Thom) Samson LPS 876) grown on chicken feather as a sole of carbon, nitrogen, and energy source [7]. In this paper, we report the biochemical characterization, including the effect of some surfactants and bleaching agents on enzyme stability, its compatibility with various commercial liquid and solid detergents and a study of an efficient stabilization method toward heat inactivation, of the keratinase produced by *Paecilomyces lilacinus* growing on hair waste substrate.

A wash performance was also done with particular emphasis on its potential application as an enzyme ingredient for the formulation of laundry detergents.

2. Material and Methods

2.1. Microorganism and Identification as a Keratinolytic Fungus. Paecilomyces lilacinus (Thom) Samson LPS 876, a nonpathogenic fungal strain locally isolated from alkaline forest soils, was used. It was selected from Spegazzini Institute Fungal Culture Collection (La Plata National University, Argentina) after a preliminary screening for keratinolytic fungal strains on feather meal agar containing (per liter) the following: defatted chicken feather meal, 15 g; NaCl, 0.5 g; K_2HPO_4, 0.3 g; KH_2PO_4, 0.4 g; agar, 15 g, pH 7.2. The strain selected was punctual streaked and incubated at 28°C for 15 days. The growth of the colony and the clear zone formation around it were daily studied. The ability to degrade keratin was determined according to the presence or absence of hydrolysis halo [8].

2.2. Culture Conditions and Enzyme Production. Production of protease by *P. lilacinus* was carried out in a minimal mineral medium containing (per liter) the following: 10 g hair waste, 0.496 g NaH_2PO_4, 2.486 g K_2HPO_4, 0.016 g $FeCl_3·6 H_2O$, 0.013 g $ZnCl_2$, 0.010 g $MgCl_2$, and 0.11 mg $CaCl_2$. Hair waste, obtained from a local tannery, was washed extensively with tap water, dried at 60°C for 2 days, and then kept at room temperature until used. In all cultures, it was a sole carbon, nitrogen, and energy source. The pH was adjusted to 7.0 previous to sterilization [9]. Cultures were performed at 28°C and 200 rpm for 10 days in an orbital shaker, in 500 mL Erlenmeyer flasks containing 200 mL of medium, inoculated with 2×10^6 conidia per mL. Samples of 5 mL were withdrawn at regular intervals, centrifuged (3,000 ×g, 10 min, 4°C), and the supernatant was used for pH, protein content, and enzyme activities determinations.

2.3. Protein Determination. The protein content was determined by Bradford's method using bovine albumin fraction V (SIGMA) as a standard [10].

2.4. Protease Activity. Proteolytic activity was measured as described by Liggieri et al. [11] with some modifications. Reaction mixture containing 100 μL of appropriately diluted enzyme preparation and 250 μL of 1% (w/v) azocasein solution in 0.1 M Tris-HCl buffer (pH: 9) was incubated for 30 min at 37°C. Reaction was stopped by precipitation of the residual substrate with 1 mL of trichloroacetic acid (TCA, 10%). The mixture was kept at room temperature for 15 min and then centrifuged at 3,000 ×g (10 min, 20°C). One mL of 1 M NaOH was added to 1 mL of the supernatant, and absorbance was measured at 440 nm. Measurement was made in triplicate using a blank with a heat inactivated enzyme solution. One unit of proteolytic activity (U_C) was defined as the amount of enzyme that, under test conditions, causes an increase of 0.1 units in the absorbance at 440 nm per minute. Azocasein was synthesized as described by Riffel et al. [12].

2.5. Keratinase Activity. Keratinolytic activity was assayed as described by Joshi et al. [13] with some modifications: 800 μL 0.1 M Tris-HCl buffer (pH: 9) was added to 30 mg of azokeratin, and mixture was stirred for 15 min at room temperature until the azokeratin was completely suspended. Appropriately diluted enzyme preparation (100 μL) was added and incubated for 25 min at 37°C with individual magnetic stirring. Reaction was then stopped by the addition of 200 μL of TCA (10%) and centrifuged at 3,000 ×g (10 min, 20°C). The absorbance of the supernatant was measured at 440 nm. A blank was prepared using heat-inactivated enzyme preparation. One unit of keratinase activity (U_K) was defined as the amount of enzyme that, under test conditions, causes an increase of 0.01 units in the absorbance at 440 nm per minute. Azokeratin was synthesized as described by Joshi et al. [13] using defatted feather meal as keratin source.

The relationship between keratinolytic and proteinolytic activity (called K : C ratio) is widely used as a parameter for evaluation of the keratinolytic potential of proteases [14, 15].

K:C ratio of *P. lilacinus'* keratinase was compared with those of several commercial proteases such as proteinase-K (Promega), Alcamax (Cergen), and papain (FLUKA). Stock solutions of these commercial enzymes, in concentration of 1 mg/mL, were prepared in distilled water. They were diluted adequately, and both enzyme activities (keratinase and protease) were determined as described above.

2.6. Biochemical Characterization.

A supernatant of a 5-day-old culture was used as crude enzyme preparation for the biochemical characterization of the keratinases produced by *P. lilacinus* (2.5 U_C/mL). It is worth to mention that, for practical reasons, protease activity was used in our research to represent keratinase activity, since keratinolytic (azokeratin) and proteolytic (azocasein) activities are directly related [16], a fact that was also later confirmed for *P. lilacinus* keratinase (see below).

2.7. Effect of pH and Temperature on Enzyme Stability and Activity.

The effect of pH and temperature on enzyme activity and stability was studied using a Response Surface Methodology (RSM) based on the use of a matrix of experiments by which the simultaneous variations of the factors can be studied. Uniform shell design proposed by Doehlert was selected for design the response surface [17]. The main advantage of this procedure lies in the possibility of extending this uniform net in any direction and increasing the number of factors in the study. The real values of the independent variables were coded based on a linear functionality between codified (Z) and actual values (X) according to:

$$X : \frac{Z * \Delta X}{\Delta Z} + X_0, \qquad (1)$$

where X_0 is the real value of the central point and ΔX and ΔZ are the difference between the highest and lowest values of real and coded numbers, respectively.

Multiple regression analysis based on the least square method was performed using Mathcad 2001 software [17]. For both determinations, the central values (zero level) for the experimental designs were pH 7.5 and temperature 40°C.

The pH and thermal stability were evaluated incubating the enzyme preparation for 2 h at each chosen experimental condition. The residual protease activity was determined under standard conditions and expressed as percentage of residual activity relative to a control (measured at 0 h of incubation). Temperature varied between 20 and 60°C and pH between 3.0 and 12.0, using a mixture of buffers (Glicine, MES and Tris, 20 mM each).

The protective effect of $CaCl_2$ and $MgCl_2$ (5 mM, each) and sorbitol (10% w/v) on heat inactivation was also studied. The crude enzyme was incubated at 50–60°C with and without the chemicals mentioned above, and residual enzyme activity was measured at regular intervals under standard assay conditions.

The effect of pH and temperature on enzyme activity was determined at each condition set by the Doehlert's design. In this case, temperature varied between 20 and 60°C and pH between 6.0 and 12.0 using the same mixture of buffers described above.

2.8. Effect of Inhibitors, Metal Ions, and Organic Solvents on Enzyme Activity.

In order to investigate the effect of different inhibitors of proteases, metal ions, and organic solvents on enzyme activity, the crude enzyme was preincubated for 1 h at room temperature with different reagents. Residual enzyme activity was determined and expressed as percentage relative to a reaction control (no addition). The different reagents tested were phenylmethanesulphonyl fluoride (PMSF, 2 mM), iodoacetate (10 mM), ethylenediaminetetraacetate (EDTA, 5 mM), 1,10-Phenantroline (1 mM), and Pepstatin A (100 μg/mL), Ca^{2+}, Mg^{2+}, Zn^{2+}, and Hg^{2+} (1 mM each). The solvents tested were DMSO, ethanol, methanol, and isopropanol (1% v/v each).

2.9. Effect of Surfactants and Bleaching Agents on Enzyme Stability.

The suitability of the crude protease of *P. lilacinus* as a detergent additive was determined by testing its stability in presence of some surfactants such as SDS (sodium dodecyl sulphate), Triton X-100, and Tween 20, and bleaching agents such as hydrogen peroxide (H_2O_2) and sodium perborate. The crude protease was incubated with different concentrations of these additives for 1 h at room temperature (22°C), and then the residual proteolytic activity was measured under standard conditions against a control without any additives, which was taken as 100%.

2.10. Detergent Compatibility.

The compatibility of the protease activity in crude extract with commercial solid and liquid laundry detergents (locally available) was also studied. The solid detergents tested were Ariel (Procter & Gamble), Drive, Skip, and Ala matic (Unilever), and the liquid ones were Ace and Ariel (Procter & Gamble); also a prewashed liquid named Mr Musculo (SC Johnson & son) was tested.

Solid detergents were diluted in tap water to give a final concentration of 7 mg/mL, and liquid detergents and prewashed were diluted 100-fold to simulate washing conditions [18]. The endogenous enzymes contained in laundry detergents were inactivated by heating the diluted detergents for 1 h at 65°C prior the addition of an aliquot of crude protease. The corresponding reactions mixture were incubated in each detergents mentioned above for 1 h at different temperatures (30–50°C), and the remaining activities were determined under standard conditions. The enzyme activity of a control, incubated under the similar conditions without detergent, was taken as 100%.

2.11. Evaluation of Washing Performance.

Clean cotton cloth pieces (2.5 cm × 2.5 cm) were soiled with blood: 100 μL of blood without pretreatment was applied to cloth piece and then dried. The stained cloth pieces were subjected to wash treatments with commercial solid detergent (Skip, a solid detergent available in Argentineans' market) diluted

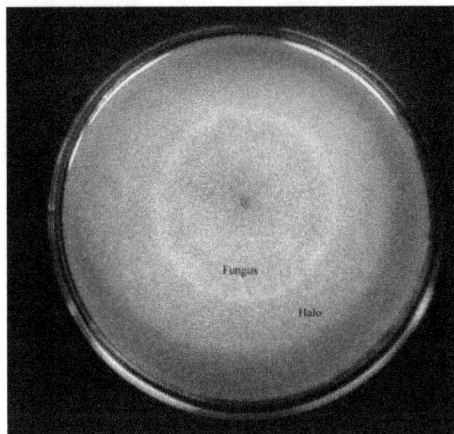

FIGURE 1: Qualitative test in feather meal agar plates for keratin degrading enzyme activity. A degradation halo surrounding the colony of *P. lilacinus* is observed.

in tap water at 7 mg/mL, supplemented with and without crude enzyme. When the wash treatment was with the supplementation of the crude enzyme, endogenous enzymes contained in laundry detergents were inactivated by heating the diluted detergents for 1 h at 65°C prior to the addition of an aliquot of crude protease.

Two stained cloth pieces were taken in separate flasks, with 50 mL as final volume, as indicated above: flask with tape water, only, flask with tap water and commercial detergent at final concentration of 7 mg/mL, and flask with tap water, commercial detergent, and crude enzyme of *P. lilacinus* ($62.5\ U_C/50\ mL$).

Each flask was incubated at two temperatures: 30 and 40°C for 30 and 60 min under agitation (200 rpm). After incubation, cloth pieces were taken out, rinsed with water, and dried. Visual examination of various pieces showed the effect of the crude enzyme in the removal of stains [19].

2.12. Statistical Analysis. All analyses were performed at least in triplicate, and data were expressed as means ± standard deviations.

3. Results and Discussion

3.1. Identification of P. lilacinus as Keratinolytic Fungus. A series of fungal strains of the Spegazzini Institute Fungal Culture Collection (La Plata National University, Argentina) were preliminary screened for their keratinolytic potential using feather meal agar. In our case, *P. lilacinus* (Thom) Samson LPS 876 was selected because, after 15 days of incubation at 28°C, a hydrolysis halo was observed indicating the keratinolytic capability of this strain (Figure 1). The use of this technique in a preliminary screening with a similar purpose was reported by Wawrzkiewicz et al. [20]. Where, among the 16 strains of dermatophytes tested, only *Trichophyton verrucosum* showed tiny fungal colonies surrounded by a wide clear zone of solubilized keratin.

3.2. Grow Profile of P. lilacinus on Hair Waste Medium. P. lilacinus LPS 876 grew well in a minimal mineral medium containing salts and hair waste as sole carbon, nitrogen, and energy source. As can be seen in Figure 2, extracellular enzyme activities (protease and keratinase) were associated with both an increment in soluble protein concentration as well as with a continuous increase in pH values in culture broth. These facts were reported by several authors for other microorganisms with high keratinolytic activity growing on this kind of substrates. The increment in pH value has been pointed out as an important indicator of the keratinolytic potential in microorganisms because of high level of deamination, with the concomitant ammonium accumulation in culture medium [21]. Moreover, Korniłłowicz-Kowlaska and Bohacz [22] concluded that substrate mass loss, a release of peptides and ammonia, sulfate excretion, and substrate alkalinization should be recognized as homogenized (due to mutual correlation) assessment parameters of the keratinolytic activity of fungi.

Enzyme activities were evaluated during the culture time course. Maximum protease and keratinase activities concurred at around 111–117 h of cultivation, (Figure 2). The ratio observed between both enzyme activities (K : C ratio) at different culture times was quite constant (11.28 ± 1.06). Therefore, the presence of a single enzyme responsible of both activities was tentatively postulated. Using the same enzyme assay conditions, it is assumed that a protease with K : C ratio higher than 5 is a true keratinase [23]. Generally, reported keratinases have a K : C ratio ranged from 5 to 20 [14, 15, 24].

The hydrolysis of azokeratin and azocasein by the proteases produced by *P. lilacinus* growing on hair waste was compared with commercial enzymes. The K : C was chosen as criterion for enzyme specificity for keratinous substrates (Table 1). As can be seen, our crude enzyme preparation was superior for hydrolysis of keratin substrate compared with other commercially available proteases. Similar results were reported by Cheng et al. [25] for the keratinase of *Bacillus licheniformis* and by Gradišar et al. [14] for the keratinases of *P. marquandii* and *Doratomyces microsporus*.

3.3. Effect of pH and Temperature on Enzyme Stability. In general, all detergent compatible enzymes are alkaline thermostable in nature with a high pH optima because the pH of laundry detergents is generally in the range of 9–12 and varying thermostability at laundry temperatures, (50−60°C) [26–28]. For the study of the effect of pH and temperature a RSM was used. The pH and temperature values used in Doehlert's design for enzyme stability determination are shown in Table 2. The central point was replicated three times in order to determine the experimental error. Data presented in Table 2 were converted into second-order polynomial equation.

Statistical analysis of the results revealed that, in the range studied, the two variables, as well as their interactions, have a significant effect on protease (keratinase) stability.

The following regression equation was obtained to calculate the percentage of residual enzyme activity (% R.A.)

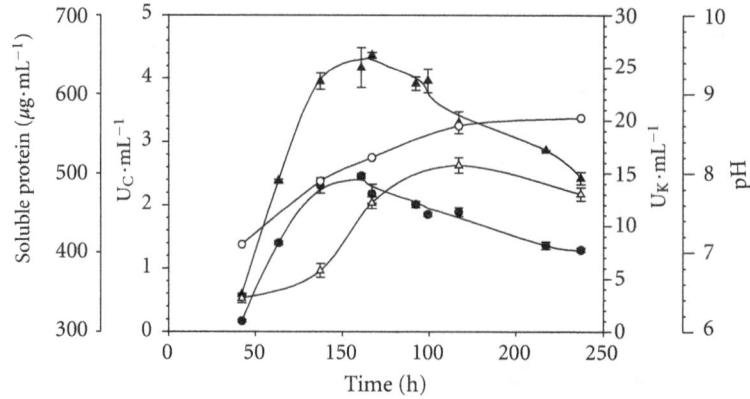

FIGURE 2: Time course of *P. lilacinus* culture using a minimal mineral medium containing $10 \, \text{g L}^{-1}$ of hair waste (pH: 7). (•) proteolytic activity; (▲) keratinolytic activity; (△) soluble protein; (○) pH.

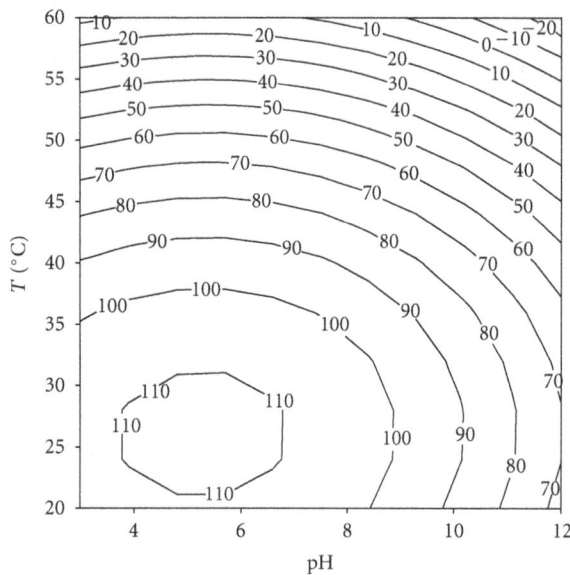

FIGURE 3: Response surface plot for the effect of pH and temperature on enzyme stability.

TABLE 1: Specific proteolytic activity (U_C/mg), keratinolytic activity (U_K/mg), and K : C ratio of crude extract and commercial proteases.

	Proteolytic activity (U_C/mg)	Keratinolytic activity (U_K/mg)	K : C ratio
Enzyme preparation	15.67	159.3	10.17
Proteinas K	33.9	218.7	6.45
Alcamax	4.6	27.4	5.96
Papain	7.6	3.9	0.51

after 2 h of incubation:

$$(\%) \, \text{R.A.} : 91.33 - 18.12 * \text{pH} - 55.43 * T - 18.83 \\ * \text{pH}^2 - 46.15 * T^2 + 0.72 * \text{pH} * T, \tag{2}$$

TABLE 2: Actual values (experimental data) and residual activity attained in Doehlert's design for pH and temperature stability.

	pH	Temperature (°C)	Residual activity (%)
	Experimental data		
A	7.5	40	93.4
A	7.5	40	89.1
A	7.5	40	92.2
B	12	40	45.1
C	9.75	60	3.25
D	5.25	60	4.5
E	3	40	104.4
F	5.25	20	106.4
G	9.75	20	104.4

where pH and T are given as codified data. The r^2 value was equal to 0.91, indicating that only 9% of the total variation was not explained by the model. The contour graph of the proteolytic activity observed as a response to the interaction of pH versus temperature is shown in Figure 3. These results indicate that the enzyme is stable in a wide range of pH and temperatures, preserving more than 60% of its activity after 2 h of incubation at pH 11 and 45°C. A serine proteinase purified from *P. lilacinus* (Thom) Samson VKM F-3891 displayed 40% of residual activity after 3 h incubation at 60°C [29]. However, in other cases, such as the serine proteinase from *Aspergillus chrysogenum*, enzyme stability in the alkaline range is substantially reduced.

Studies on thermostability of the enzyme at 50, 55, and 60°C revealed that heat inactivation displays a typical first order kinetic (Figure 4(a)). The enzyme exhibited half lives of 62, 29, and 10 min at 50, 55, and 60°C, respectively. Addition of metal ions such as $CaCl_2$ and $MgCl_2$ (5 mM) individually, improved the thermostability of the enzyme (Figures 4(b) to 4(d)). It can be seen in Figure 4(b), the apparent half-life of the enzyme increased by 1.3-fold, 1.2-fold, and 1.1-fold by the addition of Mg^{2+}, Ca^{2+}, and sorbitol, respectively. Similar results were observed at 55 and 60°C

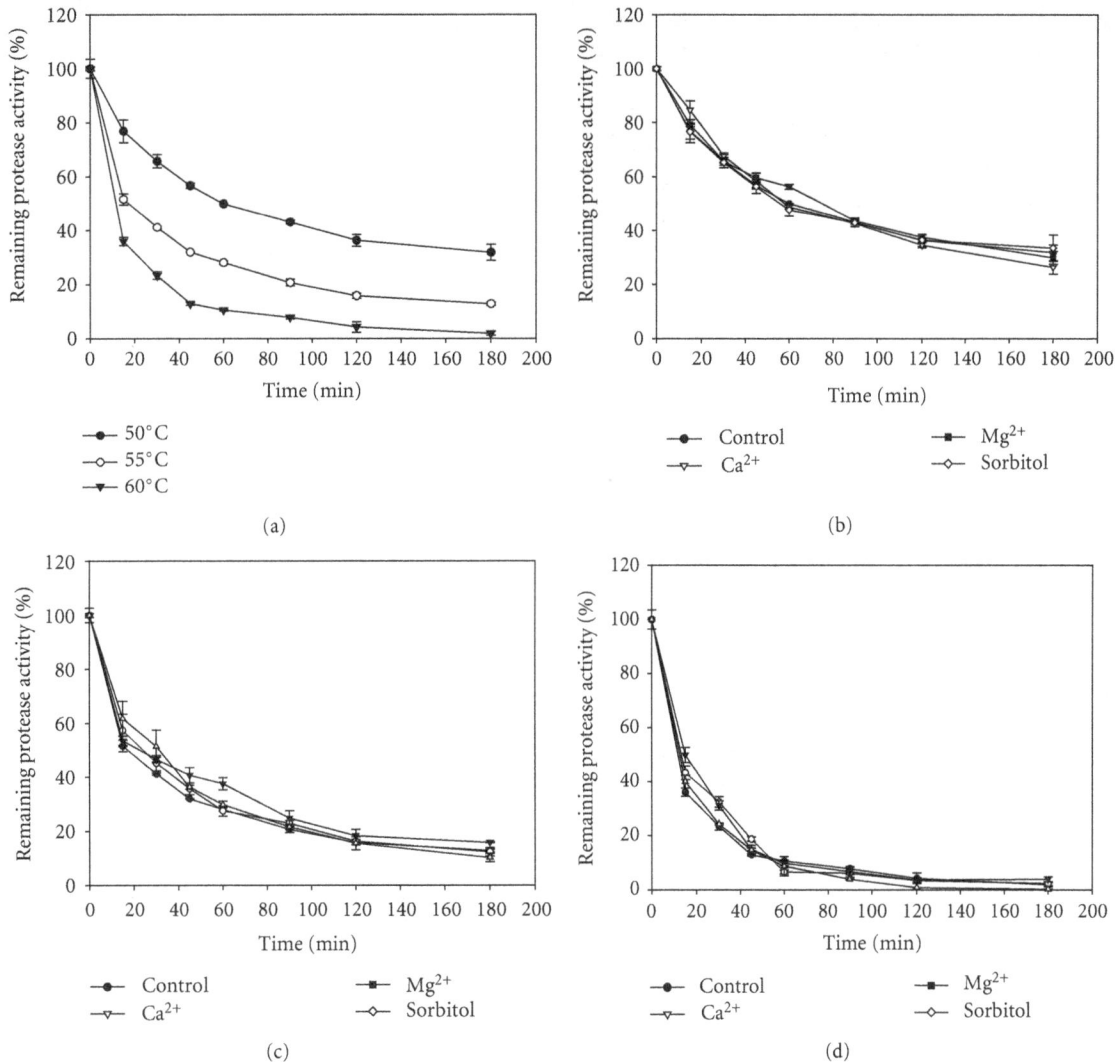

FIGURE 4: (a) Effect of temperature on protease stability (•) 50°C; (○) 55°C; (▼) 60°C. (b) Effect of stabilizers on heat inactivation at 50°C. (c) Effect of stabilizers on heat inactivation at 55°C. (d) Effect of stabilizers on heat inactivation at 60°C. For (b), (c), and (d) the original activity before preincubation was taken as 100%. Values are means of three independent determinations.

where sorbitol increased half-life by 1.2-fold. In the case of Mg^{2+}, this metal increased the apparent half-life by 1.1-fold and 1.3 fold at 55 and 60°C, respectively. In general, all chemicals tested here produced a slight increase in the thermal stability of the enzyme.

It had been reported that the addition of Ca^{2+} or polyhydric alcohols, such as glycerol and polyethylene glycol caused an increase in thermal stability of alkaline proteases. The addition of sorbitol improved the thermal stability for an alkaline protease from *B. cereus* BG1, which increased its thermal stability by approximately 2-fold at 60°C [30]. Kelkar and Deshpande [31] studied the influence of various polyols on the thermostability of pullulan-hydrolysing activity from *Sclerotium rolfsii*. The half-life of the enzyme activity at 60°C was about 30 min, and in presence of xylitol and sorbitol (3 M) they reported a significant enhancement in the thermostability of the enzyme with retention of 100%

activity after incubation for 7 h at 60°C. Ghorbel et al. [30] reported an alkaline protease that, in the presence of 10 mM Ca^{2+}, retained 100, 93, and 26% of its initial activity after heating for 15 min at 55, 60, and 70°C, respectively. However, the enzyme was completely inactivated when incubated at 55°C for 15 min in the absence of calcium. On the contrary, the enzyme reported here in presence of 5 mM Ca^{2+} retained about 57 and 43% of its initial activity after heating 15 min at 55 and 60°C, respectively, but the enzyme in absence of calcium retained more than 25% of its initial activity after 60 min of incubation at 55°C.

Several reports showed that the addition of various additives such as polyols and PEG could enhance enzymes' thermal stability [2, 18]. The increase in the thermal stability by adding these such as additives was probably due to the reinforcement of the hydrophobic interactions among nonpolar amino acids inside the enzyme molecules and thus

TABLE 3: Actual values (experimental data) and enzyme activity ($U_C \cdot mL^{-1}$) attained in Doehlert's design for the study of the effect of pH and Temperature on enzyme activity.

	pH	Temperature (°C)	$U_C \cdot mL^{-1}$
		Experimental data	
A	9	40	6.5
A	9	40	6.4
A	9	40	6.5
B	12	40	6.6
C	10.5	60	5
D	7.5	60	11.6
E	6	40	6
F	7.5	20	1.2
G	10.5	20	1.2

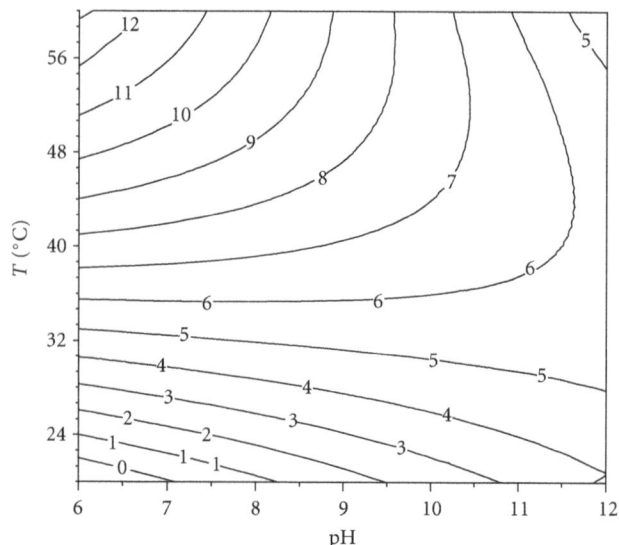

FIGURE 5: Response surface plot for the effect of pH and temperature on enzyme activity.

increased their resistance to inactivation, since it had been reported that polyols modify the structure of water and/or strengthen hydrophobic interactions among nonpolar amino acids inside the protein molecules [32].

The effect of Ca^{2+} on the improvement of thermal stability against heat inactivation may be explained by the strengthening of interactions inside proteases molecules and by binding of this metal to the autolysis site. The activity of *B. mojavensis* protease was enhanced not only by Ca^{2+} but by Fe^{2+} and Mn^{2+}. It is believed that metal ions protect the enzyme against thermal denaturation and play a vital role in maintaining the active conformation of the enzyme at higher temperatures [33].

3.4. Effect of pH and Temperature on Enzyme Activity. The pH and temperature values used in Doehlert's design for determination of the effect of pH and temperature on enzyme activity are given in Table 3.

The following regression equation was obtained to calculate the enzyme activity (E.A):

$$EA \left(U \cdot mL^{-1} \right) : 6.48 - 0.928 * pH + 4.12 * T - 0.188$$

$$* pH^2 - 2.23 * T^2 - 3.835 * pH * T, \tag{3}$$

where pH and T are given as codified data. The r^2 value was equal to 0.89, indicating that only 11% of the total variation was not explained by the model. The contour graph of the keratinolytic activity observed as a response to the interaction of pH versus temperature is shown in Figure 5. These results indicate an optimal pH and temperature of 6.0 and 60°C, respectively. Kotlova et al. [29] reported a thiol-dependent serine proteinase from *P. lilacinus* (Thom) Samson VKM F-3891 with an optimal pH into the alkaline range. Such pH dependence is also reported by Bonants et al. [34] for another proteinase isolated from a different strain of the fungus, *P. lilacinus* (Thom) Samson (CBS 143.75). These proteinases seem to belong to the subgroup of subtilisin-like enzymes with an optimum pH in the alkaline range (pH 10–12).

In our case, the protease present in the enzyme extract belongs to the group of enzymes, with an optimum pH in a neutral range (pH 6–8) at an assay temperature of 60°C. Although optimal temperature for keratin degradation is 60°C, the stability of the enzyme under this condition is not so high. Because of that, a temperature of 55°C, where the enzyme retains more than 90% of the optimal activity, is proposed for practical applications.

3.5. Effect of Protease Inhibitors, Metal Ions, and Organic Solvents on Enzyme Activity. The effect of various chemical reagents and metal ions on enzyme activity with azocasein as substrate is shown in Table 4. The enzyme activity was strongly inhibited by PMSF (93% of inhibition), a well-known inhibitor of serine proteinases, in particular subtilisins serine proteinases [35]. It was slightly inhibited by Pepstatin A but neither by iodoacetate, EDTA, nor by 1,10-Phenantroline. These facts suggest that in our case, the enzyme produced by *P. lilacinus* corresponds to a serine protease. Actually, most keratinases described until now are classified into this category [36]. The stability of the enzyme in presence of EDTA is advantageous for its use as a detergent additive. An enzyme, which is to be used as a detergent additive, should not have requirement for a metal cofactor. This is because detergent formulations contain high amounts of chelating agents, which specifically bind to and chelate metal ions making them unavailable in the detergent solution. The chelating agents remove the divalent cations responsible for water hardness and also assist in stain removal.

Among the metal ions tested, Hg^{2+} strongly inhibited proteolytic activity (94% of inhibition), whereas Ca^{2+} and Mg^{2+} caused slight activation. Hg^{2+} is recognized as an oxidant agent of thiol-groups, and the enzyme inhibition by this ion suggests the presence of an important-SH group (such as free cysteine) at/or near the active site [37]. In

TABLE 4: Effect of protease inhibitors, metal ions, detergents, and solvents on protease activity (data are given as Residual activity (%) ± SD).

Chemical none	Concentration	Residual activity (%) 100
Inhibitor		
PMSF	2 mM	7.0 ± 0.0
Iodoacetate	10 mM	95.1 ± 4.7
EDTA	5 mM	99.6 ± 6.3
1,10-Phenantroline	1 mM	100 ± 0.2
Pepstatin A	100 μg/mL	87.5 ± 5.5
Metal ion		
Mg^{2+}	1 mM	105.0 ± 1.2
Zn^{2+}	1 mM	92.8 ± 1.4
Ca^{2+}	1 mM	102.9 ± 0.5
Hg^{2+}	1 mM	6.0 ± 0.6
Detergents		
Triton X-100	0.5% (v/v)	97.7 ± 2.7
Tween 20	0.5% (v/v)	98.5 ± 2.0
SDS	0.5% (v/v)	75.9 ± 3.4
Bleaching agent		
	1% (w/v)	140 ± 2.3
H_2O_2	2%	137 ± 0.3
	3%	122.7 ± 4.0
	0.2% (w/v)	99.7 ± 2.4
Sodium perborate	0.5%	97.6 ± 0.9
	1.0%	90.8 ± 2.0
Solvent		
DMSO	1% (v/v)	106.0 ± 1.6
Ethanol	1% (v/v)	105.4 ± 4.0
Methanol	1% (v/v)	117.0 ± 2.9
Isopropanol	1% (v/v)	98.9 ± 1.7

addition, the strong inactivation by Hg^{2+} is typical for proteinases belonging to the thermitase and proteinase K subgroups [38]. This feature makes our *P. lilacinus* keratinase substantially different to the true bacillary subtilisins, as well as to the serine proteinase from *P. lilacinus* isolated by Bonants et al. [34], which is not inactivated by Hg^{2+}. Based on the presence of functionally important sulfhydryl groups, our keratinase resembles proteinase K and bacillary thiol-dependent subtilisins much more than other fungal serine proteases.

Divalent metal ions such as Ca^{2+} and Mg^{2+} slightly activated the enzyme. This fact could be explained because of that they could act as salt or ions bridges stabilizing the enzyme under its active conformation, and thus they might protect the enzyme against thermal denaturation [39, 40].

Crude enzyme preparation showed to be highly stable in presence of different organic solvents such as DMSO, ethanol, methanol, and isopropanol (Table 4), a positive fact considering the potential practical application.

3.6. Effect of Surfactants and Bleaching Agents on Enzyme Stability. The above-mentioned characteristics of our *P. lilacinus* protease suggested its potential use in different applications like laundry detergent formulation. In order to be effective during washing, a good detergent protease must be compatible and stable with all commonly used detergent compounds such as surfactants, bleaching agents, and other additives, which might be present in detergent formulation [1]. In our case, a crude protease extract was incubated 60 min at room temperature in presence of several additives, and then the residual protease activity was assayed under standard conditions.

As can be seen in Table 4, crude protease was highly stable in presence of nonionic surfactants. It retained near 100% of its initial activity in presence of 0.5% Triton X-100 and 0.5% Tween 20. In presence of 0.5% of SDS, a strong anionic surfactant, it exhibited moderated stability (75%) after 1 h of incubation. SDS is known to be a strong denaturant of proteins including alkaline proteases. It could unfold most proteins through the interaction between the charged head group of SDS and the positively charged amino acid side chains of proteins and between the alkyl chain of SDS and the nonpolar parts on the surface as well as in the interior of proteins [41]. The retention of protease activity by our enzyme preparation in presence of SDS was higher than that of a protease from *Aspergillus clavatus* ES1, which retained only 33% of its activity under the same stability assay conditions [42].

On the other hand, our enzyme preparation showed excellent stability toward bleaching agents such as H_2O_2 and sodium perborate (Table 4). It showed an stability similar to proteases from *B. licheniformis* NH1, which retained 85% and 80% of its activity after incubation with 0.5% H_2O_2 [19] and resulted in being more stable than an alkaline protease from *B. licheniformis* RP1 [43] which is less stable against bleaching agents; it's just retained 68% and 48% of its activity after 1 h incubation at 40°C in presence of 2% H_2O_2 and 0.2% sodium perborate, respectively. Bleaching agents inactivate proteins oxidatively, being Met the primary site for oxidative inactivation. All subtilisins contain a Met residue next to the catalytic Ser residue, so that many of them tend to undergo oxidative inactivation in presence of a bleaching agent such as hydrogen peroxide. Thus, many of available alkaline proteases exhibited low activity and stability toward oxidants, which are common ingredients in modern bleach-based detergents. To overcome these shortcomings, several attempts have been made to enhance enzyme stability by protein engineering [44]. That is why it is important to obtain enzymes with high stability against surfactants and oxidants for practical applications. Detergent applications for keratinases have been also suggested [36]. These include removal of keratinous dirt that are often encountered in the laundry, such as collar of shirts and used as additives for cleaning up drains clogged with keratinous waste.

3.7. Detergent Compatibility. All the commercial detergents contain hydrolytic enzymes, and these enzymes-based detergents known as "green chemicals" find a wide range of

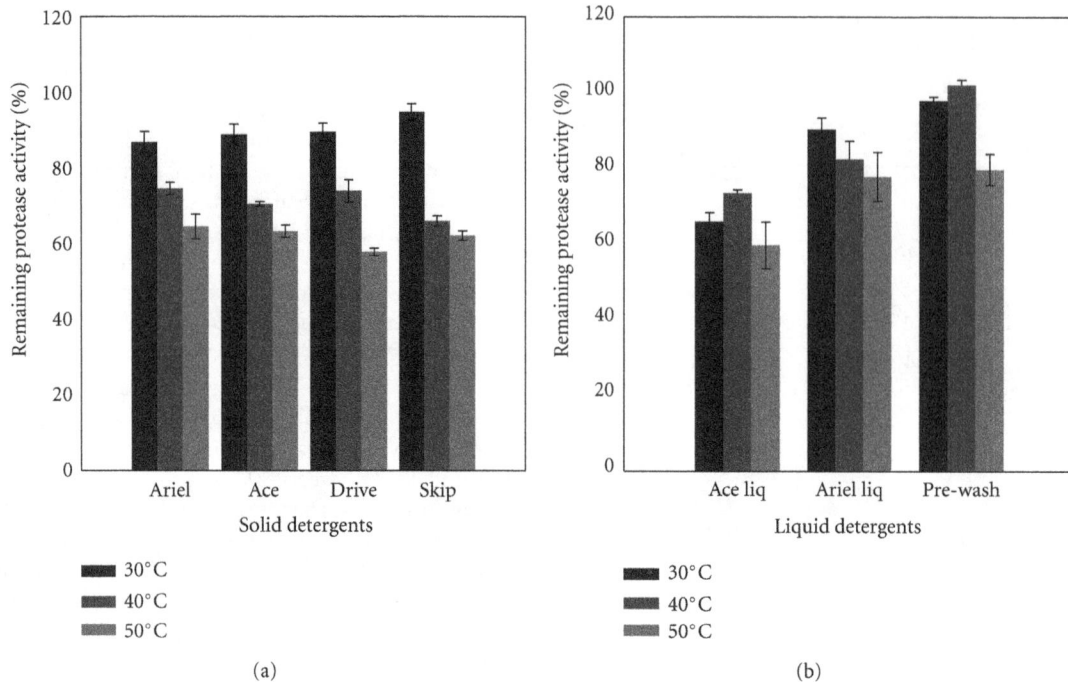

FIGURE 6: Stability of the crude enzyme in the presence of various commercial solid (a) and liquid detergents (b). CE was incubated in each detergent mentioned for 1 h at different temperatures (30–50°C), and the remaining activities were determined under standard conditions. The enzyme activity of a control, incubated under similar conditions without detergent, was taken as 100%.

applications in laundry, dishwashing, textile, and other related industries [45].

In order to check the compatibility with liquid and solid detergents, the crude enzyme was preincubated in presence of various commercial laundry detergents for 1 h at 30, 40, and 50°C. Solid detergents were diluted in tap water to a final concentration of 7 mg/mL, and the liquid ones were diluted 100-fold to simulate washing conditions. As can be seen in Figure 6(a) the crude enzyme was very stable towards all solid detergents tested, even at 50°C after 1 h of incubation, it retained more than 60% of its activity in presence of Ariel, Ace, and Skip. In presence of Drive, it retained about 57% of its activity being more stable than an alkaline protease from *Vibrio fluvialis* strain VM10 reported by Venugopal and Saramma [46], which retained just 42% of its activity in presence of Ariel as well as an alkaline serine protease from *Bacillus sp.* SSR1 reported by Singh et al. [47] which retain lees than 40% of its activity after 1 h of incubation in presence of Ariel at 40°C. Interestingly, it was more stable than the commercial protease named Maxacal, which retained less than 60% after 1 h of incubation in presence of Ariel at 40°C [47]. Similarly, proteases from *B. mojavensis* A21 [48] and from *B. licheniformis* RP1 [43] are shown to retain more than 40 and 80% of their activity in presence of Dixan after 1 h of incubation at 50°C, respectively.

In presence of liquid detergents, the crude enzyme retained more than 75% and 50% of its initial activity in presence of Ariel and Ace, respectively, after 1 h of incubation at 50°C (Figure 6(b)).

From the results presented here about the compatibility and stability whit commercial detergents at different temperatures, it can be concluded that our protease will be more effective at temperature from 30 to 40°C for long washing cycles (60 min) and at 50°C for short washing cycles (10–30 min). But with Ariel liq. long washing cycles could be done at 55°C too, because it retained about 78% of its original stability.

3.8. Wash Performance Analysis. The wash performances of the protease present in the crude extract were assessed by its ability to remove blood stain from white cotton cloth (Figure 7). Enzyme in combination of the commercial detergent Skip was tried. The visual comparison of the washed cloth revealed that washing with distilled water at temperatures of 30 and 40°C removed some amount of blood stain from the cotton cloth (Figures 7(a), 7(d), 7(g), and 7(j)). As can be seen, the replacement of the enzymes present in the commercial detergent by the crude enzyme gave a complete blood stain elimination at both temperatures and times, such as the endogenous enzymes from the commercial detergent has done (Figures 7(c), 7(f), 7(i), and 7(l)). These results show the efficiency of *P. lilacinus* protease in proteinaceous stain removal efficiency.

Abidi et al. [49] showed also the significant improvement of the supplementation of proteolytic preparation of *Botrytis cinerea*, in a laundry detergent (Henkel-Alki), in the elimination of blood, egg yolk, and chocolate stains on fabric. Jellouli et al. [50] and Savitha et al. [51] reported similar results in

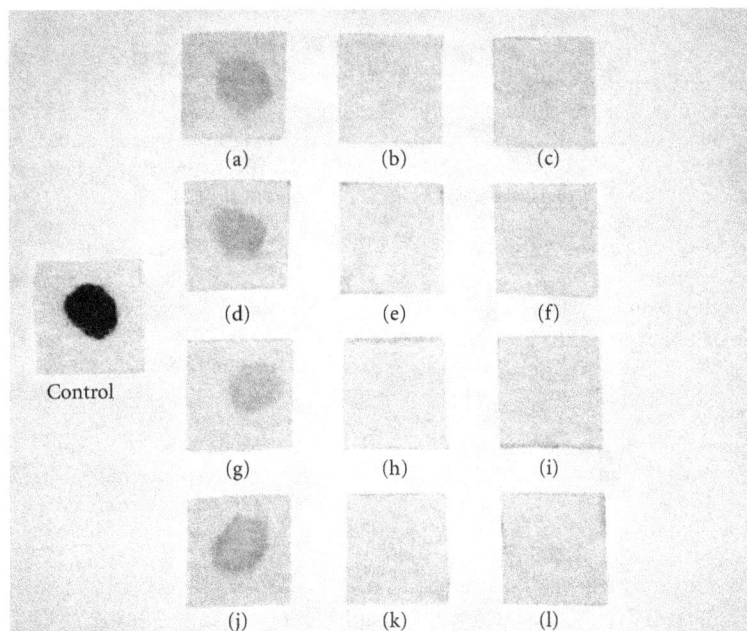

FIGURE 7: Washing performance analysis of the *P. lilacinus* enzyme preparation in the presence of the commercial detergent Skip. Analysis were done at 30°C ((a)–(f)) and 40°C ((g)–(l)), for 30 m ((a)–(c) and (g)–(i)) and 60 m ((d)–(f) and (j)–(l)). Cloth washed with tap water ((a), (d), (g), and (j)); ((b), (e), (h), and (k)) cloth washed with Skip; ((c), (f), (i), and (l)) cloth washed with Skip added with crude enzyme of the *P. lilacinus* protease.

their wash performance tests. Therefore, *P. lilacinus* crude extracts containing protease activity could be considered as a potential candidate for use as cleaning additive in detergents to facilitate the release of proteinaceous stains.

4. Conclusions

A locally isolated *P. lilacinus* strain produces an extracellular protease with keratinase activity when grown on hair waste, the main solid-wastes produced in tanneries, as substrate in submerged cultures.

The protease was characterized, and it exhibited remarkable stability toward surfactants, bleaching agents, and detergent additives like EDTA and sodium perborate. This property of the enzyme is very essential for its application as detergent additive. Our study shows that the extracellular proteolytic enzyme produced by this strain could have an industrial application in detergent industries. Moreover, the enzyme was compatible with most of the laundry detergents tested and showed a good washing performance.

Acknowledgment

This research work was supported by CONICET (Argentina National Council of Scientific and Technical Research).

References

[1] R. Gupta, Q. Beg, and P. Lorenz, "Bacterial alkaline proteases: molecular approaches and industrial applications," *Applied Microbiology and Biotechnology*, vol. 59, no. 1, pp. 15–32, 2002.

[2] M. B. Rao, A. M. Tanksale, M. S. Ghatge, and V. V. Deshpande, "Molecular and biotechnological aspects of microbial proteases," *Microbiology and Molecular Biology Reviews*, vol. 62, no. 3, pp. 597–635, 1998.

[3] K. Tsuchiya, Y. Nakamura, H. Sakashita, and T. Kimura, "Purification and characterization of a thermostable alkaline protease from alkalophilic *Thermoactinomyces* sp. HS 682," *Bioscience, Biotechnology, and Biochemistry*, vol. 56, no. 2, pp. 246–250, 1992.

[4] A. M. Wolff, M. S. Showell, M. G. Venegas et al., "Laundry performance of subtilisin proteases," in *Subtilisin Enzymes: Practical Protein Engineering*, R. Bott and C. Betzel, Eds., pp. 113–120, Springer, New York, NY, USA, 1996.

[5] D. A. D. Parry and A. C. T. North, "Hard α-keratin intermediate filament chains: substructure of the N-and C-terminal domains and the predicted structure and function of the C-terminal domains of type I and type II chains," *Journal of Structural Biology*, vol. 122, no. 1-2, pp. 67–75, 1998.

[6] B. C. Galarza, I. Cavello, C. A. Greco, R. Hours, M. M. Schuldt, and C. S. Cantera, "Alternative technologies for adding value to bovine hair waste," *Journal of the Society of Leather Technologies and Chemists*, vol. 94, no. 1, pp. 26–32, 2010.

[7] I. Cavello, S. Cavalitto, and R. Hours, "Biodegradation of a keratin waste and the concomitant production of detergent stable serine proteases from *Paecilomyces lilacinus*," *Applied Biochemistry and Biotechnology*, vol. 167, no. 5, pp. 945–958, 2012.

[8] S. Sangali and A. Brandelli, "Isolation and characterization of a novel feather-degrading bacterial strain," *Applied Biochemistry and Biotechnology A*, vol. 87, no. 1, pp. 17–24, 2000.

[9] B. C. Galarza, L. M. Goya, C. S. Cantera, M. L. Garro, H. E. Reinos, and L. M. I. López, "Fungal biotransformation of bovine hair part 1: isolation of fungus with keratinolytic

activity," *Journal of the Society of Leather Technologies and Chemists*, vol. 88, no. 3, pp. 93–98, 2004.

[10] M. M. Bradford, "A rapid and sensitive method for the quantitation of microgram quantities of protein utilizing the principle of protein dye binding," *Analytical Biochemistry*, vol. 72, no. 1-2, pp. 248–254, 1976.

[11] C. Liggieri, M. C. Arribére, S. A. Trejo, F. Canals, F. X. Avilés, and N. S. Priolo, "Purification and biochemical characterization of asclepain c I from the latex of *Asclepias curassavica* L.," *Protein Journal*, vol. 23, no. 6, pp. 403–411, 2004.

[12] A. Riffel, F. Lucas, P. Heeb, and A. Brandelli, "Characterization of a new keratinolytic bacterium that completely degrades native feather keratin," *Archives of Microbiology*, vol. 179, no. 4, pp. 258–265, 2003.

[13] S. G. Joshi, M. M. Tejashwini, N. Revati et al., "Isolation, identification and characterization of a feather degrading bacterium," *International Journal of Poultry Science*, vol. 6, no. 9, pp. 689–693, 2007.

[14] H. Gradišar, J. Friedrich, I. Krizaj et al., "Similarities and specificities of fungal keratinolytic proteases: comparison of keratinases of paecilomyces marquandii and doratomyces microsporus to some known proteases," *Applied and Environmental Microbiology*, vol. 71, no. 7, pp. 3420–3426, 2005.

[15] S. Sangali and A. Brandelli, "Feather keratin hydrolysis by a Vibrio sp. strain kr2," *Journal of Applied Microbiology*, vol. 89, no. 5, pp. 735–743, 2000.

[16] A. P. F. Corrêa, D. J. Daroit, and A. Brandelli, "Characterization of a keratinase produced by *Bacillus* sp. P7 isolated from an Amazonian environment," *International Biodeterioration and Biodegradation*, vol. 64, no. 1, pp. 1–6, 2010.

[17] S. F. Cavalitto and C. F. Mignone, "Application of factorial and Doehlert designs for optimization of protopectinase production by a *Geotrichum klebahnii* strain," *Process Biochemistry*, vol. 42, no. 2, pp. 175–179, 2007.

[18] A. Haddar, A. Sellami-Kamoun, N. Fakhfakh-Zouari, N. Hmidet, and M. Nasri, "Characterization of detergent stable and feather degrading serine proteases from *Bacillus mojavensis* A21," *Biochemical Engineering Journal*, vol. 51, no. 1-2, pp. 53–63, 2010.

[19] N. Hmidet, N. El-Hadj Ali, A. Haddar, S. Kanoun, S. K. Alya, and M. Nasri, "Alkaline proteases and thermostable α-amylase co-produced by *Bacillus licheniformis* NH1: characterization and potential application as detergent additive," *Biochemical Engineering Journal*, vol. 47, no. 1–3, pp. 71–79, 2009.

[20] K. Wawrzkiewicz, T. Wolski, and J. Łobarzewski, "Screening the keratinolytic activity of dermatophytes in vitro," *Mycopathologia*, vol. 114, no. 1, pp. 1–8, 1991.

[21] A. Riffel and A. Brandelli, "Keratinolytic bacteria isolated from feather waste," *Brazilian Journal of Microbiology*, vol. 37, no. 3, pp. 395–399, 2006.

[22] T. Korniłłowicz-Kowalska and J. Bohacz, "Biodegradation of keratin waste: theory and practical aspects," *Waste Management*, vol. 31, no. 8, pp. 1689–1701, 2011.

[23] R. Sharma and R. Gupta, "Thermostable, thiol activated keratinases from *Pseudomonas aeruginosa* KS-1 for prospective application in prion decontamination," *Research Journal of Microbiology*, vol. 5, no. 10, pp. 954–965, 2010.

[24] S. Yamamura, Y. Morita, Q. Hasan, K. Yokoyama, and E. Tamiya, "Keratin degradation: a cooperative action of two enzymes from *Stenotrophomonas* sp.," *Biochemical and Biophysical Research Communications*, vol. 294, no. 5, pp. 1138–1143, 2002.

[25] S. W. Cheng, H. M. Hu, S. W. Shen, H. Takagi, M. Asano, and Y. C. Tsai, "Production and characterization of keratinase of a feather-degrading *Bacillus licheniformis* PWD-1," *Bioscience, Biotechnology, and Biochemistry*, vol. 59, no. 12, pp. 2239–2243, 1995.

[26] H. Takami, T. Akiba, and K. Horikoshi, "Production of extremely thermostable alkaline protease from *Bacillus* sp. no. AH-101," *Applied Microbiology and Biotechnology*, vol. 30, no. 2, pp. 120–124, 1989.

[27] S. H. Bhosale, M. B. Rao, V. V. Deshpande, and M. C. Srinivasan, "Thermostability of high-activity alkaline protease from *Conidiobolus coronatus* (NCL 86.8.20)," *Enzyme and Microbial Technology*, vol. 17, no. 2, pp. 136–139, 1995.

[28] U. C. Banerjee, R. K. Sani, W. Azmi, and R. Soni, "Thermostable alkaline protease from *Bacillus brevis* and its characterization as a laundry detergent additive," *Process Biochemistry*, vol. 35, no. 1-2, pp. 213–219, 1999.

[29] E. K. Kotlova, N. M. Ivanova, M. P. Yusupova, T. L. Voyushina, N. E. Ivanushkina, and G. G. Chestukhina, "Thiol-dependent serine proteinase from *Paecilomyces lilacinus*: purification and catalytic properties," *Biochemistry*, vol. 72, no. 1, pp. 117–123, 2007.

[30] B. Ghorbel, A. Sellami-Kamoun, and M. Nasri, "Stability studies of protease from *Bacillus cereus* BG1," *Enzyme and Microbial Technology*, vol. 32, no. 5, pp. 513–518, 2003.

[31] H. S. Kelkar and M. V. Deshpande, "Effect of polyols on the thermostability of pullulan-hydrolysing activity from *Sclerotium rolfsii*," *Biotechnology Letters*, vol. 13, no. 12, pp. 901–906, 1991.

[32] J. F. Back, D. Oakenfull, and M. B. Smith, "Increased thermal stability of proteins in the presence of sugars and polyols," *Biochemistry*, vol. 18, no. 23, pp. 5191–5196, 1979.

[33] D. R. Durham, D. B. Stewart, and E. J. Stellwag, "Novel alkaline- and heat-stable serine proteases from alkalophilic *Bacillus* sp. strain GX6638," *Journal of Bacteriology*, vol. 169, no. 6, pp. 2762–2768, 1987.

[34] P. J. M. Bonants, P. F. L. Fitters, H. Thijs, E. den Belder, C. Waalwijk, and J. W. D. M. Henfling, "A basic serine protease from *Paecilomyces lilacinus* with biological activity against *Meloidogyne hapla* eggs," *Microbiology*, vol. 141, no. 4, pp. 775–784, 1995.

[35] R. J. Siezen and J. A. M. Leunissen, "Subtilases: the superfamily of subtilisin-like serine proteases," *Protein Science*, vol. 6, no. 3, pp. 501–523, 1997.

[36] R. Gupta and P. Ramnani, "Microbial keratinases and their prospective applications: an overview," *Applied Microbiology and Biotechnology*, vol. 70, no. 1, pp. 21–33, 2006.

[37] D. J. Daroit, A. Simonetti, P. F. Hertz, and A. Brandelli, "Purification and characterization of an extracellular β-glucosidase from *Monascus purpureus*," *Journal of Microbiology and Biotechnology*, vol. 18, no. 5, pp. 933–941, 2008.

[38] M. Ballinger and J. A. Wells, "Subtilisin," in *Handbook of Proteases*, A. Barrett, N. D. Rawlings, and J. F. Woessner, Eds., pp. 289–294, 1998.

[39] S. Balaji, M. S. Kumar, R. Karthikeyan et al., "Purification and characterization of an extracellular keratinase from a hornmeal-degrading *Bacillus subtilis* MTCC (9102)," *World Journal of Microbiology and Biotechnology*, vol. 24, no. 11, pp. 2741–2745, 2008.

[40] R. Sareen and P. Mishra, "Purification and characterization of organic solvent stable protease from *Bacillus licheniformis* RSP-09-37," *Applied Microbiology and Biotechnology*, vol. 79, no. 3, pp. 399–405, 2008.

[41] D. E. Otzen, "Protein unfolding in detergents: effect of micelle structure, ionic strength, pH, and temperature," *Biophysical Journal*, vol. 83, no. 4, pp. 2219–2230, 2002.

[42] M. Hajji, S. Kanoun, M. Nasri, and N. Gharsallah, "Purification and characterization of an alkaline serine-protease produced by a new isolated *Aspergillus clavatus* ES1," *Process Biochemistry*, vol. 42, no. 5, pp. 791–797, 2007.

[43] A. Sellami-Kamoun, A. Haddar, N. E. H. Ali, B. Ghorbel-Frikha, S. Kanoun, and M. Nasri, "Stability of thermostable alkaline protease from *Bacillus licheniformis* RP1 in commercial solid laundry detergent formulations," *Microbiological Research*, vol. 163, no. 3, pp. 299–306, 2008.

[44] D. A. Estell, T. P. Graycar, and J. A. Wells, "Engineering an enzyme by site-directed mutagenesis to be resistant to chemical oxidation," *The Journal of Biological Chemistry*, vol. 260, no. 11, pp. 6518–6521, 1985.

[45] A. K. Mukherjee, M. Borah, and S. K. Rai, "To study the influence of different components of fermentable substrates on induction of extracellular α-amylase synthesis by *Bacillus subtilis* DM-03 in solid-state fermentation and exploration of feasibility for inclusion of α-amylase in laundry detergent formulations," *Biochemical Engineering Journal*, vol. 43, no. 2, pp. 149–156, 2009.

[46] M. Venugopal and A. V. Saramma, "Characterization of alkaline protease from *Vibrio fluvialis* strain VM10 isolated from a mangrove sediment sample and its application as a laundry detergent additive," *Process Biochemistry*, vol. 41, no. 6, pp. 1239–1243, 2006.

[47] J. Singh, N. Batra, and R. C. Sobti, "Serine alkaline protease from a newly isolated *Bacillus* sp. SSR1," *Process Biochemistry*, vol. 36, no. 8-9, pp. 781–785, 2001.

[48] N. Fakhfakh-Zouari, A. Haddar, N. Hmidet, F. Frikha, and M. Nasri, "Application of statistical experimental design for optimization of keratinases production by *Bacillus pumilus* A1 grown on chicken feather and some biochemical properties," *Process Biochemistry*, vol. 45, no. 5, pp. 617–626, 2010.

[49] F. Abidi, F. Limam, and M. M. Nejib, "Production of alkaline proteases by *Botrytis cinerea* using economic raw materials: assay as biodetergent," *Process Biochemistry*, vol. 43, no. 11, pp. 1202–1208, 2008.

[50] K. Jellouli, O. Ghorbel-Bellaaj, H. B. Ayed, L. Manni, R. Agrebi, and M. Nasri, "Alkaline-protease from *Bacillus licheniformis* MP1: purification, characterization and potential application as a detergent additive and for shrimp waste deproteinization," *Process Biochemistry*, vol. 46, no. 6, pp. 1248–1256, 2011.

[51] S. Savitha, S. Sadhasivam, K. Swaminathan, and F. H. Lin, "Fungal protease: production, purification and compatibility with laundry detergents and their wash performance," *Journal of the Taiwan Institute of Chemical Engineers*, vol. 42, no. 2, pp. 298–304, 2011.

11

Family-Specific Degenerate Primer Design: A Tool to Design Consensus Degenerated Oligonucleotides

Javier Alonso Iserte,[1] **Betina Ines Stephan,**[1] **Sandra Elizabeth Goñi,**[1] **Cristina Silvia Borio,**[1] **Pablo Daniel Ghiringhelli,**[2] **and Mario Enrique Lozano**[1]

[1] *LIGBCM-Área Virosis Emergentes y Zoonóticas, Universidad Nacional de Quilmes, B1876BXD Buenos Aires, Argentina*
[2] *LIGBCM-Área Virosis de Insectos, Universidad Nacional de Quilmes, B1876BXD Buenos Aires, Argentina*

Correspondence should be addressed to Javier Alonso Iserte; jiserte@unq.edu.ar

Academic Editor: Goetz Laible

Designing degenerate PCR primers for templates of unknown nucleotide sequence may be a very difficult task. In this paper, we present a new method to design degenerate primers, implemented in family-specific degenerate primer design (FAS-DPD) computer software, for which the starting point is a multiple alignment of related amino acids or nucleotide sequences. To assess their efficiency, four different genome collections were used, covering a wide range of genomic lengths: *Arenavirus* (10×10^4 nucleotides), *Baculovirus* (0.9×10^5 to 1.8×10^5 bp), *Lactobacillus* sp. (1×10^6 to 2×10^6 bp), and *Pseudomonas* sp. (4×10^6 to 7×10^6 bp). In each case, FAS-DPD designed primers were tested computationally to measure specificity. Designed primers for *Arenavirus* and *Baculovirus* were tested experimentally. The method presented here is useful for designing degenerate primers on collections of related protein sequences, allowing detection of new family members.

1. Introduction

The polymerase chain reaction (PCR), one of the most important analytical tools of molecular biology, allows a highly sensitive detection and specific genotyping of environmental samples, specially important in the metagenomic era [1]. A large list of genome typing applications includes arbitrarily primed PCR [2] (AP-PCR), random amplified primed DNAs [3] (RAPDs), PCR restriction fragment length polymorphism [4] (PCR-RFLP), and direct amplification of length polymorphism [5] (DALP). All of these techniques require a high quality and purity of the specific target template, because any available DNA could be substrate for the amplification step. In view of this, genotyping procedures of large genomes or complex samples are more reliable if they are based on DNA amplification using specific oligonucleotides. Therefore, primer design is crucial for efficient and successful amplification.

Several primer design programs are available (e.g., OLIGO [6], OSP [7, 8], Primer Master [9], PRIDE [10],

Primer3 [11], among others). Regardless of each computational working strategy, all of these use a set of common criteria (e.g., G/C content, melting temperature, etc.) to evaluate the quality of primer candidates in a specific target region selected by the user. Alternative programs are aimed at more specific purposes, such as selection of primers that bind to conserved genomic regions based on multiple sequence alignments [12, 13], primer design for selective amplification of protein-coding regions [14], oligonucleotide design for site-directed mutagenesis [15], and primer design for hybridization [16]. Usually, the design of truly specific primers requires the information of the complete nucleotide sequence. This is the starting point for most of the programs described in the literature. However, the need of designing specific primers is not always accompanied by the complete knowledge of the target genome sequence.

A primer, or more generally any DNA sequence, is called specific if it represents a unique sequence and is called degenerate if it represents a collection of unique sequences. For example, the amino acid sequence "YHP" could be coded by "TATCATCCC," "TACCATCCA," or "TACCAC-CCG," among others; all of these are unique sequences that can be summarized in a "degenerate" nucleotide sequence "TAYCARCCN," using IUPAC code. Operatively, the use of a degenerate primer implies the use of a population of specific primers that cover all the possible combinations of nucleotide sequences coding for a given protein sequence. Also, primers including modified bases can be used. Some modified bases can match different bases.

Although the increase in degeneracy rises the chance of unspecific annealing of the designed primers, it also increases the probability of finding unknown divergent variants of a sequence family. This dual behavior must be taken into account during the design. Algorithmic search of primers that include degenerated positions is usually defined as the degenerate primer design (DPD) problem. In recent years, several methods were developed to solve DPD problem. Each one has a specific scope or is designed to solve a variant of the problem, but all of them aim to minimize the number of degenerations of the resulting primers.

The DPD problem was expressed in different ways by many researchers. Linhart and Shamir [17] presented the maximum coverage DPD problem (MC-DPD), with the goal of finding a primer that covers the maximum number of input sequences. The selection of primers is constrained by limiting the maximum degeneracy. They also stated the minimum degeneracy DPD problem (MD-DPD), in which the objective is finding a primer with the minimum degeneracy that covers all the input sequences. To solve MC-DPD they have developed the HYDEN program [18]. Wei et al. [19] developed the DePiCt program that uses hierarchical clustering of protein blocks to design the primers. Rose et al. [20] developed a method for hybrid degenerate-nondegenerate primers, where the 3′ region is degenerated and its 5′ region is a consensus clamp. It was implemented in CODEHOP [21] and iCODEHOP [22] programs and was used to search new members of protein families and for identification and characterization of viral genomes. Balla and Rajasekaran [23] described a method for a variant of MD-DPD that tolerates mismatch errors, implemented in the minDPS program. The programs PT-MIPS and PAMPS address mainly the problem of multiple degenerate primer design. The aim of these programs is finding the minimum number of degenerate primers that cover all the input sequences, taking into account that none of them may be more degenerated than an input value.

In this study a new method for solving the DPD problem is proposed, in which the focus is shifted away from the global minimum degenerated primer in favor of maximizing a score value which contains degeneracy but weighted by its proximity to the 3′ end of the primer. This minimizes the degeneracy at that end while allowing more freedom in the remaining positions. Hereby, the best scoring primers may not be the less degenerated, but take into account a biological restraint that is not so heavily considered in other methods. The 3′ end is the essential anchoring site because it is where the polymerase initiates its activity. From a strategic point of view, a decision must be made whether or not to allow degeneracy at this end. The presence of degeneracy at the 3′ end probably assures a greater diversity of sequences to

be detected. However, at the same time, it diminishes the proportion of primer specific for a given sequence. Therefore, we decided to be very strict in the search of conserved regions and minimize the amount of degeneracy incorporated at this end. If the input set of sequences is sufficiently large, it is highly probable that a region identified as conserved among all known sequences will likewise be conserved in any new member of the family.

2. Scoring and Primer Search Strategy

The method presented here can be used starting with DNA or protein sequence alignments (Figure 1(a)). If the input was DNA, sequences were aligned to obtain one global degenerate DNA consensus. If the input was a protein alignment, each protein of the alignment is backtranslated into a degenerate DNA sequence. All the degenerate DNA sequences were combined in one global degenerate DNA consensus. This consensus sequence covers all the putative input sequences that could be the origin of each protein sequence (Figure 1(b)). Also, the consensus sequence may code for amino acids that were not detected in the known sequences. This is inevitable given the kind of degeneracy of the genetic code.

Then, the degenerate consensus sequence was analyzed using an overlapping window-based strategy. The window length corresponds to the required oligonucleotide length, and each window corresponds to a putative primer. For each candidate primer a score is calculated. In the first place, for each position of a candidate primer a position score (Sp_i) was calculated using (1):

$$Sp_i = 1 - \log_{10}(ND_i), \tag{1}$$

where ND_i is the degeneracy value at the position i of the oligonucleotide ($1 \leq i \leq n$, where n is the length of the primer). ND_i is 1 for "A, C, G or T," 2 for "K, M, R, S, W or Y," 3 for "B, D, H or V," and 4 for "N." This expression takes a value of 1 for nondegenerate bases and decreases for more degenerated bases. On the other hand, it is known that in PCR reactions, the 3′ end of the primer is more important than the 5′ end. The region of the 3′ end of the primer must be as little degenerated as possible. Therefore, a good annealing at this end is imperative in order to minimize unspecific amplifications. Considering this, the value of Sp_i is multiplied by a weighting value (Wp_i) defined by a straight line function that increases as it comes closer to the 3′ end (2):

$$Wp_i = pA + \frac{i \times (N_y - pA)}{N_x}, \tag{2}$$

where i is the position from the 5′ end along the oligonucleotide ($1 \leq i \leq n$, where n is the length of the primer) and pA, N_y, and N_x are user adjustable parameters defining the straight line function. pA is the axis intersection and $(N_y - pA)/N_x$ is the slope. Default values for pA, N_y, and N_x are 0, 1, and 1, respectively. Changing them will permit them to be more or less strict about including degenerations closer to the 3′ end of the primer. Increasing pA or N_x, or

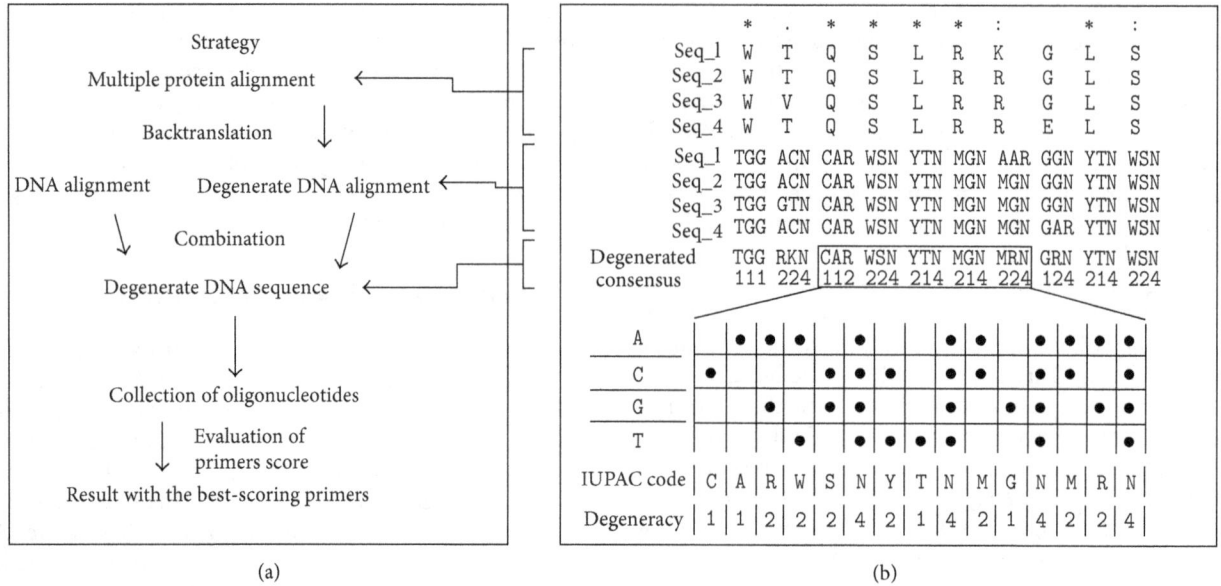

FIGURE 1: Minimum degenerated sequence generation. (a) Diagram of the general strategy used. (b) Sample protein alignment showing an example for the steps of the strategy diagram. Each sequence is computationally backtranslated to hypothetical nucleic acid sequences. IUPAC codes were used to show ambiguous positions. These sequences are piled up in order to get the degenerated consensus sequence. Numbers below this indicate the degeneration value of each position.

decreasing N_y, results in lesser stringency on the designed primer. Finally, to obtain a scaled global score (S_g), the result of $Wp_i \times Sp_i$ is divided by the maximum possible score (M_s, (3)). Global normalized score (S_g) was calculated according to (4). In this way, S_g value varies from 0 to 1. Maximum score is obtained when the value of the Sp_i is 1 for each position. Therefore, ND_i must also be 1 too, and this only happened with nondegenerated primers:

$$M_s = n \times pA + \frac{(n+1) \times n \times (N_y - pA)}{2 \times N_x}, \qquad (3)$$

$$S_g = \frac{\sum_{i=1}^n Sp_i \times Wp_i}{M_s}. \qquad (4)$$

3. Methods

3.1. Alignment and Sequence Comparison Tools.
For global alignment of protein sequences, the program ClustalW 1.83 [24] was used with default parameters. Local alignments of proteins against genomes were made using stand-alone Blast 2.2.13 [25] with default parameters. Oligonucleotide match searches were made with specifically developed tools written in C language.

3.2. Sequence Data.
Several sets of sequences were used in the tests of the program, for designing and comparison of the primer sequences against genomes. All sequences GenBank's accession numbers are presented in Table 1.

3.3. Filtering Primers.
In addition to the scoring process, FAS-DPD can optionally filter the primers individually

according to common criteria: melting point temperature (estimated using Santalucia's method [26]), $G + C$ content, $5'$ versus $3'$ stability, presence of tandem repeats of the same base occurring at $3'$ end or any place in the sequence, presence of a degenerated position at the $3'$ end, and formation of homodimer structures. Also, primer pairs can be filtered according to amplification product size, melting point temperature compatibility, $G + C$ content compatibility, and formation of heteroduplex structures.

3.4. PCR Amplification.
The PCR conditions used in all experiments follow a common protocol. The reaction mix contained 1X Taq DNA polymerase buffer (Productos Biológicos, Argentina), 0.2 mM dNTPs, 0.5 μM of each primer, 20 pM template, and different concentration of $MgCl_2$ and dimethyl sulfoxide (DMSO) in different reactions. The $MgCl_2$ was used from 2 mM to 3 mM, and DMSO was used from 0% (v/v) to 5% (v/v). The reactions were performed in a total volume of 10 μL, and the thermal profile consisted of an initial denaturation step of 94°C for 2 min, followed by 35 cycles of denaturation/annealing/extension steps. The denaturation step was at 92°C for 10 seconds, the temperature of the annealing step was not the same in all experiments, varying from 45°C to 60°C, and the time was always 15 seconds (see Figure 4). The extension step was at 72°C; the time of this step was 15 seconds. In all cases, one of the primers is specific for the template, while the other primer was designed by the method described in this work. The last step was a final extension of 5 minutes at 72°C. For Junin Virus, the template used was a plasmid containing a copy of cDNA of JUNV S genomic segment. For *Baculovirus*, the template was a plasmid containing a fragment of *Anticarsia gemmatalis*

TABLE 1: List of sequences used in the test of FAS-DPD. Accession numbers and brief description are presented.

Acc. number	Sequence description	Acc. number	Sequence description
	Arenaviral sequences		
AY129248.1	Machupo v. st. Carvallo	U41071.1	Sabia v.
AF485260.1	Machupo v. st. Carvallo	EU260463.1	Chapare v. st. 810419
AY924206.1	Machupo v. st. MARU-216606	AY081210.1	Allpahuayo v. CLHP-2098
AY924202.1	Machupo v. st. Chicava	AY012686.1	Allpahuayo v. from Peru
AY624355.1	Machupo v. st. Chicava	AY012687.1	Allpahuayo v. st. CLHP-2472
AY924205.1	Machupo v. st. 9301012	AF485262.1	Pirital v. st. VAV-488
AY619645.1	Machupo v. st. Mallele	AF277659.1	Pirital v.
AY924203.1	Machupo v. st. 9430084	M16735.1	Pichinde v.
AY924208.1	Machupo v. st. MARU 249121	AF485261.1	Parana v. st. 12056
AY924204.1	Machupo v. st. 200002427	AF512829.1	Parana v. st. 10256
AY924207.1	Machupo v. st. MARU 222688	AF512831.1	Flexal v. st. BeAn 293022
AY571959.1	Machupo v. st. 9530537	AF485257.1	Flexal v. st. Pinheiro
AY746353.1	Junin v. st. Candid-1	AF512831.1	Flexal v. st. BeAn 293022
AY358023.2	Junin v. st. XJ13	AF512830.1	Latino v. st. MARU 10924
AY619641.1	Junin v. st. Rumero	AF485259.1	Latino v. st. Maru 10924
D10072.2	Junin v. st. MC2	U34248.1	Oliveros v.
M20304.1	Tacaribe v.	AY847350.1	LCM v. st. Armstrong 53b
AF485256.1	Amapari v. st. BeAn 70563	M20869.1	LCM v. st. Armstrong 53b
AF512834.1	Amapari v. st. BeAn 70563	EU136038.1	Dandenong v. is. 0710-2678
AF512832.1	Cupixi v. st. BeAn 119303	DQ328874.1	Mopeia v. st. Mozambique
AY129247.1	Guanarito v. st. INH-95551	DQ328877.1	Ippy v. st. Dak-An-B-188-d
AF485258.1	Guanarito v. st. INH-95551	X52400.1	Nigeria Lassa v.
AY497548.1	Guanarito v. st. CVH-960101	AY628206.1	Lassa v. st. Weller
AY924392.1	Bear Canyon v. st. AV 98470029	AY628201.1	Lassa v. st. Macenta
AY924391.1	Bear Canyon v. st. AV A0070039	AY628205.1	Lassa v. st. Z148
AF512833.1	Bear canyon v. st. A0060209	J04324.1	Lassa v. st. Josiah
DQ865244.1	Catarina v. st. AV A0400135	AY772168.1	Mopeia Lassa reassortant 29
DQ865245.1	Catarina v. st. AV A0400212	AY628203.1	Lassa v. st. Josiah
EU123328.1	Skinner Tank v. st. AV D1000090	AF181853.1	Lassa v. st. LP
EU123331.1	North American arenav. st. AV 96010024	AY628207.1	Lassa v. st. Pinneo
EU123330.1	North American arenav. st. AV 96010151	AY628208.1	Lassa v. st. Acar-3080
AF228063.1	Whitewater Arroyo v. st. 9310135,	AF181854.1	Lassa v. st. 803213
AF485264.1	Whitewater Arroyo v. st. 9310141	AY342390.1	Mobala v. st. ACAR-3080-MRC5-P2
EU123329.1	North American arenav. st. AV D1240007	M33879.1	Mopeia v. st. AN-21366
AF485263.1	Tamiami v. st. CDC W-10777	AY772170.1	Mopeia v. st. AN-20410
AF512828.1	Tamiami v. st. W 10777		
	Baculoviral sequences		
AP006270.1	*Adoxophyes honmai* nucleopolyhedrovirus DNA	X77048.1	*Cryptophlebia leucotreta* granulosis
AF547984.1	*Adoxophyes orana* granulovirus	X79569.1	*Cryptophlebia leucotreta* granulosis
NC_005839.2	*Agrotis segetum* granulovirus	NC_002816.1	*Cydia pomonella* granulovirus
L22858.1	*Autographa californica* nucleopolyhedrovirus clone C6	NC_003083.1	*Epiphyas postvittana* NPV
L33180.1	*Bombyx mori* nuclear polyhedrosis virus isolate T3	NC_002654.2	*Helicoverpa armigera*
NC_005137.2	*Choristoneura fumiferana* DEF MNPV	AF081810.1	*Lymantria dispar*
NC_004778.3	*Choristoneura fumiferana* MNPV	NC_003529.1	*Mamestra configurata* NPV-A
AY864330.1	*Chrysodeixis chalcites* NPV	U75930.2	*Orgyia pseudotsugata* MNPV
AY456389.1	*Chrysodeixis chalcites* NPV	AF499596.1	*Phthorimaea operculella* granulovirus
AY456390.1	*Chrysodeixis chalcites* NPV	NC_002593.1	*Plutella xylostella* granulovirus
AY545786.1	*Chrysodeixis chalcites* NPV	NC_004323.1	*Rachiplusia ou* MNPV
AY545787.1	*Chrysodeixis chalcites* NPV	NC_002169.1	*Spodoptera exigua* MNPV
AY229987.1	*Cryptophlebia leucotreta* granulovirus	NC_003102.1	*Spodoptera litura* NPV

TABLE 1: Continued.

Acc. number	Sequence description	Acc. number	Sequence description
AY096241.1	*Cryptophlebia leucotreta* granulovirus	NC_007383.1	*Trichoplusia ni* SNPV
AY096242.1	*Cryptophlebia leucotreta* granulovirus		
	Pseudomonas sp. sequences		
NC_007492.2	*Pseudomonas fluorescens* Pf0-1	NC_004578.1	*Pseudomonas syringae*
NC_005773.3	*Pseudomonas syringae*	NC_002947.3	*Pseudomonas putida*
NC_004129.6	*Pseudomonas fluorescens*	NC_002516.2	*Pseudomonas aeruginosa*
NC_007005.1	*Pseudomonas syringae*		
	Lactobacillus sp. sequences		
NC_005362.1	*Lactobacillus johnsonii*	NC_002662.1	*Lactococcus lactis* subsp.
NC_007576.1	*Lactobacillus sakei* subsp.	NC_004567.1	*Lactobacillus plantarum*

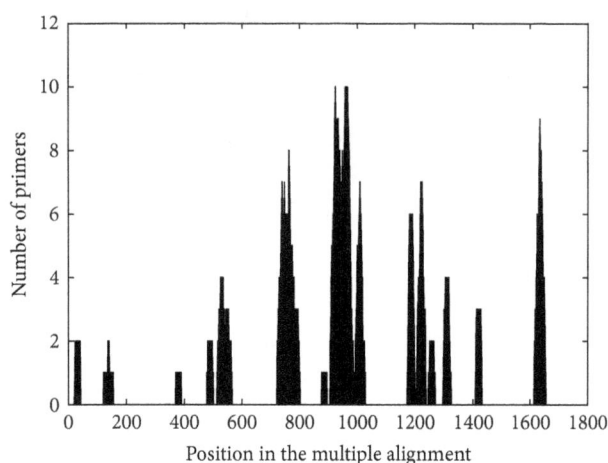

FIGURE 2: Primer distribution along one ORF. A collection of the best scoring primers for the nucleoprotein of *Arenavirus*, comprised of 50 primers for the genomic sequence and 50 for the antigenomic sequence, were represented in the corresponding alignment position. The height of each point indicates the cumulative number of primers corresponding at this position. The alignment was made with 71 arenavirus N protein sequences.

MNPV p74 gene. Sensitivity of the PCR assay was determined by dilution of cloned fragments from Junin virus [27] and *Baculovirus* template.

4. Results

4.1. Distribution of Generated Primers. The distribution of the resulting primers along the input sequence was analyzed. For this, the best one hundred primers obtained from a protein alignment were selected. For each position in the alignment, the number of the selected primers that correspond to this position was recorded (Figure 2). The test was repeated for different protein alignments.

The selected primers were located around a few hot spots in the alignment. This behavior indicates that there are generally few regions in a sequence alignment useful for degenerate primer design. Many primers found by the program are almost identical, shifting one or two bases

between them, and located for most cases in a 30–40 base run. Similar results were obtained with all proteins tested.

4.2. Intragenomic Specificity and Score Analysis. Because it is possible that the best primers are not the less degenerated substrings in the collection of candidates, their specificity was tested. Also, it was necessary to get a more precise understanding of the score assigned by FAS-DPD in terms of specificity. To achieve this, the primers were compared with the complete genome sequences used to design them, looking for unspecific perfect matches.

For this task, a wide range of genome sizes was covered. Four collections of complete genome sequences were used: *Arenavirus* (genome in 10^4 bases order), *Baculovirus* (genome in 10^5 bases order), *Lactobacillus* (genome in 10^6 bases order), and *Pseudomonas* (genome in 10^6 bases order). For each set, a randomly selected genome was used as reference. Each annotated ORF of this genome was used to search related ORFs in the other genomes of the collection using the local Blast tool. The expected value of Blast was used to decide when two ORFs were related. When an ORF of the reference genome had a related one in all other genomes, all of them were aligned with ClustalW and used in further analysis.

Each resulting alignment was used as input for FAS-DPD to search primers. For each genome polarity the best fifty nonoverlapping primers were selected. This selection was made to avoid concentration of overrepresented, hot-spot-derived, high score primers. This allowed us to find a balanced set of primers, with high and low scores.

In order to find the relationship between the score calculated for each primer and its specificity, all the primers were compared with all the oligonucleotides of the same size derived from each genome, searching for perfect matches (Figure 3). The results were similar for the four systems despite their differences in genome size.

There is an inverse correlation between primer score and the number of unspecific perfect matches. But this correlation is not linear. The quantity of unspecific perfect matches of primers with a minimal score of 0.85 and their target genome was generally zero. The number of unspecific perfect matches grew enormously with lower primer scores.

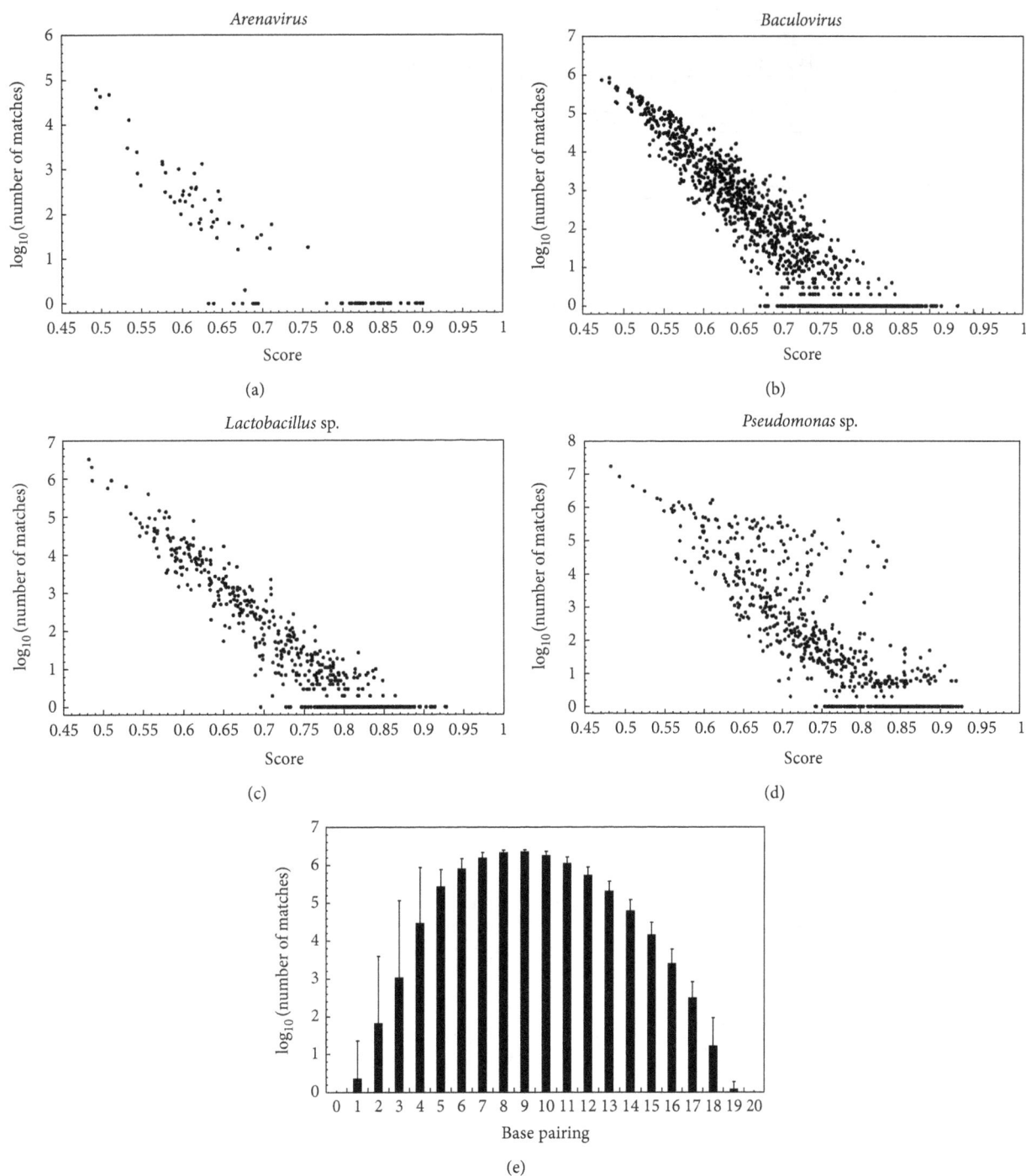

FIGURE 3: Specificity of primers. Primers designed for all ORFs shared among each model organism used were compared against the complete set of genomes for perfect matches with oligonucleotides of the same length. Each point represents the number of perfect matches (in $\log_{10}$ scale) of a primer in relation to its score. The length of the primers was 20 nucleotides. (a) *Arenavirus* genomes: 71 for S (small) RNA, 24 for L (large) RNA. (b) 22 *Baculovirus* genomes. (c) 5 *Lactobacillus* sp. genomes. (d) 7 *Pseudomonas* sp. genomes. (e) A set of primers for *Lactobacillus* sp. with scores between 0.85 and 0.90 were tested for nonperfect matches that could anneal unspecifically in PCR. Each bar represents the number of matches against the complete set of *Lactobacillus* genomes. The number below the bar indicates how many bases are shared.

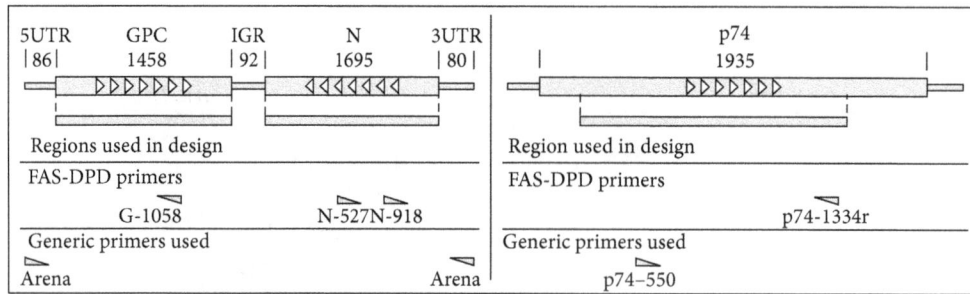

(a)

Primers (FAS-DPD/specific)	Product size	Annealing temperature (°C)	Sensitivity (copies/reaction)	Score
N527/Arena	605 bp	54.2	20	0.849
N918/Arena	986 bp	57.7	2×10^2	0.868
GR1058/Arena	947 bp	46.8	20	0.86
p74-1334r/p74–550	896 bp	58.7	2×10^5	0.828

(b)

FIGURE 4: Experimental challenge of designed primers. (a) Genomic organization of the *Arenaviruses* S RNA and P74 ORF. *Arenavirus* shows an ambisense coding strategy of the GPC and N ORFs and three noncoding regions: 5′ untranslated region (5UTR), intergenic region (IGR), and 3′ untranslated region (3UTR). The location of each designed primer (GR1058, N918, N537, and p74-1334r) and specific primers (Arena, p74-550) is also shown. (b) The results obtained with each pair of primers tested and characteristics of reaction are shown.

4.3. Experimental Challenge. In addition to theoretic tests to determine the usefulness of FAS-DPD designed primers, experimental challenges were performed using *Arenavirus* and *Baculovirus* as models. The assay consisted in performing PCRs using a pair of primers, including a degenerated FAS-DPD designed primer and a standard nondegenerated primer (this allowed testing individually each designed primer), optimizing the reaction conditions and measuring its sensitivity.

For arenavirus, the primers were designed using sequences of 71 different GenBank records for the nucleoprotein (N protein) and the glycoprotein precursor (GPC protein). From the lists of the highest scored primers, three were randomly selected and synthesized for experimental evaluation, one for GPC (GR1058: RCNWHRTTNYCRAARCAYTT, score: 0.8596) and two for N (N527: GGNRYNSWNCCRAAYTGRTT, score: 0.8494; N918: NANRTTYTCRTANGGRTTNC, score: 0.8437) (Figure 4(a)).

Amplification reactions were performed using each of these primers together with the Arena primer CGCAC-CGGGGATCCTAGGC) as nondegenerated counterpart. The latter is a generic primer for *Arenaviruses* that matches almost perfectly with the nineteen bases of 3′ end of the genomic RNA sequence and with the nineteen bases of 3′ end of the antigenomic RNA sequence of all known arenaviruses. The reaction template was a cDNA corresponding to the Junin virus small RNA segment which encodes the N and GPC proteins.

For *Baculovirus*, one primer (p74-1334r: BYRWRNC-CVWRNGGRTCSCA, score: 0.8281) was designed using 57 sequences of p74 different *Baculovirus*. As its counterpart, a specific primer for *Anticarsia gemmatalis MNPV* was used

[28] (p75-550r: GGcGTGGACGACGTGC). The reaction template was the *Anticarsia gemmatalis MNPV* p74 isolate 2D [29] gene cloned in a plasmid.

PCRs were assayed with different sets of conditions, and the sensitivity was measured. Sensitivity achieved with arenavirus primers was high. Twenty copies/μL or less of specific template were detected. For *Baculovirus* the detection was not as sensible as for arenavirus, but it can be considered as a good sensitivity; 2×10^4 copies/μL of specific template were detected. This difference can be explained taking into account that the divergence observed for baculovirus sequences is greater than for arenavirus. Therefore, the score for the p74-1334r primer was lower than that of *Arenavirus*.

4.4. Increment of Degeneration of FAS-DPD Designed Primers in relation to Minimum Degenerated Substring. The aim of FAS-DPD is to design universal degenerated primers that are not necessarily the less degenerated sequences of the collection of candidates. In order to know how much degeneration FAS-DPD designed primers acquire, another test was performed. Given an alignment of homologous ORFs, the degeneration was calculated for the highest scoring primer selected with FAS-DPD and for the minimum degenerated substring of the same length. Then, the ratio of these two values was obtained. The comparison was made with the complete set of ORF alignments used before (*Arenavirus, Baculovirus, Pseudomonas,* and *Lactobacillus*) (Figure 5). In more than 90% of the cases the increase of degeneration value is at most fourfold (e.g., changing "...A..." to "...N..." or "...A...A..." to "...R...W..."). Therefore, these primers

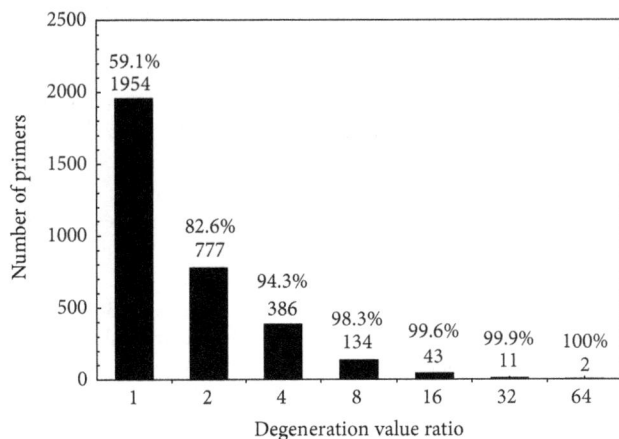

FIGURE 5: Comparison of FAS-DPD designed primers and minimum degenerated substrings. Collection of primers with the highest score designed for all the ORFs shared by all the genomes used were compared against the minimum degenerated subsequence of the same length for each ORF in order to know how much more degenerated they are. The number below each bar indicates the ratio of degeneration between the designed primer and the minimum degeneration substring. The number above each bar indicates the amount of primers that correspond with the ratio mentioned before. The percentages are cumulative with respect to increasing degeneration ratios and referred to the total number of primers used in the test.

have only up to two more degenerated positions than the substring with minimum degeneration.

It is important to note that, in general, there is not only one minimum degeneracy substring for each ORF. The decision of which primer is better must not only take into account the degeneration value. The position of degenerated bases in the sequence is crucial. The ratio of greater increase of degeneration found was 64; this corresponds to only less than 0.1% of primers. This result shows that FAS-DPD primers are more degenerated than the less degenerated substring, but this increase of degeneration is slight and does not imply a high compromise of the specificity.

5. Discussion

In this work we presented a new algorithm, implemented in the FAS-DPD software, as an alternative strategy to solving DPD problems. FAS-DPD was designed to use multiple alignments of proteins or nucleic acids as input data and constructs a consensus degenerate sequence from that, which is then used to design the putative primers.

The experimental background knowledge from molecular biology teaches us that in the real world the $3'$ ends of primers are key determinants of a successful amplification. FAS-DPD takes into account this property and incorporates special considerations in the global score calculation becoming more strict for the $3'$ end than for the $5'$ end.

The specificity of the set of primers designed with FAS-DPD was computationally tested with several collections of whole genomes, ranging from 10^4 bp to 10^6 bp. The

restriction to higher lengths was due to the lack of whole genome collections for genus of bigger sizes with several individuals. In all genome collections assayed the results showed the same behavior; there is a relationship between the score value and the number of unspecific perfect matches. This analysis allows us to suggest a cut-off score (0.85) for primers that could be more successful.

PCRs were successfully performed on arenaviral and baculoviral models. For arenavirus, the designed GPC or N primers were used with the universal Arena primer [30]. For *Baculovirus*, the designed p74 primer was used with a specific p74 primer [28]. Each reaction was tested in different conditions in order to optimize its yield.

FAS-DPD software is licensed under GNU General Public License Version 3 and is available at http://www.github.com/javieriserte/fas-dpd.

In general, the results suggest that FAS-DPD could be used to design generalized degenerate primers for detection of known or unknown members of gene families or organism families, including different types of pathogens. Also, this tool would allow a more efficient search for enzymes and other proteins with commercial or biotechnological importance, making for a faster and cheaper research process.

References

[1] K. Nelson, *Metagenomics as a Tool to Study Biodiversity*, ASM Press, Washington, DC, USA, 2008.

[2] J. Welsh and M. McClelland, "Fingerprinting genomes using PCR with arbitrary primers," *Nucleic Acids Research*, vol. 18, no. 24, pp. 7213–7218, 1990.

[3] J. G. K. Williams, A. R. Kubelik, K. J. Livak, J. A. Rafalski, and S. V. Tingey, "DNA polymorphisms amplified by arbitrary primers are useful as genetic markers," *Nucleic Acids Research*, vol. 18, no. 22, pp. 6531–6535, 1990.

[4] W. C. Nichols, S. E. Lyons, J. S. Harrison, R. L. Cody, and D. Ginsburg, "Severe von Willebrand disease due to a defect at the level of von Willebrand factor mRNA expression: detection by exonic PCR-restriction fragment length polymorphism analysis," *Proceedings of the National Academy of Sciences of the United States of America*, vol. 88, no. 9, pp. 3857–3861, 1991.

[5] E. Desmarais, I. Lanneluc, and J. Lagnel, "Direct amplification of length polymorphisms (DALP), or how to get and characterize new genetic markers in many species," *Nucleic Acids Research*, vol. 26, no. 6, pp. 1458–1465, 1998.

[6] W. Rychlik and R. E. Rhoads, "A computer program for choosing optimal oligonucleotides for filter hybridization, sequencing and in vitro amplification of DNA," *Nucleic Acids Research*, vol. 17, no. 21, pp. 8543–8551, 1989.

[7] L. Hillier and P. Green, "OSP: a computer program for choosing PCR and DNA sequencing primers," *PCR Methods and Applications*, vol. 1, no. 2, pp. 124–128, 1991.

[8] P. Li, K. C. Kupfer, C. J. Davies, D. Burbee, G. A. Evans, and H. R. Garner, "PRIMO: a primer design program that applies base quality statistics for automated large-scale DNA sequencing," *Genomics*, vol. 40, no. 3, pp. 476–485, 1997.

[9] V. Proutski and E. C. Holmes, "Primer Master: a new program for the design and analysis of PCR primers," *Computer Applications in the Biosciences*, vol. 12, no. 3, pp. 253–255, 1996.

[10] S. Haas, M. Vingron, A. Poustka, and S. Wiemann, "Primer design for large scale sequencing," *Nucleic Acids Research*, vol. 26, no. 12, pp. 3006–3012, 1998.

[11] S. Rozen and H. Skaletsky, "Primer3 on the WWW for general users and for biologist programmers," *Methods in molecular biology*, vol. 132, pp. 365–386, 2000.

[12] A. Gibbs, J. Armstrong, A. M. Mackenzie, and G. F. Weiller, "The GPRIME package: computer programs for identifying the best regions of aligned genes to target in nucleic acid hybridisation-based diagnostic tests, and their use with plant viruses," *Journal of Virological Methods*, vol. 74, no. 1, pp. 67–76, 1998.

[13] M. D. Gadberry, S. T. Malcomber, A. N. Doust, and E. A. Kellogg, "Primaclade: a flexible tool to find conserved PCR primers across multiple species," *Bioinformatics*, vol. 21, no. 7, pp. 1263–1264, 2005.

[14] C. E. López-Nieto and S. K. Nigam, "Selective amplification of protein-coding regions of large sets of genes using statistically designed primer sets," *Nature Biotechnology*, vol. 14, no. 7, pp. 857–861, 1996.

[15] A. Turchin and J. F. Lawler, "The primer generator: a program that facilitates the selection of oligonucleotides for site-directed mutagenesis," *BioTechniques*, vol. 26, no. 4, pp. 672–676, 1999.

[16] D. Hyndman, A. Cooper, S. Pruzinsky, D. Coad, and M. Mitsuhashi, "Software to determine optimal oligonucleotide sequences based on hybridization simulation data," *BioTechniques*, vol. 20, no. 6, pp. 1090–1097, 1996.

[17] C. Linhart and R. Shamir, "Degenerate primer design: theoretical analysis and the HYDEN program," *Methods in Molecular Biology*, vol. 402, pp. 221–244, 2007.

[18] C. Linhart and R. Shamir, "The degenerate primer design problem," *Bioinformatics*, vol. 18, supplement 1, pp. S172–S180, 2002.

[19] X. Wei, D. N. Kuhn, and G. Narasimhan, "Degenerate primer design via clustering," *IEEE Computer Society Bioinformatics Conference*, vol. 2, pp. 75–83, 2003.

[20] T. M. Rose, J. G. Henikoff, and S. Henikoff, "CODEHOP (COnsensus-DEgenerate Hybrid Oligonucleotide Primer) PCR primer design," *Nucleic Acids Research*, vol. 31, no. 13, pp. 3763–3766, 2003.

[21] T. M. Rose, "CODEHOP-mediated PCR: a powerful technique for the identification and characterization of viral genomes," *Virology Journal*, vol. 2, article 20, 2005.

[22] R. Boyce, P. Chilana, and T. M. Rose, "iCODEHOP: a new interactive program for designing COnsensus-DEgenerate Hybrid Oligonucleotide Primers from multiply aligned protein sequences," *Nucleic Acids Research*, vol. 37, no. 2, pp. W222–W228, 2009.

[23] S. Balla and S. Rajasekaran, "An efficient algorithm for minimum degeneracy primer selection," *IEEE Transactions on Nanobioscience*, vol. 6, no. 1, pp. 12–17, 2007.

[24] J. D. Thompson, D. G. Higgins, and T. J. Gibson, "CLUSTAL W: improving the sensitivity of progressive multiple sequence alignment through sequence weighting, position-specific gap penalties and weight matrix choice," *Nucleic Acids Research*, vol. 22, no. 22, pp. 4673–4680, 1994.

[25] S. F. Altschul, W. Gish, W. Miller, E. W. Myers, and D. J. Lipman, "Basic local alignment search tool," *Journal of Molecular Biology*, vol. 215, no. 3, pp. 403–410, 1990.

[26] J. SantaLucia, "A unified view of polymer, dumbbell, and oligonucleotide DNA nearest-neighbor thermodynamics," *Proceedings of the National Academy of Sciences of the United States of America*, vol. 95, no. 4, pp. 1460–1465, 1998.

[27] A. S. Parodi, D. J. Greenway, H. R. Rugiero et al., "Concerning the epidemic outbreak in Junin," *El Día médico*, vol. 30, no. 62, pp. 2300–2301, 1958.

[28] M. F. Bilen, M. G. Pilloff, M. N. Belaich et al., "Functional and structural characterisation of AgMNPV ie1," *Virus Genes*, vol. 35, no. 3, pp. 549–562, 2007.

[29] J. V. de Castro Oliveira, J. L. C. Wolff, A. Garcia-Maruniak et al., "Genome of the most widely used viral biopesticide: *Anticarsia gemmatalis* multiple nucleopolyhedrovirus," *Journal of General Virology*, vol. 87, no. 11, pp. 3233–3250, 2006.

[30] S. E. Goñi, J. A. Iserte, B. I. Stephan, C. S. Borio, P. D. Ghiringhelli, and M. E. Lozano, "Molecular analysis of the virulence attenuation process in Junín virus vaccine genealogy," *Virus Genes*, vol. 40, no. 3, pp. 320–328, 2010.

12

Characterization of Myelomonocytoid Progenitor Cells with Mesenchymal Differentiation Potential Obtained by Outgrowth from Pancreas Explants

Marc-Estienne Roehrich[1] and Giuseppe Vassalli[1, 2]

[1] Department of Cardiology, Centre Hospitalier Universitaire Vaudois (CHUV), Avenue du Bugnon, 1011 Lausanne, Switzerland
[2] Molecular Cardiology Laboratory, Fondazione Cardiocentro Ticino, via Tesserete 48, 6900 Lugano, Switzerland

Correspondence should be addressed to Giuseppe Vassalli, giuseppe.vassalli@chuv.ch

Academic Editor: Gabriel A. Monteiro

Progenitor cells can be obtained by outgrowth from tissue explants during primary ex vivo tissue culture. We have isolated and characterized cells outgrown from neonatal mouse pancreatic explants. A relatively uniform population of cells showing a distinctive morphology emerged over time in culture. This population expressed monocyte/macrophage and hematopoietic markers (CD11b$^+$ and CD45$^+$), and some stromal-related markers (CD44$^+$ and CD29$^+$), but not mesenchymal stem cell (MSC)-defining markers (CD90$^-$ and CD105$^-$) nor endothelial (CD31$^-$) or stem cell-associated markers (CD133$^-$ and stem cell antigen-1; Sca-1$^-$). Cells could be maintained in culture as a plastic-adherent monolayer in culture medium (MesenCult MSC) for more than 1 year. Cells spontaneously formed sphere clusters "pancreatospheres" which, however, were nonclonal. When cultured in appropriate media, cells differentiated into multiple mesenchymal lineages (fat, cartilage, and bone). Positive dithizone staining suggested that a subset of cells differentiated into insulin-producing cells. However, further studies are needed to characterize the endocrine potential of these cells. These findings indicate that a myelomonocytoid population from pancreatic explant outgrowths has mesenchymal differentiation potential. These results are in line with recent data onmonocyte-derivedmesenchymal progenitors (MOMPs).

"M.-E. Roehrich passed away whilst this article was in press."

1. Introduction

The pancreas is a complex organ consisting of three principal cell types: endocrine islets, exocrine acini, and ducts. Evidence of differentiation of new β-cells from pancreatic nonislet cells suggests the existence of pancreatic nonendocrine stem/progenitor cells [1, 2]. New β-cells may also result from replication of preexisting β-cells [3], or from progenitor cells originating from the ductal epithelium [4–6] or the exocrine tissue of the pancreas [7–9]. Pancreatic progenitor cells express key transcription factors involved in the embryological development of endocrine cells such as pancreatic and duodenal homeobox factor 1 (Pdx1),

neurogenin 3 (Ngn3) and paired box 4 (Pax4), or embryonic markers such as Oct-4 and Nanog, or nestin [10]. Pancreatic progenitor cells have been prospectively isolated by fluorescence-activated cell sorting (FACS) using specific antibodies that recognize cell-surface epitopes expressed by stem/progenitor cells in other tissues, such as CD133, CD117 (c-kit/stem cell factor receptor), ATP-binding cassette (ABC) G2, and mesenchymal stem cell (MSC) markers [11–15].

An alternative method for the isolation of tissue-resident progenitor cells is the explant outgrowth approach. This method does not rely on positive cell selection. Within tissue explants, progenitor cells are located in close proximity to stem cell niches, which regulate stem and progenitor cell function [16]. These cells may become activated during *ex vivo* primary tissue culture, migrate across chemotactic gradients towards the surface of the tissue explant, are shed by it, and form a monolayer. Primary tissue cultures of the adult or embryonic pancreas have been described extensively [17]. In contrast, data on pancreas explant cell outgrowths are limited. Using the explant outgrowth technique, Schneider et al. [18] isolated stellate cells from pancreata of rats with cerulein pancreatitis. Bläuer et al. [19] designed a new explant outgrowth system that allowed for the isolation of pancreatic acinar cells at the gas-liquid interphase. Carlotti et al. [20] reported that the cell outgrowth from isolated human islets was comprised of adherent fibroblastoid cells that expressed MSC and pericyte markers, as well as nestin and vimentin, but not genes for endocrine hormones. When cultured under appropriate conditions, these cells differentiated into adipocytes and osteoblasts lineages and expressed insulin, glucagons, and somatostatin genes. Several other studies attempting to generate β-cells from precursor cells from endocrine or exocrine pancreatic explants documented the presence of plastic-adherent mesenchymal cells in cell cultures [21–25]. While early studies suggested that epithelial-to-mesenchymal transition by β-cells might be responsible for the occurrence of these mesenchymal cells [21], this assumption was recently refuted based on lineage tracing experiments [26–29].

MSCs are multipotent precursor cells for stromal cells, which are capable of differentiating into multiple ectodermal, mesodermal, and endodermal tissues [30]. As such, they have been considered a source of cells for therapeutic approaches for various conditions, including type-1 diabetes. Experimental evidence suggests bone marrow (BM), or adipose tissue-derived MSCs are capable of differentiating into insulin-producing cells *in vitro* and contribute to the restoration of normoglycemia in animal models of diabetes *in vivo* [31, 32]. Human mesenchymal stromal cells that differentiate and mature to hormone-expressing cells *in vivo* have been referred to as islet-derived precursor cells (IPCs) [33]. Recent evidence suggests MSCs may act as trophic mediators to attenuate β-cell death and activate endogenous regenerative mechanisms [34–41].

The present study aimed to characterize the mouse pancreas explant cell outgrowth during *ex vivo* tissue culture. Unlike Carlotti et al. [20] who studied islet outgrowths, we used whole pancreas explants. We reproducibly obtained a population of cells that exhibited a relatively uniform morphology and a stable cell-surface marker profile. The latter was characterized by expression of monocyte/macrophage and hematopoietic markers (CD11b and CD45), pericyte/perivascular markers (neuron-glial antigen 2 [NG2] proteoglycan and, to a lesser extent, CD146) [42], and certain MSC and/or endothelial progenitor cell (EPC) markers (CD29 and CD44), but not MSC-defining (CD90 and CD105) and endothelial (CD31) markers. The isolated myelomonocytoid population was propagated for up to 5 passages and was maintained in culture as a monolayer for more than 1 year with no major morphologic or immunophenotypic changes. Plastic-adherent cells spontaneously formed spherical clusters that detached

from plastic, which is considered a feature of stemness [43]. They were capable of differentiating along multiple mesenchymal lineages (fat, cartilage, and bone) although this was not demonstrated with single-cell cloning. These findings indicate that pancreas explant cell outgrowths can give rise to a myelomonocytoid population endowed with mesenchymal differentiation potential. These findings are inline with recent data on monocyte-derived mesenchymal progenitors (MOMPs) [44].

2. Materials and Methods

2.1. Cell Isolation and Culture. Pancreatic explants were obtained from neonatal (1-2 days of age) male C57Bl/6 mice (from Charles River Laboratories, France) or C57BL/6-Tg(CAG-EGFP)1Osb/J transgenic mice expressing enhanced green fluorescent protein (EGFP) from an immediate-early CMV promoter (gift of T. Pedrazzini, CHUV, Lausanne). Tissue explants were rinsed abundantly with heparinized saline and then cut into small pieces that were placed in Corning Costar 6-well culture plates (Sigma) with no extracellular matrix (EMC) protein coating. Explants were cultured in MesenCult (MesenCult MSC Basal Medium [Mouse] supplemented with serum-containing MesenCult MSC Stimulatory Supplements [Mouse], both from Stem Cell Technologies). After 2 weeks, tissue explants were removed from the culture plates, while the cell outgrowth was left in place. When adherent cells formed a nearly confluent monolayer, they were detached from plastic with PBS-EDTA, collected, and seeded onto new plates. In separate experiments ($n = 2$), cells were cultured in Dulbecco-modified Eagle medium supplemented with 10% fetal calf serum (DMEM-10% FCS), with or without granulocyte-macrophage colony-stimulating factor (GM-CSF). In a separate experiment, cells were cultured using a MethoCult (Stem Cell Technologies)-based 3D system.

2.2. Flow Cytometric Analyses. For flow cytometric analyses ($n = 6$), cells were gently detached from plastic with PBS-EDTA, filtered through a 70-μm filter, centrifuged, and resuspended in reagent A Leucoperm B4FO9B (AbD Serotec) for 15 min at RT. Then, PBS was added, and cells were centrifuged, resuspended in reagent B, incubated with primary antibody (see Supplementary Table 1 available online at doi: 10.1155/2012/429868) for 30 min at RT, and washed with PBS. When needed, cells were resuspended in reagent B and incubated with mouse anti-rat Alexa 488 (1 : 25 dilution) for 30 min at RT. Flow cytometric analyses were performed using a FACSCalibur system (BD Bioscience) and the CellQuest software. Gates used to resolve antigen-expressing cells were set using appropriate isotype-specific control antibodies.

2.3. Immunocytochemistry. Immunocytochemistry was performed as previously described [45]. Briefly, cells were seeded on Lab-Tek Chamber-Slides (Nunc) and fixed with 1% paraformaldehyde (PFA). For immunostaining of NG2 proteoglycan, a polyclonal rabbit anti-NG2 antibody (Chemicon/Millipore) followed by a goat-anti-rabbit secondary Ab

Characterization of Myelomonocytoid Progenitor Cells with Mesenchymal Differentiation Potential Obtained by Outgrowth from Pancreas Explants

107

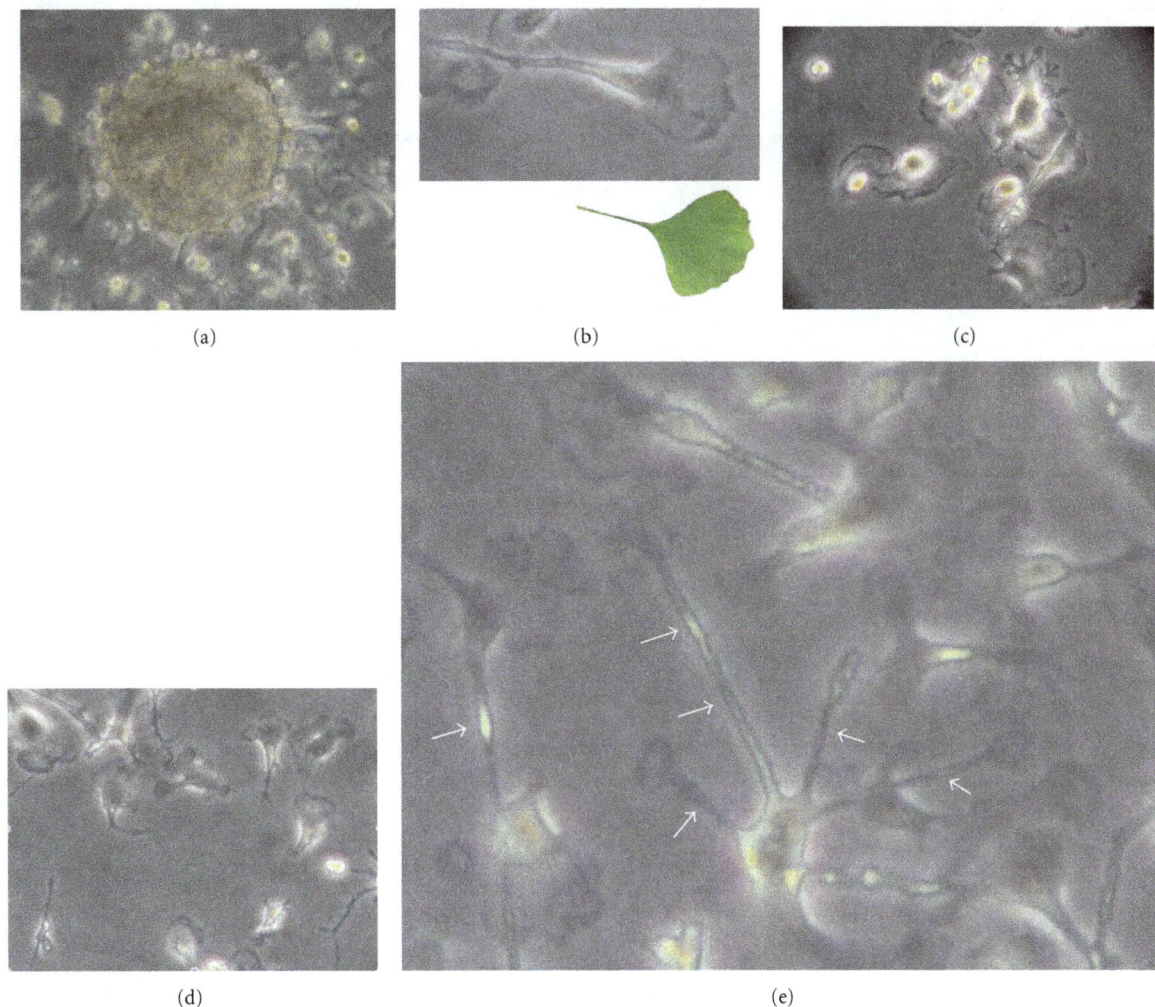

(a)

(b)

(c)

(d)

(e)

FIGURE 1: (a) Phase photomicrograph of a pancreatic explant with outgrowing cells in the primary *ex vivo* tissue culture. (b) High magnification view of an expanded pancreas-derived cell showing a characteristic *Gingko biloba* leaf-like shape (insert). (c and d) Expanded pancreas-derived cells showing refringent nuclei and thin cytoplasmic processes. (e) Expanded pancreas-derived cells showing a "stellate" pattern of cytoplasmic processes (arrows) with knobs on their extremities.

labeled with Alexa 488 (Invitrogen; 1 : 400 dilution) was used. Nuclei were stained with DAPI.

2.4. Sphere Formation and Clonogenicity. Free-floating spherical clusters formed spontaneously from monolayers of plastic-adherent cells plated on Corning Costar 6-well plates ($n = 2$ experiments). To assess whether spherical cell clusters were clonally derived, mixtures of pancreatic cell outgrowths from C57BL/6 wildtype (WT) and from C57BL/6-Tg(CAG-EGFP) 1Osb/J transgenic mice were cultured at varying cell ratios. Spheres were analyzed for green fluorescent areas under the fluorescence microscope after 3 weeks ($n = 1$ experiment).

2.5. Differentiation Assays and Cell Staining. To induce adipogenic, osteogenic, and chondrogenic differentiation, pancreatic outgrowth-derived cells were cultured for 3 months in MesenCult and then changed to NH AdipoDiff, OsteoDiff,

and ChondroDiff Media (all from Miltenyi), respectively, for 17 days. Adipogenic differentiation was detected by incubating cells with 1% PFA for 10 min, followed by Oil red-O solution for 15 min, and three PBS washes. Osteogenic differentiation was detected by incubating cells with 1% PFA for 10 min, followed by 2% Alizarin red for 5 min, and three PBS washes. Chondrogenic differentiation was detected by staining cells with Alcian-blue. For detection of pancreatic β-cells, cells cultured in MesenCult supplemented with 1.27 μM dexamethasone for 18 days were incubated with 1% PFA for 10 min, and stained with the zinc-chelating agent, dithizone (Merck) [46–49], for 15 min according to the manufacturer's instructions.

3. Results

3.1. Cell Morphology and Culture. Using the explant outgrowth approach, cells shed by cultured pancreatic explants from neonatal mice were observed on day 3-4, initially as a

(a)

(b)

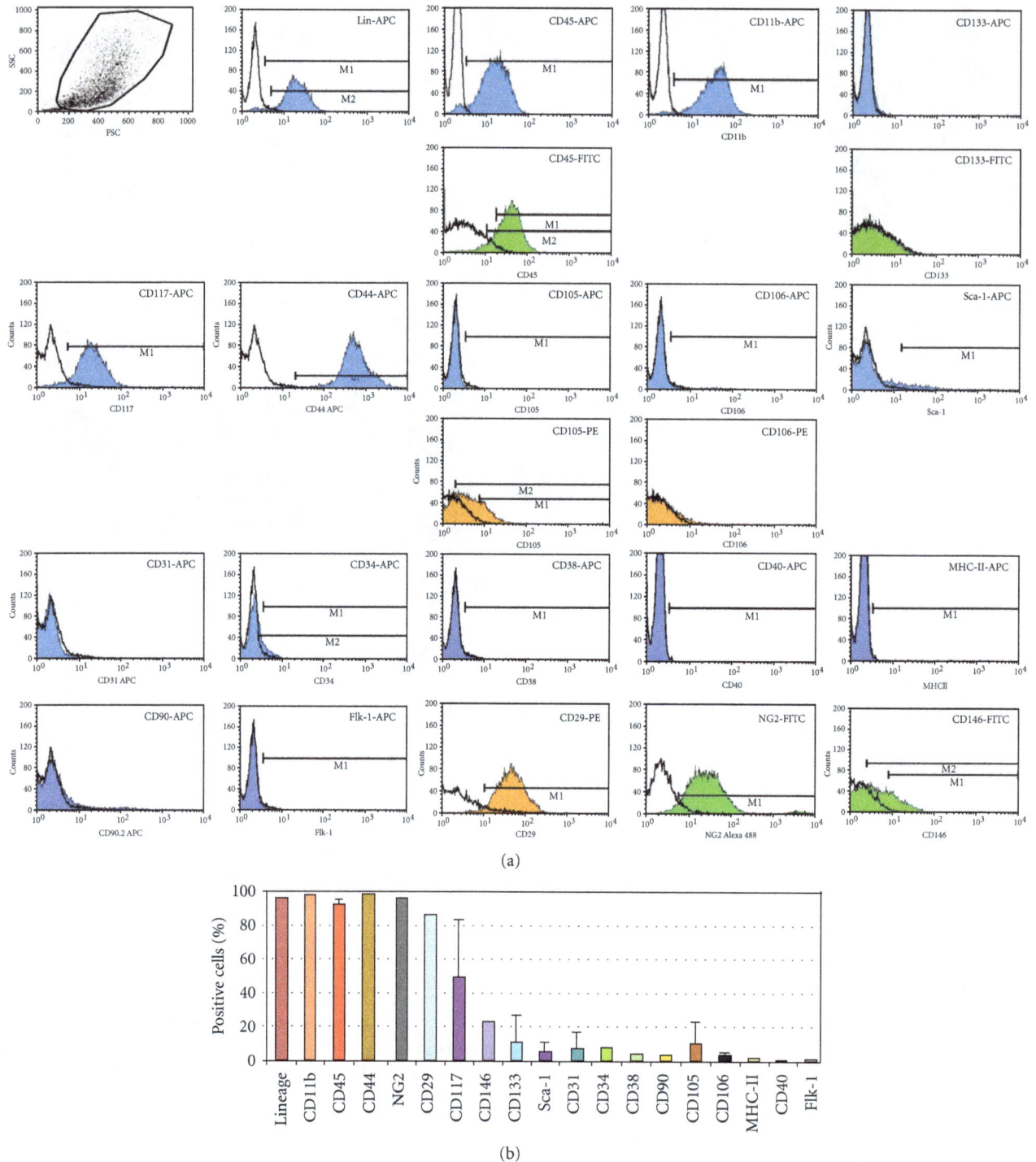

FIGURE 2: Flow cytometric analyses of the cell-surface marker profile of pancreas outgrowth-derived cells. (a) Representative analysis of cell-surface marker expression of cells cultured 2 months in MesenCult. Blue, green, and orange colors indicate APC, FITC, and PE fluorochromes, respectively. Selected markers (CD45, CD105, CD106, and CD133) were determined with two different fluorochromes. (b) Mean percentages (±SD) of cells expressing the indicated cell-surface markers (*n* = 6 analyses; 3–5 samples for each marker, excepted for a subset of markers [NG2, CD29, CD146, CD34, CD38, MHC-II, CD40, and Flk-1] for which a single measure is available).

heterogeneous population of plastic-adherent cells showing both spindle-shaped and round morphologies (Figure 1(a)). After 4–6 weeks, cells acquired a relatively uniform morphology characterized by one or multiple thin cytoplasmic processes carrying a knob on their extremities (this distinctive cell morphology resembled a *gingko biloba* leaf; Figures 1(b) and 1(d)). Cells with multiple thin processes exhibited a stellate shape, whereby processes from neighboring cells

Characterization of Myelomonocytoid Progenitor Cells with Mesenchymal Differentiation Potential Obtained by Outgrowth from Pancreas Explants

109

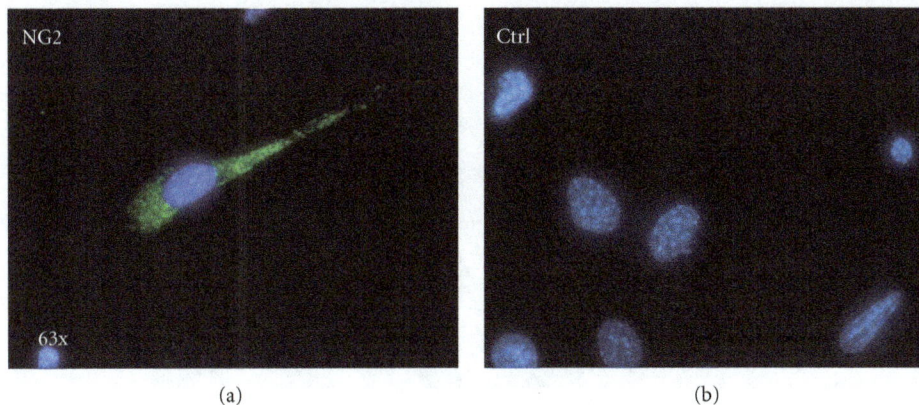

FIGURE 3: NG2 expression by pancreas outgrowth-derived cells. (a) NG2 immunostaining using an Alexa 488-labeled secondary antibody (green); nuclear staining with DAPI (blue). (b) Control (secondary antibody only).

appeared to establish inter-cellular contacts (Figure 1(e)). Cells grew slowly and were propagated for up to 5 passages. They could be maintained in culture for more than 1 year.

3.2. Cell-Surface Marker Profile. Cells cultured for 2 months in MesenCult were analyzed by flow cytometry. They expressed lineage (Lin) differentiation antigens, the common leukocyte antigen CD45, the monocytic marker CD11b, the hematopoietic marker c-kit (CD117), the pericyte/mesoangioblast markers NG2 proteoglycan, and CD146 [24], CD44 (a receptor for hyaluronic acid considered a MSC and EPC marker), and CD29 (integrin β1/fibronectin receptor; Figure 2). NG2 expression was demonstrated immunocytochemically (Figure 3). Stem cell antigen-1 (Sca-1), CD34 (an hematopoietic and EPC marker), CD133 (a stem cell marker), CD31 (an endothelial marker), CD90 (THY1 T-cell antigen; a MSC marker), and CD105 (endoglin; a MSC marker) were expressed by small cell subsets. CD38, CD40, Flk-1, and major histocompatibility complex (MHC) class II molecules were not expressed. In cells cultured in MesenCult for 2 months and then changed to DMEM-10% FCS for 3 weeks, CD45$^+$, CD11b$^+$, or c-kit$^+$ subsets were reduced by approximately half compared with cells maintained in MesenCult (Figure 4). Cell culturing in DMEM-10% FCS supplemented with GM-CSF did not alter CD45, CD11b, and c-kit expression (data not shown).

3.3. Nonclonal Sphere Formation. The isolated population formed spherical clusters that detached from plastic and floated free in the medium. Spheres collected and plated onto new plates disaggregated and gave rise to a monolayer of sphere-derived cells, which were able to form a second generation of spheres. This procedure could be repeated for 3 cycles, at least. We tested the clonality of first-generation spheres by culturing mixtures of cells derived from GFP transgenic and WT mice. While spheres formed by GFP cells were entirely green fluorescent, and those formed by WT cells were non-fluorescent, those formed by mixtures of GFP and WT cells included both green fluorescent and non-fluorescent areas (Figure 5), indicating nonclonality.

3.4. 3D-Cell Culture System in MethoCult. Cells placed in a 3D-culture MethoCult system formed long, dendritic-like filaments after 4 to 8 days in culture (Figure 6).

3.5. Cell Differentiation Potential. Under appropriate conditions, pancreas outgrowth-derived cells differentiated along osteogenic, chondrogenic, and adipogenic lineages, as evidenced by Alizarin-red (Figure 7(a)), Alcian-blue (Figure 7(b)), and oil red-O staining (Figures 7(c) and 7(d)), respectively. Dithizone staining was positive for a subset of cells cultured in 1.27 μM dexamethasone/MesenCult for 18 days (Figures 7(e) and 7(f)).

4. Discussion

The main finding of the present study is that cells obtained by outgrowth from murine pancreas explants in MesenCult give rise to a population of myelomonocytoid cells endowed with mesenchymal differentiation potential. These cells also stain positive with dithizone, a zinc-chelating agent commonly used to detect insulin-producing cells [46–49]. The endocrine differentiation potential of these cells is being addressed in an ongoing study. In the present study, we focus on their phenotype and MSC-like characteristics.

Monocyte-derived cells include macrophages, fibrocytes, dendritic cells, osteoclasts, and adipocytes. Monocytes, unlike macrophages and fibrocytes, do not express CD105 [50–52]. Because the isolated population lacks CD105 expression, it appears to have a monocytoid phenotype. This population meets only part of the minimal criteria for defining MSC established by the International Society for Cellular Therapy [53]: plastic-adherence in standard culture conditions and capacity to differentiate into osteoblasts, adipocytes, and chondroblasts *in vitro* (in the present study, multilineage differentiation potential was not demonstrated with single-cell cloning). Regarding the cell-surface marker profile, the minimal criteria for defining MSC, namely, expression of CD105 and CD90 but not CD45, are not met. However, other MSC/stromal markers (CD44 and CD29) and pericyte/perivascular markers (NG2 and CD146) are

Complete MesenCult medium

(a)

(b)

(c)

DMEM-FCS 10%

(d)

(e)

(f)

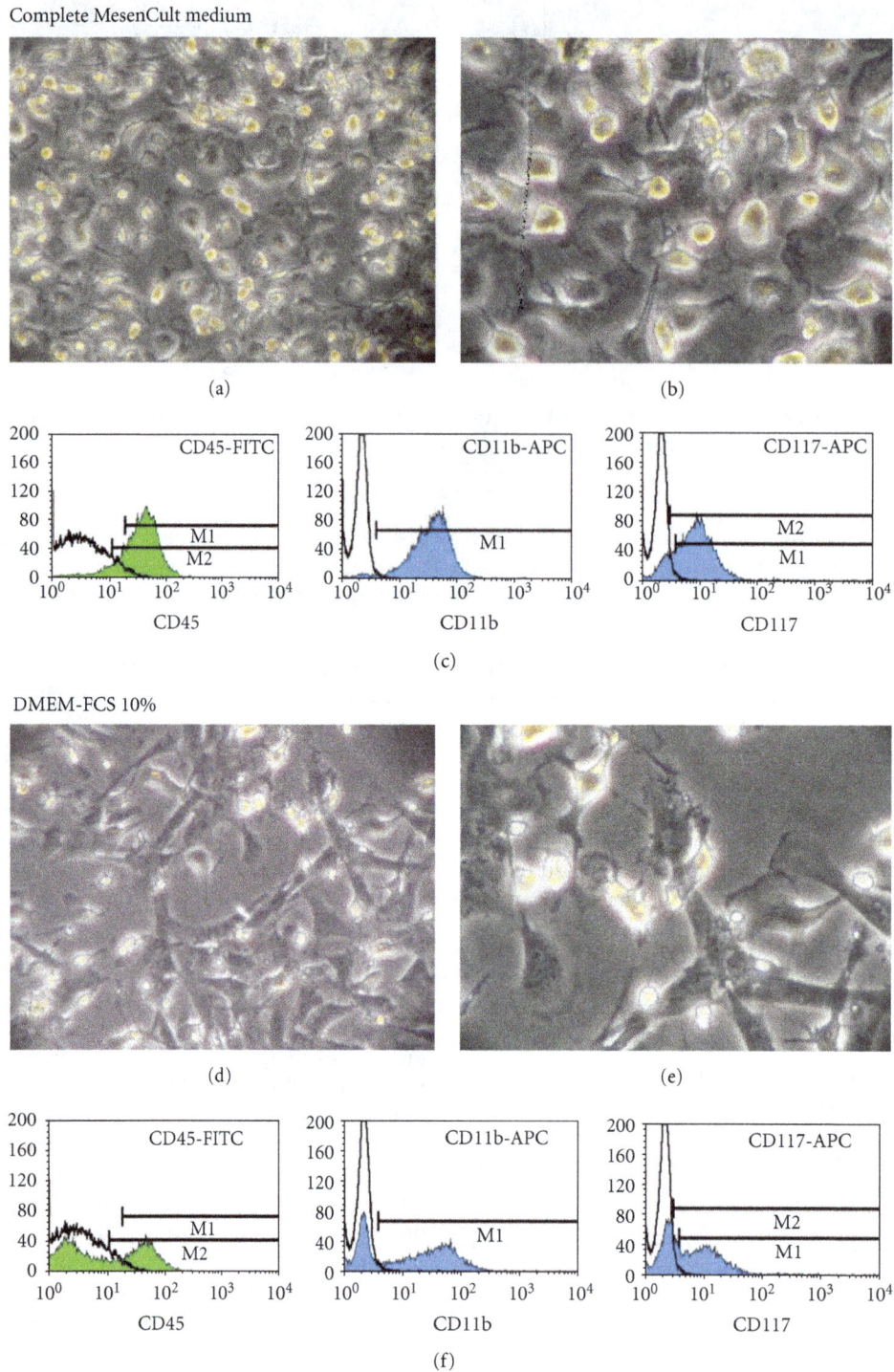

FIGURE 4: Effects of the culture medium on cell morphology and antigen expression. (a–c) Cells cultured in MesenCult for 1 month display highly refringent nuclei, *gingko biloba* leaf-like shapes, and CD45, CD11b, and CD117 expression. (d–f) Cells initially derived in MesenCult and then changed to DMEM-FCS 10% show less refringent nuclei, rhomboid shapes, and decreased subsets of $CD45^+$, $CD11b^+$, or $CD117^+$ cells.

expressed [42]. In this regard, it has been shown that human MSCs in several organs originate from pericytes/perivascular cells and express NG2 [54]. For comparison, previous studies have shown that human islet outgrowths are positive for multiple MSC and pericyte markers ($CD105^+$, $CD90^+$, $CD44^+$, $CD29^+$, $NG2^+$, and $CD146^+$) but negative for CD45.

Pancreas-derived cells cultured in the 3D-MethoCult system exhibit a dendritic or oligodendrocytic-like morphology

Characterization of Myelomonocytoid Progenitor Cells with Mesenchymal Differentiation Potential Obtained by Outgrowth from Pancreas Explants

111

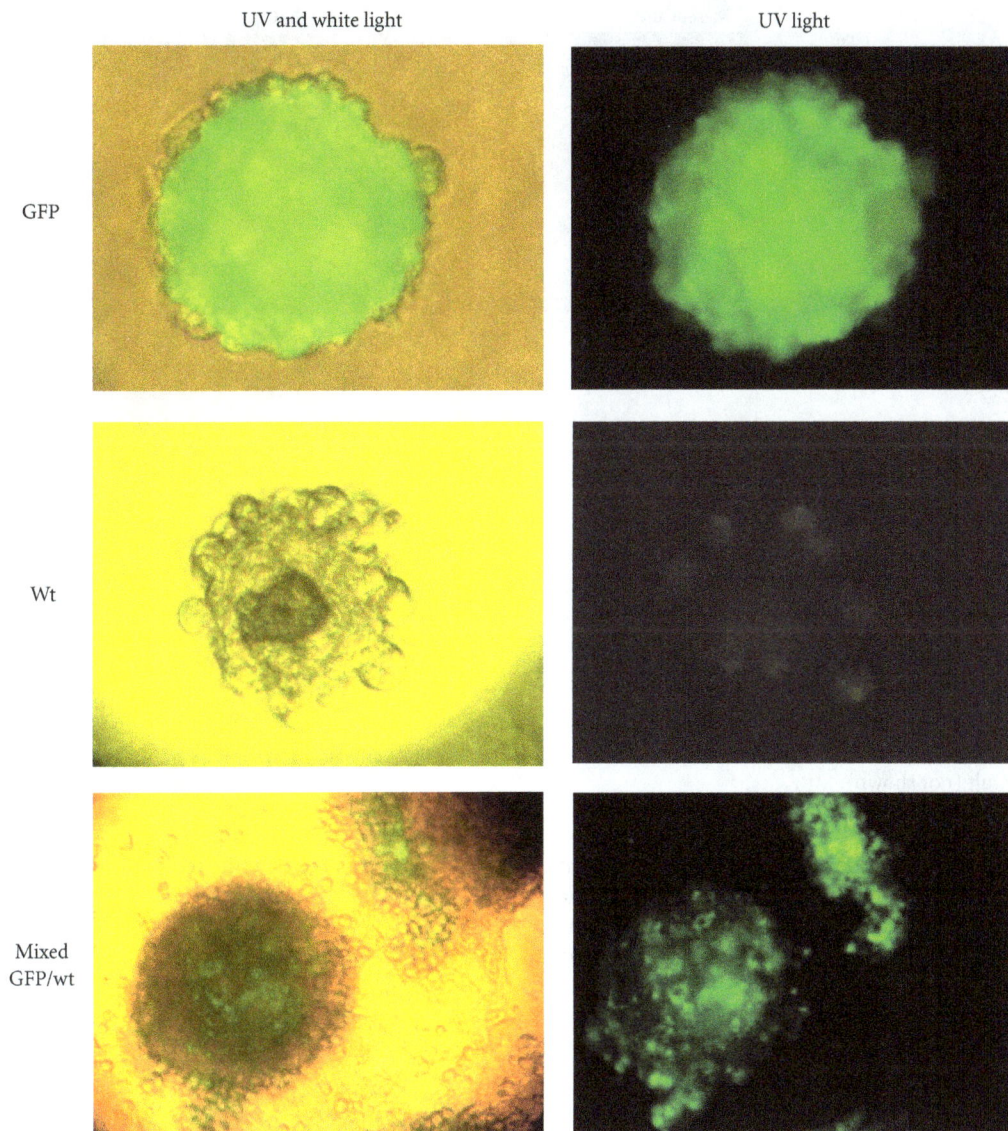

FIGURE 5: Photomicrographs and UV photomicrographs of sphere clusters formed by pancreas outgrowth-derived cells from either WT or GFP-transgenic mice, or by mixtures of the two. The latter show a patchy white/green pattern under the UV light, indicating that spheres are nonclonal.

characterized by multiple branched filaments. Microglia have a CD11b$^+$CD45^{lo} phenotype and can be distinguished from primary macrophages on the basis of their CD45 expression level [55]. Differentiation of mouse BM-derived stem cells toward microglia-like cells has been reported [56].

The isolated myelomonocytoid population appears to have an advantage in terms of survival or growth compared with other cells present in the cellular outgrowth from pancreatic explants. These cells may die off and be taken over by the myelomonocytoid component that persists after extended periods of time. The underlying mechanism is unclear. Because MesenCult is a commercially available medium that has been optimized for growth of MSC, the emergence of a myelomonocytoid population over time is somewhat surprising. In a previous study [15], we used this medium to expand mouse cardiac-derived MSC,

which displayed a stable phenotype (Lin$^-$, Sca-1$^+$, CD90$^+$, CD105$^+$, CD45$^-$, and CD31$^-$) for more than 25 passages. This observation indicates that MesenCult can preserve the phenotype of cultured MSC for extended periods of time, at least under certain circumstances.

The mesenchymal differentiation potential of the myelomonocytoid population may appear at odds with the established CD45$^-$ MSC phenotype [53]. However, Sordi et al. [41] recently showed that mesenchymal cells emerging from human pancreatic culture did not result from an epithelial to mesenchymal transition but represented the expansion of a pool of resident MSC located in the periacinar, perivascular, and periductal space. Using a GFP$^+$ BM transplant model, they showed that mesenchymal cells emerging from pancreatic endocrine or exocrine tissue culture originated mainly from the CD45$^+$ BM compartment.

MesenCult   MethoCult

Day 4

(a)   (b)

Day 8

(c)

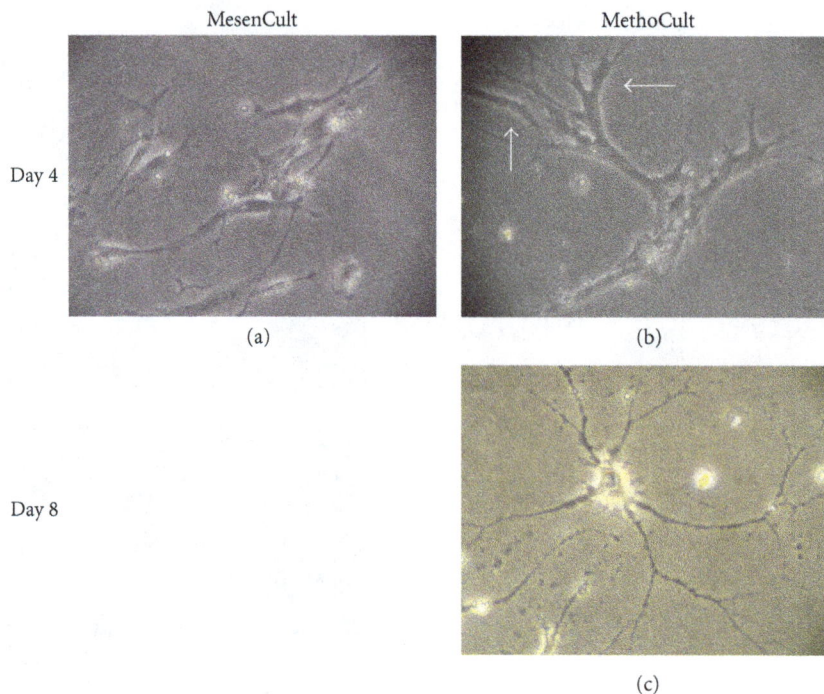

FIGURE 6: Pancreas-derived cells (5 weeks after plating; passage 3) were seeded at very low density (10^3 cells/35 mm plate) in either MesenCult (a) or MethoCult (b and c). Pictures were taken 4–8 days later. Thin dendritic-like cell extensions, or filaments, were seen in MethoCult (c), but not in MesenCult (not shown).

Athough these cells expressed negligible levels of islet-specific genes, they improved islet function and neovascularization after transplantation with a minimal islet mass in a mouse model. Kaiser et al. [57] showed that a small population of BM MSC originated from the CD45+CD34+ fraction, whereas the majority was obtained from the CD45−CD34− fraction. MSC from either fraction could be differentiated into adipocytes, osteocytes, and chondroblasts. Additional studies confirmed that MSC can express CD45 under certain conditions [58–60]. In some cases, CD45 expression was dramatically downregulated during *in vitro* culture [57, 58]. As mentioned above, our data suggest that MesenCult may preserve CD45 expression in cultured pancreas-derived myelomonocytoid cells, whereas partial downregulation of CD45 expression was observed in a standard culture medium (DMEM-10% FCS).

Pancreas-derived myelomonocytoid cells form spherical clusters, which is considered a feature of stemness [43]. However, these spheres are nonclonal. Similar findings have been reported for neural stem cells, as colonies formed by these cells can grow clonal or nonclonal [61].

The origin of the isolated myelomonocytoid population remains unclear. It might originate from blood monocytes trapped in the intravascular compartment of tissue explants, as recently shown for CD45+ cells from cardiac explant outgrowths [62]. It should be mentioned, however, that we were not able to isolate myelomonocytoid cells from murine BM-derived cells using the same culture conditions. Alternatively, this population might originate from pancreas-resident monocytes, monocyte-derived cells, or MSC. In this regard, Freisinger et al. [63] showed that clonally isolated,

adipose-derived MSC cultured in appropriate differentiation media gave rise to cells expressing monocyte/macrophage and early hematopoietic markers.

Our findings are in general agreement with recent reports on multipotent monocytes. Zhao et al. [64] isolated a subset of adult pluripotent stem cells (CD14+, CD34+, and CD45+) from human peripheral blood monocytes. These cells in appearance resembled fibroblasts, expanded in the presence of macrophage colony-stimulating factor (M-CSF), and could be differentiated into mature macrophages and T lymphocytes, as well as into epithelial, endothelial, neuronal, and liver cells in the presence of appropriate growth factors. Kuwana et al. [44] and Kuwana and Seta [65] described human blood monocyte-derived multipotent cells (MOMCs; CD14+, CD34+, CD45+, and type-I collagen+) that exhibited a fibroblast-like morphology and contained progenitors with the capacity to differentiate into bone, cartilage, fat, skeletal muscle, cardiac muscle, neuron, and endothelium [65–67]. Romagnani et al. [68] described circulating clonogenic, multipotent CD14+ CD34lo cells that proliferated in response to stem cell growth factors. Ungefroren and Fändrich [69] reported that the programmable cell of monocytic origin (PCMO) is a potential adult stem/progenitor cell source for the generation of islet cells. Hur et al. [48] recently showed that human peripheral blood monocytes could be differentiated into insulin-producing cells using the hemato-sphere culture technique. Collectively, these data suggest that blood monocytes and monocyte-derived cells, although not considered classic adult stem cells, may represent versatile progenitor cells capable of generating multiple types of cells.

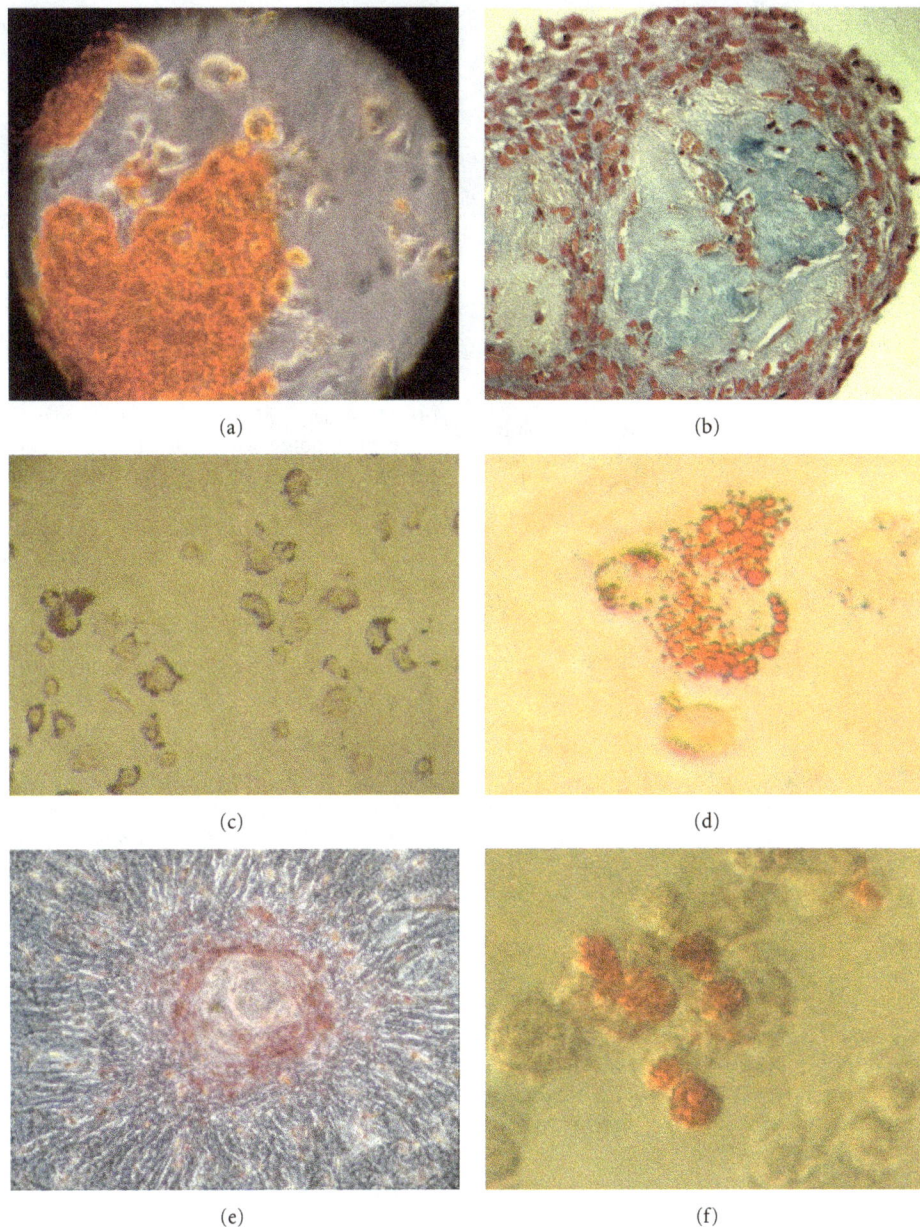

FIGURE 7: Multilineage differentiation of pancreas-derived cells. (a) Osteogenic differentiation (Alizarin-red staining). (b) Chondrogenic differentiation (Alcian-blue staining). (c and d) Adipogenic differentiation (Oil red-O staining; low/high magnification views). (e and f) Cells cultured for 2 months in MesenCult showed areas of positive staining with dithizone (red) as evidence of zinc-rich insulin-producing cells (low/high magnification views).

Owing to their mesenchymal differentiation potential, pancreas outgrowth-derived myelomonocytoid cells are of potential interest to cell therapy applications even though this aspect was not directly addressed by the present study. Sordi et al. [41] reported beneficial effects of pancreatic MSC in diabetic mice, as mentioned above. When cotransplanted with a minimal islet mass, these cells improved neovascularization and islet function. This effect was not due to MSC differentiation into insulin-secreting cells, but to MSC-mediated protective effects on transplanted islets. Moreover, Johansson et al. [70] showed that MSC within composite endothelial cell-MSC-pancreatic islets improved endothelial cell proliferation and sprouting *in vitro*. It therefore could be speculated that pancreas-derived myelomonocytoid cells endowed with MSC potential might exert trophic effects on pancreatic islets via paracrine mechanisms, as reported for pancreatic MSC by Sordi et al. [41]. Further studies are warranted to test this hypothesis and to define the origin and the endocrine potential of the pancreas-derived myelomonocytoid population.

Acknowledgments

Support by the Fondation Vaudoise de Cardiologie (Lausanne, Switzerland), the Cecilia Augusta Foundation, the METIS Foundation Sergio Mantegazza, the "Fondazione per la ricerca sulla trasfusione e sui trapianti" (Lugano, Switzerland), and the Swiss Heart Foundation (Berne, Switzerland) is gratefully acknowledged.

References

[1] J. Domínguez-Bendala, L. Inverardi, and C. Ricordi, "Regeneration of pancreatic beta-cell mass for the treatment of diabetes," *Expert Opinion on Biological Therapy*, vol. 12, no. 6, pp. 731–741, 2012.

[2] T. Guo and M. Hebrok, "Stem cells to pancreatic β-cells: new sources for diabetes cell therapy," *Endocrine Reviews*, vol. 30, no. 3, pp. 214–227, 2009.

[3] Y. Dor, J. Brown, O. I. Martinez, and D. A. Melton, "Adult pancreatic β-cells are formed by self-duplication rather than stem-cell differentiation," *Nature*, vol. 429, no. 6987, pp. 41–46, 2004.

[4] S. Bonner-Weir, M. Taneja, G. C. Weir et al., "In vitro cultivation of human islets from expanded ductal tissue," *Proceedings of the National Academy of Sciences of the United States of America*, vol. 97, no. 14, pp. 7999–8004, 2000.

[5] M. Reichert and A. K. Rustgi, "Pancreatic ductal cells in development, regeneration, and neoplasia," *The Journal of Clinical Investigation*, vol. 121, pp. 4572–4578, 2011.

[6] X. Xu, J. D'Hoker, G. Stangé et al., "β cells can be generated from endogenous progenitors in injured adult mouse pancreas," *Cell*, vol. 132, no. 2, pp. 197–207, 2008.

[7] V. K. Ramiya, M. Maraist, K. E. Arfors, D. A. Schatz, A. B. Peck, and J. G. Cornelius, "Reversal of insulin-dependent diabetes using islets generated in vitro from pancreatic stem cells," *Nature Medicine*, vol. 6, no. 3, pp. 278–282, 2000.

[8] R. M. Seaberg, S. R. Smukler, T. J. Kieffer et al., "Clonal identification of multipotent precursors from adult mouse pancreas that generate neural and pancreatic lineages," *Nature Biotechnology*, vol. 22, no. 9, pp. 1115–1124, 2004.

[9] R. M. Baertschiger, D. Bosco, P. Morel et al., "Mesenchymal stem cells derived from human exocrine pancreas express transcription factors implicated in beta-cell development," *Pancreas*, vol. 37, no. 1, pp. 75–84, 2008.

[10] V. M. Schwitzgebel, D. W. Scheel, J. R. Conners et al., "Expression of neurogenin3 reveals an islet cell precursor population in the pancreas," *Development*, vol. 127, no. 16, pp. 3533–3542, 2000.

[11] H. T. Lin, S. H. Chiou, C. L. Kao et al., "Characterization of pancreatic stem cells derived from adult human pancreas ducts by fluorescence activated cell sorting," *World Journal of Gastroenterology*, vol. 12, no. 28, pp. 4529–4535, 2006.

[12] Y. Oshima, A. Suzuki, K. Kawashimo, M. Ishikawa, N. Ohkohchi, and H. Taniguchi, "Isolation of mouse pancreatic ductal progenitor cells expressing CD133 and c-Met by flow cytometric cell sorting," *Gastroenterology*, vol. 132, no. 2, pp. 720–732, 2007.

[13] A. Suzuki, H. Nakauchi, and H. Taniguchi, "Prospective isolation of multipotent pancreatic progenitors using flow-cytometric cell sorting," *Diabetes*, vol. 53, no. 8, pp. 2143–2152, 2004.

[14] T. Sugiyama, R. T. Rodriguez, G. W. McLean, and S. K. Kim, "Conserved markers of fetal pancreatic epithelium permit prospective isolation of islet progenitor cells by FACS," *Proceedings of the National Academy of Sciences of the United States of America*, vol. 104, no. 1, pp. 175–180, 2007.

[15] H. Immervoll, D. Hoem, P. Ø Sakariassen, O. J. Steffensen, and A. Molven, "Expression of the "stem cell marker" CD133 in pancreas and pancreatic ductal adenocarcinomas," *BMC Cancer*, vol. 8, article 48, 2008.

[16] Y. C. Ya-Chieh Hsu and E. Fuchs, "A family business: stem cell progeny join the niche to regulate homeostasis," *Nature Reviews Molecular Cell Biology*, vol. 13, pp. 103–114, 2012.

[17] F. Esni, Y. Miyamoto, S. D. Leach, and B. Ghosh, "Primary explant cultures of adult and embryonic pancreas," *Methods in Molecular Medicine*, vol. 103, pp. 259–271, 2005.

[18] E. Schneider, A. Schmid-Kotsas, J. Zhao et al., "Identification of mediators stimulating proliferation and matrix synthesis of rat pancreatic stellate cells," *American Journal of Physiology*, vol. 281, no. 2, pp. C532–C543, 2001.

[19] M. Bläuer, I. Nordback, J. Sand, and J. Laukkarinen, "A novel explant outgrowth culture model for mouse pancreatic acinar cells with long-term maintenance of secretory phenotype," *European Journal of Cell Biology*, vol. 90, pp. 1052–1160, 2011.

[20] F. Carlotti, A. Zaldumbide, C. J. Loomans et al., "Isolated human islets contain a distinct population of mesenchymal stem cells," *Islets*, vol. 2, no. 3, pp. 164–173, 2010.

[21] M. C. Gershengorn, A. A. Hardikar, C. Wei, E. Ceras-Raaka, B. Marcus-Samuels, and B. M. Raaka, "Epithelial-to-mesenchymal transition generates proliferative human islet precursor cells," *Science*, vol. 306, no. 5705, pp. 2261–2264, 2004.

[22] A. Lechner, A. L. Nolan, R. A. Blacken, and J. F. Habener, "Redifferentiation of insulin-secreting cells after in vitro expansion of adult human pancreatic islet tissue," *Biochemical and Biophysical Research Communications*, vol. 327, no. 2, pp. 581–588, 2005.

[23] L. Ouziel-Yahalom, M. Zalzman, L. Anker-Kitai et al., "Expansion and redifferentiation of adult human pancreatic islet cells," *Biochemical and Biophysical Research Communications*, vol. 341, no. 2, pp. 291–298, 2006.

[24] M. Eberhardt, P. Salmon, M. A. von Mach et al., "Multipotential nestin and Isl-1 positive mesenchymal stem cells isolated from human pancreatic islets," *Biochemical and Biophysical Research Communications*, vol. 345, no. 3, pp. 1167–1176, 2006.

[25] R. Gallo, F. Gambelli, B. Gava et al., "Generation and expansion of multipotent mesenchymal progenitor cells from cultured human pancreatic islets," *Cell Death and Differentiation*, vol. 14, no. 11, pp. 1860–1871, 2007.

[26] R. A. Morton, E. Geras-Raaka, L. M. Wilson, B. M. Raaka, and M. C. Gershengorn, "Endocrine precursor cells from mouse islets are not generated by epithelial-to-mesenchymal transition of mature beta cells," *Molecular and Cellular Endocrinology*, vol. 270, no. 1-2, pp. 87–93, 2007.

[27] F. Atouf, H. P. Cheol, K. Pechhold, M. Ta, Y. Choi, and N. L. Lumelsky, "No evidence for mouse pancreatic β-cell epithelial-mesenchymal transition in vitro," *Diabetes*, vol. 56, no. 3, pp. 699–702, 2007.

[28] L. G. Chase, F. Ulloa-Montoya, B. L. Kidder, and C. M. Verfaillie, "Islet-derived fibroblast-like cells are not derived via epithelial-mesenchymal transition from Pdx-1 or insulin-positive cells," *Diabetes*, vol. 56, no. 1, pp. 3–7, 2007.

[29] H. A. Russ, Y. Bar, P. Ravassard, and S. Efrat, "In vitro proliferation of cells derived from adult human β-cells revealed by cell-lineage tracing," *Diabetes*, vol. 57, no. 6, pp. 1575–1583, 2008.

[30] D. S. Krause, N. D. Theise, M. I. Collector et al., "Multi-organ, multi-lineage engraftment by a single bone marrow-derived stem cell," *Cell*, vol. 105, no. 3, pp. 369–377, 2001.

[31] C. Moriscot, F. de Fraipont, M. J. Richard et al., "Human bone marrow mesenchymal stem cells can express insulin and key transcription factors of the endocrine pancreas developmental pathway upon genetic and/or microenvironmental manipulation in vitro," *Stem Cells*, vol. 23, no. 4, pp. 594–603, 2005.

[32] K. Timper, D. Seboek, M. Eberhardt et al., "Human adipose tissue-derived mesenchymal stem cells differentiate into insulin, somatostatin, and glucagon expressing cells," *Biochemical and Biophysical Research Communications*, vol. 341, no. 4, pp. 1135–1140, 2006.

[33] B. Davani, L. Ikonomou, B. M. Raaka et al., "Human islet-derived precursor cells are mesenchymal stromal cells that differentiate and mature to hormone-expressing cells in vivo," *Stem Cells*, vol. 25, no. 12, pp. 3215–3222, 2007.

[34] A. Ianus, G. G. Holz, N. D. Theise, and M. A. Hussain, "In vivo derivation of glucose-competent pancreatic endocrine cells from bone marrow without evidence of cell fusion," *The Journal of Clinical Investigation*, vol. 111, no. 6, pp. 843–850, 2003.

[35] A. Lechner, Y. G. Yang, R. A. Blacken, L. Wang, A. L. Nolan, and J. F. Habener, "No evidence for significant transdifferentiation of bone marrow into pancreatic beta-cells in vivo," *Diabetes*, vol. 53, no. 3, pp. 616–623, 2004.

[36] A. I. Caplan and J. E. Dennis, "Mesenchymal stem cells as trophic mediators," *Journal of Cellular Biochemistry*, vol. 98, no. 5, pp. 1076–1084, 2006.

[37] R. H. Lee, M. J. Seo, R. L. Reger et al., "Multipotent stromal cells from human marrow home to and promote repair of pancreatic islets and renal glomeruli in diabetic NOD/scid mice," *Proceedings of the National Academy of Sciences of the United States of America*, vol. 103, no. 46, pp. 17438–17443, 2006.

[38] V. S. Urbán, J. Kiss, J. Kovács et al., "Mesenchymal stem cells cooperate with bone marrow cells in therapy of diabetes," *Stem Cells*, vol. 26, no. 1, pp. 244–253, 2008.

[39] E. J. Estrada, F. Valacchi, E. Nicora et al., "Combined treatment of intrapancreatic autologous bone marrow stem cells and hyperbaric oxygen in type 2 diabetes mellitus," *Cell Transplantation*, vol. 17, no. 12, pp. 1295–1304, 2008.

[40] M. Zhao, S. A. Amiel, S. Ajami et al., "Amelioration of streptozotocin-induced diabetes in mice with cells derived from human marrow stromal cells," *PLoS ONE*, vol. 3, no. 7, Article ID e2666, 2008.

[41] V. Sordi, R. Melzi, A. Mercalli et al., "Mesenchymal cells appearing in pancreatic tissue culture are bone marrow-derived stem cells with the capacity to improve transplanted islet function," *Stem Cells*, vol. 28, no. 1, pp. 140–151, 2010.

[42] C. L. Maier, B. R. Shepherd, T. Yi, and J. S. Pober, "Explant outgrowth, propagation and characterization of human pericytes," *Microcirculation*, vol. 17, no. 5, pp. 367–380, 2010.

[43] E. Pastrana, V. Silva-Vargas, and F. Doetsch, "Eyes wide open: a critical review of sphere-formation as an assay for stem cells," *Cell Stem Cell*, vol. 8, no. 5, pp. 486–498, 2011.

[44] M. Kuwana, Y. Okazaki, H. Kodama et al., "Human circulating CD14+ monocytes as a source of progenitors that exhibit mesenchymal cell differentiation," *Journal of Leukocyte Biology*, vol. 74, no. 5, pp. 833–845, 2003.

[45] A. Meinhardt, A. Spicher, M. E. Roehrich, I. Glauche, P. Vogt, and G. Vassalli, "Immunohistochemical and flow cytometric analysis of long-term label-retaining cells in the adult heart," *Stem Cells and Development*, vol. 20, no. 2, pp. 211–222, 2011.

[46] Z. A. Latif, J. Noel, and R. Alejandro, "A simple method of staining fesh and cultured islets," *Transplantation*, vol. 45, no. 4, pp. 827–830, 1988.

[47] A. Shiroi, M. Yoshikawa, H. Yokota et al., "Identification of insulin-producing cells derived from embryonic stem cells by zinc-chelating dithizone," *Stem Cells*, vol. 20, no. 4, pp. 284–292, 2002.

[48] J. Hur, J. M. Yang, J. I. Choi et al., "New method to differentiate human peripheral blood monocytes into insulin producing cells: human hematosphere culture," *Biochemical and Biophysical Research Communications*, vol. 418, pp. 765–769, 2012.

[49] A. Rezania, M. J. Riedel, R. D. Wideman et al., "Production of functional glucagon-secreting α-cells from human embryonic stem cells," *Diabetes*, vol. 60, no. 1, pp. 239–247, 2011.

[50] W. Gorczyca, Z. Y. Sun, W. Cronin, X. Li, S. Mau, and S. Tugulea, "Immunophenotypic pattern of myeloid populations by flow cytometry analysis," *Methods in Cell Biology*, vol. 103, pp. 221–266, 2011.

[51] D. Pilling, T. Fan, D. Huang, B. Kaul, and R. H. Gomer, "Identification of markers that distinguish monocyte-derived fibrocytes from monocytes, macrophages, and fibroblasts," *PLoS ONE*, vol. 4, no. 10, Article ID e7475, 2009.

[52] S. J. Curnow, M. Fairclough, C. Schmutz et al., "Distinct types of fibrocyte can differentiate from mononuclear cells in the presence and absence of serum," *PLoS ONE*, vol. 5, no. 3, Article ID e9730, 2010.

[53] M. Dominici, K. Le Blanc, I. Mueller et al., "Minimal criteria for defining multipotent mesenchymal stromal cells. The International Society for Cellular Therapy Position Statement," *Cytotherapy*, vol. 8, no. 4, pp. 315–317, 2006.

[54] M. Crisan, S. Yap, L. Casteilla et al., "A perivascular origin for mesenchymal stem cells in multiple human organs," *Cell Stem Cell*, vol. 3, no. 3, pp. 301–313, 2008.

[55] A. L. Ford, A. L. Goodsall, W. F. Hickey, and J. D. Sedgwick, "Normal adult ramified microglia separated from other central nervous system macrophages by flow cytometric sorting: phenotypic differences defined and direct ex vivo antigen presentation to myelin basic protein-reactive CD4+ T cells compared," *Journal of Immunology*, vol. 154, no. 9, pp. 4309–4321, 1995.

[56] A. Hinze and A. Stolzing, "Differentiation of mouse bone marrow derived stem cells toward microglia-like cells," *BMC Cell Biology*, vol. 12, article 35, 2011.

[57] S. Kaiser, B. Hackanson, M. Follo et al., "BM cells giving rise to MSC in culture have a heterogeneous CD34 and CD45 phenotype," *Cytotherapy*, vol. 9, no. 5, pp. 439–450, 2007.

[58] F. Deschaseaux, F. Gindraux, R. Saadi, L. Obert, D. Chalmers, and P. Herve, "Direct selection of human bone marrow mesenchymal stem cells using an anti-CD49a antibody reveals their CD45med,low phenotype," *British Journal of Haematology*, vol. 122, no. 3, pp. 506–517, 2003.

[59] Y. Koide, S. Morikawa, Y. Mabuchi et al., "Two distinct stem cell lineages in murine bone marrow," *Stem Cells*, vol. 25, no. 5, pp. 1213–1221, 2007.

[60] I. Rogers, N. Yamanaka, R. Bielecki et al., "Identification and analysis of in vitro cultured CD45-positive cells capable of multi-lineage differentiation," *Experimental Cell Research*, vol. 313, no. 9, pp. 1839–1852, 2007.

[61] B. L. Coles-Takabe, I. Brain, K. A. Purpura et al., "Don't look: growing clonal versus nonclonal neural stem cell colonies," *Stem Cells*, vol. 26, no. 11, pp. 2938–2944, 2008.

[62] D. R. Davis, Y. Zhang, R. R. Smith et al., "Validation of the cardiosphere method to culture cardiac progenitor cells from myocardial tissue," *PLoS ONE*, vol. 4, no. 9, Article ID e7195, 2009.

[63] E. Freisinger, C. Cramer, X. Xia et al., "Characterization of hematopoietic potential of mesenchymal stem cells," *Journal of Cellular Physiology*, vol. 225, no. 3, pp. 888–897, 2010.

[64] Y. Zhao, D. Glesne, and E. Huberman, "A human peripheral blood monocyte-derived subset acts as pluripotent stem cells," *Proceedings of the National Academy of Sciences of the United States of America*, vol. 100, no. 5, pp. 2426–2431, 2003.

[65] M. Kuwana and N. Seta, "Human circulating monocytes as multipotential progenitors," *Keio Journal of Medicine*, vol. 56, no. 2, pp. 41–47, 2007.

[66] H. Kodama, T. Inoue, R. Watanabe et al., "Neurogenic potential of progenitors derived from human circulating CD14+ monocytes," *Immunology and Cell Biology*, vol. 84, no. 2, pp. 209–217, 2006.

[67] H. Kodama, T. Inoue, R. Watanabe et al., "Cardiomyogenic potential of mesenchymal progenitors derived from human circulating CD14+ monocytes," *Stem Cells and Development*, vol. 14, no. 6, pp. 676–686, 2005.

[68] P. Romagnani, F. Annunziato, F. Liotta et al., "CD14+CD34low cells with stem cell phenotypic and functional features are the major source of circulating endothelial progenitors," *Circulation Research*, vol. 97, no. 4, pp. 314–322, 2005.

[69] H. Ungefroren and F. Fändrich, "The programmable cell of monocytic origin (PCMO): a potential adult stem/progenitor cell source for the generation of islet cells," *Advances in Experimental Medicine and Biology*, vol. 654, pp. 667–682, 2010.

[70] U. Johansson, I. Rasmusson, S. P. Niclou et al., "Formation of composite endothelial cell-mesenchymal stem cell islets: a novel approach to promote islet revascularization," *Diabetes*, vol. 57, no. 9, pp. 2393–2401, 2008.

13

Display of the Viral Epitopes on *Lactococcus lactis*: A Model for Food Grade Vaccine against EV71

Nadimpalli Ravi S. Varma,[1] **Haryanti Toosa,**[2] **Hooi Ling Foo,**[2,3]
Noorjahan Banu Mohamed Alitheen,[2,4] **Mariana Nor Shamsudin,**[2,5] **Ali S. Arbab,**[1]
Khatijah Yusoff,[2,6] **and Raha Abdul Rahim**[2,4]

[1] *Cellular and Molecular Imaging Laboratory, Department of Radiology, Henry Ford Hospital, Detroit, MI 48202, USA*
[2] *Institute of Bioscience, Faculty of Biotechnology and Biomolecular Sciences, Universiti Putra Malaysia, 43400 Serdang, Selangor, Malaysia*
[3] *Department of Bioprocess Technology, Faculty of Biotechnology and Biomolecular Sciences, Universiti Putra Malaysia, 43400 Serdang, Selangor, Malaysia*
[4] *Department of Cell and Molecular Biology, Faculty of Biotechnology and Biomolecular Sciences, Universiti Putra Malaysia, 43400 Serdang, Selangor, Malaysia*
[5] *Department of Medical Microbiology and Parasitology, Faculty of Medicine and Health Sciences, 43400 Serdang, Selangor, Malaysia*
[6] *Department of Microbiology, Faculty of Biotechnology and Biomolecular Sciences, Universiti Putra Malaysia, 43400 Serdang, Selangor, Malaysia*

Correspondence should be addressed to Raha Abdul Rahim; raha@biotech.upm.edu.my

Academic Editor: Yu Hong Wei

In this study, we have developed a system for display of antigens of Enterovirus type 71 (EV71) on the cell surface of *L. lactis*. The viral capsid protein (VP1) gene from a local viral isolate was utilized as the candidate vaccine for the development of oral live vaccines against EV71 using *L. lactis* as a carrier. We expressed fusion proteins in *E. coli* and purified fusion proteins were incubated with *L. lactis*. We confirmed that mice orally fed with *L. lactis* displaying these fusion proteins on its surface were able to mount an immune response against the epitopes of EV71. This is the first example of an EV71 antigen displayed on the surface of a food grade organism and opens a new perspective for alternative vaccine strategies against the EV71. We believe that the method of protein docking utilized in this study will allow for more flexible presentations of short peptides and proteins on the surface of *L. lactis* to be useful as a delivery vehicle.

1. Introduction

Enterovirus 71 infection manifests most frequently as the childhood illness known as hand-foot- and-mouth disease (HFMD) and is considered to be clinically indistinguishable from HFMD caused by Coxsackie A16 (CA16). However, the former has the propensity to cause neurological disease during acute infection, a feature not observed in CA16 infections [1]. Children under 5 years of age are partichltularly susceptible to the more severe forms of EV71-associated neurological disease, including aseptic meningitis, brainstem or cerebellar encephalitis, and acute flaccid paralysis. Several large epidemics of severe EV71 infection in young children, including numerous cases of fatal brainstem encephalitis, have recently been reported in South East Asia and Western Australia [2–6] raising concern that there may be an increase in both the prevalence and virulence of EV71. Two candidate vaccines against EV71 utilizing a formalin-inactivated whole virus and a DNA vaccine expressing VP1 have previously been developed [7]. In addition, both recombinant and subunit vaccine strategies optimized as a neutralizing antibody had been shown to provide some protection against EV71 lethal challenges in neonatal mice [8].

The use of a live, food grade organism that is noninvasive and nonpathogenic as antigen delivery vehicle is a promising vaccine strategy. This strategy could overcome potential problems due to the use of live attenuated enteroviral strains, which may have the risk of reversion and residual virulence. The immunogenicity by *L. lactis* expressing several bacterial and viral antigens has been documented [9–11]. One of the main factors inhibiting their use in a live vaccine delivery is the lack of expression vectors with strong promoters. To overcome these problems associated with high expression of proteins in *L. lactis*, we have chosen the *E. coli* expression host due to the availability of a wide variety of expression vectors and that recombinant proteins produced in *E. coli* can be easily purified. In this work, we expressed and purified individually the fusion proteins (viral epitopes fused with cell wall binding anchor protein) and successfully anchored the epitopes on the outer surface of *L. lactis* to be presented as a surface displayed antigen. Preliminary immunological studies have demonstrated the generation of specific antibody responses in mice orally fed with *L. lactis* displaying epitopes of EV71.

2. Materials and Methods

2.1. Microorganisms. Escherichia coli TOP10 (Invitrogen, Carlsbad, CA, USA) was used as a cloning host. *E. coli* BL21 (DE3) F$^-$ *ompT hsd*S$_B$ (r$_B^-$ m$_B^-$) *gal dcm* (DE3) plysS (Cam^R) was used as the *E. coli* expression host. *L. lactis* MG1363 [12], was used to display the viral epitopes.

2.2. Culture Conditions. Lactococcal cells were grown at 30°C in M17 broth (Oxoid, USA) (Tryptone (5 g/L), Soya peptone (5 g/L), Lab-Lemco (5 g/L), Yeast extract (2.5 g/L), Ascorbic acid (0.5 g/L), Magnesium sulphate (0.5 g/L), and Di-sodium-glycerophosphate (19 g/L)) or M17 agar with 0.5% glucose as standing culture. *E. coli* cells were grown at 37°C with agitation in Luria-Bertani (LB) (Oxoid) broth (Tryptone (10 g/L), Yeast extract (5 g/L), and NaCl (10 g/L)). Whenever required, a total concentration of 50 μg/mL ampicillin was used for the recombinant *E. coli* cultures.

2.3. Plasmids. pCR 2.1 (Amp^R, Km^R, Invitrogen) is an *E. coli* vector used for subcloning the nucleotides from 1 to 201 (VP1$_{1-201nt}$) and 103 to 300 (VP1$_{103-300nt}$) of VP1 gene of EV71; pSVac (Amp^R) is an *E. coli* expression plasmid harbouring the N-acetylmuramidase (*acmA'*) gene fragment [13]. pRSETC (Amp^R, Invitrogen) is pUC-derived expression vector designed for high level protein expression of cloned genes under T7 promoter in *E. coli*.

2.4. Construction of Plasmids pSVacVP1$_{1-201}$ and pSVacVP1$_{103-300}$. Plasmids pSVacVP1$_{1-201}$ and pSVacVP1$_{103-300}$ were previously constructed [13]. The N-terminal region A1 represents VP1$_{1-201}$ and N-terminal region A3 represents VP1$_{103-300}$.

2.5. Expression Studies. The *E. coli* BL21 (DE3) plysS cells containing the recombinant vectors were grown overnight at 37°C with shaking at 250 rpm. The cells were subcultured into a fresh 10 mL of LB medium containing ampicillin (50 μg/mL) and chloramphenicol (35 μg/mL) grown to an OD$_{600}$ of 0.6 before being induced with 1 mM of isopropyl-β-D-thiogalactopyranoside (IPTG) for 3 h. Sodium dodecyl sulfate-polyacrylamide gel electrophoresis (SDS-PAGE) was performed according to Laemmli, 1970 [14], using 10%–12.5% (w/v) polyacrylamide gels. The *E. coli* cultures were harvested by centrifugation at 2,000 g for 10 min. The cell pellets were resuspended in 100 μL of 2X sample buffer (0.125 M Tris, 4% SDS, 0.2 M DDT, 0.02% bromophenol blue, and 20% glycerol) prior to boiling at 95°C for 5 min before centrifugation at 10,000 g. A volume of 10 μL of the supernatant was loaded onto the gel. Semidry blotter (Hoeffer, Pharmacia Biotech, UK) was utilized to transfer the electrophoresed protein bands from the SDS-PAGE to polyvinylidene difluoride (PVDF) membrane. The membrane was then incubated in 1% (w/v) blocking solution (Roche Diagnostics GmbH, Mannheim, Germany) for 1 h at room temperature with gentle agitation. Then the membrane was incubated with primary rabbit anti-VP1 (Professor Dr. Mary Jane Cardosa, Universiti Malaysia Sarawak, Malaysia). The conjugated membrane was then washed with TSBT (Roche) three times before incubation with peroxidase-labelled goat anti-rabbit IgG secondary antibody (50 mU/mL in 0.5% (w/v) blocking solution (Roche)) for 60 min at room temperature with gentle agitation. After washing with TSBT, the membrane was exposed to film for 20 min and visualized.

2.6. Purification of Recombinant Protein Fragment and Binding to L. lactis. The cell cultures (10 mL) were harvested after 2–3 h induction with IPTG. The cells were resuspended in 400 μL of PBS pH 7.4 (20 mM K$_2$HPO$_4$, 5 mM KH$_2$PO$_4$, and 150 mM NaCl) and then lysed by a combination of lysozyme (10 mg/mL) and glass beads (Sigma, St. Louis, MO, USA). The crude homogenates were centrifuged at 10,000 g for 15 min and the supernatant was applied into the Ni^{2+} affinity column (Qiagen GmbH, Germany). The recombinant proteins were eluted with 250 mM imidazole buffer and each of their concentrations calculated based on Bradford method using the Bio-Rad protein assay kit (Bio-Rad, USA). Three mL of exponentially grown *L. lactis* MG1363 were centrifuged and gently resuspended in 600 μL of fresh M17 broth. Then, 200 μL of purified AcmA/VP1$_{1-67aa}$ and AcmA/VP1$_{35-100aa}$ were separately added to 600 μL of the cells and incubated at 30°C for 2 h. The mixture was then centrifuged again at 2,000 g for 10 min and the cell pellets were washed with 1 mL of PBS three times. The binding of the purified recombinant proteins were then analysed by immunofluorescence microscopy.

2.7. Immunofluorescence Microscopy. The control *L. lactis* cells and cells mixed with either the AcmA/VP1$_{1-67aa}$ or AcmA/VP1$_{35-100aa}$ fusion proteins were initially placed on chamber slides precoated with poly-L-lysine followed by incubation for 15 min before being fixed with 4%

paraformaldehyde. Cells were also incubated with 3% bovine serum albumin (BSA) in PBS for 30 min at room temperature to block nonspecific binding and washed with PBS. The fixed cells were then labeled with primary (rabbit anti-VP1) antibodies. The slide was then washed with PBS and incubated with rhodamine labeled goat anti-rabbit secondary antibody (diluted at 1 : 200 in 1% BSA) at room temperature for 1 h. This was followed by washing with PBS three times, air-dried, and mounted in an antifading agent (Fluoroguard, (Bio-Rad, Hercules, CA, USA)). The labeled slides were then analysed by Confocal Microscope (Bio-Rad MRC 1024 Confocal Laser Scanning Microscope, Bio-Rad) (excitation 550 nm; emission 570 nm). Cells were observed under a 40x objective. Images were taken and analyzed with Bio-Rad laser sharp software (Bio-Rad).

2.8. Stability Assay. Stability of anchored protein on cell surface was analyzed for a period of 5 days. In brief, *L. lactis* cells were added to fusion protein and incubated at 30°C for 2 h. The mixture was centrifuged and washed with PBS. ELISA was carried out on the *L. lactis* cells displaying fusion protein at every 24 h up to 120 h to determine the stability. The lithium chloride stability assay was performed to further test the stability of the anchored proteins [15]. *L. lactis* cells incubated with fusion proteins were harvested and treated with $100\,\mu$L of 8 M LiCl solution at 30°C for 30 min. After treatment, cells were analyzed by ELISA for the detection of the presence of fusion proteins on the cell surface of *L. lactis*.

2.9. Immunogenicity Studies. Specific pathogen-free female 2-week-old BALB/c mice were used. The mice were housed in microisolator cages with free access to water and feed. Three groups of 5 mice were orally fed with $500\,\mu$L of *L. lactis* cells (10^9 cells) displaying either AcmA/VP1$_{1-67aa}$ fusion protein or AcmA/VP1$_{35-100aa}$ fusion protein or with both of the fusion proteins. The first control mice group was immunized with the $500\,\mu$L (10^9 cells) of *L. lactis* cells. The second control mice group received $500\,\mu$L of PBS. All the mice were fed using oral gavage tube without anesthesia and received the booster dose (same as the initial immunization dose) on days 7, 14, and 21. Blood samples were collected from a tail vein of the immunized mice at 0, 7, 14, 21, 28, and 35 days, and the collected blood was incubated at 37°C for 1 h. The sera were separated from red blood cells by centrifugation at 4,500 g for 10 min and stored at 4°C. For long term storage, serum samples were kept at −20°C.

2.10. Western Blot for the Detection of Antigen-Specific Serum Antibody. Purified fusion proteins (AcmA/VP1$_{1-67aa}$ and AcmA/VP1$_{35-100aa}$) and total protein extractions of *L. lactis* and *E. coli* BL21 (DE3) pLysS (pRSETC) cells were separated by 12.5% SDS-PAGE and electroblotted on a PVDF (Millipore Corp., Billerica, MA, USA) membrane. The membrane was then incubated in 1% (w/v) BSA in DBT (Amresco, Solon, OH, USA) for 1 h, followed by incubation for 1 h in 10 mL of DBT (Amresco) containing $10\,\mu$L of the respective sera collected at day 21 (7 days after the 2nd booster dose) from the immunized mice. After washing with the DBT, the conjugated

membrane was incubated with goat anti-mouse antibody conjugated HRP (50 mU/mL in TBS, (Amresco) for 1 h, washed with DBT, and developed using 4-chloronaphthol (Amresco).

2.11. Analysis of Antigen-Specific Serum Antibody by ELISA. ELISA plate wells were coated with purified recombinant VP1 protein (complete VP1 protein; $1\,\mu$g/mL in coating buffer 0.015 M Na$_2$CO$_3$, 0.03 M NaHCO$_3$, and pH 9.6). ELISA plates coated with EV71 virus were also used to analyse the serum of immunized mice. (The purified recombinant VP1 protein and EV71 virus coated plates were obtained from Professor Dr. Mary Jane Cardosa, Universiti Malaysia Sarawak, Malaysia). The wells were blocked with 2% BSA in PBS for 1 h. A volume of $100\,\mu$L of serially diluted hyperimmune mouse sera (1 : 1000; 1 : 10,000, and 1 : 100,000 dilutions) were added to the wells and incubated for 1 h. The serum from blood collected at 0, 7, 14, 21, and 28 days from all groups of immunized mice was analyzed at 1 : 1000; 1 : 10,000, and 1 : 100,000 dilutions. The wells were then washed six times with 1x PBS before incubation with the secondary antibody ($100\,\mu$L of HRP conjugated anti-rabbit antibodies (Roche, Switzerland) diluted at 1 : 500 in 0.5% BSA in 1x PBS) at room temperature for 1 h. After incubation, the unconjugated secondary antibody was removed by washing with 1x PBS (6 times, 10 min each). Then, $100\,\mu$L of substrate (BM Blue, Roche) was added to each well. After color development the reaction was stopped by adding $50\,\mu$L of 1 M H$_2$SO$_4$ and absorbance was measured by ELISA reader at OD$_{450}$.

3. Results

3.1. Construction of pSVacVP1$_{1-201nt}$, pSVacVP1$_{103-300nt}$ and Expression of AcmA/VP1$_{1-67aa}$ and AcmA/VP1$_{35-100aa}$. Two fragments of the N-terminal region of VP1 were amplified and subcloned separately into plasmid pSVac [13]. Total protein extracts of *E. coli* BL2 (DE3) containing the recombinants AcmA/VP1$_{1-67aa}$ and AcmA/VP1$_{35-100aa}$ and *E. coli* BL21 (DE3) were analyzed by SDS-PAGE and Western blot. The SDS-PAGE protein profile showed the presence of 28 kDa and 25 kDa bands, which approximately corresponded to the expected size of the recombinant AcmA/VP1$_{1-67aa}$ and AcmA/VP1$_{35-100aa}$. Western blot analysis using anti-VP1 confirmed that the two bands were immunoreactive to the antibody (Figure 1). This suggested that the pSVacVP1$_{1-201nt}$ and pSVacVP1$_{103-300nt}$ recombinant constructs were successfully expressed in *E. coli*.

3.2. Affinity Purification of Recombinant Fusion Proteins. In order to study the display of EV71 capsid protein (VP1$_{1-201nt}$ and VP1$_{103-300nt}$ regions of VP1 gene) on the cell wall surface of *L. lactis*, recombinant *E. coli* BL21 (DE3) pLysS cells harbouring pSVacmVP1$_{1-201}$, pSVacmVP1$_{103-300}$, pSVnpVP1$_{1-201}$, and pSVnpVP1$_{103-300}$ vectors were grown and induced with IPTG (Gibco BRL, USA). The protein fractions from the cells were purified on Ni^{2+} affinity columns, and the eluted proteins were analysed by SDS-PAGE (data not shown).

3.3. Binding of the EV71 VP1 Epitopes to the Cell Surface of Lactococcus . Purified $AcmA/VP1_{1-67aa}$ and $AcmA/VP1_{35-100aa}$ fragments were incubated with *L. lactis* and subjected to ELISA analysis and immunofluorescence staining. A positive color change was detected for the *L. lactis* cells incubated with $AcmA/VP1_{1-67aa}$, $AcmA/VP1_{35-100aa}$ fusion proteins. Immunofluoresence analysis also indicated that the display of $AcmA/VP1_{1-67aa}$ and $AcmA/VP1_{35-100aa}$ fusion proteins on the *L. lactis* cell surface was in stable conformation. It was observed that the *L. lactis* cells incubated with both fragments ($AcmA/VP1_{1-67aa}$ and $AcmA/VP1_{35-100aa}$) were efficiently stained by rhodamine labeled secondary antibody whilst the control cells remained free from staining (Figure 2). These results strongly suggest that the fusion proteins constituting the $AcmA/VP1_{1-67aa}$ and $AcmA/VP1_{35-100aa}$ expressed in *E. coli* had maintained the active binding domains and the capacity to dock-onto the outer surface of *L. lactis* cell wall.

3.4. Binding Stability of Fusion Proteins on the Surface of L. lactis. In order to apply this system for the display of foreign proteins on *L. lactis*, it is important to determine the stability of the anchorage of fusion proteins. The stability assay was conducted for 5 days, at each 24 h interval, after which the *L. lactis* cells incubated with the fusion proteins ($AcmA/VP1_{1-67aa}$ and $AcmA/VP1_{35-100aa}$) were probed with rabbit anti-VP1 antibody. This was followed by HRP conjugated anti-rabbit IgG antibody (Roche) before being analysed by ELISA reader. *L. lactis* cells without incubation with fusion proteins were used as the negative control. The fusion proteins still present on the surface of *L. lactis* even after five days of incubation (data not shown). We further tested stability of anchored protein by treating with LiCl. LiCl is commonly used to remove proteins from bacterial cell walls. We interested to observe the effect of LiCl on *L. lactis* cells displaying $AcmA/ VP1_{1-67aa}$ or $VP1_{35-100aa}$. The mode of action of LiCl is the cleavage of covalent or noncovalent bonds between the surface proteins and cell walls. We want to test the stability of anchored proteins by treating LiCl. *L. lactis* displaying fusion proteins ($AcmA/VP1_{1-67aa}$ and $AcmA/VP1_{35-100aa}$) were treated with 8 M LiCl, after the treatment of cells was analyzed by whole cell ELISA. Results showed the presence of fusion proteins on the cell surface of *L. lactis* even after treatment with LiCl, which indicates that the proteins are anchored strongly to the cell surface (data not shown).

3.5. Detection of Serum Antibody Response for $VP1_{1-67aa}$ and $VP1_{35-100aa}$ of VP1 in Mice. The sera of mice orally immunized with live *L. lactis* cells displaying $VP1_{1-67aa}$ or $VP1_{35-100aa}$ antigens were tested for VP1 specific antibodies by ELISA using purified recombinant VP1 fusion protein (complete VP1 protein) as the antigen. The antiserum from mice orally fed with *L. lactis* displaying the immunogens ($VP1_{1-67aa}$ or $VP1_{35-100aa}$ or both) clearly reacted with the fusion proteins (recombinant VP1 fusion protein of EV71) (Figure 3), whereas the antiserum from mice orally immunized with only *L. lactis* or mice orally immunized with PBS

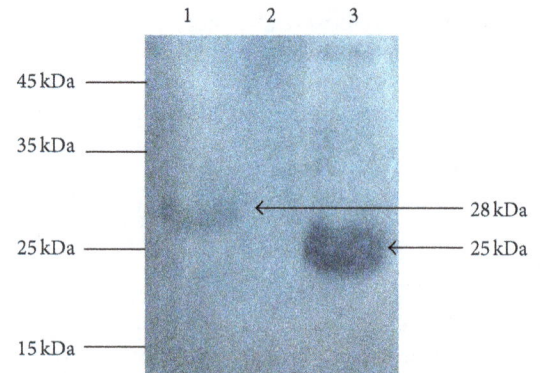

Figure 1: Western blot analyses of the over-expressed recombinant fusion proteins ($AcmA/VP1_{1-67aa}$ and $AcmA/VP1_{35-100aa}$). Lane 1, Total protein of BL21 (DE3) pLysS ($pSVacVP1_{1-201nt}$); lane 2, Total protein of BL21 (DE3) pLysS (pRSETC) as negative control; lane 3, Total protein of BL21 (DE3) pLysS ($pSVacVP1_{103-300nt}$). The arrow shows recombinant fusion proteins: $AcmA/VP1_{1-67aa}$ (~28 kDa), $AcmA/VP1_{35-100aa}$ (~25 kDa).

did not react with the recombinant VP1 fusion protein of EV71 (Figure 3).

The antibody titers of mice orally fed with *L. lactis* displaying $VP1_{1-67aa}$ were shown to have lower antibody titers after primary immunization when compared with the antibody titers of mice orally fed with *L. lactis* displaying $AcmA/ VP1_{1-67aa}$ or $VP1_{35-100aa}$ (Figure 4(a)). The antibody titers increased after the 1st booster dose in *L. lactis* displaying $VP1_{1-67aa}$ (Figure 4(a)). On the other hand, mice orally fed with *L. lactis* displaying $VP1_{35-100aa}$ gave a higher level of antibody titers in primary immunized serum as well as in all booster doses when compared to the antibody titers of mice fed with *L. lactis* displaying $VP1_{1-67aa}$ (Figure 4(b)). The highest level of antibody titers at 1 : 1000 dilution was, however, seen in the serum of mice fed with *L. lactis* displaying both epitopes when compared to *L. lactis* displaying only $VP1_{1-67aa}$ or $VP1_{35-100aa}$ (Figure 4(c)). These results indicated a better response when a combination of both epitopes were used. There was no reaction between recombinant VP1 fusion protein and the serum of mice orally immunized with PBS (Figure 4(d)). A very minor reaction was observed with the serum of mice immunized with *L. lactis* at 1 : 1000 serum dilution and lower (Figure 4(e)).

In addition, ELISA results demonstrated that the antiserum from mice orally fed with *L. lactis* displaying immunogens ($VP1_{1-67aa}$ or $VP1_{35-100aa}$ or both) of VP1 of EV71 clearly reacted in the wells coated with EV71 virus (data not shown), whereas the antiserum from mice orally fed with only *L. lactis* or mice orally given PBS did not react with the EV71 virus (data not shown). These results clearly indicated that the fusion proteins ($AcmA/VP1_{1-67aa}$ and $AcmA/VP1_{35-100aa}$) displayed on the cell surface of *L. lactis* were able to elicit an antigen-specific immune response in mice against VP1 protein. The antibody response against $VP1_{1-67aa}$ and $VP1_{35-100aa}$ antigens of EV71 in mice was also tested by Western blot analysis. Groups of five mice were orally immunized with live *L. lactis* cells displaying $AcmA/VP1_{1-67aa}$ and $AcmA/VP1_{35-100aa}$

(a)

(b)

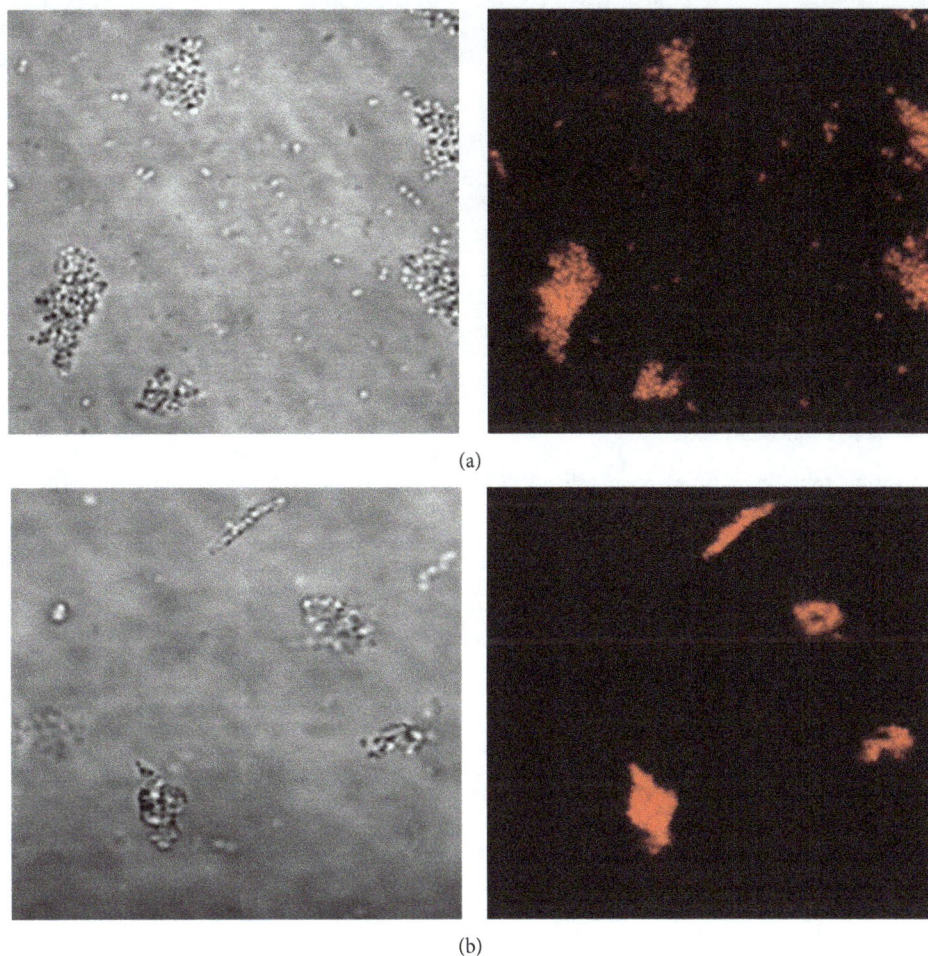

FIGURE 2: Confocal micrographs of the binding of fusion proteins to *L. lactis*: (a) bright field and fluorescence image of *L. lactis* cells incubated with AcmA/VP1$_{1-67aa}$ protein; (b) bright field and fluorescence image of *L. lactis* cells incubated with AcmA/VP1$_{35-100aa}$ protein.

and mouse sera (7 days after the second booster dose) were tested for AcmA/VP1$_{1-67aa}$ and AcmA/VP1$_{35-100aa}$ specific antibodies by Western blot analysis using AcmA/VP1$_{1-67aa}$ and AcmA/VP1$_{35-100aa}$ fusion proteins as the capturing antigens. The antisera from mice orally fed with *L. lactis* displaying either one or both of the fusion proteins of EV71 were shown to have reacted with the fusion proteins (Figure 5), whereas the antisera from mice orally fed with only *L. lactis* or PBS did not show any positive reaction (data not shown). These results clearly indicate that the fusion proteins (AcmA/VP1$_{1-67aa}$ and AcmA/VP1$_{35-100aa}$) displayed on the cell surface of *L. lactis* were able to elicit an antigen-specific immune response in the mice.

4. Discussion

A system for targeting purified anchor proteins to the cell surface of *Lactococcus* and other lactic acid bacteria (LAB) has been developed [13, 15, 16]. Since *L. lactis* is a noncolonizing commensal organism, the approach of this work was to append the surface of the organism *in vitro* with antigens prior to immunization to enhance antibody response. Our objectives were to study the capability of the purified anchor protein AcmA that has gone through the *E. coli* system to attach and deliver specific antigens such as those of VP1$_{1-67aa}$ and VP1$_{35-100aa}$ fragments onto the surface of *L. lactis* in order to elicit an immune response in the host. *L. lactis* has been reported to successfully express and target tetanus toxin model antigen into the cytoplasm, cell wall, and extracellular medium that elicited immune and protective responses [11]. In addition, interleukin-10 secreted by *L. lactis* was shown to have biological activity in mice [17]. Dieye et al. 2003 [18] also reported that the presentation of infectious bursal disease virus antigens (VP2) utilizing *Lactococcus* as a delivery vehicle showed a partial protection of the cell wall bound Nuc-VP2 against proteolysis as opposed to secreted Nuc-VP2. Recently, Ramasamy et al. [19] reported their work on the immunogenicity of a malaria parasite antigen displayed by *Lactococcus lactis* in oral immunizations. However, lactococcal system for vaccine delivery is hindered due to low levels of expression recombinant protein in *Lactococcus* and the use of antibiotic markers in recombinant *Lactococcus* often makes the bacteria resistance to antibiotics. In addition, we cannot control the expression of antigens when we directly make recombinant *Lactococcus* for vaccine delivery. We need

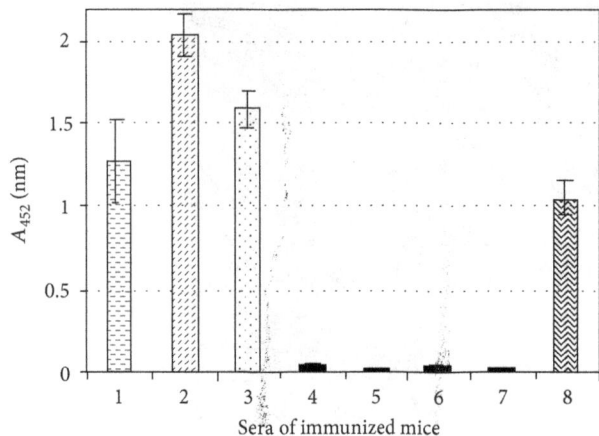

FIGURE 3: Analysis of serum from immunized mice using VP1 coated ELISA plates. 1: serum of mice immunized with *L. lactis* displaying VP1$_{1-67aa}$; 2: serum of mice immunized with *L. lactis* displaying VP1$_{1-67aa}$ and VP1$_{35-100aa}$; 3: serum of mice immunized with *L. lactis* displaying VP1$_{35-100aa}$; 4: serum of control mice immunized with *L. lactis*; 5: serum of control mice immunized with PBS; 6: serum of the rabbit; 7: serum of preimmunized mice; 8: rabbit anti-VP1 antibodies as positive control. Sera from immunized mice (Balb/c) were taken after 3rd booster immunization. Testing of sera was done at sera dilution 1:10,000. Note: the absorbance value shown was after the deduction of the background value obtained from purified His-tag protein.

alternative strategy to overcome some of these problems associated with *Lactococcus*. We selected *E. coli* as expression host to produce the fusion proteins (antigen/anchor) to overcome low expression associated with *Lactococcus*. *E. coli* have a number of commercially established high protein expression vectors and *E. coli* can easily be grown in a bioreactor and the recombinant proteins can be purified using simple purification systems such as fast protein liquid chromatography (FPLC). In addition, specific concentration of proteins (antigen) can be calculated, mixed with the appropriate number of *L. lactis*, where we can control dose vaccine by controlling a number of antigens and *Lactococcus* molecules. Since recombinant plasmids are not introduced into *Lactococcus* which eliminates antibiotic marker as selective pressure, this, therefore eliminates the worry of antibiotic resistant genes contaminating the environment when using recombinant vaccines. A number of advantages with *E. coli* make them an attractive host for the expression of fusion proteins (EV71 epitopes fused with cell wall binding domain of AcmA). AcmA is an autolysin which plays a key role in *Lactococcus* growth and propagation. AcmA naturally expressed in *Lactococcus* and expressed AcmA travels to cell wall and binds to the cell wall. Once it binds to cell, it starts the lysis of cell wall to release intracellular proteases into the media to digest the proteins into micronutrients which requires their cell survival. We utilized this natural phenomenon of the AcmA protein for the cell wall binding of EV71 epitopes. Cell wall binding domain of AcmA has three repeated regions of lysin motif (LysM) domains. The LysM domain is about 40 amino acids long and present in a number of surface

associated proteins in a wide range of bacteria. The LysM domain has a $\beta\alpha\alpha\beta$ structure and conserved asparate or glutamate in this shallow groove assumed to be involved in the binding with peptidoglycan and the mechanism of AcmA binding to cell wall was unknown.

In this study, the N-terminal fragments of VP1 of EV71 were subcloned into pSVac to allow for the expression of C-terminal fusion proteins. The sequences of VP1$_{1-67aa}$ and VP1$_{35-100aa}$ at the N-terminal region of the VP1 protein of EV71 were chosen as antigens to be displayed on *Lactococcus*. VP1 protein of EV71 has high immunogenicity and antigenicity [20–22], and it has been a major candidate for the development of vaccines [20]. The studies by Hovi and Roivainen [23] showed that a highly conserved region of 42–52 amino acids close to the N-terminus of VP1 was involved in immunogenicity and that antibodies against this region can be used as a group reagent recognizing Enteroviruses. Peptide antibodies against 42–52 amino acid motif were shown to be capable of precipitating purified poliovirus particles, indicating that this region is exposed and involved in immunogenicity [24]. To create an N-terminal epitope for surface display, VP1 gene was truncated into VP1$_{1-201nt}$ and VP1$_{103-300nt}$ regions. The VP1$_{1-201nt}$ region represented amino acids 1 to 67, and VP1$_{103-300nt}$ region represented amino acid sequences 35 to 100, both from the N-terminal. The truncation of VP1 protein was done to increase the solubility of fusion protein and keep the structure small to avoid the possibility of masking the cell wall binding domains of AcmA. In this vector construct, the foreign genes were cloned upstream of the *acmA* gene fragment, thus allowing for a free C-terminal fusion for binding to the cell wall surface of Lactococcal cells. The AcmA/VP1$_{1-67aa}$ and AcmA/VP1$_{35-100aa}$ fusion proteins were then purified and targeted to the cell surface of *L. lactis*, and the recombinant *Lactococci* was used to immunize BALB/c mice by oral administration. Both the VP1$_{1-67aa}$ and VP1$_{35-100aa}$ could be docked onto the surface of *L. lactis*.

The AcmA repeat cell wall anchor has been previously used for the surface expression of the *Bacillus licheniformis* alpha-amylase and *E. coli* beta-lactamase [25], and the mechanism by which the acmA encoded attachment domains interact with the cell wall components has been suggested to be covalent in nature [26, 27]. Our main concern was the folding and stability of the fusion proteins after they were expressed in *E. coli* and purified. The expression of foreign genes in *E. coli* has been well documented [28, 29]. Observations from immunofluorescence studies showed that the purified AcmA proteins from *E. coli* cells had maintained their capability to anchor onto the surface of *Lactococcus* cells and are stably docked for at least 5 days [13]. Free proteins that may have been detached from the Lactococcal cell carrier presumably will not be able to survive the gastrointestinal tract to render any immunological reaction [30, 31]. Immunogenicity results indicated an immunogenic reaction in the test mice where the production of specific antibodies against VP1$_{1-67aa}$ and VP1$_{35-100aa}$ was observed by ELISA and Western blot analyses. The VP1$_{1-67aa}$ and VP1$_{35-100aa}$ antigens carrying *Lactococcus* represents the first step towards the development of a new strategy for vaccination against EV71 and perhaps

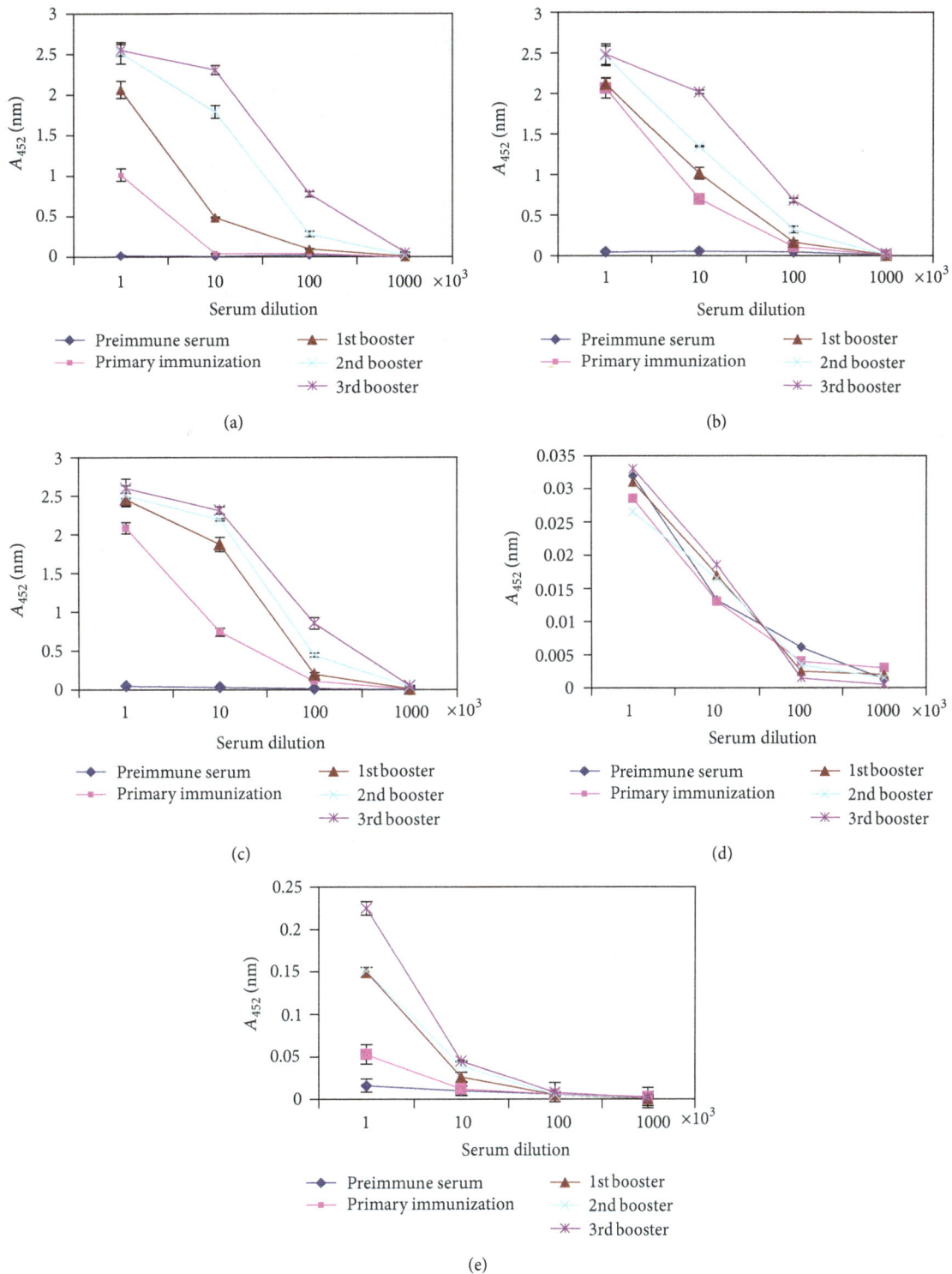

FIGURE 4: Determination of antibody titers by ELISA. (a) Serum from mice immunized with *L. lactis* displaying VP1$_{1-67aa}$, (b) serum from mice immunized with *L. lactis* displaying VP1$_{35-100aa}$, (c) serum from mice immunized with *L. lactis* displaying both epitopes (VP1$_{1-67aa}$ and VP1$_{35-100aa}$), and (d) serum from control mice immunized with PBS. (e) Serum from control mice immunized with *L. lactis*. Antibodies were measured using complete VP1 protein coated ELISA plates. Sera from mice (Balb/c) were taken before and after each immunization with *L. lactis* displaying VP1$_{1-67aa}$.

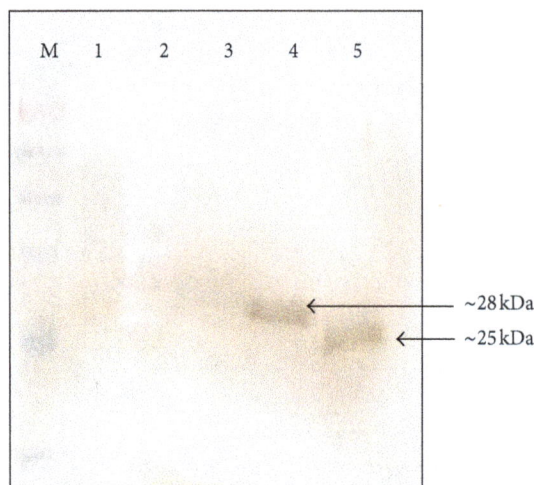

FIGURE 5: Detection of serum antibody response against VP1$_{1-67aa}$ and VP1$_{35-100aa}$ epitopes of EV71 in mice immunized with *L. lactis* displaying AcmA/VP1$_{35-100aa}$. Lane 1: BSA protein; lane 2: *E. coli* (pRSET) total proteins; lane 3: total proteins of *L. lactis* MG1363; lane 4: purified AcmA/VP1$_{1-67aa}$ proteins; lane 5: purified AcmA/VP1$_{35-100aa}$ protein; lane M: protein marker (Fermentas, Hanover, MD, USA). The arrow shows recombinant fusion proteins: AcmA/VP1$_{1-67aa}$ (~28 kDa) and AcmA/VP1$_{35-100aa}$ (~25 kDa).

other viral infections. Such a delivery system, utilizing lactic acid bacteria for oral administration of vaccine through food and water, would be very attractive because of its safety, low cost, and nonimmunosuppressing properties. In conclusion, a cell surface display system in which the AcmA cell wall binding protein of *L. lactis* was used as an anchoring motif was studied. Fusion proteins of up to 79 amino acids long were successfully displayed on the *L. lactis* outer membrane. Furthermore, the strains developed in this study were shown to be capable of inducing immunogenicity in orally fed mice. We believe that the method of protein docking utilized in this study will allow for more flexible presentations of short peptides and polypeptides on the surface of *L. lactis* to be useful as a delivery vehicle.

Authors' Contribution

N. R. S. Varma, R. Abdul Rahim, K. Yusoff, and H. L. Foo developed the concept and designed the experiments. N. R. S. Varma carried out the experiments and the analysis of the data. H. Toosa performed the purification of Acma-a1 epitopes from *E. coli*, B. M. Alitheen N. M. N. Shamsudin, and A. S. Arbab supported the project.

Acknowledgments

This work was supported by a grant from the Ministry of Science, Technology and Innovations, Malaysia (Grant no. 09-02-04-006 BTK/ER/024). The authors would like to thank Professor Dr. Mary Jane Cardosa (University Malaysia Sarawak, Malaysia) for the gift of VP1 gene of EV71, as well as Shahrul Ezhar Abdul Rahman, Nur Adeela Yasid, and Lalita (Universiti Putra Malaysia) for their help.

References

[1] A. M. Q. King, F. Brown, P. Christian, T. Hovi, and T. Hyypia, "Picornaviridae," in *Virus Taxonomy. Seventh Report of the International Committee For the Taxonomy of Viruses*, M. H. V. Van Regenmortel, C. M. Fauquet, D. H. L. Bishop, and C. H. Calisher, Eds., pp. 657–673, Academic Press, New York, NY, USA, 2000.

[2] M. J. Cardosa, S. Krishnan, P. H. Tio, D. Perera, and S. C. Wong, "Isolation of subgenus B adenovirus during a fatal outbreak of enterovirus 71-associated hand, foot, and mouth disease in Sibu, Sarawak," *The Lancet*, vol. 354, no. 9183, pp. 987–991, 1999.

[3] L. Y. Chang, Y. C. Huang, and T. Y. Lin, "Fulminant neurogenic pulmonary oedema with hand, foot, and mouth disease," *The Lancet*, vol. 352, no. 9125, pp. 367–368, 1998.

[4] K. Komatsu, Y. Shimizu, Y. Takeuchi, H. Ishiko, and H. Takada, "Outbreak of severe neurologic involvement associated with *Enterovirus* 71 infection," *Pediatric Neurology*, vol. 20, pp. 17–23, 1999.

[5] L. C. S. Lum, K. T. Wong, S. K. Lam et al., "Fatal enterovirus 71 encephalomyelitis," *Journal of Pediatrics*, vol. 133, no. 6, pp. 795–798, 1998.

[6] P. McMinn, K. Lindsay, D. Perera, Hung Ming Chan, Kwai Peng Chan, and M. J. Cardosa, "Phylogenetic analysis of enterovirus 71 strains isolated during linked epidemics in Malaysia, Singapore, and Western Australia," *Journal of Virology*, vol. 75, no. 16, pp. 7732–7738, 2001.

[7] M. Y. Liau, R. J. Chiang, S. Y. Li et al., "Development of vaccines against enterovirus 71," in *APEC Enteroviral Watch Program for Children, International Scientific Symposium Proceedings*, pp. 81–82, 2000.

[8] C. K. Yu, C. C. Chen, C. L. Chen et al., "Neutralizing antibody provided protection against enterovirus type 71 lethal challenge in neonatal mice," *Journal of Biomedical Science*, vol. 7, no. 6, pp. 523–528, 2000.

[9] L. Chamberlain, J. M. Wells, K. Robinson, K. Schofield, and R. W. F. Le Page, "Mucosal immunization with recombinant *Lactococcus lactis*," in *Gram-Positive Bacteria as Vaccine Vehicles For Mucosal Immunization*, G. Pozzi and J. M. Wells, Eds., pp. 83–106, Landes Bioscience, Austin, Tex, USA, 1997.

[10] K. Q. Xin, Y. Hoshino, Y. Toda et al., "Immunogenicity and protective efficacy of orally administered recombinant *Lactococcus lactis* expressing surface-bound HIV Env," *Blood*, vol. 102, no. 1, pp. 223–228, 2003.

[11] J. M. Wells, P. W. Wilson, P. M. Norton, M. J. Gasson, and R. W. F. Le Page, "*Lactococcus lactis*: high-level expression of tetanus toxin fragment C and protection against lethal challenge," *Molecular Microbiology*, vol. 8, no. 6, pp. 1155–1162, 1993.

[12] M. J. Gasson, "Plasmid complements of *Streptococcus lactis* NCDO 712 and other lactic streptococci after protoplast-induced curing," *Journal of Bacteriology*, vol. 154, no. 1, pp. 1–9, 1983.

[13] A. R. Raha, N. R. S. Varma, K. Yusoff, E. Ross, and H. L. Foo, "Cell surface display system for *Lactococcus lactis*: a novel development for oral vaccine," *Applied Microbiology and Biotechnology*, vol. 68, no. 1, pp. 75–81, 2005.

[14] U. K. Laemmli, "Cleavage of structural proteins during the assembly of the head of bacteriophage T4," *Nature*, vol. 227, no. 5259, pp. 680–685, 1970.

[15] T. Bosma, R. Kanninga, J. Neef et al., "Novel surface display system for proteins on non-genetically modified gram-positive bacteria," *Applied and Environmental Microbiology*, vol. 72, no. 1, pp. 880–889, 2006.

[16] M. L. Roosmalen, R. Kanninga, M. E. Khattabi et al., "Mucosal vaccine delivery of antigens tightly bound to an adjuvant particle made from food-grade bacteria," *Methods*, vol. 38, pp. 144–149, 2006.

[17] L. Steidler, W. Hans, L. Schotte et al., "Treatment of murine colitis by *Lactococcus lactis* secreting interleukin-10," *Science*, vol. 289, no. 5483, pp. 1352–1355, 2000.

[18] Y. Dieye, A. J. W. Hoekman, F. Clier, V. Juillard, H. J. Boot, and J. C. Piard, "Ability of *Lactococcus lactis* to export viral capsid antigens: a crucial step for development of live vaccines," *Applied and Environmental Microbiology*, vol. 69, no. 12, pp. 7281–7288, 2003.

[19] R. Ramasamy, S. Yasawardena, A. Zomer, G. Venema, J. Kok, and K. Leenhouts, "Immunogenicity of a malaria parasite antigen displayed by *Lactococcus lactis* in oral immunisations," *Vaccine*, vol. 24, no. 18, pp. 3900–3908, 2006.

[20] N. W. Cheng, C. L. Ya, F. Cathy, S. L. Nan, R. S. Shih, and S. H. Mei, "Protection against lethal enterovirus 71 infection in newborn mice by passive immunization with subunit VP1 vaccines and inactivated virus," *Vaccine*, vol. 20, no. 5-6, pp. 895–904, 2001.

[21] G. S. Page, A. G. Mosser, J. M. Hogle, D. J. Filman, R. R. Rueckert, and M. Chow, "Three-dimensional structure of poliovirus serotype 1 neutralizing determinants," *Journal of Virology*, vol. 62, no. 5, pp. 1781–1794, 1988.

[22] T. J. Smith, E. S. Chase, T. J. Schmidt, N. H. Olson, and T. S. Baker, "Neutralizing antibody to human rhinovirus 14 penetrates the receptor- binding canyon," *Nature*, vol. 383, no. 6598, pp. 350–354, 1996.

[23] T. Hovi and M. Roivainen, "Peptide antisera targeted to a conserved sequence in poliovirus capsid protein VP1 cross-react widely with members of the genus *Enterovirus*," *Journal of Clinical Microbiology*, vol. 31, no. 5, pp. 1083–1087, 1993.

[24] M. Roivainen, L. Piirainen, T. Rysa, A. Narvanen, and T. Hovi, "An immunodominant N-terminal region of VP1 protein of poliovirion that is buried in crystal structure can be exposed in solution," *Virology*, vol. 195, no. 2, pp. 762–765, 1993.

[25] G. Buist, *AcmA of Lactococcus lactis, a cell-binding major autolysin [Ph.D. thesis]*, University of Groningen, Groninge, The Netherlands, 1997.

[26] K. J. Leenhouts, G. Buist, and J. Kok, "Anchoring of proteins to lactic acid bacteria," *Antonie van Leeuwenhoek*, vol. 76, no. 1–4, pp. 367–376, 1999.

[27] A. Steen, G. Buist, K. J. Leenhouts et al., "Cell wall attachment of a widely distributed peptidoglycan binding domain is hindered by cell wall constituents," *Journal of Biological Chemistry*, vol. 278, no. 26, pp. 23874–23881, 2003.

[28] F. Baneyx, "Recombinant protein expression in *Escherichia coli*," *Current Opinion in Biotechnology*, vol. 10, no. 5, pp. 411–421, 1999.

[29] H. Schwab, "Principles of genetic engineering for *Escherichia coli*," in *Biotechnology: Genetic Fundamentals and Genetic Engineering*, H. J. Rehm, G. Reed, A. Puhler, and P. Stadler, Eds., pp. 375–419, VCH Verlagsgesellschaft MbH, Weinheim, Germany, 2nd edition, 1993.

[30] R. I. Walker, "New strategies for using mucosal vaccination to achieve more effective immunization," *Vaccine*, vol. 12, no. 5, pp. 387–400, 1994.

[31] A. L. Mora and J. P. Tam, "Controlled lipidation and encapsulation of peptides as a useful approach to mucosal immunizations," *Journal of Immunology*, vol. 161, no. 7, pp. 3616–3623, 1998.

14

Production of Pectinolytic Enzymes by the Yeast *Wickerhanomyces anomalus* Isolated from Citrus Fruits Peels

María A. Martos,[1] **Emilce R. Zubreski,**[1] **Oscar A. Garro,**[2] **and Roque A. Hours**[3]

[1] *Facultad de Ciencias Exactas, Químicas y Naturales, Universidad Nacional de Misiones, Felix de Azara 1552, N3300LQH Posadas, Argentina*
[2] *Universidad Nacional del Chaco Austral, Comandante Fernández 755, H3700LGO Presidencia Roque Sáenz Peña, Argentina*
[3] *Centro de Investigación y Desarrollo en Fermentaciones Industriales (CINDEFI, UNLP, CONICET La Plata), Facultad de Ciencias Exactas, Universidad Nacional de la Plata, Calle 47 y 115, B1900ASH La Plata, Argentina*

Correspondence should be addressed to María A. Martos; amartos@arnet.com.ar

Academic Editor: Triantafyllos Roukas

Wickerhamomyces anomalus is pectinolytic yeast isolated from citrus fruits peels in the province of Misiones, Argentine. In the present work, enzymes produced by this yeast strain were characterized, and polygalacturonase physicochemical properties were determined in order to evaluate the application of the supernatant in the maceration of potato tissues. *W. anomalus* was able to produce PG in liquid medium containing glucose and citrus pectin, whose mode of action was mainly of endo type. The supernatant did not exhibit esterase or lyase activity. No others enzymes, capable of hydrolyzing cell wall polymers, such as cellulases and xylanases, were detected. PG showed maximal activity at pH 4.5 and at temperature range between 40°C and 50°C. It was stable in the pH range from 3.0 to 6.0 and up to 50°C at optimum pH. The enzymatic extract macerated potato tissues efficiently. Volume of single cells increased with the agitation speed. The results observed make the enzymatic extract produced by *W. anomalus* appropriate for future application in food industry, mainly for the production of fruit nectars or mashed of vegetables such as potato or cassava, of regional interest in the province of Misiones, Argentine.

1. Introduction

Enzymes hydrolyzing pectic substances, which contribute to the firmness and structure of plant cells, are known as pectinolytic enzymes or pectinases. Based on their mode of action, these enzymes include polygalacturonase (PG), pectinesterase (PE), and lyases (pectinlyase (PL) and pectate-lyase (PAL)). PG, PL, and PAL are depolymerizing enzymes, which split the α-(1,4)-glycosidic bonds between galacturonic monomers in pectic substances either by hydrolysis (PG) or by β-elimination (PL, PAL). PG catalyzes the hydrolytic cleavage of the polygalacturonic acid chain while PL performs a transeliminative split of pectin molecule, producing an unsaturated product. PE catalyzes the de-esterification of the methoxyl group of pectin, forming pectic acid [1, 2]. There are two types of PGases with different technological applications: exopolygalacturonases (exo-PG) that break down the distal groups of the pectin molecule, reducing chain length relatively slowly, and endopolygalacturonases (endo-PG) which act randomly on all the links in the chain, reducing molecular dimensions and viscosity more rapidly [3].

Pectinolytic enzymes play an important role in food technology, mainly in the processing of fruit juices and wines and in the maceration of plant tissue. Maceration is a process by which organized tissue is transformed into a suspension of intact cells, resulting in pulpy products used in the food industry for the production of fruit nectars as pears, peaches, apricots, strawberries, and vegetables mashed such as potatoes, carrots, red pepper, and others that are used in babies and seniors foods [4]. For such purposes, only the intercellular cementing material that holds together cells and some portion of primary plant cell walls should be removed without damage to adjacent secondary cell walls, to help avoid cell lysis, keeping nutritional properties of food [5].

For this reason, cellulases in the enzyme mixture are undesirable [6].

The stability of pectinases is affected by both physical parameters (pH and temperature) and chemical parameters (inhibitors or activators). Enhancing the stability and maintaining the desired level of activity over a long period are two important points considered for an efficient application of these enzymes [7].

Pectinases used in the food industry are commercially produced by *Aspergillus niger*. Commercial preparations of fungal origin contain a complex mixture of different enzymes with pectinolytic activity, including PGases, lyases, the undesirable PE, and others enzymes. Yeasts have advantages compared to filamentous fungi, because they are unicellular, the growth is relatively simple, and usually yeasts do no secret PE [8].

A yeast isolated from citrus fruit peels in the province of Misiones (Argentine) and identified as *Wickerhamomyces anomalus*, recent reclassification of the species *Pichia anomala* [9], produced pectinolytic enzymes in liquid medium containing glucose and citrus pectin as carbon and energy sources and inductor, respectively. In the present work, enzymes produced by this wild yeast strain were characterized, and physicochemical properties of polygalacturonase were determined by the study of the effect of temperature and pH on its activity and stability, in order to evaluate the application of the supernatant in the maceration of potato tissues.

2. Materials and Methods

2.1. Microorganism. W. anomalus was isolated from citrus fruit peels in the province of Misiones (Argentine).

2.2. Culture Media

YM Medium. Yeast extract (Sigma), 5 g/L; tryptone (Difco-Becton Dickinson & Co.), 5 g/L; glucose (Britania), 10 g/L; agar (Britania), 15 g/L, pH 5.0.

YNB Medium. Yeast Nitrogen Base (YNB, Difco-Becton Dickinson & Co.), 6.7 g/L; glucose (Britania), 5 g/L; citrus pectin (Parafarm), 5 g/L; pH 5.0.

Citrus pectin was washed with a 70% (v/v) ethanol-HCl (0.05 N) solution to remove soluble sugars [10].

All components of media were autoclaved (121°C, 15 min) except in the case of YNB solution which was sterilized separately by filtration through a cellulosic filter paper (0.22 μm, Sartorius).

2.3. Production of Pectinolytic Enzymes in Submerged Fermentation. Five hundred millilitre Erlenmeyers flasks with 95 mL of YNB medium were inoculated with 5 mL of an appropriate dilution of a suspension of the microorganism (DO_{620} = 0.96), grown in *YM medium* (30°C, 24 h). The Erlenmeyers flasks were incubated at 30°C for 10 h on a rotary shaker at 180 rpm. The biomass was separated by centrifugation at 4000 rpm, for 10 min at 5°C. The culture medium supernatant was frozen at –18°C and used as source of extracellular enzymes (named enzymatic extract, EE). The assay batch cultures were run in triplicate and mean values were calculated.

2.4. Enzyme Assays

Polygalacturonase (PG). PG activity was assayed by measuring the reducing groups released by dinitrosalicylic acid method [11]. A calibration curve was made using galacturonic acid (GA, Sigma) as standard. One unit of PG was defined as the amount of enzyme which releases 1 μmol of GA per minute.

Xylanase and Cellulose. Xylanase and cellulase activities were assayed as was PG activity except for the use of xylan (Sigma) and carboxymethylcellulose (Sigma), respectively, as substrates. Xylose (Sigma) and glucose (Sigma) were used as standard for xylanase and cellulase, respectively.

Pectinlyase (PL). PL was assayed by monitoring the increase in absorbance at 235 nm of citrus pectin (Sigma) solution, as described by [12]. One unit of PL activity was defined as the amount of enzyme which produces an increase of one unit of absorbance in the conditions of the assay.

Pectatelyase (PAL). PAL was assayed as was PL activity except for the use of PGA (Sigma) as the substrate.

Pectinesterase (PE). PE activity was determined by color change of a pH indicator (bromocresol green) added to the reaction mixture, due to carboxyl groups being released during the reaction. As a substrate, it was used 0.5% (w/v) citrus pectin (Sigma) in water, pH 5.0 [13].

2.5. Mode of Action of PG. The endo- or exo-mode of action of PG was determined by measuring the formation of reducing groups and the changes in viscosity of 5 g/l PGA (Sigma) solution in AcB (0.2 M, pH 5.0), at 37°C.

For thin-layer chromatography (TLC) analysis of PGA degradation products, heat inactivated samples were spotted (10 μL) on aluminium sheets (silica gel 60 F254, Merck) and the chromatography performed by using the ascending method with n-butanol : acetic acid : water (9 : 4 : 7, v/v/v) as the solvent system. Detection was accomplished by spraying the dried plate with 3% (w/v) phosphomolybdic acid dissolved in 10% (v/v) sulfuric acid in ethanol followed by heating at 105°C for 5 min. GA was used as standard [14].

An endo-PG is characterized by a strong reduction in viscosity (e.g., 50%) with a concomitantly low release of reducing groups and the first products are oligomers, whereas an exo-PG has to hydrolyse greater than 20% of the glycosidic linkages to obtain an equivalent viscosity reduction and the first degradation products are monomers or dimmers [15–17].

2.6. Effect of pH on Polygalacturonase Activity and Stability. The effect of pH on PG activity was determined by incubating the reaction mixture at pH values ranging from 3.5 to 6.0, under standard enzyme assay conditions.

The pH stability of the enzyme was evaluated by measuring the residual activity, under standard enzyme assay conditions, after incubating the EE without substrate for 24 h at 4°C at various pH from 2.0 to 8.0. The buffers employed in these measurements were citrate/phosphate buffer, for pH 2.0–3.5 and 6.0–8.0 and AcB (0.2 M) for pH 4.0–5.5. All the experiments were conducted in triplicate and the results show the mean values of the activities.

2.7. Effect of Temperature on Polygalacturonase Activity and Stability.

The effect of reaction temperature on PG activity was tested by incubating the reaction mixture at temperatures from 15°C to 60°C, at pH optimum, under standard enzyme assay conditions.

The thermostability of the enzyme was determined by measuring the residual activity, under standard enzyme assay conditions, after incubating the EE without substrate at temperatures from 45°C to 60°C at pH optimum.

All the experiments were conducted in triplicate and the results show the mean values of the activities.

2.8. Assay of Maceration Activity.

The effect of reaction time and shaking on potato maceration and final yield of single cells were evaluated.

Potatoes, purchased from a local market, were used for tests of maceration with the EE of *W. anomalus*. Potatoes were peeled and cut into pieces measuring 3-4 mm on each side. Enzymatic maceration was carried out in 125 mL Erlenmeyer flasks containing 5 mL of AcB (0.2 M) at pH optimum, 3 g of vegetable tissue, and 5 mL of EE. Flasks were incubated, at optimum temperature, up to 300 min in a shaker at different agitation speed. The whole content of the flasks was filtered through a 20 mesh screen into a 10 mL graduated conical test tube. Suspension of single cells was kept at 5°C for 4 h. The volume of single cells decanted was measured. Residual undegraded plant tissue was dried at 80°C until constant weight and then weighed. As a control, blanks were prepared with heat-denatured enzymes. Microscopic observations of the maceration process were also done [6]. Each experience was run in triplicate and mean values were calculated.

3. Results and Discussion

3.1. Extracellular Enzyme Activities.

W. anomalus was able to produce PG (~51 U/mL) in liquid medium containing YNB, glucose, and citrus pectin. The supernatant (10x) did not exhibit esterase or lyase activity. No other enzymes, capable of hydrolyzing cell wall polymers, such as cellulases and xylanases, were detected.

These results are in agreement with the observation of several authors who reported that the most common enzyme found to be secreted by pectinolytic yeasts is PG [17, 18]. Schwan et al. [16] reported that four yeast strains isolated from cocoa fermentations (*Kluyveromyces marxianus, K. thermotolerans,* and *Saccharomyces cerevisiae* var. *chevalieri*) showed extracellular PG activity, and neither PE nor lyases were detected in culture filtrates. Eight wine yeast strains of *Saccharomyces* sp. produced PG but none of them produced

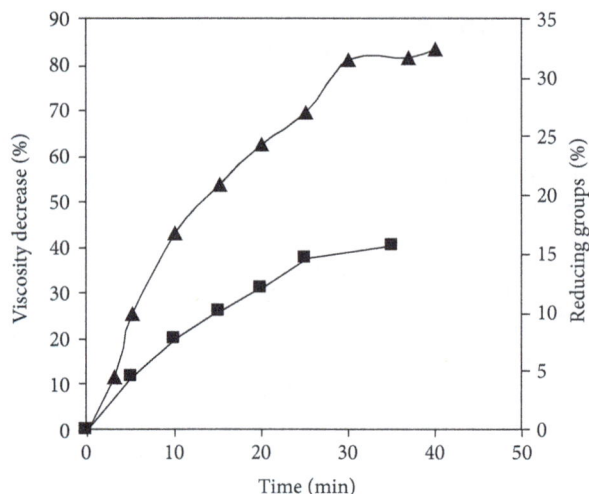

FIGURE 1: Degradation of polygalacturonic acid with the extract enzymatic of *W. anomalus*. Symbols: (-▲-) viscosity decrease, (-■-) reducing groups.

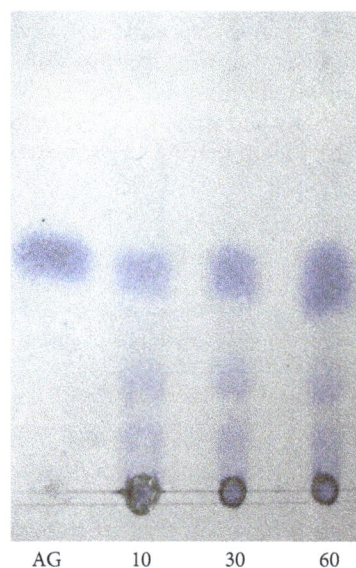

FIGURE 2: Thin-layer chromatography of the degradation products during enzymatic digestion of PGA solution with the extract enzymatic of *W. anomalus*. Numbers below each line indicate the reaction time.

PL or PAL [19]. Masoud and Jespersen [20] reported that yeasts predominant during coffee processing (six strains of *Pichia anomala*, four strains of *P. kluyveri,* and two strains of *Hanseniaspora uvarum*) were found to secrete PG but no PL or PE was found to be produced by the yeasts examined.

3.2. Mechanism of Action of PG.

Figure 1 shows the decrease in viscosity and increase in reducing groups as a function of time of a PGA solution by the EE of *W. anomalus* and Figure 2, shows the thin-layer chromatography of the degradation products during enzymatic digestion.

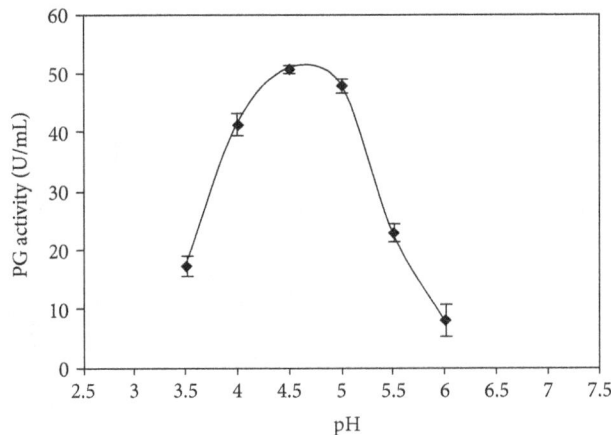

FIGURE 3: Effect of pH on PG activity produced by *W. anomalus*.

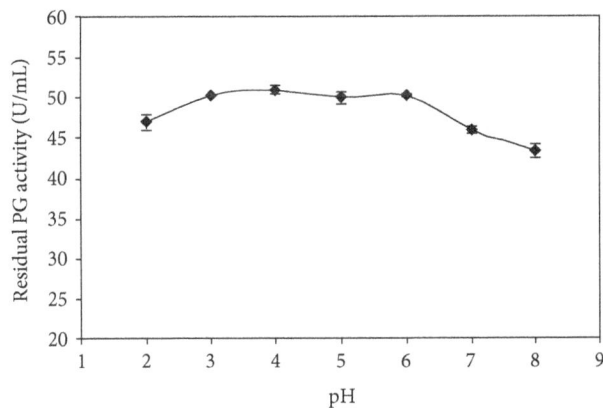

FIGURE 4: Effect of pH on PG stability produced by *W. anomalus*.

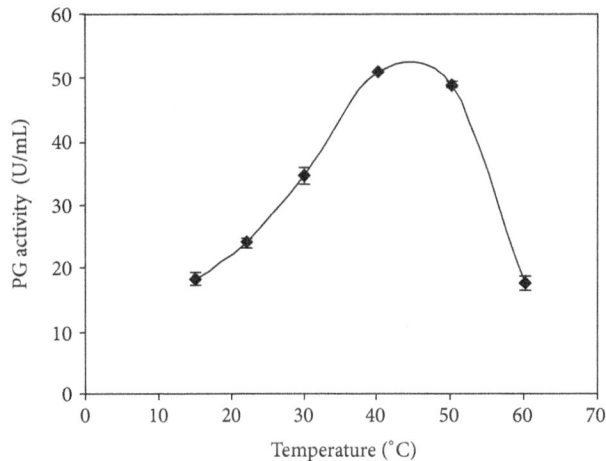

FIGURE 5: Effect of temperature on PG activity produced by *W. anomalus*.

Figure 4 shows that the enzyme was stable at a pH range from 3.0 to 6.0, after incubation time of 24 h at 5°C. Analysis of variance ($P < 0.05$) revealed no significant differences between these values. The enzyme retained 92% and 85% of its activity at pH 2.0 and 8.0, respectively.

Blanco et al. [18] reported that yeasts PGases exhibit an optimum pH in the acidic region between 3.5 and 5.5. This is in accordance with that reported for PGases produced by *S. cerevisiae* IM1-8b; *S. cerevisiae* 1389 [26]; *K. wickerhamii* [24]; *S. cerevisiae* UCLMS-39, yeast isolated from wine ecosystems [3]; *K. wickerhamii* strain 185 and *K. marxianus* strain 166, both yeasts from tropical fruits [8], and PGases from *K. marxianus* CCT 3172, *P. anomala* S16, and *P. kluyveri* S13Y4, yeasts predominant during coffee processing [20].

3.4. Effect of Temperature on PG Stability. The effect of temperature on PG activity and stability produced by *W. anomalus* is shown in Figures 5 and 6, respectively.

Figure 5 shows that PG activity was higher in a temperature range between 40°C and 50°C at pH 4.5. Analysis of variance revealed no significant differences between these values ($P < 0.05$). The value of PG activity at 40°C was 1.5 times higher than the value obtained at 30°C.

PGases isolated from different microbial sources differ markedly from each other with respect to their physicochemical properties; most have optimal temperature range of 30°C–50°C [1]. PG produced by *K. marxianus* [16], *K. marxianus* CCT 3172, and *P. anomala* S16 [20] exhibited maximum activity at 40°C and that of *K. Wickerhamii* [24] and *P. kluyveri* S13Y4 [20] at 50°C.

Figure 6 shows that in the absence of substrate, PG was stable at 45°C and 50°C during 8 h of incubation, at optimum pH. At 55°C, the enzymatic activity decreased and retained 78% and 54% of the initial activity after 30 min and 1 h of incubation, respectively. At 60°C thermal inactivation rate was higher and after 1 h of incubation the residual activity was only 24%.

Figure 1 shows that the viscosity of the substrate decreased 50% when only 9% of the glycosidic bonds were split. The TLC analysis of the products of PGA hydrolysis indicates that mono-, di-, and tri-galacturonanos and higher oligosaccharides were produced from the initial stages of the hydrolysis and accumulated throughout the incubation period. PG did not seem able to attack dimers and trimers as these products were accumulated throughout the incubation period (Figure 2). From these results, it can be deduced that PG of *W. anomalus* acts by an endo-splitting mechanism, so it is an endo-PG (EC 3.2.1.15) [17, 21].

Pectinolytic enzymes from yeasts are mainly endo-PG [22]. This observation is in agreement with those reported for *K. marxianus* [16], *S. cerevisiae* 1389 and *S. cerevisiae* IMI-89 [23], *K. wickerhamii* [24] and *Thermoascus aurantiacus* CBMAI-756 [25] which acted by an endo mechanism.

3.3. Effect of pH on PG Activity and Stability. The effect of pH on PG activity and stability produced by *W. anomalus* is shown in Figures 3 and 4, respectively.

PG secreted by *W. anomalus* exhibited maximal activity at pH 4.5. At pH 4.0 and 5.0, PG activity values were 81% and 94%, respectively (Figure 3).

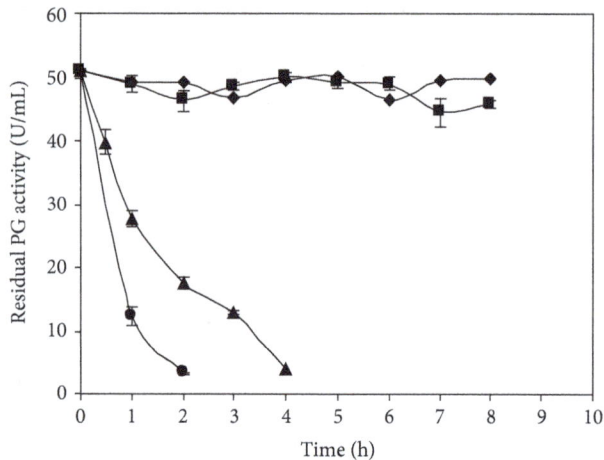

FIGURE 6: Effect of temperature on PG stability produced by *W. anomalus*. Symbols: (◆) 45°C, (■) 50°C, (▲) 55°C, (•) 60°C.

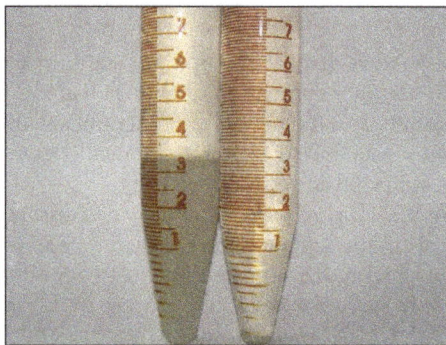

FIGURE 7: Decanted free cells after enzymatic maceration of potato tissue with the EE of *W. anomalus*. Left: negative control.

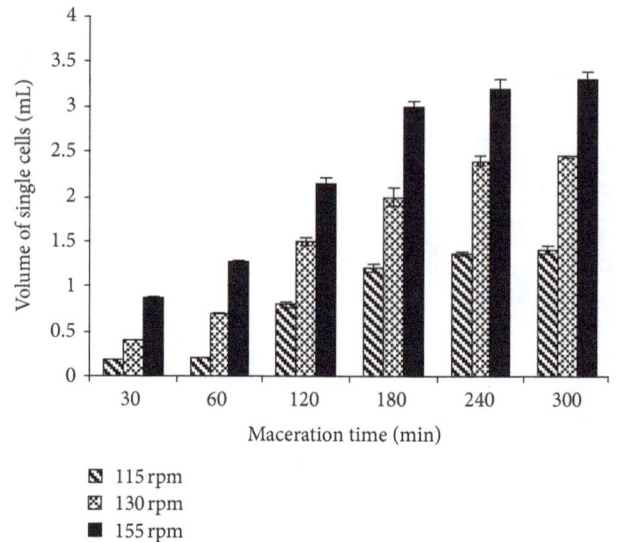

FIGURE 8: Volume of single cells as a function of reaction time at different agitation speeds, during maceration of potato tissues with enzymatic extract of *W. anomalus*.

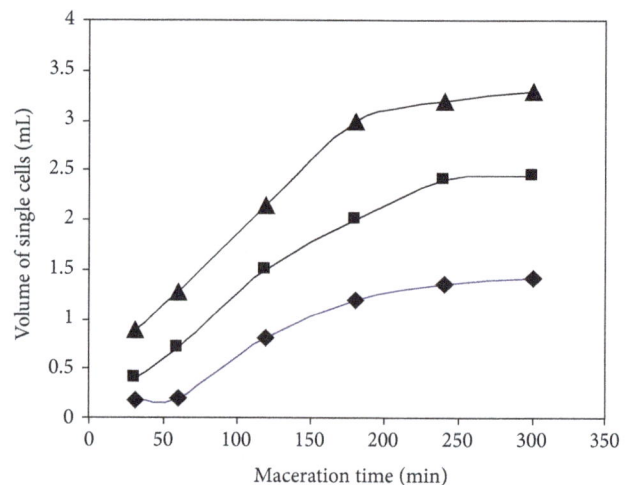

FIGURE 9: Kinetics of maceration process from potato tissues with the enzymatic extract of *W. anomalus*. Symbols (rpm): 115 (◆), 130 (■), 155 (▲).

The termoestabilidad of PG produced by *W. anomalus* was similar to that reported for PGases from other yeasts like *S. cerevisiae* IM1-8b and *S. cerevisiae* 1389 which were quite stable in the 20–50°C temperature range but were inactivated (80%) within 5 min at 55°C [23].

The knowledge of enzyme deactivation and stability is important to maintain the desired level of enzyme activity over a long period of time and improve its stability for an efficient application in an industrial process [25]. Besides after any application, the enzyme has to be inactivated, so the knowledge of thermal inactivation has great importance [27].

3.5. Assay of Maceration Activity. Figure 7 shows a photograph of conical tubes containing the decanted material (released cells) after enzymatic maceration of potato tissue with EE of *W. anomalus* at 45°C (within the range of PG stability) and pH 4.5 (optimum pH of PG). Effect of shaking on single-cell production is shown in Figure 8 and kinetics of single-cell production is present in Figure 9.

Figure 7 shows that the EE macerated plant tissue efficiently, and maceration blanks yielded negligible values. Blanks with the inactivated enzyme showed that, in all cases,

the effect was caused mainly by maceration activity of PG present in the EE and not by mechanical (shear) effects only.

Volume of single cells increased with reaction time at the three agitation speeds tested (115, 130, and 155 rpm) (Figure 8).

The rate of maceration at 115 rpm was low, and it increased at higher agitation speed (Figure 9). The rate of maceration at 155 rpm was higher at all times tested and yielded large amounts of single cells. At this agitation speed, volume of single cells increased almost linearly up to 180 min, suggesting that longer reaction times may not be necessary to achieve maximum maceration yields.

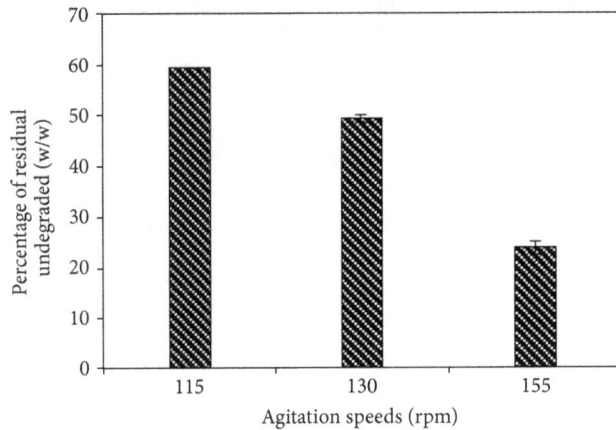

FIGURE 10: Residual undegraded of potato tissue as a function of agitation speeds after 300 min of time reaction with the enzymatic extract of *W. anomalus*.

The percentage of residual undegraded plant tissue produced at the end of the maceration process (300 min) at three agitation speeds is shown in Figure 10.

Figure 10 shows that it was obtained a 59.6%, (w/w), 49% (w/w), and 24.9% (w/w) of undegraded residue at 115, 130, and 155 rpm, respectively. Therefore the efficiency of the enzyme in macerating potato was higher at 155 rpm (about 75% of plant material was converted into cell free) and the yield of single cells produced was high. It would be important because the main purpose of enzymatic maceration is to maximize conversion of plant tissues into single cells. Consequently, amounts of the plant material that resisted enzymatic reaction should be minimized to optimize yield [6]. In contrast, at 115 rpm most of the initial material remained as an insoluble residue and the yield single cell volume was low.

The ability to release pectin from protopectin, leading to the maceration of plant tissues, depends on two factors: the chemical structure of the substrate and the ability of the enzyme to reach and degrade the specific site where the reaction takes place. Once partial depolymerization of the middle lamella had occurred, and a shear force was needed to transform the plant material into a suspension of loose cells [6].

P. anómala produces PG with maceration activity of potato tissues. Several PPases from bacterial, yeast or fungal origins have been isolated and characterized [28, 29].

4. Conclusions

The results showed that *W. anomalus* has a pectolytic system which consists essentially of an enzyme with polygalacturonase activity, whose mode of action is mainly of endo-type. Other enzymes such as PE, lyases, cellulases, and xylanases were not detected. PG exhibited an optimum pH in the acidic region and was stable up to 50°C, suited to most fruit and vegetable processing applications. The enzyme was responsible of the maceration of potato tissue observed.

This yeast is able to produce only an enzyme with polygalacturonase activity, making the downstream processing easier if pure enzyme is needed since separation from other enzymes is not required. Therefore polygalacturonase like the one characterized in this study could be a potential candidate for different applications in food industry, mainly in the softening of vegetables for the preparation of babies and seniors foods or for the production of dehydrated mashed cassava of regional interest in the province of Misiones, Argentine.

References

[1] R. S. Jayani, S. Saxena, and R. Gupta, "Microbial pectinolytic enzymes: a review," *Process Biochemistry*, vol. 40, no. 9, pp. 2931–2944, 2005.

[2] C. Tari, N. Gögus, and F. Tokatli, "Optimization of biomass, pellet size and polygalacturonase production by *Aspergillus sojae* ATCC 20235 using response surface methodology," *Enzyme and Microbial Technology*, vol. 40, no. 5, pp. 1108–1116, 2007.

[3] M. Fernández-González, J. F. Úbeda, T. G. Vasudevan, R. R. Cordero Otero, and A. I. Briones, "Evaluation of polygalacturonase activity in *Saccharomyces cerevisiae* wine strains," *FEMS Microbiology Letters*, vol. 237, no. 2, pp. 261–266, 2004.

[4] W. Pilnik and A. G. J. Voragen, "The significance of endogenous and exogenous pectic enzymes in fruit and vegetable processing," in *Food Enzymology*, vol. 1, pp. 303–336, 1991.

[5] J. A. V. Costa, E. Colla, G. Magagnin, L. Oliveria dos Santos, M. Vendruscolo, and T. E. Bertolin, "Simultaneous amyloglucosidase and exo-polygalacturonase production by *Aspergillus niger* using solid-state fermentation," *Brazilian Archives of Biology and Technology*, vol. 50, no. 5, pp. 759–766, 2007.

[6] T. Nakamura, R. A. Hours, and T. Sakai, "Enzymatic maceration of vegetables with protopectinases," *Journal of Food Science*, vol. 60, pp. 468–472, 1995.

[7] S. N. Gummadi and T. Panda, "Purification and biochemical properties of microbial pectinases: a review," *Process Biochemistry*, vol. 38, no. 7, pp. 987–996, 2003.

[8] E. Geralda da Silva, M. de Fátima Borges, C. Medina, R. Hilsdorf Piccoli, and R. Freitas Schwan, "Pectinolytic enzymes secreted by yeasts from tropical fruits," *FEMS Yeast Research*, vol. 5, no. 9, pp. 859–865, 2005.

[9] C. P. Kurtzman, C. J. Robnett, and E. Basehoar-Powers, "Phylogenetic relationships among species of *Pichia*, *Issatchenkia* and *Williopsis* determined from multigene sequence analysis, and the proposal of *Barnettozyma* gen. nov., *Lindnera* gen. nov. and *Wickerhamomyces* gen. nov.," *FEMS Yeast Research*, vol. 8, no. 6, pp. 939–954, 2008.

[10] S. F. Cavalitto, J. A. Arcas, and R. A. Hours, "Pectinase production profile of *Aspergillus foetidus* in solid state cultures at different acidities," *Biotechnology Letters*, vol. 18, no. 3, pp. 251–256, 1996.

[11] G. L. Miller, "Use of dinitrosalicylic acid reagent for determination of reducing sugar," *Analytical Chemistry*, vol. 31, no. 3, pp. 426–428, 1959.

[12] P. Albersheim, H. Neukom, and H. Deuel, "Splitting of pectin chain molecules in neutral solutions," *Archives of Biochemistry and Biophysics*, vol. 90, no. 1, pp. 46–51, 1960.

[13] C. Vilariño, J. F. Del Giorgio, R. A. Hours, and O. Cascone, "Spectrophotometric method for fungal pectinesterase activity determination," *LWT—Food Science and Technology*, vol. 26, no. 2, pp. 107–110, 1993.

[14] J. C. Contreras Esquivel and C. E. Voget, "Purification and partial characterization of an acidic polygalacturonase from *Aspergillus kawachii*," *Journal of Biotechnology*, vol. 110, no. 1, pp. 21–28, 2004.

[15] S. Yoshitake, T. Numata, T. Katsuragi, R. A. Hours, and T. Sakai, "Purification and characterization of a pectin-releasing enzyme produced by *Kluyveromyces wickerhamii*," *Journal of Fermentation and Bioengineering*, vol. 77, no. 4, pp. 370–375, 1994.

[16] R. F. Schwan, R. M. Cooper, and A. E. Wheals, "Endopolygalacturonase secretion by *Kluyveromyces marxianus* and other cocoa pulp-degrading yeasts," *Enzyme and Microbial Technology*, vol. 21, no. 4, pp. 234–244, 1997.

[17] A. R. García, M. I. Balbín, J. C. Cabrera, and A. Castelvi, "Actividad endopoligalacturonasa de un preparado de la levadura Kluyveromyces marxianus aislada de la pulpa de café," *Cultivos Tropicales*, vol. 23, no. 1, pp. 67–72, 2002.

[18] P. Blanco, C. Sieiro, and T. G. Villa, "Production of pectic enzymes in yeasts," *FEMS Microbiology Letters*, vol. 175, no. 1, pp. 1–9, 1999.

[19] F. Radoi, M. Kishida, and H. Kawasaki, "Endo-polygalacturonase in *Saccharomyces wine* yeasts: effect of carbon source on enzyme production," *FEMS Yeast Research*, vol. 5, no. 6-7, pp. 663–668, 2005.

[20] W. Masoud and L. Jespersen, "Pectin degrading enzymes in yeasts involved in fermentation of coffea arabica in East Africa," *International Journal of Food Microbiology*, vol. 110, no. 3, pp. 291–296, 2006.

[21] F. M. Rombouts and W. Pilnik, "Pectic enzymes," in *Microbial Enzymes and Bioconversions*, A. H. Rose, Ed., pp. 227–282, Academic Press, London, UK, 1980.

[22] D. B. Pedrolli and E. C. Carmona, "Purification and characterization of the exopolygalacturonase produced by *Aspergillus giganteus* in submerged cultures," *Journal of Industrial Microbiology and Biotechnology*, vol. 37, no. 6, pp. 567–573, 2010.

[23] P. Blanco, C. Sieiro, A. Díaz, and T. G. Villa, "Differences between pectic enzymes produced by laboratory and wild-type strains of *Saccharomyces cerevisiae*," *World Journal of Microbiology and Biotechnology*, vol. 13, no. 6, pp. 711–712, 1997.

[24] S. Moyo, B. A. Gashe, E. K. Collison, and S. Mpuchane, "Optimising growth conditions for the pectinolytic activity of *Kluyveromyces wickerhamii* by using response surface methodology," *International Journal of Food Microbiology*, vol. 85, no. 1-2, pp. 87–100, 2003.

[25] E. S. Martins, D. Silva, R. S. R. Leite, and E. Gomes, "Purification and characterization of polygalacturonase produced by thermophilic *Thermoascus aurantiacus* CBMAI-756 in submerged fermentation," *Antonie van Leeuwenhoek*, vol. 91, no. 3, pp. 291–299, 2007.

[26] P. Blanco, C. Sieiro, A. Diaz, and T. G. Villa, "Production and partial characterization of an endopolygalacturonase from *Saccharomyces cerevisiae*," *Canadian Journal of Microbiology*, vol. 40, no. 11, pp. 974–977, 1994.

[27] C. Tari, N. Dogan, and N. Gogus, "Biochemical and thermal characterization of crude exo-polygalacturonase produced by *Aspergillus sojae*," *Food Chemistry*, vol. 111, no. 4, pp. 824–829, 2008.

[28] S. F. Cavalitto, R. A. Hours, and C. F. Mignone, "Growth and protopectinase production of *Geotrichum klebahnii* in batch and continuous cultures with synthetic media," *Journal of Industrial Microbiology and Biotechnology*, vol. 25, no. 5, pp. 260–265, 2000.

[29] N. L. Rojas, S. F. Cavalitto, C. F. Mignone, and R. A. Hours, "Role of PPase-SE in *Geotrichum klebahnii*, a yeast-like fungus able to solubilize pectin," *Electronic Journal of Biotechnology*, vol. 11, no. 1, pp. 1–8, 2008.

15

Biotechnological Tools for Environmental Sustainability: Prospects and Challenges for Environments in Nigeria—A Standard Review

Chukwuma S. Ezeonu,[1] Richard Tagbo,[1] Ephraim N. Anike,[2] Obinna A. Oje,[3] and Ikechukwu N. E. Onwurah[3]

[1] Industrial Biochemistry and Environmental Biotechnology Unit, Chemical Sciences Department, Godfrey Okoye University, P.M.B. 01014, Enugu, Nigeria
[2] Pure and Industrial Chemistry Unit, Chemical Sciences Department, Godfrey Okoye University, P.M.B. 01014, Enugu, Nigeria
[3] Pollution Control and Biotechnology Unit, Department of Biochemistry, University of Nigeria, Nsukka, Enugu State, Nigeria

Correspondence should be addressed to Chukwuma S. Ezeonu, chuksmaristos@yahoo.com

Academic Editor: Manuel Canovas

The environment is a very important component necessary for the existence of both man and other biotic organisms. The degree of sustainability of the physical environment is an index of the survival and well-being of the entire components in it. Additionally, it is not sufficient to try disposing toxic/deleterious substances with any known method. The best method of sustaining the environment is such that returns back all the components (wastes) in a recyclable way so that the waste becomes useful and helps the biotic and abiotic relationship to maintain an aesthetic and healthy equilibrium that characterizes an ideal environment. In this study, the method investigated includes biological method of environmental sustainability which seeks to investigate the various biotechnological tools (biotools) in current use and those undergoing investigations for future use.

1. Introduction

Biotechnological tools are those processes of bioscientific interests that use the chemistry of living organisms through cell manipulation to develop new and alternative methods aimed at cleaner and more effective ways of producing traditional products and at the same time maintain the natural and aesthetic beauty of the environment. Biotechnology is the current trend in production processes across the world, as opposed to the conventional chemical synthesis of products. The reason is due to the fact that biotechnological methods are ecofriendly while the latter method adds pollutants and waste into our environment. A lot of problems associated with conventional methods of pollutant treatment by incineration or landfills have given the impetus on the need for alternative, economical, and reliable biological methods of pollution treatments.

Chen et al. [1] enumerated vividly that environmental biotechnology refers to the utilization of microorganisms to improve environmental quality. Although the field of environmental biotechnology has been around for decades, starting with the activated sludge and anaerobic digestion in the early 20th century, the introduction of new technologies from modern microbiology and molecular biology has enabled engineers and scientists to tackle the more contemporary environment problems such as detoxification of hazardous wastes through the use of living organisms.

As the earth's human population has increased, natural ecosystems have declined and changes in the balance of natural cycles have had a negative impact on both humans and other living systems. Thus, there is abundant scientific evidence that humanity is living unsustainably, and returning human use of natural resources to within limits will require a major collective effort [2]. Given the challenges of population increase and its attendant problems of pollution increase, biotechnology remains the most reliable means of environmental sustenance. The world is currently endangered; government and people of many counties are concerned about this endemicity of pollutants (most of which are recalcitrant) in our otherwise aesthetic environment. Africa generally and Nigeria in particular have not imbibed maximally the benefit of using biotechnology in maintenance of the beautiful environment. This paper will address the issues relating to the use of biotechnological methods vis-à-vis biotools in solving the problems of environmental degradation, with a view to encourage the adoption of these biotechnological methods in Nigeria, Africa, and other countries where waste has been a menace to the environments.

2. Environmental Sustainability

Sustainability is the capacity to endure. The word sustainability is derived from the Latin *sustinere* (tenere, to hold; sus, up). In ecology the word describes how biological systems remain diverse and productive over times. For humans it is the potential for long-term maintenance of well-being, which in turn depends on the well-being of the natural world and the responsible use of natural resources (http://en.wikipedia.org/wiki/Environmental_Sustainability_Index) [3]. Environmental sustainability is the process of making sure current processes of interaction with the environment are pursued with the idea of keeping the environment as pristine as naturally possible based on ideal-seeking behaviours. An "unsustainable situation" occurs when natural capital (the sum total of nature's resources) is used up faster than it can be replenished. Sustainability requires that human activity only uses nature's resources at a rate at which they can be replenished naturally. Theoretically, the long-term result of environmental degradation is the inability to sustain human life. Such degradation on a global scale could imply extinction for humanity [4].

A healthy environment is one that provides vital goods and services to humans as well as other organisms within its ecosystem. This can be achieved in two ways and include discovering ways of reducing negative human impact and enhancing the well-being and vitality of all living organisms (plants and animals) in the environment. Daly [5] suggested three broad criteria for ecological sustainability: renewable resources should provide a sustainable yield (the rate of harvest should not exceed the rate of regeneration); for non-renewable resources there should be equivalent development of renewable substitutes; waste generation should not exceed the assimilative capacity of the environment.

It is important to also clearly define what the environment is to the humans who are the focus and are adversely affected positively or negatively according to their activities within their surroundings. Thus, Bankole [6] reported that "Environment" refers to the physical surroundings of man, of which he is part and on which he depends for his activities, like physiological functioning, production, and consumption. His physical environment stretches from air, water, and land to natural resources like metals, energy carriers, soil, and plants, animals, and ecosystems. For urbanized man, a large part of his environment is man-made. But even then, the artificial environments (buildings, roads) and implements (clothes, automobiles) are the result of an input of both labour and natural resources.

3. Environmental Sustainability Index (ESI)

This is a composite index tracking 21 elements of environment sustainability covering natural resource endowments, past and present pollution levels, environmental management efforts, contributions to protection of the global commons, and a society's capacity to improve its environmental performance over time [7].

The Environmental Sustainability Index was developed and published between 1999 and 2005 by Yale University's Centre for Environmental Law and Policy in collaboration with Columbia University's Centre for International Earth Science Information Network (CIESIN), and the World Economic Forum. The ESI developed to evaluate environmental sustainability relative to the paths of other countries. Due to a shift in focus by the terms developing the ESI, a new index was developed, the Environmental Performance Index (EPI) that uses the outcome-oriented indicators, then works as a benchmark index that can be more easily used by policy makers, environmental scientists, advocates and the general public [8].

4. The Nigerian Physical Environment

4.1. The Niger Delta Environment. Nigeria has one of the worst environmental records in the world. In late 1995, Nigeria's execution of eight environmental activists, notable Nobel Peace Prize nominee Ken Saro-Wiwa, made international headlines and brought world-wide recognition of the serious environmental degradation of Nigeria [9]. Today, the oil-rich Niger Delta region of Nigeria is always the first point of reference when analysing the Nigerian environment. This stems from the fact that despite the sacrifice of Ken Saro-Wiwa and others, much has not changed in terms of making the environment pollution free. In fact, it has even gotten worse with time and recent developments. During the 1990s, the Niger Delta locals learned that extortion pays. Villagers found that by sabotaging oil installations to collect oil spill compensation from shell (an oil firm) they could earn more than by marginal subsistence farming on degraded lands. Thus, sabotaging and spills became a new dimension of increasing crude oil pollutants. Attacks on oil facilities and pipelines became even more relentless, and the Niger River delta was an increasingly bloody place. Environmental degradation from crude oil productions continued, and by 1999 the United Nations declared the delta the most threatened in the world [9]. In early 2006, conditions worsened in the delta. The number of kidnapping of oil workers increased as did attacks on oil facilities. The Niger Delta is made up of six states of the south-south region, namely, Bayelsa, Akwa Ibom, Cross River, Delta, Rivers, and Edo.

This lingering problem caused the former President of Nigeria, Alhaji Umaru Musa Yar'adua, to create a Ministry of the Niger Delta so as to oversee the well-being of the environment and the people of this region. Currently more is being done in the Niger Delta to bring about reduction in vandalization of oil pipelines as well as an amnesty for the irate youths in this part of Nigeria in order to find a lasting solution to both the social and environmental problem of the Niger Delta area of Nigeria.

Crude oil spill affects germination and growth of some plants [10], it also affects the overall production of crop (e.g., *Zea mays*) due to its negative impact on the chlorophyll content which is a marker of the yield of plants [11]. Severe crude oil spill in Cross River state, Nigeria, has forced some farmers to migrate out of their traditional home, especially those that depend solely on agriculture. The negative impact of oil spillages remains the major cause of depletion of the Niger Delta of Nigeria vegetation cover and the mangrove ecosystem [12].

4.2. The Nigerian Environment: Case Study of Solid Waste Generation in Cities.
The Nigerian cities such as Aba, Enugu, Onitsha, Kano, Ibadan, and Lagos are characterised by huge mounds of solid waste dumps generated from households, industries, markets, schools, and street trading. This can be attributed to migration, population increase, urbanization, constructions, and industrialization coupled with inefficient, improper and some times nondisposal of wastes. Solid waste dumps are indiscriminately formed on streets, homes, road side, markets, and other places where human activities take place in the cities.

Solid wastes can be broadly grouped into two as it relates to the concept of this write up. These two categories are the following.

(a) The Biodegradables (Biowastes).
These include those solid wastes generated, which could be decomposed by microorganisms and does not constitute major sources of pollution for a long period of time. They are paper products (such as printing papers, waste books, newspapers, carton, toilet paper, card boards), and wastes of plant origin (fruits, stems, roots, vegetables, leaves, food remains and garden solid wastes, etc.), wastes of animal origin (faecal matter, carcass, droppings, and poultry waste products). These groups of solid waste even though they are easily degraded by microorganism in minimal time, give off offensive odour and constitute nuisance to the aesthetic environment more than the nonbiodegradable solid wastes. They can also constitute a good habitat for the thriving of pathogenic microorganisms which could easily pollute fresh food product and sources of fresh water in the urban cities in Nigeria.

(b) Nonbiodegradable (Rubbish/Garbage).
These groups of solid wastes are not degradable or hardly degraded by microorganisms. Hence, other means of treatment such as incineration, land refill, and recycling are currently employed in Nigeria as ways of disposing them. Examples of this group of solid wastes are solid wastes of metallurgical and smelting industries (abandoned vehicles, motor cycles, vehicle part and scrap metals, iron, zinc, aluminium sheets and other metals, machine parts); solids wastes of construction industries (sand, gravel, bitumen wastes, concrete and waste building materials); solid waste of plastic industries (plastic buckets, cable insulators, tyres, chairs, tables, cellophane bags, plastic bottles, cutleries, sachet water containments, etc.) and glass products. These might not give out offensive odour, but they are even worse nuisance to the environment since their disposal has become a "Herculean" and near-impossible task in Nigeria.

Solid waste management activities include prevention (pollution prevention from sources), source reduction (pollution minimization in waste generating activities at point of good production), and treatment (safe disposal of non-recyclable residues, recycling, transport of waste to land refills).

The major problem is that Nigeria is yet to develop efficient ways of waste disposal which are eco-friendly and which could be recycled back into the environment without constituting nuisance to the environment or affecting the health of the biotic components of the ecosystem.

4.3. Environmental Degradation due to Mining Activities in Nigeria.
The natural topography of many cities and country side in Nigeria had been destroyed as a result of commercial activities involved in the exploration and exploitation of numerous minerals that abound in the country. Places like Jos, Bauchi, Nasarawa, and Enugu states have been worst affected by environmental degradation which had defaced the beautiful landscape of the natural environment.

No consistent mining regulatory law is enforced in the country. The exploration of tin in the Plateau (Jos) started as early as 1808 by the British colonialists, and in the 1970s Nigeria produced an average of 10,000 tons of tin ore annually. Output fell to 3,000 tons in the 1980s and dropped again to 500 tons in the 1990s. Nigeria now earns less than 0.5% of its foreign exchange from tin [13]. For over 70 years Jos tin mining industry was mostly controlled by overseas companies. But when the company was nationalised in 1972, no one took responsibility for clearing up the mess left behind. In places on the plateau such as Bukuru, Rayfield, Barkin Ladi, Mangu, Anglo Jos, Zawan, Du, Shen, Gyel and Shere Hills, ugly gashes left over from past mining activities can be seen everywhere. Alarmingly, effluents from nearby industries have seeped deep into mines-turned-water holes. Farmers use water from dams which resulted from tin mining activities for irrigation. The top soil also washes into streams in neighbouring village water that is used for drinking and other domestic purposes.

In addition, locals use soil left over from the abandoned mining sites—containing naturally found radioactive heavy metals to build houses. Environmentalists fear that people living in these houses risk being exposed to unhealthy levels of radiation [13].

Tin mining has also displaced many people from fertile agricultural land. The mining sites are located in the best areas in terms of the terrain and the flatness of the land. The people are now compelled to farm on rocky land. Government on its part instead of reclaiming the lands and resettling the people only asked people whose lands are destroyed to move out of these danger zones without compensation or arrangements to resettle them properly.

Thus, the tin areas have environment whose topography is made up of dams (which claim lives of both human and animals annually) as well as "a lunar landscape of steep-sided mounds with multicoloured ponds or lakes" [13].

Nigerian environmentalists have agreed that mining activities such as tin-Jos, Coal-Enugu, and others have done great damages to the environment which will need a concerted effort especially adoption of better mining practices in order to remediate.

4.4. Erosion, Desertification, and Deforestation: Loss of Biodiversity in the Nigerian Environment. Erosion problem is also a major environmental threat in Nigeria as sheet and gully erosion have wrecked untold havoc in several states such as Abia, Adamawa, Anambra, Delta, Ebonyi, Edo, Enugu, Gombe, Jigawa, Kogi, Ondo, Ogun, and Lagos. In Lagos state and other coastal areas, coastal erosion has destroyed properties and valuable lands were washed away. Most of the flooding and erosion seen in cities are as a result of poor drainage system.

Places such as the eastern states of Nigeria (Anambra, Imo, Abia, Enugu, and Ebonyi) have regions prone to erosion. This has resulted in the entire loss of farm land and buildings. The situation is so pathetic that a whole clan in a southern part of Anambra state was forced to take refuge in a primary school. Places like Agulu and most part of Aguata and Orumba Local government areas are highly endangered with erosion invasion. The Northern part of the country has ecosystem characterised by the Savannah clime. Starting from the North Central region encompassing Benue to Katsina states in the farthest part of the North made up of the southern savannah, Northern savannah, Sudan, and Sahel savannah characterized by low foliage and little trees. The environment is marked by constant grazing and building of huts which affect the type of plant survival as well as desert encroachment from the Niger and the Chad republics at the furthest part of the country.

Deforestation is a serious problem in Nigeria, which currently has one of the highest rates of forest loss (3.3 Percent) in the world. Since 1990, the country has lost some 1 million hectares or 35.7 percent of its forest covers [9]. Worse Nigeria's most biodiverse ecosystems—its old-growth forests is that are disappearing at an even faster rate. Between 1990 and 2005, the country lost a staggering 79% of these forests and since 2000 Nigeria has been losing an average of 11 percent of its primary forests per year—double the rate of the 1990s. These figures mark Nigeria as having the highest deforestation rate of natural forest on the planet. As its forests fall, Nigeria has seen wildlife populations plummet downward from poaching and habitat loss, increasing desertification. It appears that Nigeria's swift economic development has exacted a high toll on its people and environment [9].

The problems of environmental degradation have continued to plague Nigeria, and they have defied proffered solution mainly due to improper applications and also the lack of proper waste control and environmental maintenance. The major causes of environmental degradation problems were identified by the Vision 2010 Committee set up by the Federal Government. Aina and Salau [14] enumerated some of these problems as follows:

(i) poverty as a cause consequence of environmental exploitation, with the poor scavenging marginal lands to eke out a living;

(ii) bush burning for farming and ever-increasing depletion of young forests for fuel wood.

(iii) uncontrolled logging accentuated by lack of restocking in many parts of the country. This practice is linked with the loss of precious biological diversity (nature's gene bank of raw materials for future development);

(iv) gas flaring, Crude oil spill and the resultant problem of ecosystem destabilization, heat stress, acid rain and acid precipitation-induced destruction of fresh water fishes and forests in the coastal areas of the country. Nigeria alone accounted for about 28% of the world's total gas flared;

(v) a general inability of the agencies responsible for the environment to enforce laws and regulations, particularly with respect to urban planning and development, prospecting for minerals and adherence to industrial standards, sitting of public and residential quarters in flood-prone areas, unsettled dump site improperly reclaimed and converted to plots.

5. The Need for Pollution Prevention

Most of the pollutants in the environment are directly or indirectly the product of industrial activities/production. Awareness of the deleterious effect of pollutants in the environment is on the increase. Government, environmentalists, and communities for a long time have been frowning at the degradation of the environment due to man-made pollutants especially those that are by-product of industries. Industries on the other hand are under pressure by their communities to minimize the pollutants they generate. This has placed the manufacturing industries at high cost of revenue for pollution treatment as well as Billion of Naira for research into eco-friendly ways of manufacturing processes which minimizes pollution generation. Most of the pollution released from industrial processes includes discharge into the environment, namely: air, land and water. The best points of pollution prevention involves, source reduction (by using raw materials more efficiently); pollution control (substituting less harmful substances for hazardous materials); pollution management (eliminating toxic substances from the production process).

By implementing pollution prevention practices, companies often reduce their operational waste disposal, and compliance costs (http://www.p2.org/about/nppr_p2.cmf/).

6. Biotechnology: The Hope for Environmental Sustainability

As earlier stated, man's activities in his environment involve a lot of chemical synthesis in the process of converting the natural products in his environment into other forms convenient for his consumption. In the quest for converting wood into timber, use of fruits in juice production, use of herb for drug synthesis, conversion of petrochemical substances into polythene products, the environment correspondingly becomes littered with substances not needed in the cause of production. In the process of creating products, man also creates problems either consciously or unconsciously vis-à-vis pollution. As a result, the most acceptable solution to the generated wastes in the environment is such that will conveniently integrate them back into the environment. That method involves the use of microorganisms—usually yeasts, bacteria, or fungi as whole cell usage production system or in the form of industrial enzymes. In many cases these microorganisms or their products are integrated into the substrates which give us the products, desired in the industries, examples of these are bioleaching (biomining), biodetergent, biotreatment of pulp, biotreatment of wastes (bioremediation), biofiltrations, aquaculture treatments, biotreatment of textiles, biocatalysts, biomass fuel production, biomonitoring, and so forth. These are biotools (biotechnological tools), which could solve the problem of pollution and help sustain the environment. This is so because when the products or their constituents are discarded, they go back into the ecosystem. As such, they become reconverted into organic components of the environments. Moreover, their production is strictly biological instead of chemical (synonymous to pollution introduction).

These biotechnology tools have long been used in many developed countries in the world such as the United States, Finland, Sweden, Germany, Japan, and others. Africa is still lagging from being integrated into these environmental sustainability best practices. Nigeria is the focus on how to begin to make use of these biotools for the improvement of the badly degraded environment.

7. Biotechnology

Biotechnology is defined as a set of scientific techniques that utilize living organisms or parts of organisms to make, modify, or improve products which could be plants or animals. It is also the development of specific organisms for specific application or purposes and may include the use of novel technologies such as recombinant DNA, cell fusion, and other new bioprocesses [15].

Biotechnology is not new; it has been employed for centuries in the production of fermented foods such as gari, bread, yoghurt, and cheese and beverages such as wine and beer [16]. Thus, it is a natural phenomenon in use even in Africa (Nigeria) though its principle was not well understood. CTA [16] report illustrated the denomination of "green," "red" and "white" biotechnology according to its

TABLE 1: Classification of biotechnologies. Modification from: Disilva [17].

Red	Medical
Yellow	Food biotechnology
Green	Agriculture
Blue	Aquatic
White	Gene-based industry
Grey	Fermentation
Brown	Arid
Gold	Nanotechnology/bioinformatics
Purple	Intellectual
Dark	Bioterrorism/warfare

uses and applications. Moreover, Disilva [17] has a different classification as shown in Table 1.

Despite the that classification for convenience, using the CTA [16] classification encompasses all others. As a result "Green biotechnology" encompasses a wide range of techniques that consists of culturing plant tissues and/or organs, followed by the multiplication of the relevant plants with desirable characteristics. Genetically identical plantlets are thus available for distribution to farmers, horticulturalists, forestry growers, and nurseries all the year round. It also includes the transformation of plants, crop species, and varieties through genetic engineering techniques, leading to what are known as "genetically modified" (GM) crops. In addition, green or agricultural biotechnology also applies to techniques used in livestock husbandry (nutrition and reproduction). Green biotechnology should therefore not only be equated with advances in genetic engineering.

"Red biotechnology" encompasses the genetic engineering technique that has been used since the mid-1970s to produce drugs and vaccines in microorganisms, animal cells, and more recently in plants. For example, insulin, human and bovine growth hormones, interferon, cell growth factors, antihepatitis B vaccine, and others are being produced in this way. A wide range of diagnostic techniques and veterinary vaccines are produced using red or medical biotechnology [16].

"White biotechnology" refers to a wide range of processes resulting in fermented products and chemicals (e.g., enzymes, biofuels such as ethanol and bioplastics) as well as to the technologies used in recycling waste water, industrial effluents, and solid wastes. These "bioremediation" processes contribute to the abatement of pollution. The extraction of metals from ores with the help of microorganisms (biomining) is also part of white or environmental biotechnology [16].

8. Environmental Biotechnology

Environmental biotechnology is "the integration of natural sciences and engineering in order to achieve the application of organisms, cells, parts thereof and molecular analogues for the protection and restoration of the quality of our environment" [18].

The nomenclature of "white biotechnology" is alluded to both industrial and environmental biotechnology. Biotechnological tools for environmental sustainability are qualified to be greatly associated as major component of the "White biotechnology."

Biotechnological processes to protect the environment have been used for almost a century now, even longer than the term "biotechnology" exists [18]. Municipal sewage treatment plants and filters to purify town gas were developed around the turn of the century. They proved very effective although, at the time, little was known about the biological principles underlying their function. Since that time, our knowledge base has increased enormously [18].

Biotechnological techniques to treat waste before or after it has been brought into the environment are components of environmental biotechnological tools. Biotechnology can also be applied industrially for use in developing products and processes that generate less waste and use less nonrenewable resources and consume less energy. In this respect biotechnology is well positioned to contribute to the development of a more sustainable society through a sustainable environment. Recombinant DNA technology has improved the possibilities for the prevention of pollution and holds a promise for a further development of bioremediation [18]. What this means for environmental biotechnology is that it is futuristic and limitless in application and usage.

9. Biotechnological Tools for Environmental Sustainability

Biotools for the sustenance of the environment are those biotechnological processes that make use of bioproducts as well as microorganisms for pollution reduction, production of environmental friendly products as well as general maintenance of the pristine (natural) environment for the benefit of man and other ecosystem components. It is an aspect of environmental biotechnology concerned with prevention of processes capable of causing an unsustainable environment for man and ecocomponents. Some of the biotools in use will be briefly and concisely enumerated here, and it is by no means exhaustive due to current and future addition to the body of knowledge in the environmental biotechnology field. The discussion will centre on the current or future projection of the usage of these tools elsewhere and the need for Nigeria and other countries which are hitherto not adapting to their usage due to environmental and technological limitations to break such barriers and begin in earnest to adopt their usage. Some of the biotools are as enumerated in the following.

9.1. Biodetergents and Biosolvent Research/Production. FAIR [19] stated that solvents and detergents are important in a number of industries, and most are derived from petroleum. There is increasing concern that prolonged contact with solvents causes health problems for workers in the factories where solvents and detergents are produced and for those using such substances in domestic and commercial laundry in their everyday work. The aim for research into biodetergents is to create biological substitutes for solvents and detergents derived from petroleum. Apart from detergents, research is also ongoing into developing substitutes of biological origin for solvents used in the production of paints, offset-printing ink, and so forth. These new solvents will be made by mixing together several common bioliquids: such as bioethanol, terpenes, vegetable oils, fatty acids, methyl esters, and derivatives of related compounds [19].

The key to success is mixing the right compounds together in the right proportions. To do this, researchers are developing mathematical models, which will allow the "recipes" for the new biosolvents to be optimised. The mixtures being developed are designed to meet the criteria established by the companies participating in the project. Those small- and medium-sized enterprises from France, the United Kingdom, The Netherlands, Denmark, and Belgium are paint producers, manufacturers of ink for use in offset printing, and producers of detergent. Their activities are examples of only three industries where solvents are used: the application of biosolvents to other industrial processes will also be promoted as part of this project [19].

9.1.1. Benefits for Farmers, Workers, and the Environment. FAIR [19] SMEs research release reported that most solvents and detergents in current use are derived from petroleum, as a result there are two major drawbacks to these solvents and detergents compared to the biological counterparts. First, one day the crude oil from which they are derived will no longer be available since they are nonrenewable natural resources and again they present a major waste disposal problem as they do not break down readily when disposed of hence pollute the environment.

Presently [19], the success of the new biosolvents and detergents are limited since they are more expensive than the traditional products. At the moment, conventional solvents are cheap and the replacement bioproducts could not compete on price. But as governments start to use taxation to stimulate the use of renewable resources, economic viability of biosolvents should improve. Increased use of biosolvents and biodetergents will stimulate demand for the raw materials used to manufacture the new products and so help farmers by creating a new outlet for their produce. The new solvents and detergents being developed will have a much less detrimental effect than existing products on the health of workers making and using them. Agricultural product, job creation, and environmental sustainability are the benefits accruable due to this biotool innovation. Development of the product will help lessen dependence on nonrenewable resources (prevents nonrenewable environmental exploitation), safeguard human health, and protect the environment from chemical pollutants seeping into aquatic body and atmosphere.

Nigeria can take a cue from this giant stride and begin its own processing of the natural and agricultural product as well as make use of the fertile land to develop bio products friendly to the environment. It is a viable project friendly to the environment. It is a viable project which can be adopted since it requires less energy consumption and minimal technological effort. Hence, Nigerian government, environmentalists, industrialists, and research institution can adopt this novel production of biodetergent and biosolvents for its environmental sustainability.

9.2. Bioremediation. Bioremediation is the use of biological systems for the reduction of pollution from air or from aquatic or terrestrial systems [18], it also involves extracting a microbe from the environment and exposing it to a target contaminant so as to lessen the toxic component [20]. Thus, the goal of bioremediation is the employment of biosystems such as microbes, higher organisms like plants (phytoremediation) and animals to reduce the potential toxicity of chemical contaminants in the environment by degrading, transforming, and immobilizing these undesirable compounds.

Biodegradation is the use of living organisms to enzymatically and otherwise attack numerous organic chemicals and break them down to lesser toxic chemical species. Biotechnologists and bioengineers classify pollutants with respect to the ease of degradation and types of processes that are responsible for this degradation, sometimes referred to as treatability [20].

Biodegradation with microorganisms is the most frequently occurring bioremediation option. Microorganisms can break down most compounds for their growth and/or energy needs. These biodegradation processes may or may not need air. In some cases, metabolic pathways which organisms normally use for growth and energy supply may also be used to break down pollutant molecules. In these cases, known as cometabolisms, the microorganism does not benefit directly. Researchers have taken advantage of this phenomenon and used it for bioremediation purposes [18].

A complete biodegradation results in detoxification by mineralising pollutants to carbon dioxide (CO_2), water (H_2O), and harmless inorganic salts [18]. Incomplete biodegradation (i.e., mineralization) will produce compounds that are usually simpler (e.g., cleared rings, removal of halogens), but with physical and chemical characteristics different from the parent compound. In addition, side reactions can produce compounds with varying levels of toxicity and mobility in the environment [20].

Biodegradation may occur spontaneously, in which case the expressions "intrinsic bioremediation" or "natural attenuation" are often used [18]. In many cases the natural circumstances may not be favourable enough for natural attenuation to take place due to inadequate nutrients, oxygen, or suitable bacteria. Such situations may be improved by supplying one or more of the missing/inadequate environmental factors. Extra nutrients [18] were disseminated to speed up the break down of the oil spilled on 1000 miles of Alaskan shoreline by the super tanker Exxon Valdez in 1989.

According to Vallero [20], there are millions of indigenous species of microbes living at any given time within many soil environments. The bioengineer simply needs to create an environment where those microbes are able to use a particular compound as their energy source. Biodegradation processes had been observed empirically for centuries, but putting them to use as a distinct field of

bioremediation began with the work of Raymond et al. [21]. This seminal study found that the addition of nutrients to soil increases the abundance of bacteria that was associated with a proportional degradation of hydrocarbons, in this case petroleum by-products [21].

9.2.1. Life Chemical Dynamics (Biochemodynamics) of Bioremediation. Bioremediation success [20] depends on the following:

(1) the growth and survival of microbial populations; and

(2) the ability of these organisms to come into contact with the substances that need to be degraded into less toxic compounds;

(3) sufficient numbers of microorganisms to make bioremediation successful;

(4) the microbial environment must be habitable for the microbes to thrive.

Sometimes, concentrations of compounds can be so high that the environment is toxic to microbial populations. Therefore, the bioengineer must either use a method other than bioremediation or modify the environment (e.g., dilution, change of pH, pumped Oxygen, adding organic matter, etc.) to make it habitable. An important modification is the removal of non-aqueous-phase liquids (NAPLs) since the microbes' biofilm and other mechanisms usually work best when the microbe is attached to a particle; thus, most of the NAPLs need to be removed, by vapour extraction [20]. Thus, low permeability soils, like clays, are difficult to treat, since liquids (water, solutes, and nutrients) are difficult to pump through these systems. Usually bioremediation works best in soils that are relatively sandy, allowing mobility and greater likelihood of contact between the microbes and the contaminant [20]. Therefore, an understanding of the environmental conditions sets the stage for problem formulation (i.e., identification of the factors at work and the resulting threats to health and environmental quality) and risk management (i.e., what the various options available to address these factors are and how difficult it will be to overcome obstacles or to enhance those factors; that make remediation successful). In other words, bioremediation is a process of optimization by selecting options among a number of biological, chemical and physical factors these include correctly matching the degrading microbes to conditions, understanding and controlling the movement of the contaminant (microbial food) so as to come into contact with microbes, and characterizing the abiotic conditions controlling both of these factors [20]. Optimization can vary among options, such as artificially adding microbial populations known to break down the compounds of concern. Only a few species can break down certain organic compounds [20]. Two major limiting factors of any biodegradation process are toxicity to the microbial population and inherent biodegradability of the compound. Numerous bioremediation projects include in situ (field treatment) and ex situ (sample/laboratory treatment) waste treatment using biosystems [20].

TABLE 2: Environmental process and bioremediation procedures involved.

Environmental condition	Biosystem/microbes used	Bioremediation benefit
Waste water and industrial effluents	Sulphur-metabolising bacteria	(1) Microorganisms in sewage treatment plants remove common pollutants (heavy metals and sulphur compounds) from waste water before it is discharged into rivers or sea. (2) Production of animal feed from fungal biomass after penicillin production in penicillin industries. (3) Useful biogas (methane, etc.) production from anaerobic waste water treatment.
Drinking and process water	Organic degrading microbes (Bacteria, fungi, and algae)	(1) Reclamation and purification of waste waters for reuse and provision of portable recyclable drinking water for the public consumption and for livestock use. (2) Remove wastes for organic fertilizer agric use.
Air and waste gases	Bacteria, fungi	Biofilter application of pollutant purifying bacteria. Application of bioscrubbers, immobilized microorganism in inert matrix and nutrient film trickling devices for better air and gas purification. For example, bioscrubber-based system for removal of nitrogen and sulphur oxides from flue gas of blast furnaces in place of limestone gypsum process, and elimination of styrene from the waste gas of polystyrene processing industries by a fungi biofilter model.
Soil and land treatment	*Pseudomonas* spp., *Bacillus* spp., Fungi, *Rhodococcus*, *Acinetobacter*, *Mycobacterium*	Both in situ (in its original place) and ex situ (somewhere else) are commercially exploited for the cleanup of soil and groundwater. Use of microorganisms (bioaugmentation, ventilation, and/or adding nutrient solution (biostimulation) that is, petroleum decontamination, can involve use of plants (phytoremediation). Bacteria in association with roots of plants (Rhizobacterium), and so forth. Use of bioreactors for ex situ treatment with introduction of suitable microbes and environmental factors.
Solid waste	Bacteria, fungi, and so forth	Composting or anaerobic digestion of domestic and garden wastes helps in recovery of high-value biogas and useful organic compost without the toxic components. Free breakdown of solid waste by microbial biota for recyclable waste, an acceptable alternative to incineration.

Table 2 shows the application of bioremediation in various environmental processes.

10. A Practical Application of Microorganism in Crude Oil Bioremediation

According to Onwurah [22] many microorganisms can adapt their catabolic machinery to utilize certain environmental pollutants as growth substrates, thereby bioremediating the environment. Some microorganisms in carrying out their normal metabolic function may fortuitously degrade certain pollutants as well. This process termed cometabolism obviously requires adequate growth substrates. Diazotrophs, such as *Azotobacter vinelandii*, beyond their ability to fix atmospheric nitrogen also have the capacity, in some case, to cometabolise petroleum hydrocarbons [10].

Onwurah [22] carried out a bioremediation study that involved two bacteria, a hydrocarbonoclastic and diazotrophic bacteria. The hydrocarbonoclastic was tentatively identified as *Pseudomonas* sp. and designated as $NS_{50}C_{10}$

by the Department of Microbiology, University of Nigeria, Nsukka. The diazotrophic bacteria was *Azotobacter vinelandii*, which was isolated from previously crude oil-contaminated soil [10]. This study describes the mineral media and procedure for isolation and multiplication of the bacteria to the required cell density. Crude oil spill was simulated by thoroughly mixing 50, 100, and 150 mg fractions of crude oil with 100 g batches of a composite soil sample in beakers. The soil samples were taken from a depth of 0–50 cm from the Zoological garden, University of Nigeria, Nsukka. The mixing was conducted using a horizontal arm shaker adjusted to a speed of 120 rpm for 30 minutes. The contaminated soil samples, in beakers, were inoculated with optimal combinations (cell density) of $NS_{50}C_{10}$ and *A. vinelandii*. Water was added to the crude oil-contaminated soil samples (both inoculated and those not inoculated to a saturation point but not in excess), and then the samples were left to stand undisturbed for seven days. $NS_{50}C_{10}$ was applied first, followed by *A. vinelandii*, 12 hours later. At the seventh day of soil treatment, 20 sorghum grains (previously soaked overnight in distilled water) were planted

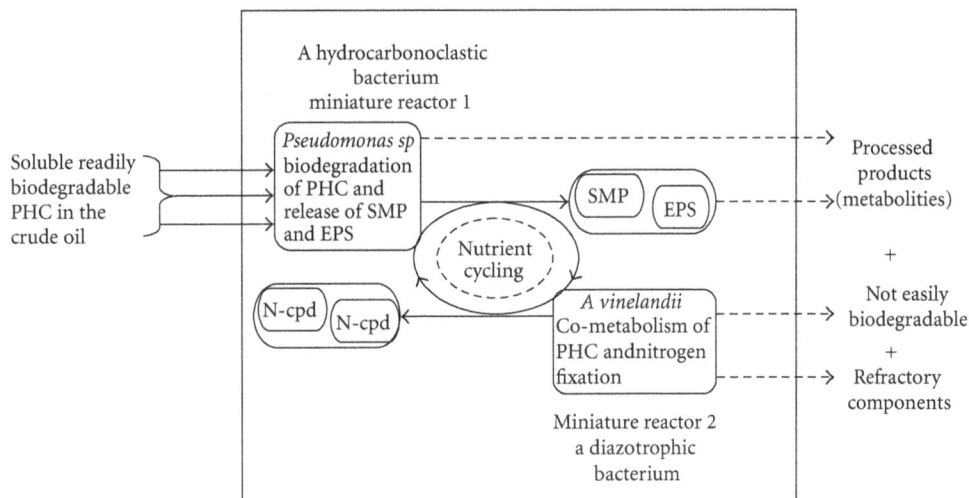

FIGURE 1: Simplified bioremediation conceptual model of *Pseudomonas* sp. and *A. vinelandii* operating as a unit of two miniature sequencing bioreactors, in situ (SMP: soluble microbial products; N-cpd: fixed nitrogen compounds; EPS: exopolysaccharide; PHC = petroleum hydrocarbons) [22].

in each soil sample followed by irrigation to aid germination. Seven days after the planting of the sorghum grains, the soil from each beaker was carefully removed. The number of germinated seed per batch of soil sample was noted, the length of radicules was measured, and the mean length was taken from each batch.

The results of this experiment showed that *Pseudomonas* sp. grew well on agar plates containing a thin film of crude oil as the only carbon source, while *A. vinelandii* did not. However, cell-free extract of *Azotobacter vinelandii* fixed atmospheric nitrogen as ammonium ion (NH_4^+) under appropriate condition. The specific growth rate values in contaminated soil samples inoculated with both normal $NS_{50}C_{10}$ and *A. vinelandii* (consortium) were highest in all cases. By adding an aerobic, free living diazotroph *A. vinelandii* with the *Pseudomonas sp.* ($NS_{50}C_{10}$), an improvement on bioremediation of soil over that of the pure $NS_{50}C_{10}$ alone was achieved to the order of 51.96 to 82.55%. This innovative application that uses the synergetic action of several microorganisms to clean up oil-polluted soil has potential application for the bioremediation of oil-contaminated soil in the Niger delta region.

The method described above is the biotechnological application known as *bioaugmentation* which is the addition of selected organisms to contaminated soils (sites) in order to supplement the indigenous microbial population and speed up degradation. Figure 1 presents a model of the process involved in this bioremediation technique.

This bioremediation method by the authors has been applied in bioremediation especially in Niger delta areas of Nigeria. The authors also serve in the capacity of industrial consultants in the specialized field of crude oil pollution clean-up procedures using this specific biotool (bioremediation).

10.1. Biofiltration.
This is a pollution control technique employing the use of living material to capture and biologically degraded process pollutants. Common uses of biofiltration processes are for processing waste water, capturing harmful chemicals or silt from surface runoff, and microbiotic oxidation of contaminants in air (http://www.biofilter.com/).

In multimedia-multiphase bioremediation, waste streams containing volatile organic compounds (VOCs) may be treated with combinations of phases, that is, solid media, gas, and liquid flow in complete biological systems. These systems are classified as three basic types: biofilters, biotrickling filters, and bioscrubbers (http://www.biofilter.com/). Biofilms of microorganisms (bacteria and fungi) are grown on porous media in biofilters and biotrickling systems. The application of this biotechnological tool includes the following.

10.1.1. Control of Air Pollution.
When applied to air filtration and purification, biofilters use microorganisms to remove air pollution (http://www.biofilter.com/). The air flows through a packed bed, and the pollutant transfers into a thin biofilm on the surface of the packing material. Microorganisms, including bacteria and fungi, are immobilized in the biofilm and degrade the pollutant. Trickling filters and bioscrubbers rely on a biofilm and the bacterial action in their recirculating waters (http://www.biofilter.com/). The air or other gas containing the VOCs is passed through the biologically active media, where the microbes break down the compounds to simpler compounds, eventually to carbon dioxide (if aerobic), methane (if anaerobic), and water. The major difference between biofiltration and trickling systems is how the liquid interfaces with the microbes. The liquid phase is stationary in a biofilter (Figure 2), but liquids move through the porous media of a biotrickling system (i.e., the liquid "trickles").

A particular novel biotechnological method in biofilteration (Figure 3) uses compost as the porous media. Compost contains numerous species of beneficial microbes that are

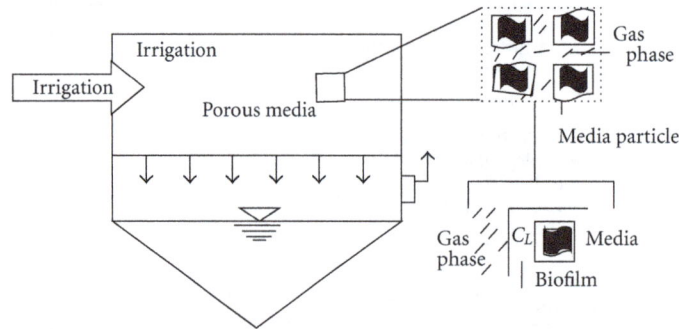

FIGURE 2: Schematic of packed bed biological control system to treat volatile compounds. Air containing gas phase pollutants (CG) traverse porous media. The soluble fraction of the volatilized compounds in the air steam partition into the biofilm (CL) according to Henry's Law. CL = {CG/H} where H is Henry's Law constant. Adapted and modified from Vallero [23].

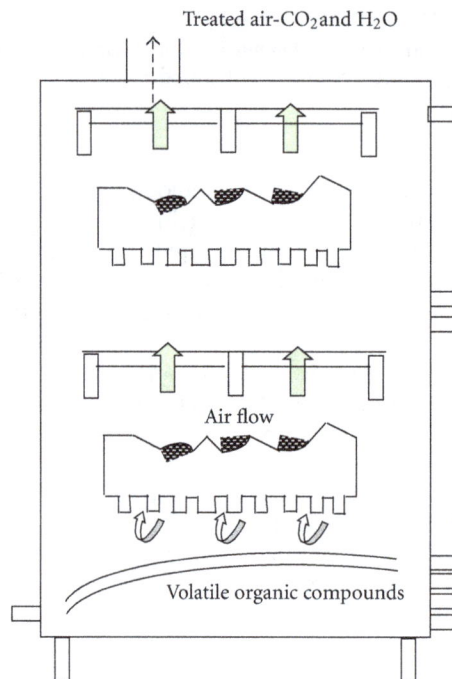

FIGURE 3: Biofiltration without a liquid phase used to treat vapour phase pollutants. Air carrying the volatilized contaminants upward through porous media (e.g., compost) containing microbes acclimated to break down the system can be heated to increase the partitioning to the gas phase. Microbes in the biofilm surrounding each individual compost particle metabolize the contaminants into simpler compounds, eventually converting them into carbon dioxide and water vapour. Modified from Vallero, [20].

already acclimated to organic wastes. Industrial compost biofilters have achieved removal rates at the 99% level [20]. Biofilters are also the most common method for removing VOCs and odorous compounds from air streams.

In addition to a wide assortment of volatile chain aromatic organic compounds, biological systems have successfully removed vapour-phase inorganics, such as ammonia, hydrogen sulfide, and other sulfides including carbon disulfide, as well as mercaptans. The operational key is the biofilm. The gas must interface with the film. Compost has been a particularly useful medium in providing this partitioning [20]. Industries employing the biofiltration technology include food and animal products, offgas from waste water treatment facilities, pharmaceuticals,

wood products manufacturing, paints, and coatings application and manufacturing and resin manufacturing and application. Compounds treated are typically mixed VOCs and various sulfur compounds, including hydrogen sulfide (http://www.biofilter.com/). Maintaining proper moisture condition is an important factor in biofiltration. The air normally humidifies before it enters the bed with a watering (spray) system, humidified chamber, bioscrubber, or biotrickling filter. Properly maintained, a natural organic packing media peat, vegetable mulch, bark, or wood chips may last for several years. However, engineered combined natural organic and synthetic component packing materials will generally last much longer, up to 10 years. A number of companies offer these types or proprietary packing materials and multiyear guarantees, not usually provided with a conventional compost or wood chip bed biofilter. For large volumes of air, a biofilter may be the only cost effective solution (http://www.biofilter.com/). There is no secondary pollution (unlike the case of incineration where additional CO_2, CO, and NO gases are produced from burning fuel(s) and degradation products form additional biomass, carbon dioxide and water).

10.1.2. Water Treatment. Trickling filters have been used to filter water for various end uses for almost two centuries. Biological treatment has been used in Europe to filter surface water for drinking purposes since the early 1900s and is now receiving more interest worldwide (http://www.biofilter.com/). Media irrigation water, although many systems recycle part of it to reduce operating costs, has a moderately high biochemical oxygen demand (BOD) and may require treatment before disposal. Biofilters are being utilized in Columbia falls, Montana at Plum Creed Timber Company's fibreboard plant (http://www.biofilter.com/).

Biofiltration is one of the most effective water treatment technologies. Its application includes water filtration in farms, livestock operations, city municipal, industrial, and household applications. Some of the organisations which have supported the development or application biofiltration of water (http://www.biofilter.com/) over the past 14 years include the following:

(i) Prairie Farm Rehabilitation Administration (PFRA),

(ii) The National Research Council (NRC),

(iii) The Saskatchewan Research Council (SRC),

(iv) Napier University (Scotland),

(v) Agriculture and Agro-food Canada.

Biofiltration is ideal for well, lake, pond, river, and dug out water. Biofilters remove the following substances from air and water: iron and iron bacteria, parasites, colour, cysts, manganese, pesticides, arsenic, lead, mercury, turbidity, dissolved organic carbon (dissolved organic material in water), tannins [26].

A good number of research and practical work has been and is being carried out by Nigerian scientists and academics in the area of biofiltration of waste water. Bearing in mind that good water is a very essential commodity which is not readily available in most part of the country, it is therefore of great necessity to look at economic feasible ways to treat water for the benefit of the citizens. The sources of portable water for most Nigerian cities are government treated tap water and commercially treated drinking water as well as domestic water by water service private firms. Rural communities make do with water from ponds, streams, rivers, rain, and spring which are prone to contamination by water-borne diseases such as typhoid and diarrhoea which is common in those communities. Some of the applications of this important biotool in Nigeria are as follows.

(1) Asamudo et al. [27] demonstrated the effectiveness of using the fungus *Phanerochaete chrysosporium* in the biofiltration of textile effluent, polycyclic aromatic hydrocarbons (PAH), and pulp and paper effluents. The microorganism was capable of producing extracellular enzymes such as manganese peroxidase, cellulases, and lignin peroxidases, in achieving total remediation of these effluents.

(2) Ezeronye and Okerentugba [28] carried out a study to demonstrate the effectiveness of a yeast biofilter composed of a mixed culture of *Saccharomyces* spp., *Candida* spp., *Schizosaccharomyces* spp. and *Geotrichum candidum* in the treatment of fertilizer factory effluents and 98% treatment efficiency was achieved. The biochemical oxygen demand (BOD) of the effluent was reduced from a range of 1200–1400 to 135–404 mg/L. Besides, ammonia nitrogen (NH_3-N) and nitrate-nitrogen (NO_3-N) were reduced from 1000–10 mg/L and 100–17.6 mg/L, respectively.

(3) Ogunlela and Ogunlana [29] developed a system using lava stones and oyster shells biofilter substrates for the oxidation of ammonia in a recirculatory aquaculture system. The effluent was treated using the biofilter, and chemical analyses were carried out once a week for four consecutive weeks. The results at the end of the fourth week indicated that the ammonia and nitrite concentrations were 0.0374 mg/L and 0.292 mg/L, respectively, which were below the permissible limits of 0.05 mg/L and 0.3 mg/L for ammonia and nitrite, respectively.

One of the most recent innovations in the use of this biotool in Nigeria was by Rabah et al. [30]. Their work describes the use of yeast biofilters in the treatment of abattoir waste water. Thus, Nono (locally fermented milk product) and Kunun-zaki (a refreshing drink made from millet) samples were obtained at the minimarket of the main campus of the Usmanu Danfodiyo University, Sokoto, Nigeria, in sterile sample bottles and transported in an icebox to the laboratory for the isolation of yeasts. Wastewater was collected from an abattoir in Sokoto, Nigeria, using sterile two litre capacity sample bottles and transported in an icebox to the laboratory. The wastewater was collected from three points in the abattoir: at the point where the wastewater leaves the slaughter hall (Point A, PA), midway through the drainage channel (Point B, PB), and the point where the wastewater drained to the surrounding soil (Point C, PC). A total of three samples were collected from each point at different times.

The biofilter was constructed using Perspex glass with a length of 18.0 cm, width of 10.8 cm, and a depth of 10.5 cm. The filter has upper and lower compartments separated by a perforated partition made up of the same Perspex glass. It also has a tap for the collection of filtered wastewater. Potato peels were ground to smaller particles, wetted, and placed on the perforated partition. The yeast biomass was inoculated on the peels and left for one week at ambient laboratory temperature ($28 \pm 2°C$) to allow the cells to grow. Then the abattoir wastewater was introduced into the filter bed and left to stand for a minimum period of 14 days. The filtered wastewater was collected from the lower chamber of the filter through a tap fitted to the chamber.

The yeast species isolated from the Nono and Kunun-zaki and identified for use as biofilters in the biofiltration process were identified as *Candida krusei, Candida morbosa, Torulopsis dattila, Torulopsis glabrata,* and *Saccharomyces chevalieri.* Also the results of the physicochemical qualities of the abattoir wastewater before and after biofiltration process from the three sampling points (PA, PB, and PC) revealed that there was a considerable reduction in pH, nitrate (NO_3), dissolved oxygen (DO), biochemical oxygen demand (BOD), and chemical oxygen demand (COD) after the biofiltration of the wastewater collected from the three sampling points. It was also observed that the concentrations of other compounds in the wastewater varied with the sampling points probably due to contamination from human activities in the abattoir such as dumping of cow dung and pieces of bones in the wastewater channels. According to Rabah et al. [30], the results generally indicated that the yeast biofilter was fairly effective in the bioremediation process. The biofilter had a percentage efficiency of 42.5%.

10.2. Biomining.
Bacteria leaching is now used throughout the world as an additional technique for extracting metals from ores. Metals which can be extracted in this way include copper, uranium, cobalt, lead, nickel, and gold [31].

Biomining is a generic term [32] that describes the processing of metal-containing ores and concentrates of metal containing ores using microbiological technology. Biomining has application as an alternative to more traditional physical-chemical methods of mineral processing. Commercial practices of biomining can be broadly categorized in two, namely, mineral biooxidation and bioleaching. Both processes use naturally occurring microorganisms to extract metals from sulphide bearing minerals. Minerals biooxidation refers to the process when it is applied to enhance the extraction of gold and silver, whereas bioleaching usually refers to the extraction of base metals, such as Zinc, Copper, and Nickel.

Collectively, minerals biooxidation and bioleaching are commercially proven, biohydrometallurgical or biomining processes that are economic alternatives to smelting, roasting, and pressure oxidation to treat base and precious metals associated with sulphide minerals [32].

Metals are essential physical components of the ecosystem, whose biologically available concentrations depend primarily on geological and biological processes [33]. Elevated levels of metals at specific sites can create a significant environmental and health problem when the release of metals through geological processes of decomposition and anthropogenic processes far exceeds that of natural processes of metal cycling. Metal contamination of both aqueous and terrestrial environments is of great concern, due to the toxicity and persistence of metals in the ecosystem and their threat to animal and human health [34]. Bacteria play an important role in the geochemical cycle of metals in the environment, and their capabilities and mechanisms in transforming toxic metals are of significant interest

in the environmental remediation of contaminated sites. Microorganisms [34] colonize and shape the Earth in many ways, and their ability to adsorb and transform metals can shade light on solving pollution problems and proposing solutions in the clean up of contaminated site.

10.2.1. Extraction Role of Microbes in Biomining.
Although many undiscovered microbial communities are involved in biomining, some of the popular and discovered bacteria responsible are: *Leptospirillum ferrooxidans, Acidithiobacillus thiooxidans,* and *Acidithiobacillus ferrooxidans* [31]. There is a good understanding of the exact role of microbes in biomining, thanks to today's sophisticated instrumentation that can examine materials at the atomic level. Given the fact that many microbes float freely in the solution around the minerals, many microbes attach to the mineral particles forming a biofilm [31]. The microbes, whether they are freely floating or whether they are in the biofilm, continuously devour their food sources—iron (chemically represented as Fe^{2+}) and sulphur. The product of the microbial conversion of iron is "ferric iron," chemically represented as "Fe^{3+}". According to Brierley [32], ferric iron is a powerful oxidizing agent, corroding metal sulphide minerals (e.g., pyrite arsenopyrite, chalcocite, and sphalerite) and degrading them into dissolved melts, such as copper, zinc, and more iron, the latter being the food source for the microbes. The sulphide portion of the mineral is converted by the microbes to sulphuric acid.

Uranium occurs in oxidation states ranging from U (III) to U (VI), with the most stable species, U (VI) and U (IV), existing in the environment [34]. U (VI) is predominant in the oxic surface waters, and UO_2^{2+} (uranyl) always forms stable, soluble complexes with ligands such as carbonate, phosphate, and humic substances [34]. In natural waters the solubility of U (VI) usually increases several orders of magnitude at higher pH values, due to complexation with carbonate or bicarbonate. By contrast, U (IV) is commonly found in the anoxic conditions and is present primarily as an insoluble uranite (UO_2). Therefore, reduction of the soluble uranyl to the insoluble uranite seems to be an effective means to immobilize uranium in the anoxic environment to decrease the potential release of the mobile species [34].

More research interests in the bioreduction of U (VI) are demonstrated in the dissimilatory metal-reducing bacteria (DMRB) under anaerobic conditions [34]. Lovley et al. [35] first demonstrated the occurrence of dissimulatory U (VI) reduction by the Fe (III) reducing bacteria *Geobacter met-allireducens* and *Alteromonas putrifaciens* (later, *Shewanella putrefacians*), which could conserve energy for anaerobic growth via the reduction of U (VI). Soluble U (VI) is more readily reduced to U (IV) by *G. metallireducens* and other Fe (III) reducing microorganisms than are insoluble Fe (III) oxides, and once produced, U (IV) can be reoxidized to U (VI) with the reduction of Fe (III) to Fe (II) [36].

10.2.2. Microbial Gold Mining.
In some precious-metal deposits gold [32] occurs as micrometer-sized particles that are occluded, or locked, within sulphide minerals, principally

pyrite (an iron sulphide mineral) and arsenopyrite (an arsenic containing iron sulphide mineral). According to Brierley [32], to effectively recover the precious metals, the sulfides must be degraded (oxidized) to expose the precious metals. Once the sulphides are sufficiently degraded to expose the gold and silver, a dilute solution of cyanide is used to dissolve the precious metals. If the occluded gold and silver [32] are not exposed by breaking down the sulphide minerals, the cyanide cannot help in the release of the metals and recovery will be low. The ferric iron that is produced by the microorganism is the chemical agent that breaks down (oxidizes) the sulfide mineral. The microorganisms can be thought of as the manufacturing facility for producing the ferric iron. Microorganisms in the ore are destroyed by lime. Cyanide leaching can be accomplished in another heap or the oxidized and lime-conditioned ore can be ground and cyanide leached in a mill. The residue slurry is rinsed with fresh water, neutralized with lime, subjected to solid/liquid separation, and the solid residue is cyanide leaching to extract the gold. Gold recoveries are in the 95–98% range.

Advantages of biomining [32] using organisms include the following.

(1) Biomining microorganisms do not need to be genetically modified; they are used in their naturally occurring form.

(2) Unlike humans, animals, and plants, microorganisms reproduce by doubling; that is, when there is abundant food (iron and sulfur) for biomining microbes and optimal conditions (sufficient oxygen, carbon dioxide and a sulfuric acid environment), a microbe will simply divide. Thus, in heap of minerals biooxidation for pretreating gold ores, there are about one million microbes per gram of ore.

(3) High altitudes have no effect on the biomining microorganisms. However, additional air must be supplied to give the organisms an optimal performance.

(4) Biomining using microorganisms does not produce dangerous waste products. Base metals, for example, zinc and copper are recycled and neutralized with lime/limestone.

(5) The biomining microbes cannot escape from the heap or bioreactor to cause environmental problems. These microbes exist in the environment only where conditions are suitable (i.e., sources of iron and sulfur are oxidized, air and a sulfuric acid environment).

Biomining as a biotool has not been explored in Nigeria. Though Nigeria has many solid minerals in different states of the country, some of the minerals are tin (found in Plateau, Nassarawa, Kaduna, Bauchi and Gombe states), gold (found in Oyo, Osun and Ondo states), copper (Edo and Benue states), tantalite (Gombe, Plateau, Kaduna, and Nasarawa states), and uranium (Bauchi state) among others. The procedures, equipments, nonawareness/interest by government as well as competition with the physicochemical methods of extraction of these minerals are the greatest limitation

in the exploitation and usage of this biotool in mining of minerals from their ores. Providing information to Nigeria scientists, ministries and government agencies is the solution to this limitation. Thus, the essence of the suggestion here is to create awareness in this regard.

10.3. Biomonitoring. In a broad sense, biological monitoring involve any component that makes use of living organisms, whole or part as well as biological systems to detect any harmful, toxic, or deleterious change in the environment. There are various components employed in biomonitoring of contaminants in the environment. They include biomarkers (biological markers), biosensors, and many others.

Biomonitoring or biological monitoring is a promising, reliable means of quantifying the negative effect of an environmental contaminant.

Biological Markers. A biomarker is an organism or part of it, which is used in soliciting the possible harmful effect of a pollutant on the environment or the biota [37]. Biological markers (biomarkers) are measurement in any biological specimen that will elucidate the relationship between exposure and effect such that adverse effects could be prevented [38]. The use of chlorophyll production in *Zea mays* to estimate deleterious effect of crude oil contaminants on soils is a typical plant biomarker of crude oil pollution [11]. When a contaminant interacts with an organism, substances like enzymes are generated as a response. Thus, measuring such substances in fluids and tissue can provide an indication or "marker" of contaminant exposure and biological effects resulting from the exposure. The term biomarker includes any such measurement that indicates an interaction between an environmental hazard and biological system [39]. It should be instituted whenever a waste discharge has a possible significant harm on the receiving ecosystem. It is preferred to chemical monitoring because the latter does not take into account factors of biological significance such as combined effects of the contaminants on DNA, protein, or membrane. Onwurah et al. [37] stated that some of the advantages of biomonitoring include the provision of natural integrating functions in dynamic media such as water and air, possible bioaccumulation of pollutant from 10^3 to 10^6 over the ambient value, and/or providing early warning signal to the human population over an impending danger due to a toxic substance. Microorganisms can be used as an indicator organism for toxicity assay or in risk assessment. Tests performed with bacteria are considered to be most reproducible, sensitive, simple, economic, and rapid [40] (Table 3).

10.4. Biosensor. A biosensor is an analytical device consisting of a biocatalyst (enzyme, cell, or tissue) and a transducer, which can convert a biological or biochemical signal or response into a quantifiable electrical signal [46]. A biosensor could be divided into two component analytical devices comprising of a biological recognition element that outputs a measurable signal to an interfaced transducer [24]. Biorecognition typically relies on enzymes, whole cells,

TABLE 3: Biomarkers and their applications

Biomarker type	Uses	Reference
Chlorophyll content *Zea mays* L.	Detection of level of hydrocarbon contamination of agricultural soil	[11]
Sensitivity of *Nitrobacter* sp.	Based on the effect of crude oil on oxidation of nitrite to nitrate	[41]
Azotobacter sp.	Used in evaluating the effect of oil spill in aquatic environment	[42]
Algae/plant steranes and bacteria hopanes	Steranes formed as components of crude oil and hopanes used to determine the source rock that generated a crude oil	[43]
Ethoxyresorufin-O-deethylase (EROD) in fish in vivo	Indicates exposure of fish to planar-halogenated hydrocarbons (PAHs) by receptor-mediated induction of cytochrome P-450-dependent monooxygenase exposed to PAHs and similar contaminants	[44, 45]

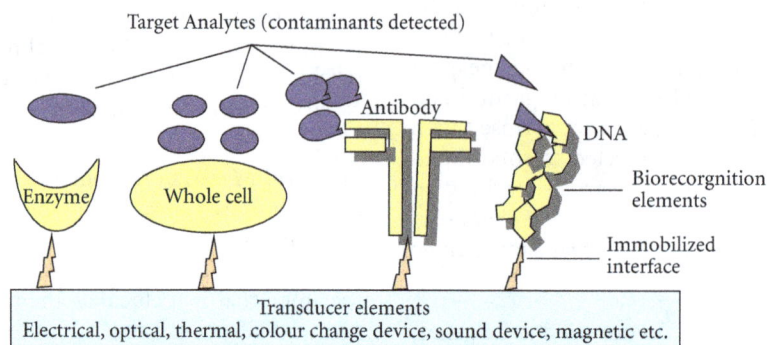

FIGURE 4: Anatomy of a Biosensor. The interaction between the target analyte and the biorecognition element creates a signalling event detectable by the interfaced transducer element. Modified from source: Ripp et al. [24].

antibodies, or nucleic acids, whereas signal transduction exploits electrochemical (amperometric, chronoamperometric, potentiometric, field-effect transistors, conductometric, capacitive), optical (absorbance, reflectance, luminescence, chemiluminescence, bioluminescence, fluorescence, refractive index, light scattering), piezoelectric (mass sensitive quartz crystal microbalance), magnetic, or thermal (thermistor, pyroelectric) interfaces [24]. The biocatalyst component of most biosensors is immobilized on to a membrane or within a gel, such that the biocatalyst is held in intimate contact with the transducer and may be reused. Biosensors are already of major commercial importance, and their significance is likely to increase as the technology develops [46]. Biosensors are still emerging biotechnology for the future in environmental biomonitoring since they have specific limitations. Biosensors on a general sense are often employed for continuous monitoring of environmental contamination or as bioremediation process monitoring and biocontrol tools to provide informational data on what contaminants are present, where they are located, and a very sensitive and accurate evaluation of their concentrations in terms of bioavailability. Ripp et al. [24] explained that bioavailability measurements are central to environmental monitoring as well as risk assessment because they indicate the biological effect of the chemical, whether toxic, cytotoxic, genotoxic, mutagenic, carcinogenic, or endocrine disrupting, rather than mere chemical presence as is achieved with analytical instruments. As the name suggests they are biological instruments that detest and signal the presence of harmful contaminants in the environment. There are different types based on the biological components on which

their sensitivities are based (Figure 4). Some of them, though not exhaustive are the following.

10.4.1. Enzyme-Based Biosensors.

Leyland Clark in the 1960s used an enzyme biosensor which consists of glucose oxidase enzyme immobilized on an oxygen electrode for blood glucose sensing. This historical application of enzyme-based biosensor has found a world-wide lucrative application in medical diagnosis. Nevertheless, enzyme-based biosensor gradually gained application in environmental monitoring. According to Ripp et al. [24], enzymes act as organic catalysts, mediating the reactions that convert substrate into product. Since enzymes are highly specific for their particular substrate, the simplest and most selective enzyme-based biosensors merely monitor enzyme activity directly in the presence of the substrate. A novel example of biosensors of enzyme origin which has found application in the environment is the sulfur/sulfate-reducing bacterial cytochrome C_3 reductases that reduce heavy metals. Michel et al. [47] immobilized cytochrome C_3 on a glassy carbon electrode and monitored its redox activity amperometrically in the presence of chromate [Cr (IV)] with fair sensitivity (lower detection limit of 0.2 mg/L) and rapid response (several minutes) (Figure 5).

When tested under simulated groundwater conditions, the biosensors reacted with several other metal species, albeit at lower sensitivities, and were affected by environmental variables such as pH, temperature, and dissolved oxygen. Similarly operated enzyme-based biosensors for ground

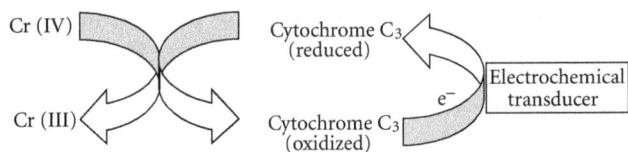

FIGURE 5: Enzymatic biosensor using cytochrome C_3 as the recognition element. Upon exposure to chromate [Cr (VI)], electrode-immobilized cytochrome C_3 reduces Cr (VI) to Cr (III). The current produced by the electrochemical regeneration of reduced cytochrome C_3 is proportional to the amount of oxidized cytochrome C_3 and, therefore, the Cr (IV) concentration. Ripp et al. [24].

water contaminant perchlorate using perchlorate reductase as the reduction enzyme (detection limit of $10 \, \mu g/L$) [48], organophosphate pesticides using parathion hydrolase or organophosphorus hydrolase as recognition enzymes (detection down to low μM concentrations) [49], and environmental estrogens using tyrosinase as the recorgnition enzyme (detection down to $1 \, \mu M$) [50] have also been designed.

Another type of enzyme biosensor relies on enzyme activation upon interaction with the target of interest. Heavy metals, for example, in the form of cofactors-inorganic ions that binds to and activate the enzyme can be detected based on this integral association. Metalloenzymes such as alkaline phosphatase, ascorbate oxidase, glutamine synthetase, and carbonic anhydrase require association of a metal ion cofactor with their active sites for catalytic activity and can thus be used as recognition element for heavy metal [24]. Alkaline phosphatase, for example, can be applied in this regard as a biosensor for zinc [Zn (II)] or ascorbate oxidase for biosensing copper (II) with detection limits down to very low part-per-billion levels [51]. Various immobilization techniques are adopted in the attachment of the enzyme to the transducing element [52]; they include adsorption, covalent attachment, entrapment in polymeric matrices such as sol-gels or Langmuir-Blodgett films, or direct cross-linking using polymer networks or antibody/enzyme conjugates. Immobilization provides the biosensor longevity and with recent integration of redox active carbon-based nanomaterials (nanofibers, nanotubes, nanowires, and nanoparticles) as transducers and their unique ability to interact with biological material, a promising advancement in enzyme biosensor design and sensitivity is in sight.

Optical transducers (absorption, reflectance, luminescence, chemiluminescence, evanescent wave, surface plasma resonance) are also commonly employed in enzyme-based biosensor [24]. This can be as simple as optically registering a pH change using a pH reactive dye; for example, bromocresol purple can be immobilized with an acetylcholinesterase-based biosensor to monitor pH changes related to this enzyme's activity upon exposure to pesticides. Acetylcholinesterase hydrolysis releases protons (H^+), resulting in a decrease in pH, which in turn instigates a decrease in the absorption spectra of bromocresol purple [24].

Andreou et al. [53] incorporated such a biosensor successfully on the distal end of a fibre optic cable for facile interrogation of water samples for pesticide residue. A great application of optical biosensor is in the Luminol, widely used as an electrochemiluminescent indicator. It reacts with the acetylcholinesterase/choline oxidase hydrogen peroxide by-product to yield luminescent light signals that have also been used to quantify pesticide concentrations.

10.4.2. Antibody-Based Biosensors (Immunosensors). These types of biosensors make use of antibodies as recognition elements (immunosensors). They are used widely as environmental monitors because antibodies are highly specific, versatile, and bind stably and strongly to target analytes (antigens) [24]. Antibodies can be highly effective detectors for environmental contaminants, and advancements in techniques such as phage display for the preparation and selection of recombinant antibodies with novel binding properties assures their continued environmental application. Perhaps the best introduction to antibody-based biosensing is the Automated Water Analyzer Computer Supported System (AWACSS) environmental monitoring system developed for remote, unattended, and continuous detection of organic pollutants for water quality control [54]. AWACSS uses an optical evanescent wave transducer and fluorescently labelled polyclonal antibodies for multiplexed detection of targeted groups of contaminants, including endocrine disruptors, pesticides, industrial chemicals, pharmaceuticals, and other priority pollutants, without requisite sample preprocessing. Antibody binding to a target sample analyte occurs in a short 5-minute preincubation step, followed by microfluidic pumping of the sample over the transducer element, which consists of an optical waveguide chip impregnated with 32 separate wells of immobilized antigen derivatives [24]. As the antibody/analyte complexes flow through these wells, only antibodies with free binding sites can attach to the well surface (in what is referred to as a binding inhibition assay). Thus, antibodies with both of their binding sites bound with analyte will not attach to the surface and will pass through the detector. A semiconductor laser then excites the fluorophore label of bound antibodies, allowing for their quantification, with high fluorescence signals indicating high analyte concentrations. A fibre optic array tied to each well permits separation and identification of signals by the well, thereby yielding a simultaneous measurement of up to 32 different sample contaminants. The instrument has been used for groundwater, wastewater, surface water, and sediment sample testing with detection limits for most analytes in the ng/L range within assay times of approximately 18 minutes [24]. Another design by Glass et al. [55], similar to the above but less refined benchtop flow-through immunosensor (KinExA) was demonstrated to detect analytes successively based on a replaceable flow cell containing fluorescently labeled antibody. Their time of assay was approximately 26 minutes, with detection limits at picomolar concentrations.

Although not as elaborate as the AWACSS, a multitude of other antibody-based biosensors have been applied as environmental monitors, traditionally serving as biosensors for pesticides and herbicides, but their target analytes have broadened considerably over the past several years to include heavy metals, polycyclic aromatic hydrocarbons (PAHs), polychlorinated biphenyls (PCBs), explosives (TNT and RDX), phenols, toxins such as microcystin, pharmaceutical compounds, and endocrine disruptors [56].

10.4.3. DNA-Based Biosensors. The principle underlying the DNA-based biosensor is the ability of a transducer to monitor a change in the nucleic acid's structure occurring after exposure to a target chemical. These structural changes are brought on either by the mutagenic nature of the chemical, resulting in mutations, intercalations, and/or strand breaks, or by the chemical's ability to covalently or noncovalently attach to the nucleic acid [24]. Immobilizing the nucleic acid as a recognition layer on the transducer surface forms the biosensor, and detection of the chemically induced nucleic acid conformational change is then typically achieved electrochemically (i.e., a change in the current) or less so through optical or other means [57].

Nucleic acid biosensors are generally nonselective and provide an overall indication of a potentially harmful (genotoxic, carcinogenic, cytotoxic) chemical or chemical mix in the test environment and, depending on the biosensor format, an estimate of concentration. Bagni et al. [58] illustrated a conventional DNA biosensor which was used to screen soil samples for genotoxic compounds, using benzene, naphthalene, and anthracene derivatives as model targets. Double-stranded DNA was immobilized on a single-use disposable screen-printed electrochemical cell operating off a handheld battery-powered potentiostat [59]. A $10\,\mu L$ drop of a preprocessed and preextracted contaminated soil sample was placed onto the working electrode for 2 minutes, and resulting electrochemical scans, based on the chemical's propensity to oxidize DNA guanine residues, were measured. The magnitude of these "guanine peaks" in relation to a reference electrode was linearly related to their concentration in solution (i.e., the higher the concentration of the target chemical, the more the damage imposed on the DNA, and the lower the electrochemical measurement of the oxidation signal). In a very discrete application of this DNA biosensor, the authors also applied it to the detection of this DNA biosensor and also to the detection of PAHs in fish bile, using the accumulation of PAH compounds in live fish to monitor for water contamination events [60].

Nucleic acid can be manipulated similarly to create target specific aptamers using a process called SELEX (systematic evolution of ligands by exponential enrichment) [24]. By iteratively incubating nucleic acid with the desired target, one can select for oligonucleotide sequences (or aptamers) with the greatest affinity for the target. Kim et al. [61] used SELEX to create an aptamer specific for 17β-estradiol and used it in an electrochemical biosensor (or aptasensor) to achieve detection of this important endocrine disruptor at levels as low as 0.1 nM. Predominant aptasensor development and application is in the clinical fields, but it is slowly and inevitably encroaching upon environmental sensing. An aptasensor for the cyanobacterial toxin microcystin (lower dertection limit of $50\,\mu g/mL$) [62] and another for zinc based on fluorophore beacon (lower detection limit of $5\,\mu M$) [63] have been reported.

Hydrazine and aromatic amine compounds in fresh and groundwater, hydroxyl radicals in uranium mine drainage waters, herbicides such as atrazine, general toxicity events in wastewater, industrially contaminated soils, and various other environmental sources have all been screened using DNA biosensors [24].

Metals are also relevant detection targets, due to their various affinities for nucleic acid. Lead, Cadmium, Nickel, Arsenic, Copper, Iron, Chromium, and others have been detected through DNA biosensing, incorporating both single- and double-stranded DNA as the sensing element, but again, nonselectivity [24]. Selectivity, though, has been demonstrated by several groups using deoxyribozymes (DNAzymes) or ribozymes (RNAzymes). These engineered catalytic oligonucleotides can mediate nucleic acid cleavages or ligation, phosphorylation, or other reactions. For example, DNAzyme biosensor for lead uses a single-stranded DNAzyme absorbed to a gold electrode [64]. The DNAzyme incorporates a methylene blue tag at concentrations as low as 62 ppb; the DNAzyme strand is cleaved, allowing the methylene blue tag to approach the transducer and transfer electrons, thereby instigating an electrochemical signal [24]. However, the rapidity (only a few minutes to detect but sample processing is often necessary), sensitivity (typically down to low part-per-billion levels), ease of use, and cost-effectiveness screen environmental sites for toxic chemical intrusions or monitoring operational endpoints of bioremediation efforts. A calorimetric DNAzyme-based biosensor for lead has also been demonstrated [65].

10.4.4. Biomimetics, BioMEMs, and Other Emerging Biosensor Technologies. The future of biosensors is clearly in the emergent technologies of Biomimetics and BioMEMs. BioMEMs (biological microelectromechanical systems) are an assortment of biomicro, bioanotechnological, and microfluidic interfaces that form lab-on-a-chip, biochip, or micrototal analysis system (μTAS) biosensors [24]. Their objectives are toward miniaturization, portability, redundancy, and a reduction in sample size, time of response, and cost. The majority of these biosensors serves biomedical rather than environmental causes, but they are slowly and inevitably being adapted for the environmental monitoring community. BioMEMs most often utilize optical transducers interfaced with enzyme, whole-cell, antibody, or nucleic acid-type receptors. Several recent examples should illustrate their various design and performance characteristics. Yakovleva et al. [66] developed a microfluidic immunosensor flow cell for the detection of atrazine in surface water. Chemiluminescently labelled antibodies directed against atrazine were combined with artificially contaminated river water and microfluidically pumped at 40 to $50\,\mu L$/minute through a 42-channel 13 mm $\times$ 3 mm silicon microchip containing a functionalized antibody affinity-capture surface [24]. Upon antibody/atrazine capture, a luminal substrate was added to mediate the chemiluminescent reaction which was monitored with a photomultiplier tube (PMT) suspended above the microchip. Islam et al. [67] have further improved this sensing strategy by essentially integrating the PMT directly on the microchip flow cell to create truly miniaturized biosensor referred to as a BBIC (bioluminescent bioreporter integrated circuit). This 1.5 mm × 1.5 mm CMOS microluminometer was designed to capture and process bioluminescent signals emanating from immobilized whole cell bioluminescent bioreporter bacteria [24]. The BBIC converts the bioluminescently derived photodiode current into a digital signal, the frequency of which is proportional to the concentration of pollutant to which the bioreporter has been exposed. In water artificially contaminated with salicylate as a model pollutant, the flow-through BBIC responded with 30 minutes to part-per-billion concentration [24].

BioMEMs also include microcantilever-based biosensors that translate a molecular recognition event into nanomechanical motion that is measured by induced bending in a microfabricated cantilever similar on a macroscale to identifying a person on a diving board based on the deflection of the diving board by their weight. Optical or piezoresistive transducers usually measure microcantilever deflections at nanometer-to-subnanometer ranges of motion, and due to their small size, several microcantilevers can be accommodated per transducer for multianalyte sensing. Alvarez et al. [68] immobilized antibodies to the pesticide DDT on a microcantilever and demonstrated real-time detection at nanomolar concentrations.

Biomimetics mimic (imitate) the attributes of naturally occurring biological materials to synthetically recreate or enhance their properties [24]. Molecularly imprinted polymers (MIPs) are one of the typical examples of Biomimetics application in biosensors which can deliver more robust, stable, and target-specific receptors. MIPs are essentially created by mixing the target analyte (or template) with a monomer. Resulting MIPs then serve as analyte-specific synthetic receptors (or artificial antibodies or enzymes) that can be associated with transducers to form sensors. Dickert et al. [69] synthesized MIP receptors for various PAH constituents, optically interrogated them with a fluorescent sensor, and demonstrated detection of individual PAHs such as pyrene down to ng/L concentrations in artificially contaminated drinking water. More recently, Xie et al. [70] molecularly imprinted the explosive 2,4,6-trinitrotoluene (TNT) onto the walls of silica nanotubes, thus implying a great future for MIP nanosensors which have faster response time and great sensitivity. Other environmentally relevant MIP sensors have been designed for various herbicides/pesticides (2,4-D, atrazine, phenylureas, CAT, DDT), aquatic toxins such as microcystin and demoic acid, and various heavy metals, with incorporation into a variety of optical, electrochemical, or piezoelectric transducer element [71].

Biosensors based on the use of whole animals or their organs represent a very unique mode of sensing. Insect antennas, for example, are covered with highly sensitive and naturally tuned receptors called sensilla that respond to chemical, physical, and mechanical signals via electrical nerve impulses. By immobilizing the antenna or even the entire insect on a transducer and measuring these induced electrical impulses (or electroantennograms), a biosensor materializes [24]. A multianalyte biosensor can be formed by adhering antennas from several different insects. The current targets for such biosensors are odourants such as those related to smoke (guaiacol and 1-octen) for early-warning fire detection or volatiles emanating from diseased plants, with detection limits in the part-per-billion range. Their parallel applications for sensing volatiles associated with environmental contaminants and even non-odour-related compounds are a potential future prospect [24]. Imagine a chip sensitive to a particular pollutant analyte attached to a fish, insects, or invertebrate and the organism released into the environment while the sensor is monitored with a computer. The possibility is as much as the imagination can go!

10.5. Biomass Fuel. Biomass is any plant or animal matter used to produce energy. Many plants and plant-derived materials can be used for energy production; the most common is wood. Other sources include food crops, grasses, agricultural residues, manure, and methane from landfills [72]. The main driving forces for adoption of biomass fuel and its encouragement is mainly due to the more efficient bioprocesses and bioproducts which are cost savings and improved product quality/performance. Environmental consideration of the quick degradation of by-products of biomass fuel is a major consideration in the development of this biotool.

The state of Texas in the United States of America is known to be an agricultural state that has adopted biomass energy production. Crops used to produce biomass energy include cotton, corn, and some soybeans—all grown in Texas [73]. In the US, the primary biomass fuels are wood, biofuels, and various waste products. Biofuels include alcohols, synfuel, and bio-diesel, a fuel made from grain and animal fats. Waste consists of municipal solid waste, landfill gas, agricultural by-products, and other material. Most biomass energy used in the US—65 percent—comes from wood [74]. Another 23 percent of biomass energy used comes from biofuels. while the remaining 12 percent comes from waste energy. Energy generated from biomass is the nation's largest source of renewable energy, accounting for 48 percent of the total in 2006. The US consumed 3,277 trillion British thermal units (Btu) of biomass energy in 2006. The next largest source of renewable energy is hydroelectric power, with 2,889 trillion Btu consumed in 2006 [75].

While cattle manure has the most potential for power use, other forms of agricultural waste have significant possibilities, too. These include poultry litter, rice straw and husk, peanut shells, cotton gin trash, and corn stover. In fact, a recent report from the Houston Advanced Research Center estimated that Texas agricultural wastes have the potential to produce 418.9 megawatts of electricity, or enough to power over 250,000 homes, based on average Texas electricity use in 2006 [76].

Plant biomass can be processed and converted by fermentation and other processes into chemicals, fuels and materials that are renewable and result in no net emissions of greenhouse gases. Also, energy, such as waste heat, can be used efficiently. This approach is called industrial ecology [77].

In the US, ethanol made from corn currently accounts for the majority of biofuel consumption in the transportation sector. In the future, however, "lignocellulosic" biofuels made from crop residue, grasses, wood products, sorghum, "energy cane," and agricultural waste are expected to supplement corn ethanol. These are commonly referred to as "cellulosic." Public and private funding for new research in cellullosic fuels is increasing. Corn ethanol requires significant amounts of fertilizers, pesticides, energy and water to grow, cellulosic biofuel production promises to be much more efficient. The production of cellulosic ethanol and other biofuels is expected to give significant increase in yield of fertilizers, pesticides and energy production. Cellulosic biofuel production promises to be much more efficient in economic usage of agricultural products.

Biologically derived products (bioproducts) are generally less toxic and less persistent than their petrochemical counterparts. Group of companies can mimic the cooperative action of organisms in natural ecosystems by clustering around the processing of a feedstock such as a biomass so the by-product of one is the starting material for another. The ability to evolve bioprocesses and bioproduction systems allows for major improvements in both economic and environmental performance. This permits a manufacturing facility to increase its profitability and capacity while maintaining or even reducing its environmental footprint [77].

In Nigeria, lots of projects are ongoing in the area of biofuel. Dependence on crude oil as a major economic stay for the country has also had its toll on other sectors such as agriculture. Despite this, Nigeria is yet to develop the refining sector of the crude oil exploration. Nigeria currently imports refined petroleum, diesel, ethanol, and kerosene from countries around the world. This has not adversely affected the price of the commodity in the country. Its negative effect is slowly being realized. As a result, it is wise to quickly look at alternative sources of energy so as to avoid the short fall of overdependence on nonrenewable sources such as the petroleum industry so as to cushion the gradual but sure effect of this activity in the economy of the country. The government is also gradually looking at ways to refine the crude oil by revitalizing the refineries and making them functional. Despite all these measures, biofuel still holds the answer to the solution of the problems in the oil sector as long as the environment is concerned. Also the future portends that better industrial management practices will preferentially make use of agro-based sources in all production process so as to put the environment first in all production practices. Two of such ongoing research in this biotechnological tool process involve: Alkali-catalysed Laboratory Production and Testing of Biodiesel Fuel from Nigerian Palm Kernel Oil by Alamu et al. [78] and BioDiesel Nigeria (BDN), a company that will provide jobs for Nigerians by farming Jatropha seeds. The company had proposed on providing employments for those living in poverty in the cities by relocating them to Jatropha farms in the north. There they will be trained to farm Jatropha trees and be paid a decent wage. Tables 4 and 5 show recorded extent of Biofuel efforts in Nigeria.

Highina et al. [80] stated that biofuel industry in Nigeria is still at its infancy, even though policy guidelines are available at the NNPC for the development of the industry, few most ground breaking achievements have been made. In 2007 the Kaduna State Government of Nigeria set up a pilot plant for bioethanol production to demonstrate the technology and its viability using local design and materials. Ahmadu Bello University, Zaria, Kaduna State Nigeria, also established a pilot plant for biodiesel production Bugaje and Mohammed [81]. Other efforts elsewhere in the country have been limited to bench-scale production of bioethanol and biodiesel from a number of feedstock. There is need to move further from this and scale up for commercial production. Ethanol is produced from fermentation of sugars and biodiesel from the reaction of plant or animal oils with an alcohol, for which methanol is universally used in commercial application today [80].

10.6. Aquaculture Treatment/Management. Catfish cultivation has assumed a commercial dimension world-wide. With the adoption of this agricultural wealth creation technique, and the uniqueness of its good protein content, it has been accepted as special delicacy among the wealthy especially in Nigeria. Most families in south-eastern and south-western Nigeria cultivate catfish on medium to commercial scale as a financial supplement for regular job. Most individual, have also engaged themselves as full-time catfish farmers in different level of this aquaculture business. Thus, some are into catfish feed production, breeding (reproduction) as well as total development and marketing of this agricultural product. Apart from the fact that there are various feeds available in the market for cultivation of catfish, various feed supplements from poultry waste to domestic food remnants as well as vegetables, nuts, fruit peels, and so forth are also integrated in the production of this fast growing aquaculture practice. In line with this encouraging agricultural venture is the attendant problem of effluent generation which can adversely affect the aesthetic environments.

Catfish pond effluent quality varies from pond to pond and from season to season. Effluent quality is usually poorest (highest concentrations of solids, organic matter, total phosphorus, and total nitrogen) in the summer when fish feeding rates and water temperatures are highest. Catfish pond effluents generally have higher concentrations of nutrients and organic matter than natural stream waters but much lower concentrations than municipal, and industrial waste water. It appears that catfish pond effluents are most likely to exceed regulatory limits for suspended solids and total phosphorus. Other measured water qualities sometimes are

TABLE 4: Bioethanol plants in Nigeria.

Name of Company	Plant location	Feed stock	Installed capacity (million litres/year)
Dura Clean	Bacita	Molasses/Cassava	4.4
AADL	Sango Ota	Cassava	10.9

used for treating agricultural, municipal and industrial waste water.

According to SRAC [82] report, there are various biological techniques adopted, which are cost efficient and eco-friendly in the treatment of aquaculture effluents. Some of such aquaculture biotechniques include the following:

Construction of Wetlands Adjacent to Ponds. Wetlands are inexpensive to build and operate, and it also eliminates the need for chemical treatment of wastewater. They also contribute stability to local hydrologic processes and are excellent wild-life habitats.

The disadvantage of wetland for treating aquaculture pond wastes is the large amount of space necessary to provide an adequate hydraulic residence time. Therefore, it will probably be necessary to integrate wetland treatment of effluents with other pond effluent management procedures to reduce the area of wetland needed. For example, a wetland centrally located on a farm, or connected to an integrated drainage system, would save on construction costs and use land efficiently. Such a system would also allow a wetland to be used to treat the overflow coming from ponds after rainfall. Pond draining could be staged so that only one pond is being drained at a time, allowing one wetland to serve numerous ponds. Effluent from a constructed wetland could even be pumped back into ponds and reused if needed.

Wetlands act as biological filters to remove pollutants from water, and natural or constructed wetland exposes the solid, semisolid waste particles for fast degradation by microorganism. The water if properly channeled can be reused, and the dried solid component of the effluent when properly treated can be used as organic fertilizer in crop production [82].

10.6.1. Treating Pond Effluents Using Grass Filter Strips. Draining effluents over grass strips filters solids from animal waste. This system may be useful for filtering catfish pond effluent. Common and coastal Bermuda, dallies, and Bahia are recommended grasses for warm climates; fescue, reed canary, and rye grasses are recommended for cool climates. Grass filter strips are highly effective in reducing the concentrations of suspended solids, biochemical oxygen demand, and ammonia, but not efficient in removing algae.

Concentrations of suspended solids, organic matter, and total nitrogen in catfish pond effluents were reduced by applying the effluent to well-established strips of either Bahia or Bermuda grass. This filtering technique was relatively easy and inexpensive and may have application if the filtered effluent is to be reused for fish production to conserve

groundwater. It could also be used to treat effluent before discharging it to receiving water [82].

10.6.2. Management Practices to Reduce the Impact of Aquaculture Effluents on the Environment

(a) Conservative Water Management Practices. This involves two major techniques, namely,

(1) reusing water for Multiple Fish Crops
(2) reducing overflow after rains by keeping pond water level below the pond drain.

Reusing Water for Multiple Fish Crops. The concentration of the substance in the effluent is dependent on the volume of water discharged as well as the mass of nutrients (organic matter) present in the ponds according to the aquacultural feeding history. Thus, reducing the concentration of potential pollutants in pond effluents is difficult, but it is relatively easy to control discharge volume. The most obvious procedure for reducing the volume of effluents from channels catfish ponds is to harvest the fish without draining the ponds. However, this practice works only if water quality does not deteriorate as the water is reused.

SRAC [82] report established that a comparison of annually drained and undrained catfish ponds showed little difference in water quality and no difference in fish production. Natural processes, such as nutrient uptake by bottom soils, microbial decomposition of organic matter, denitrification and sedimentation, continually remove potential pollutants from pond water. Operating ponds without draining makes better use of the waste assimilation capacity of ponds and saves significant amounts of water as well as reducing overall effluent volume.

Reducing Overflow after Rains by Keeping Pond Water Level below the Pond Drain. Seasonal changes in overflow volume affected the amount of waste discharged more than seasonal changes in effluent quality [82]. So, reducing overflow volume can have a dramatic impact on mass discharge of nutrients and organic matter from catfish ponds. By keeping the pond water level below the level of the drain, rainfall is captured rather than allowed to overflow and annual waste discharge is reduced by 50 to 100% depending on the weather. Specifically, waste discharges are normally greatest in winter when overflow volume are highest and not in summer when waste concentrations are highest. A study was conducted in which effluents from Georgia catfish ponds were used to determine the production of soybean irrigated with such effluents.

(b) Using Effluents for Irrigation of Soybeans. Although water discharged from aquaculture ponds is often viewed simply as a waste product, it still has value and its reuse may have multiple benefits. If ponds are located near terrestrial crops that require irrigation, pond discharge can be used for irrigation water. That use will reduce waste discharge and benefit the crop. The SRAC [82] report has it that the

TABLE 5: Proposed plants.

No.	Name of company	Project information	Budget
1	Jigawa, Benue, Anambra and Ondo States	Integrated bio-ethanol refineries and sugarcane farm	US$4 Billion
2	Nasarawa state	Integrated bioethanol refinary and cassava farm	US$27 Million
3	Casplex	Ethanol refinery and cassava farm	NA
4	Akoni	Ethanol plant	NA
5	Ekiti state	Integrated bioethanol refinary and cassava farm	US$100.7 Million

NA: not available.
Source: Agbola et al. [79].

total nitrogen available for crops varied from 0.9 to 1.2 kg/ha from each centimeter of water applied. Assuming average irrigation is 30 cm, then available nitrogen ranged from 27 to 36 kg/ha, a significant portion of the nitrogen requirement of many agronomic crops. Although the average soybean yield was 3.6 metric tons/ha, double the average yield in Georgia, the increased yield was the result of irrigation alone and not the nutrients in the irrigation water. Although the nutrient content of pond effluents may be too low to affect crop production, effluent water not useful for catfish cultivation can find application for irrigation of crops and thus reduce discharge volume. Rice irrigation with effluents of aquaculture has also been suggested.

It is important to note that even though catfish farming is the major aquaculture, there are also other aqua cultural practices like cultivation (breeding) of crawfish, shrimps, and other fresh water organisms.

The summary of the SRAC [82] publication is an examination of the impact of aquaculture pond effluent on the environment and how the use of cost reduction and yet simple management practices can help to control pollution of the environment with aquaculture effluents. These simple management practices do not require extra expense or labour. It is advocated therefore that all aquaculturists should strive to reduce the impact of their activities on the environment by adhering to the following guidelines (aquaculture management practices).

(i) Use high-quality feeds and efficient feeding practices. Feeds are the origin of all pollutants in catfish pond effluents.

(ii) Provide adequate aeration and circulation of pond water. Maintaining good dissolved oxygen levels enhances the appetite of fish and encourages good feeding conversion. Oxygen availability at bottom of ponds improves degradation of organic matter and reduces the amount of organic matter in effluent.

(iii) Minimize water exchange. Routine water exchange is of questionable value as a water quality management procedure and greatly increases effluent volume.

(iv) Operate ponds for several years without draining. Reusing water for multiple fish crops is one of the best methods of reducing waste discharge from ponds.

(v) Capture rainfall to reduce pond overflow also reduces the need for pumped water to maintain pond water levels.

(vi) Allow solids to settle before discharging water. After sieving ponds partially drained for fish harvest, hold remaining water for 2 to 3 days to allow solids settle. Better still dose not discharge this last portion of water.

(vii) Reuse water that is drained from ponds. Instead of draining ponds for fish harvest, water can be pumped to adjacent ponds and reused in the same or other ponds.

(viii) Treat effluents by using constructed wetlands.

(ix) Use effluents to irrigate terrestrial crops. Under certain conditions, the water discharged from ponds may have value as irrigation water for crops.

10.7. Biocatalysts. By using well-established tools from metabolic engineering [83] and biochemistry [84], efforts have been made on engineering microbes to function as "designer biocatalysis," in which certain desirable traits are brought together with the aim of optimizing the rate and specificity of biodegradation. Therefore, enzymes extracted from naturally occurring microorganisms, plant and animals can be used biologically to catalyse chemical reactions with high efficiency and specificity. Compared to conventional chemical processes, biocatalytic processes usually consume less energy, produce less waste, and use less organic solvents (that then require treatment and disposal) [77].

Microbial industrial production of enzyme involves a lot of aerobic steps within a submerged culture in a stirred tank reactor. The enzyme biochemistry is driven by transcription, translation, and molecular mass, number of polypeptide chains, isoelectric point, and degree of glycosylation, such as a saccharide's reaction with a hydroxyl or amino group to form a glycoside [20]. The selection of microbes as candidates for fermentation depends on process characteristics (such as viscosity or recoverability), legal approval of use, and the state of knowledge about the selected organism. Sugars comprise the principal feedstock (i.e., production process strictly biological) for microbial processes (carbon and energy sources) [20]. Feedstocks include molasses, unrefined sugar, and sulfite liquor from cellulose production plants, hydrolysates of wood and starch, or fruit juices, such as the grape juice used in wine making processes. Thus, these raw sources contain other compounds beside sugars. This can be beneficial, because vegetative materials invariably contain nitrogen, phosphorus, and potassium, important nutrients to maintain microbial growth and metabolism [20]. For the purer feed stocks, the nutrients are added to the reactor as inorganic compounds such as ammonium compounds, phosphate, and potassium chloride. Organic supplements include meal, fish meal, cotton seed, low-quality protein materials such as casein or its hydrolysates, millet, stillage, and corn steep liquor. In addition, these chemically complicated mixtures must contain micronutrients, that is trace elements and growth promoters, which are limiting factors. In general, the raw materials are dissolved or suspended in water, and then the medium is heated, filtered, and sterilized. For downstream processing (harvest, concentration, and purification) or for analytical assays during the process, additional pretreatment of the raw material can reduce unwanted side reactions [20] (Figure 6).

By imitating natural selection and evolution, the performance of naturally occurring enzymes can be improved. Enzymes can rapidly be "evolved" (this technique is called "molecular evolution") through mutation or genetic engineering and selected using high-throughput screening to catalyse specific chemical reactions and to optimize their performance under certain conditions such as elevated temperature [77].

11. Nigeria Situation

In research institutions and universities, biocatalysis have been applied in laboratory-scale experiments and primary investigation of product synthesis. Most of the detergents in the country are incorporated with various enzymes. The most significant areas of application of biocatalysis as a biotechnological tool in Nigeria are in the areas of yoghurt production by many local industries, confectionaries and bread production industries, local fermented special seasonings (Ogiri and Okpei) for soup making in the eastern part of Nigeria, produced by boiling melon seed, castor oil seeds and exposing them to microorganisms in the environment for fermentation followed by milling and packaging for consumption. Others involve fermentation of cassava for food and brewing of alcoholic drinks by pilot commercial industrial production in organizations such as Nigerian Breweries, Nigerian Distilleries, Guinness Nigeria Plc, and many other brewing industries which spread across various states in the country (Table 6).

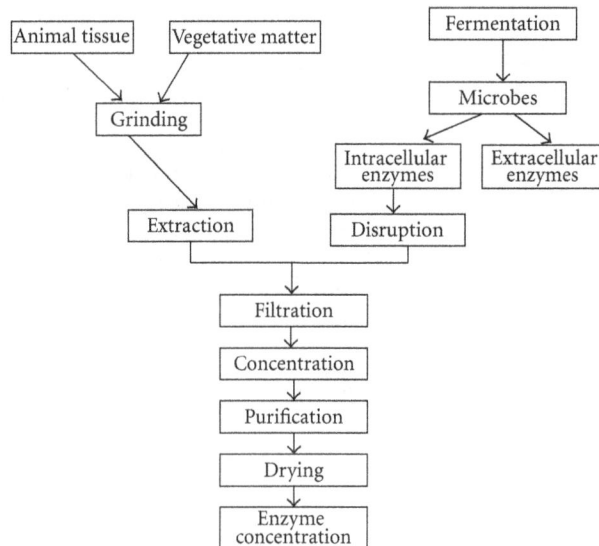

FIGURE 6: Steps in industrial fermentation (enzyme production). Source: [25].

Enzyme biotechnologies can be visualized as sets of biological reactions occurring at various scales in the environment. The activities of enzyme in reactions can lead to desirable results, such as the chemical transformation and ultimate degradation of toxic substances into harmless compounds. Biological reactions may also lead to undesirable results, such as the introduction of genetically modified organisms to an ecosystem or the generation of toxic chemicals. Here enzymes are modified to do the clearing process of these toxins.

12. Conclusion

At the backdrop of the need to meet certain challenges that affect development, thus the much needed change to bring about these developments informed the decision by the United Nation to elaborate on key agenda which most nations are expected to adhere to in order to achieve certain goals known as Millennium Development Goals (MDGs). The attainment of these goals, aiming at ensuring that participating countries, provide basic good things of life for their citizens.

The United Nations (UN) Secretary General's Special Adviser on the MDGs, Jeffrey D. Sachs, visited Nigeria recently for assessment of progress in indicators of whether Nigeria is on the part of attaining these (MDGs) and he gave his verdict. In his word as reported by Anuforo [94] "One would say Nigeria is on the path, but not well on the path. The direction is positive. The institutional innovation is exciting. But the quantitative achievement is not sufficient. So, there really need to be acceleration between 2010 and 2015."

On her part, the Senior Special Assistant to President Goodluck Jonathan on the MDGs, Mrs. Amina Az-Zubair, also gave perspective to what the Nigerian Government is doing to attain the 2007 to 2016 objectives. According

TABLE 6: Environmental friendly application of enzymes.

Industrial sector	Description	Enzyme application	Reference
Fine chemical production	Biocatalysis using selectivity of enzymes for one of the enantiomers of a chiral molecule, that is, one enantiomer of a racemate is unaffected and the other enantiomer is converted into the desired, pure chemical	Hydrolases are most prominent enzyme used in production of fine chemicals by biocatalytic resolution	Schulze and Wubbolts [85]
Biopolymers/plastics	Enzymes or whole cell systems use sugars as feedstock for product manufacturing	Microbial/enzyme emulation of fossil fuel process	
Nutritional oil production	Genetically enhanced biomass (e.g., soybeans) to yield oil with improved properties, especially functional and nutritional quality	Increasing concentration of β-conglycinin, a seed storage protein	Harlander [86]
Ethanol production	Feed stock is cellulosic biomass (e.g., corn ears and stalks, wheat straw, or switchgrass)	Recent advances in cellulose enzymes have improved efficiencies	Knauf and Moniruzzaman [87]
Leather degreasing	Developing proteases for use in soaking, dehairing, and bating processes	Proteases from *Aspergillus tamarii* and *Alcaligenes faecalis* and loosen hair without chemical assistance. Alkaline protease produced from *Rhizopus oryzae* through solid-state fermentation dehairs the skins completely; use of enzymes for dehairing; baterial cultures have keratinolytic activity	Thanikaivelan et al. [88]
Biohydrogen production	H_2 reactions catalyzed by either nitrogenase or hydrogenase enzymes	*E. coli*, *Enterobacter aerogenes*, and *Clostridium butyricum* use multienzyme systems. Can continuously produce H_2 photochemically and nonphotochemically. Nitrogenase enzymes from *Rhodopseudomonas palustris* and *Rhodobacter sphaeroides* generate H_2 under N-limited conditions	US Department of Energy, Office of Science [89]
Chemical/biological warfare agent decontamination	Enzymatic processes can speed the decomposition of organophosphate nerve agents and other warfare agents	Bacterial enzymes catalyze hydrolysis from bacteria genetically modified to express protein variants, for example phosphotriesterase and organophosphorus anhydrolase	Richardt and Blum [90]
Pulp and paper bleaching	Xylanase is applied before bleaching, replacing Cl-containing compounds in the first stage of the five-stage bleaching sequence. While rot fungus (*Phanerochaete chrysosporium*) degrades lignin in bioreactor wood chips injected with fungus and a growth medium, incubate for 2 weeks, followed by traditional chemical or mechanical processes	Enzyme replaces traditional Cl-addition. Biotechnology process reduces the amount of Cl-containing compounds by more than 10%. Bioreactor method reduces bleaching-related energy requirements by 40%, with concomitant pollution reduction	Roncero et al. [91]
Electroplating/metal cleaning	Enzymes make degreasing/metal cleaning. Fungi can be used to treat metal-laden waste	Proteases may be similar to those listed for leather degreasing *Aspergillus japonicus* used to sorp metal ions, for example, Fe (II), Ni (II), Cr (VI), and Hg (II)	Ahluwalia and Goyal [92]

Source: adapted from: [93].

to Agbaegbu [95], Az-Zubair said that the overarching objective of the countdown towards Nigerians achievement of the MDGs by 2015, on to safe water and sanitation have not improved significantly, and other environmental challenges such as erosion, coastal flooding, and climate change are growing. The MDG office reported, however, that the proportion of the population with access to safe drinking water dropped from 54 percent in 1990 to 49 percent in 2007. The proportion of the population with access to basic sanitation is said to have risen from 39 to 43 percent in the recent period. "With better implementation of the various policy frameworks and plans for water, sanitation, environment and slum upgrading, Nigeria will be on track to achieve this goal," the MDG office report said. Even so, it noted that "planning and maintenance of water supply at local level is weak. Environmental pressures range from desertification in the North to flooding, rubbish heaps and soil erosion in the coastal and Niger Delta regions, requiring a nationally coherent but localized approach."

Enduring sustenance of the environment must come from using the natural methods to remove the synthetic or deleterious activities of man in the environment. When a forest, for example, loses its trees it takes some time to regenerate but when it does it retains it natural beauty and improve the quality of other biota within it. When a lake is contaminated with pollutants, prevention of more pollution and exploitation of gases (aeration), encouragement of microbial activities restores the health of the lake. Thus, biotechnological tools are bioenvironmental technological practices aimed at encouraging without compromise with any other technology no matter how fast and efficient those other technologies could be in restoring the environment in such a way that is closer to nature if not totally natural.

The biotools enumerated by the authors are by no means exhaustive since environmental biotechnology is highly dynamic as well as futuristic. Those commonly practiced much elsewhere and very little if at all in Nigeria, which are being improved upon, are aquaculture treatment/management, biomonitoring, bioleaching, bio-catalysis, biodetergent/biosolvent production, biofiltration, bioremediation, biomass fuel production, and so forth. Therefore, improvement in the quality of human, environment anywhere can be attained by applying any of the biotechnological tools suitable in such an environment. Nigeria has already adopted some such as bioremediation, biomonitoring, biofiltration, biofuel, and biocatalysis; however, their volume of their adoption is still very low. There is therefore need to adopt these biotools much and to look at ways of encouraging the local industries to adopt biotechnological tools in their production process so as to maintain the Nigerian environment.

Developing Nations should instead of despairing on the damaged environment due to refuse heaps, mining activities, industrial production, and so forth embrace these practice (biotools) so as to improve their environment as well as to maintain and prevent continuous degradation of the environment. The Niger Delta regions of Nigeria had experienced continuous defacing of the environment due to crude oil pollution. Other nations have also had battered environments (water, soil, air) due to numerous industrial activities. Finding alternative ways of production, favourable to the environment as well as eco-friendly means of dealing with waste will go a long way in sustaining the environment.

Acknowledgments

The Library section of Godfrey Okoye University, Enugu-Nigeria, is sincerely acknowledged and of worthy mention is Mrs. Mary Ellen Chijioke whose magnanimity is unequalled and who took time to personally search and provide basic environmental biotechnology textbooks which in no small measure helped in the success of this research. The Dean Faculty of Natural and Applied Sciences, Godfrey Okoye University, Professor Emmanuel Adinna, is sincerely appreciated for the constant reminder that "this work must be done." The ebullient and amiable Vice Chancellor of Godfrey Okoye University, Enugu Nigeria, Most Rev. Fr. (Professor) Christian Anieke is most appreciated for his untiring struggles to make Godfrey Okoye -University the best in the world.

References

[1] W. Chen, A. Mulchandani, and M. A. Deshusses, "Environmental biotechnology: challenges and opportunities for chemical engineers," *AIChE Journal*, vol. 51, no. 3, pp. 690–695, 2005.

[2] Earth Policy Institute Natural Systems, Data Center, 2009, http://www.earth-policy.org/.

[3] Wikipedia, the free encyclopedia- "Sustainability", 2010, http://en.wikipedia.org/wiki/Sustainability.

[4] Independently Sustainable Regions, "Environmental Sustainability," 2010, http://www.IndependentlySustainableRegion.

[5] H. E. Daly, "Toward some operational principles of sustainable development," *Ecological Economics*, vol. 2, no. 1, pp. 1–6, 1990.

[6] O. P. Bankole, "Major environmental issues and the need for environmental statistics and indicators in Nigeria," in *Proceedings of the ECOWAS workshop on Environmental Statistics*, Abuja, Nigeria, 2008.

[7] Yale Center for Environmental Law and Policy, "2005 Environmental Sustainability Index," Yale University, 2010, http://www.yale.edu/esi/ESI2005_Main_Report.pdf.

[8] Yale Center for Environmental Law and Policy/Center for International Earth Science Information Network at Columbia University, "2008 Environmental Performance Index Report," 2008, http://www.yale.edu/epi/files/2008EPI_text.pdf.

[9] Nigeria: Environmental Profile, 2010, http://www.NigeriaEnvironmentalProfile.htm.

[10] I. N. E. Onwurah, "Restoring the crop sustaining potential of crude oil polluted soil by means of *Azotobacter* inoculation," *Plant Production Research Journal*, vol. 4, pp. 6–16, 1999.

[11] C. S. Ezeonu and I. N. E. Onwurah, "Effect of crude oil contamination on Chlorophyll content in *Zea mays L*," *International Journal of Biology and Biotechnology*, vol. 6, no. 4, pp. 299–301, 2009.

[12] E. A. Odu, "Impact of pollution in biological resources within the Niger delta," Technical Report on Environmentatl Pollution Monitoring of the Niger Delta Basin of Nigeria 6,

pp. 69–121, Environmental Consultancy Group, University of Ife, 1987.

[13] A. Raufu, "Tin Mining Wreaks havoc on "Beautiful" Nigerian City. Third World Networkonline," 1999, http://www.twnside.org.sg/title/1878-cn.htm.

[14] T. A. Aina and A. T. Salau, "The challenges of sustainable development in Nigeria," *Nigerian Environmental Study/Action Team (NEST)*, 1992.

[15] Report on National Biotechnology Policy. White House Council on competitiveness. Washington, DC, USA, 1991.

[16] African, Caribbean and Pacific (ACP), "ACP region must harness biotechnology for a better future," ACP policy brief 1, CTA, Wageningen, The Netherlands, 2005.

[17] E. J. Disilva, "The colours of biotechnological science, development and humankind," *Electronic Journal of Biotechnology*, vol. 7, no. 3, 2004.

[18] EFB, Environmental Biotechnology. European Federation of Biotechnology. Task group on public perceptions of Biotechnology Briefing paper 4, 2nd Edition, 1999, http://www.kluyver.stm.tudelft.nl/efb/home.htm.

[19] FAIR: Co-operation research for SMEs, "Integrated Production and processing chains, Quality of life and management of living resources," 2000, http://www.biodetergent.pdf.

[20] A. D. Vallero, *Environmental Biotechnology: A Biosystems Approach*, Elsevier Academic Press, Burlington, Mass, USA, 1st edition, 2010.

[21] R. L. Raymond, V. W. Jamisen, and J. O. Hudson Jr., "Final Report on Beneficial simulation of Bacterial activity in groundwater containing petroleum products," American Petroleum Institute, Washington, DC, USA, 1975.

[22] I. N. E. Onwurah, "An integrated environmental biotechnology for enhance bioremediation of crude oil contaminated agricultural land," *Journal of Biotech Research*, vol. 1, no. 2, pp. 51–60, 2003.

[23] D. A. Vallero, *Fundamentals of Air Pollution*, Academic Press, Burlington, Mass, USA, 4th edition, 2007.

[24] S. Ripp, M. L. Diclaudio, and G. S. Sayler, "Biosensors as environmental monitors," in *Environmental Microbiology*, R. Mitchell and J. Gu, Eds., pp. 213–233, Wiley-Blackwell, NJ, USA, 2nd edition, 2010.

[25] "Adaption from European commision and federal environment agency Austria," Final Report Collection of information on Enzymes, Contract No. B43040/2000/278245/MAR/E2, 2002, http://www.agronavigator.cz/attachments/enzymerepcomplete.pdf.

[26] http://www.mainstream_biofiltration_water.htm.

[27] N. U. Asamudo, A. S. Daba, and O. U. Ezeronye, "Bioremediation of textile effluent using Phanerochaete chrysosporium," *African Journal of Biotechnology*, vol. 4, no. 13, pp. 1548–1553, 2005.

[28] O. U. Ezeronye and P. O. Okerentugba, "Performance and efficiency of a yeast biofilter for the treatment of a Nigerian fertilizer plant effluent," *World Journal of Microbiology and Biotechnology*, vol. 15, no. 4, pp. 515–516, 1999.

[29] A. O. Ogunlela and A. S. Ogunlana, "Application of lava stones and oyster shells as biofilter substrates in a recirculatory aquaculture system," *Journal of Applied Sciences Research*, vol. 7, no. 2, pp. 88–90, 2011.

[30] A. B. Rabah, M. L. Ibrahim, U. J. Josiah Ijah, and S. B. Manga, "Assessment of the efficiency of a yeast biofilter in the treatment of abattoir wastewater," *African Journal of Biotechnology*, vol. 10, no. 46, pp. 9347–9351, 2011.

[31] D. Taylor, J. Gregory, M. Jones, and R. Fosbery, "The use of microorganisms to extract heavy metals from low grade

[32] C. L. Brierley, "Biomining—Extracting metals with microorganisms," Brierley Consultancy LCC, Fact sheets, 2008.

[33] H. L. Ehrlich, *Geomicrobiology*, Marcel Dekker, New York, NY, USA, 4th edition, 2002.

[34] H. Xu and J. Gu, "Sorption and transformation of toxic metals by microorganisms," in *Environmental Microbiology*, pp. 153–175, Wiley-Blackwell, NJ, USA, 2nd edition, 2010.

[35] D. R. Lovley, E. J. P. Phillips, Y. A. Gorby, and E. R. Landa, "Microbial reduction of uranium," *Nature*, vol. 350, no. 6317, pp. 413–416, 1991.

[36] K. P. Nevin and D. R. Lovley, "Potential for nonenzymatic reduction of Fe(III) via electron shuttling in subsurface sediments," *Environmental Science and Technology*, vol. 34, no. 12, pp. 2472–2478, 2000.

[37] I. N. E. Onwurah, V. N. Ogugua, N. B. Onyike, A. E. Ochonogor, and O. F. Otitoju, "Crude oils spills in the environment, effects and some innovative clean-up biotechnologies," *International Journal of Environmental Research*, vol. 1, no. 4, pp. 307–320, 2007.

[38] National Research Council (NRC), *Environmental Neurotoxicology*, National Academic Press, Washington, DC, USA, 1992.

[39] National Research Council (NRC), *Biologic Markers in Reproductive Toxicology*, National Academic Press, Washington, DC, USA, 1989.

[40] P. J. Matthews, "Toxicology for water scientists," *Journal of Environmental Management*, vol. 11, no. 1, pp. 1–16, 1980.

[41] G. C. Okpokwasili and L. O. Odokuma, "Tolerance of Nitrobacter to toxicity of some Nigerian crude oils," *Bulletin of Environmental Contamination and Toxicology*, vol. 52, no. 3, pp. 388–395, 1994.

[42] I. N. E. Onwurah, "Biochemical oxygen demand exertion and glucose uptake kinetics of Azotobacter in crude oil polluted medium," *Bulletin of Environmental Contamination and Toxicology*, vol. 60, no. 3, pp. 464–474, 1998.

[43] K. E. Peters and J. M. Moldown, *The Biomarker Guide, Interpreting Molecular Fossils in Petroleum and Ancient Sediments*, Prentice-Hall, Englewood Cliffs, NJ, USA, 1993.

[44] T. D. Bucheli and K. Fent, "Induction of cytochrome P450 as a biomarker for environmental contamination in aquatic ecosystems," *Critical Reviews in Environmental Science and Technology*, vol. 25, no. 3, pp. 201–268, 1995.

[45] J. Stegeman and M. Hahn, "Biochemistry and molecular biology of monooxygenases: current perspectives on forms, functions, and regulation of cytochrome P450 in aquatic species," in *Aquatic Toxicology: Molecular, Biochemical, and Cellular Perspectives*, D. Malins and G. Ostrander, Eds., CRC Press, Boca Raton, Fla, USA, 1994.

[46] K. Wilson and J. M. Walker, *Principles and Techniques of Practical Biochemistry*, Cambridge University Press, 4th edition, 1994.

[47] C. Michel, F. Battaglia-Brunet, C. T. Minh, M. Bruschi, and I. Ignatiadis, "Amperometric cytochrome c3-based biosensor for chromate determination," *Biosensors and Bioelectronics*, vol. 19, no. 4, pp. 345–352, 2003.

[48] B. C. Okeke, G. Ma, Q. Cheng, M. E. Losi, and W. T. Frankenberger, "Development of a perchlorate reductase-based biosensor for real time analysis of perchlorate in water," *Journal of Microbiological Methods*, vol. 68, no. 1, pp. 69–75, 2007.

[49] M. Trojanowicz, "Determination of pesticides using electrochemical enzymatic biosensors," *Electroanalysis*, vol. 14, no. 19-20, pp. 1311–1328, 2002.

ores," in *AS and A Level Biology. International Examinations Textbook*, University of Cambridge, 2nd edition, 2009.

[50] S. Andreescu and O. A. Sadik, "Correlation of analyte structures with biosensor responses using the detection of phenolic estrogens as a model," *Analytical Chemistry*, vol. 76, no. 3, pp. 552–560, 2004.

[51] I. Satoh and Y. Iijima, "Multi-ion biosensor with use of a hybrid-enzyme membrane," *Sensors and Actuators B*, vol. 24, no. 1–3, pp. 103–106, 1995.

[52] E. Lojou and P. Bianco, "Application of the electrochemical concepts and techniques to amperometric biosensor devices," *Journal of Electroceramics*, vol. 16, no. 1, pp. 79–91, 2006.

[53] V. G. Andreou and Y. D. Clonis, "A portable fiber-optic pesticide biosensor based on immobilized cholinesterase and sol-gel entrapped bromcresol purple for in-field use," *Biosensors and Bioelectronics*, vol. 17, no. 1-2, pp. 61–69, 2002.

[54] J. Tschmelak, G. Proll, J. Riedt et al., "Automated Water Analyser Computer Supported System (AWACSS) Part I: project objectives, basic technology, immunoassay development, software design and networking," *Biosensors and Bioelectronics*, vol. 20, no. 8, pp. 1499–1508, 2005.

[55] T. R. Glass, H. Saiki, T. Joh, Y. Taemi, N. Ohmura, and S. J. Lackie, "Evaluation of a compact bench top immunoassay analyzer for automatic and near continuous monitoring of a sample for environmental contaminants," *Biosensors and Bioelectronics*, vol. 20, no. 2, pp. 397–403, 2004.

[56] M. Farré, L. Kantiani, and D. Barceló, "Advances in immunochemical technologies for analysis of organic pollutants in the environment," *Trends in Analytical Chemistry*, vol. 26, no. 11, pp. 1100–1112, 2007.

[57] M. Fojta, "Electrochemical sensors for DNA interactions and damage," *Electroanalysis*, vol. 14, no. 21, pp. 1449–1463, 2002.

[58] G. Bagni, S. Hernandez, M. Mascini, E. Sturchio, P. Boccia, and S. Marconi, "DNA biosensor for rapid detection of genotoxic compounds in soil samples," *Sensors*, vol. 5, no. 6–10, pp. 394–410, 2005.

[59] A. Sassolas, B. D. Leca-Bouvier, and L. J. Blum, "DNA biosensors and microarrays," *Chemical Reviews*, vol. 108, no. 1, pp. 109–139, 2008.

[60] F. Lucarelli, L. Authier, G. Bagni et al., "DNA biosensor investigations in fish bile for use as a biomonitoring tool," *Analytical Letters*, vol. 36, no. 9, pp. 1887–1901, 2003.

[61] Y. S. Kim, H. S. Jung, T. Matsuura, H. Y. Lee, T. Kawai, and M. B. Gu, "Electrochemical detection of 17β-estradiol using DNA aptamer immobilized gold electrode chip," *Biosensors and Bioelectronics*, vol. 22, no. 11, pp. 2525–2531, 2007.

[62] C. Nakamura, T. Kobayashi, M. Miyake, M. Shirai, and J. Miyake, "Usage of a DNA aptamer as a ligand targeting microcystin," *Molecular Crystals and Liquid Crystals Science and Technology Section A*, vol. 371, pp. 369–374, 2001.

[63] M. Rajendran and A. D. Ellington, "Selection of fluorescent aptamer beacons that light up in the presence of zinc," *Analytical and Bioanalytical Chemistry*, vol. 390, no. 4, pp. 1067–1075, 2008.

[64] Y. Xiao, A. A. Rowe, and K. W. Plaxco, "Electrochemical detection of parts-per-billion lead via an electrode-bound DNAzyme assembly," *Journal of the American Chemical Society*, vol. 129, no. 2, pp. 262–263, 2007.

[65] H. Wei, B. Li, J. Li, S. Dong, and E. Wang, "DNAzyme-based colorimetric sensing of lead (Pb2+) using unmodified gold nanoparticle probes," *Nanotechnology*, vol. 19, no. 9, Article ID 095501, 2008.

[66] J. Yakovleva, R. Davidsson, M. Bengtsson, T. Laurell, and J. Emnéus, "Microfluidic enzyme immunosensors with immobilised protein a and G using chemiluminescence detection," *Biosensors and Bioelectronics*, vol. 19, no. 1, pp. 21–34, 2003.

[67] S. K. Islam, R. Vijayaraghavan, M. Zhang et al., "Integrated circuit biosensors using living whole-cell bioreporters," *IEEE Transactions on Circuits and Systems I*, vol. 54, no. 1, pp. 89–98, 2007.

[68] M. Alvarez, A. Calle, J. Tamayo, L. M. Lechuga, A. Abad, and A. Montoya, "Development of nanomechanical biosensors for detection of the pesticide DDT," *Biosensors and Bioelectronics*, vol. 18, no. 5-6, pp. 649–653, 2003.

[69] F. L. Dickert, P. Lieberzeit, and M. Tortschanoff, "Molecular imprints as artificial antibodies—a new generation of chemical sensors," *Sensors and Actuators, B*, vol. 65, no. 1, pp. 186–189, 2000.

[70] C. Xie, B. Liu, Z. Wang, D. Gao, G. Guan, and Z. Zhang, "Molecular imprinting at walls of silica nanotubes for TNT recognition," *Analytical Chemistry*, vol. 80, no. 2, pp. 437–443, 2008.

[71] L. Ye and K. Mosbach, "Molecular imprinting: synthetic materials as substitutes for biological antibodies and receptors," *Chemistry of Materials*, vol. 20, no. 3, pp. 859–868, 2008.

[72] National Renewable Energy Laboratory, "Biomass Energy Basics," 2010, http://www.nrel.gov/learning/re_biomass.html.

[73] U.S. Department of Agriculture, Texas Fact Sheet, 2010, http://www.nass.usda.gov/Statistics_by_State/Texas/index.asp.

[74] U.S. Department of Energy, Energy Information Administration, Annual Energy Review, 2006, http://www.eia.doe.gov/aer/pdf/aer.pdf.

[75] U.S. Department of Energy, Energy Information Administration, Annual Energy Review, 2006, http://www.eia.doe.gov/emeu/aer/pdf/pages/sec10_3.pdf.

[76] Houston Advanced Research Center, "Combined Heat and Power Potential using Agricultural Wastes," prepared for State Energy Conservation Office, 2008, http://www.seco.cpa.state.tx.us/zzz_re/re_biomass_chp-report 2008.pdf.

[77] Organisation for Economic Co-operation and Development (OECD), "The Application of Biotechnology to Industrial Sustainability," 2001, http://www.oecd.org/sti/biotechnology.

[78] O. J. Alamu, M. A. Waheed, and S. O. Jekayinfa, "Alkali-catalysed Laboratory production and testing of biodiesel fuel from Nigerian palm kernel oil," *Agricultural Engineering International*, vol. 9, article 07009, 2007.

[79] O. P. Agboola, O. M. Agboola, and F. Egelioglu, in *Proceedings of the World Congress on Engineering (WCE '11)*, vol. 3, London, UK, 2011.

[80] B. K. Highina, I. M. Bugaje, and B. Umar, "Liquid biofuels as alternative transport fuels in Nigeria," *Journal of Applied Technology in Environmental Sanitation*, vol. 1, no. 4, pp. 317–327, 2011.

[81] I. M. Bugaje and I. A. Mohammed, "Biofuels production for transport sector in Nigeria," *International Journal of Development Studies*, vol. 3, no. 2, pp. 36–39, 2008.

[82] Southern Regional Aquaculture Center (SRAC), "Characterisation and Management of Effluents from Aquaculture ponds in the Southeastern United States," 1999, Publication No. 470.

[83] J. Y. Lee, J. R. Roh, and H. S. Kim, "Metabolic engineering of Pseudomonas putida for the simultaneous biodegradation of benzene, toluene, and p-xylene mixture," *Biotechnology and Bioengineering*, vol. 43, no. 11, pp. 1146–1152, 1994.

[84] L. P. Wackett, M. J. Sadowsky, L. M. Newman, H.-G. Hur, and S. Li, "Metabolism of polyhalogenated compounds by a genetically engineered bacterium," *Nature*, vol. 368, no. 6472, pp. 627–629, 1994.

[85] B. Schulze and M. G. Wubbolts, "Biocatalysis for industrial production of fine chemicals," *Current Opinion in Biotechnology*, vol. 10, no. 6, pp. 609–615, 1999.

[86] S. Harlander, "Biotechnology's Possibilities for Soyfoods and Soybean Oil. United Soybean Board," 2009, http://www.soyconnection.com/newsletters/soy-connection/ health-nutrition/article.php/Biotechnology%27s +Possibilities+for+Soyfoods+and+Soybean+Oil?id=64.

[87] M. Knauf and M. Moniruzzaman, "Lignocellulosic biomass processing: a perspective," *International Sugar Journal*, vol. 106, no. 1263, pp. 147–150, 2004.

[88] P. Thanikaivelan, J. R. Rao, B. U. Nair, and T. Ramasami, "Progress and recent trends in biotechnological methods for leather processing," *Trends in Biotechnology*, vol. 22, no. 4, pp. 181–188, 2004.

[89] U.S. Department of Energy, Office of Science. Systems Biology for Energy and Environment: Biohydrogen Production, 2009, http://genomicsgtl.energy.gov/benefits/biohydrogen.shtml.

[90] A. Richardt and M. W. Blum, *Decontamination of Warfare Agents; Enzymatic Methods for the Removal of B/C Weapons*, Wiley-VCH, Weinheim, Germany, 1997.

[91] M. B. Roncero, A. L. Torres, J. F. Colom, and T. Vidal, "The effect of xylanase on lignocellulosic components during the bleaching of wood pulps," *Bioresource Technology*, vol. 96, no. 1, pp. 21–30, 2005.

[92] S. S. Ahluwalia and D. Goyal, "Microbial and plant derived biomass for removal of heavy metals from wastewater," *Bioresource Technology*, vol. 98, no. 12, pp. 2243–2257, 2007.

[93] "The third wave in Biotechnology: a primer on industrial biotechnology," 2009, http://www.bio.org/ind/background/thirdwave.asp.

[94] E. Anuforo, "Special report MDGs: Nigeria struts, frets," *The Guardian*, pp. 12–13, 2010.

[95] T. Agbaegbu, "A dream threatened," *Newswatch*, pp. 14–18, 2010.

16

Polymorphism of Myostatin Gene in Intron 1 and 2 and Exon 3, and Their Associations with Yearling Weight, Using *PCR-RFLP* and *PCR-SSCP* Techniques in Zel Sheep

Elena Dehnavi,[1] **Mojtaba Ahani Azari,**[1] **Saeed Hasani,**[1] **Mohammad Reza Nassiry,**[2] **Mokhtar Mohajer,**[3] **Alireza Khan Ahmadi,**[4] **Leila Shahmohamadi,**[1] **and Soheil Yousefi**[1]

[1] *Department of Animal Science, Gorgan University of Agricultural Sciences and Natural Resources, P.O. Box 4913815739, Gorgan, Iran*

[2] *Department of Animal Science, Ferdowsi University of Mashhad, P.O. Box 9177948978, Mashhad, Iran*

[3] *Golestan Agriculture Jahad, P.O. Box 49174, Gorgan, Iran*

[4] *Department of Animal Science, Faculty of Agrecultural Scinece and Natural Recources, Gonbad University, P.O. Box 4971799151, Gonbad, Iran*

Correspondence should be addressed to Elena Dehnavi, e.dehnavi@yahoo.com

Academic Editor: Manuel Canovas

The aim of present study was to investigate myostatin gene polymorphism and its association with yearling weight records in Zel sheep using *PCR-RFLP* and *PCR-SSCP* methods. Blood samples were collected from 200 Zel sheep, randomly, and DNA was extracted using modified salting out method. Polymerase chain reaction was carried out to amplify 337, 222, and 311 bp fragments, respectively, comprising a part of exon 3, intron 1, and intron 2 of myostatin gene. In addition, exon 3 was digested by *HaeIII* enzyme under *RFLP* method, and introns 1 and 2 were studied using *SSCP*. Under *RFLP* method, all samples showed *mm* genotype. Under *SSCP* method, intron 1 was also monomorph but intron 2 was polymorph (AA, AB, and BB). The allelic frequencies for A and B were 75.5 and 24.5%, respectively. This locus was not in Hardy-Weinberg equilibrium ($P < 0.05$), and there was no significant effect of myostatin gene on yearling weights.

1. Introduction

Considerable progress in farm animal breeding has been made in the last few decades, but achieving greater understanding in the improvement of meat quality was very slow before molecular markers became an accessible technology with wide applications in breeding methods [1].

Meat quality is one of the important economic traits in domestic animals. Determination of meat quality requires analysis and classification of fat content, composition, tenderness, water-holding capacity, color, oxidative stability, and uniformity. Meat quality is affected by several factors such as breed, genotype, feeding, fasting, preslaughter handling, stunning, slaughter methods chilling, and storage conditions [2].

Finding of main genes responsible for meat quality will benefit the producers. In recent years, a lot of works have been performed in this field to find potential genes or chromosome regions associated with the meat quality traits in different farm animals, including cattle, sheep, and chicken. Myostatin (*MSTN*) or growth differentiation factor-8 (GDF-8) is a member of the mammalian growth transforming family (TGF-beta superfamily), which plays an important role in the regulation of embryonic development and tissue homeostasis in adults [3]. They are known to block myogenesis, hematogenesis and enhance chondrogenesis as well as epithelial cell differentiation in vitro. In mice, null mutants are significantly larger than wild-type animals, with 200–300% more skeletal-muscle mass, because of hyperplasia and hypertrophy [4]. Muscular hypertrophy (mh), also known as

TABLE 1: Region, methods, primer's sequence (5′ → 3′), and length of PCR products of the ovine myostatin gene.

Region	Using method	Primer's sequence (5′ → 3′)	Length of fragment (bp)
Intron 1	PCR-SSCP	F: TAC CTT CAT CAC TCT GCC TTC C	222
		R: GGA GGA AAG AAG AGG GAC AAG	
Intron 2	PCR-SSCP	F: CAC ATT TTT CCC CCA GAA GAG	311
		R: AAG ACA GTT CAG AAA ATA GCT GG	
Exon 3	PCR-RFLP	F: CCG GAG AGA CTT TGG GCT TGA	337
		R: TCA TGA GCA CCC ACA GCG GTC	

F: forward and R: reverse.

TABLE 2: PCR conditions.

Location	Primary denaturation in 1st cycle	Denaturation		Annealing		Elongation		Final extension	Number of cycles
	°C/Sec	°C	Sec	°C	Sec	°C	Sec	°C/Sec	n
Intron 1	95/240	94	60	56.5	70	72	75	72/600	40
Intron 2	95/240	95	50	55	60	72	75	72/600	40
Exon 3	94/240	94	60	58.5	60	72	120	72/240	35

"double-muscling" in cattle, has been recognized as a physiological character for years [5] and is seen in Belgian Blue and Piedmontese cattle [6]. These animals had less bone, less fat, and 20% more muscle on an average [7]. Mutations within myostatin gene were red to muscular hypertrophy allele (mh allele) in the double muscle breeds [6]. Such a major effect of a single gene on processing yields opened a potential channel for improving processing yields of animals using knockout technology [8]. Therefore, considering of myostatin gene in farm animals is important to find better animal which opens interesting prospects for future selection programs, especially marker-assistant selection for economic traits.

In Iran, sheep meat is a major source of animal protein and investigation for meat quality and related genes is important. Zel sheep is a native Iranian meat breed and plays a great role in sheep rearing activities in the north of Iran [9]. The aim of present study was to identify genotypes of myostatin gene and their association with yearling weight records in Zel sheep using PCR-RFLP and PCR-SSCP methods in order to find effective alleles influencing meat quantity and quality traits in sheep.

2. Materials and Methods

2.1. Animals and DNA Extraction. Blood samples were randomly collected from 200 (190 ewes and 10 rams) Zel sheep (there were 230 ewes and 30 rams in this station) from Shirang's Zel Breeding Station in Fazel Abad city of Golestan province. DNA was extracted from 3 mL of blood as described by Miller et al. [10]. Quality and quantity of DNA were measured by visual and spectrophotometer methods.

2.2. PCR. Two pairs of primers were designed for each of intron 1 and 2 and exon 3 regions. The primer sequences are presented in Table 1. An aliquot of 100 ng genomic

DNA was amplified in a total volume of 15 μL PCR mix. The PCR mix consisted of 7.5 μL Master mix (Cinna clon), 2 μL forward and reverse primers (10 pmol/μL), and 4.5 μL ddH2O. Amplification conditions are shown in Table 2.

In every experiment, negative controls were used, aiming to avoid contaminations. Assays were performed in a thermal cycler (Personal Cycler-Biometra, CA, German), and the amplicons were analyzed by 1.5% agarose gel electrophoresis. The gels were stained with ethidium bromide and visualized under ultraviolet light.

2.3. Digestion Reaction. 10 μL of PCR products were incubated for 10 h at 37°C with 1 μL (10 units) of HaeIII enzyme for myostatin gene (just for exon 3, using RFLP method). Digestion products were separated by electrophoresis on 8% nondenaturing polyacrylamide gels, stained by silver nitrate staining method [11] (Figure 3).

2.4. SSCP. Genotyping of intron 1 and 2 was performed by PCR-SSCP method. PCR products (3 μL) were diluted with 13 μL of running buffer (including 800 μL formamide 99%, 100 μL loading dye, 100 μL glycerol 98%, 3 μL 0.5M EDTA, and 2 μL 10M NaOH). After heating at 95°C for 5 min, they were immediately placed on ice for 10 min. Polymorphisms were detected using 10% nondenaturing polyacrylamide gels (Figures 1 and 2). The mixture was electrophoresed for 4 h at 250 V and 10°C. DNA fragments were visualized using the silver nitrate staining method [11].

2.5. Statistical Analysis. Calculation of genotypes, alleles frequencies, mean expected and observed heterozygosities, and Chi-square test was performed using PopGene32 (ver. 1.32) [12]. Samples, which were born in 2006, 2007, and 2008 years, were used for statistical analysis. Yearling weight (YW)

Polymorphism of Myostatin Gene in Intron 1 and 2 and Exon 3, and Their Associations with Yearling Weight, Using PCR-RFLP and PCR-SSCP Techniques in Zel Sheep

161

TABLE 3: Allele and genotype frequencies, observed, expected, and average heterozygosity for intron 2 of MSTN gene.

Locus	Allelic frequencies (%)		Genotype frequencies (%)			Heterozygosity			χ^2
	A	B	AA	AB	BB	Obs.	Exp.	Ave.	
Intron 2	75.5	24.5	73.5	4	22.5	0.04	0.37	0.37	160.55*

*$P < 0.05$.

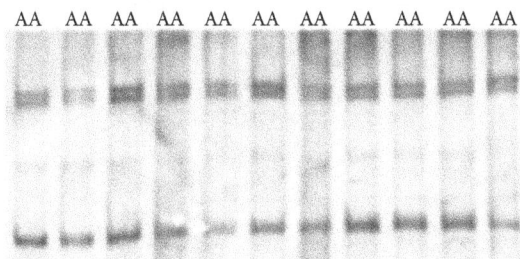

FIGURE 1: The SSCP patterns of intron 1 (222 bp), on 10% nondenatured polyacrylamide gel after silver nitrate staining.

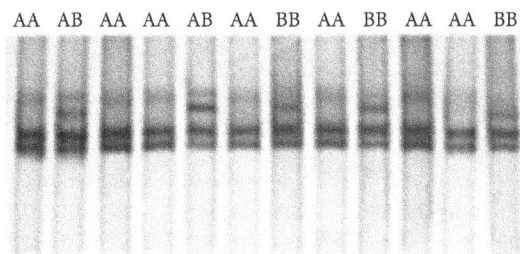

FIGURE 2: The SSCP patterns of intron 2 (311 bp), on 10% nondenatured polyacrylamide gel after silver nitrate staining. Three patterns demonstrating the 3 genotypes are presented.

was analyzed using the fixed model of SAS [13] software and by GLM procedure by the following statistical model:

$$Y_{ijkl} = \mu + S_i + D_j + G_k + e_{ijkl}, \quad (1)$$

where Y_{ijkl} is yearling weight of each animal; μ is general mean; S_i is sex effect ($i = 1$, and 2), D_j is birth year effect ($j = 1$, 2, and 3), G_k is genotype effect ($k = 1$, 2, and 3), e_{ijkl} is random error.

3. Results

3.1. Exon 3. A 337 bp fragment for exon 3 of MSTN locus was amplified. HaeIII restriction enzyme was used to digest the PCR products. The HaeIII digests the m allele but not M allele. Digestion of the m allele produced three fragments of 83, 123, and 131 bp (Figure 3). All samples were digested by HaeIII enzyme and showed the mm genotype. As a result, all of them were monomorph (Figure 1).

3.2. Intron 1 and 2. Intron 1 and 2 of myostatin gene with 222 and 311 bp lengths were amplified, respectively. Under the SSCP analysis, different conformations were detected by electrophoresis on 12% nondenaturing polyacrylamide gel

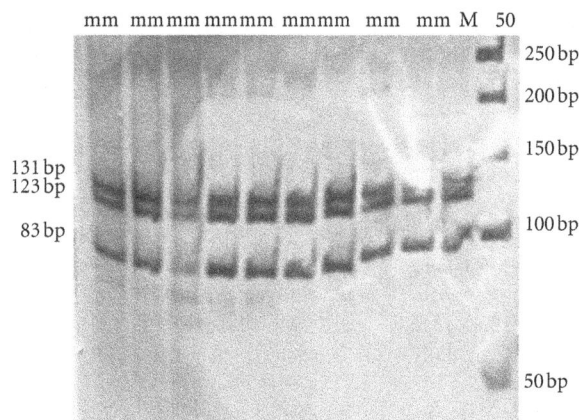

FIGURE 3: Restriction patterns of 337 bp fragments of exon 3 after digesting with HaeIII on 8% nondenatured polyacrylamide gel after silver nitrate staining. Molecular marker was M50.

TABLE 4: Least square means (LSM), standard error (SE), and probability levels for YW (kg) of intron 2 of MSTN genotypes.

Probability levels			Genotype	LSM* ± SE
AA	AB	BB		
—	0.9539	0.6367	AA	28.30ᵃ ± 0.72
0.9539	—	0.7397	AB	27.85ᵃ ± 1.61
0.6367	0.7397	—	BB	29.09ᵃ ± 0.84

*Same letters in column show no significant difference ($P > 0.05$).

(Figure 2). Results showed that intron 1 of this gene was also monomorph, and all samples showed the homozygote genotype (Figure 1). Different conformations were found in intron 2, and A and B alleles were detected with frequencies of 75.5 and 24.5%, respectively. In this population, this locus did not show Hardy-Weinberg equilibrium ($P < 0.05$) (Table 3). Observed heterozygosity for this locus was very low (0.04) showing high level of homozygosity in the herd. Results showed that there was no significant effect of genotypes of myostatin gene on yearling weights ($P > 0.05$) (Table 4). However, sex had significant effect on YW ($P < 0.01$). Yearling weight least squares means of males (31.62 ± 1.34 kg) were more than females (25.21 ± 0.57 kg).

4. Discussion

Results showed polymorphism in intron 2, but intron 1 and exon 3 were monomorph. On the contrary, Soufy et al. [14, 15] observed polymorphism for exon 3 in Sanjabi sheep and native Kermanian cattle. Intron 1 was also monomorph, and all samples showed the homozygote genotype. On the other hand, intron 2 was polymorphic and three different genotypes were detected. Three different conformational patterns (*AA*, *AB*, and *BB*) were determined with frequencies of 73.5, 4, and 22.5%. The allelic frequencies for *A* and *B* were as 75.5 and 24.5%, respectively. Similar result was observed in Iranian Baluchi sheep [16, 17]. This inconsistency may be ascribed to breed differences, population and sampling size, environmental factors, mating strategies, geographical position effect, and frequency distribution of genetic variants.

Statistical analysis showed that myostatin locus had no significant effect on *YW* ($P > 0.05$). Similar to these findings, Masoudi et al. [17] did not report any significant effect of this locus on *YW*. Although, they found significant effect of different genotypes on birth weight, they also did not observe any significant effect on weaning and six month weights. Ansary et al. [16] detected significant effect of different genotypes on daily gain from birth to 3 month of age ($P < 0.01$). This may be due to the environmental effects that exist and affect this trait. It must be pointed that mutation in intron region is classified as a silent mutation and in spite of existence of mutation in this gene any associations is not reported. However, there are reports of diseases caused by silent mutations. It also seems that introns have a role in the expression of gene and necessitate for physical instructors of DNA. But they do not have a major role in rank of amino acids and proteins' instructor [18].

In this population, this locus did not show Hardy-Weinberg equilibrium. This confirmed that factors leading to disequilibrium, especially selection, may affect the genetic structure of the population. Based on our results, the investigated population showed a low degree of genotypic variability for the *MSTN* gene. This may be explained by the conservation and breeding strategies, which have been carried out. In recent years, in this station, only a few rams have been used as sires in breeding plans. Due to small effective population size, inbreeding was high, and, as a consequence, heterozygosity and genetic variability were low. Controlled breeding might help in lowering inbreeding. In spite of low variability for genomic DNA, these data provide evidence that Iranian's Zel sheep breed have a polymorphism in intron 2 for myostatin locus. However, results showed that this locus in this population may not be useful for developing future selection programs, especially marker-assistant selection for improving weight gain and meat traits.

It can be concluded that, although *MSTN* polymorphism did not have effect on *YW*, further analysis needs to be conducted on the effect of *MSTN* genotypes on yearling weight and other body weights. Furthermore, results showed that *PCR-RFLP* and *PCR-SSCP* are appropriate tools for evaluating genetic variability.

Acknowledgments

This work was financially supported by the Jahad-e-Agriculture (Golestan and I.R. Iran). The authors also acknowledge the staffs of Shirang Research Station for their helping to provide blood samples.

References

[1] Y. Gao, R. Zhang, X. Hu, and N. Li, "Application of genomic technologies to the improvement of meat quality of farm animals," *Meat Science*, vol. 77, no. 1, pp. 36–45, 2007.

[2] K. Rosenvold and H. J. Andersen, "Factors of significance for pork quality - A review," *Meat Science*, vol. 64, no. 3, pp. 219–237, 2003.

[3] T. S. Sonstegard, G. A. Rohrer, and T. P. L. Smith, "Myostatin maps to porcine chromosome 15 by linkage and physical analyses," *Animal Genetics*, vol. 29, no. 1, pp. 19–22, 1998.

[4] A. C. McPherron, A. M. Lawler, and S. J. Lee, "Regulation of skeletal muscle mass in mice by a new TGF-β superfamily member," *Nature*, vol. 387, no. 6628, pp. 83–90, 1997.

[5] P. F. Arthur, "Double muscling in cattle: a review," *Australian Journal of Agricultural Research*, vol. 46, pp. 1493–1515, 1995.

[6] R. Kambadur, M. Sharma, T. P. L. Smith, and J. J. Bass, "Mutations in myostatin (GDF8) in double-muscled Belgian Blue and Piedmontese cattle," *Genome Research*, vol. 7, no. 9, pp. 910–916, 1997.

[7] J. Kobolak and E. Gocza, "The role of the myostatin protein in meat quality—a review," *Archives Animal Breeding*, vol. 45, no. 2, pp. 159–170, 2002.

[8] A. M. Kocabas, H. Kucuktas, R. A. Dunham, and Z. Liu, "Molecular characterization and differential expression of the myostatin gene in channel catfish (Ictalurus punctatus)," *Biochimica et Biophysica Acta - Gene Structure and Expression*, vol. 1575, no. 1–3, pp. 99–107, 2002.

[9] I. L. Mason, *A World Dictionary of Livestock Breeds, Types and Varieties*, C.A.B International, Oxford, UK, 4th edition, 1996.

[10] S. A. Miller, D. D. Dykes, and H. F. Polesky, "A simple salting out procedure for extracting DNA from human nucleated cells," *Nucleic Acids Research*, vol. 16, no. 3, p. 1215, 1988.

[11] H. Benbouza, J. M. Jacquemin, J. P. Baudin, and G. Mergeai, "Optimization of a reliable, fast, cheap and sensitive silver staining method to detect SSR markers in polyacrylamide gels," *Biotechnology, Agronomy, Society and Environment*, vol. 10, no. 2, pp. 77–81, 2006.

[12] F. C. Yeh, R. Yang, T. J. Boyle, Z. Ye, and J. M. Xiyan, *POPGENE 32, Microsoft Window-based Freeware for Population Genetic Analysis, Version 1.32*, Molecular Biology and Biotechnology Centre, University of Alberta, Edmonton, Canada, 2000.

[13] SAS Institute, *SAS User's Guide, Version 8.2*, SAS Institute, Cary, NC, USA, 2001.

[14] B. Soufy, M. R. Mohammadabadi, K. Shojaeyan et al., "Evaluation of Myostatin gene polymorphism in Sanjabi sheep by PCR-RFLP method," *Animal Science Reserches*, vol. 19, no. 1, pp. 81–89, 2009.

[15] B. Soufy, K. Shojaeian, M. M. Abadi, B. A. Zadeh, and A. Mohamadi, "Myostatin gene polymorphism in native kermanian cattle using molecular marker," in *Proceedings of the 6th National Biotechnology Congress of Iran*, Milad Tower Conference Hall, Tehran, Iran, August 2009.

[16] M. Ansary, M. Tahmoorespour, M. V. Valeh, M. R. Nassiry, and F. E. Shahroudi, "Investigation of polymorphism of GDF-8 gene and its association with average daily gain in Baluchi sheep," in *Proceedings of the 3th Congress on Animal Science*, Ferdowsi University of Mashhad, Mashhad, Iran, 2008.

[17] A. Masoudi, H. Hemrani, A. Abbasi et al., "Using PCR-SSCP techniques to study polymorphism of myostatin gene and

its association with production traits in Baluchi sheep," in *Proceedings of the 4th National Biotechnology Congress of Iran*, Kerman, Iran, August 2005.

[18] H. R. Matthews, R. Freedland, and R. L. Miesfeld, *Biochemistry a Short Course*, Wileyliss, Inc., Hoboken, NJ, USA, 1997.

17

Novel Simplified and Rapid Method for Screening and Isolation of Polyunsaturated Fatty Acids Producing Marine Bacteria

Ashwini Tilay and Uday Annapure

Food Engineering and Technology Department, Institute of Chemical Technology, Matunga, Mumbai 400 019, India

Correspondence should be addressed to Uday Annapure, udayannapure@gmail.com

Academic Editor: Manuel Canovas

Bacterial production of polyunsaturated fatty acids (PUFAs) is a potential biotechnological approach for production of valuable nutraceuticals. Reliable method for screening of number of strains within short period of time is great need. Here, we report a novel simplified method for screening and isolation of PUFA-producing bacteria by direct visualization using the H_2O_2-plate assay. The oxidative stability of PUFAs in growing bacteria towards added H_2O_2 is a distinguishing characteristic between the PUFAs producers (no zone of inhibition) and non-PUFAs producers (zone of inhibition) by direct visualization. The confirmation of assay results was performed by injecting fatty acid methyl esters (FAMEs) produced by selected marine bacteria to Gas Chromatography-Mass Spectrometry (GCMS). To date, this assay is the most effective, inexpensive, and specific method for bacteria producing PUFAs and shows drastically reduction in the number of samples thus saves the time, effort, and cost of screening and isolating strains of bacterial PUFAs producers.

1. Introduction

Microbial lipids are a diverse group of compounds with a number of vital nutraceutical and pharmaceutical applications and utilized commercially since the 1980s. These microbial lipids or polyunsaturated fatty acids (PUFAs) are obtained from various sources. Now a day, microorganism-produced (algae/fungi/bacteria) PUFAs are commercially competitive with plant and fish oils.

PUFAs are the fatty acids having more than one double bond. Eicosapentaenoic acid (EPA, 20:5, n-3) and docosahexaenoic acid (DHA, 22:6, n-3) are the important n-3 fatty acids, while arachidonic acid (AA, 20:4, n-6) is a vital n-6 fatty acid. EPA and DHA are important for prevention of arthrosclerosis, cancer, rheumatoid arthritis, psoriasis, and diseases of old age such as Alzheimer's and age-related macular degeneration [1, 2]. AA and DHA are of special importance in the brain and blood vessels and are considered essential for pre- and postnatal brain and retinal development [3]. Eicosanoids such as prostaglandins, prostacyclins, and leukotrienes derived from n-3 PUFA are also important

in new-born and infant development, modulatory vascular resistance, and wound healing [4–6]. PUFAs are either directly available as components of the diet or produced from precursors like linoleic acid (LA, C18:2 n-6) and a-linolenic acid (ALA, C18:3 n-3) [7].

Accordingly, PUFAs are highly important substances in the pharmaceutical, medical, and nutritional fields. Recent investigations have focused on microorganisms as alternative natural source for production of oil containing PUFAs. These are potentially promising lipid source because of their high growth rates in simple media and simplicity of their manipulation. As traditional sources of n-3 fatty acids such as fish oil continue to diminish, identification of alternate sources will become crucial. Marine microorganisms represent one of the less explored sources of biologically active natural products. Novel compounds with various bioactivities such as antibiotic, antitumor, cytotoxic, and anti-inflammatory, have been isolated and elucidated from this source [8].

Considering importance of PUFAs, many researchers have tried to isolate and screen marine organisms for these

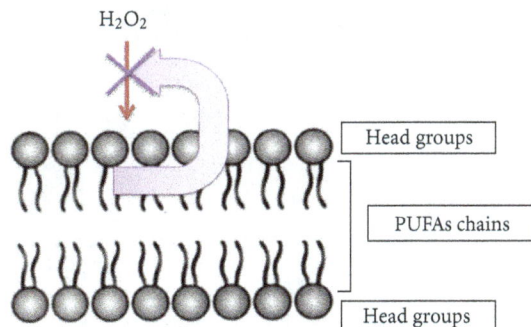

FIGURE 1: Proposed mode of action for the H_2O_2-plate assay method.

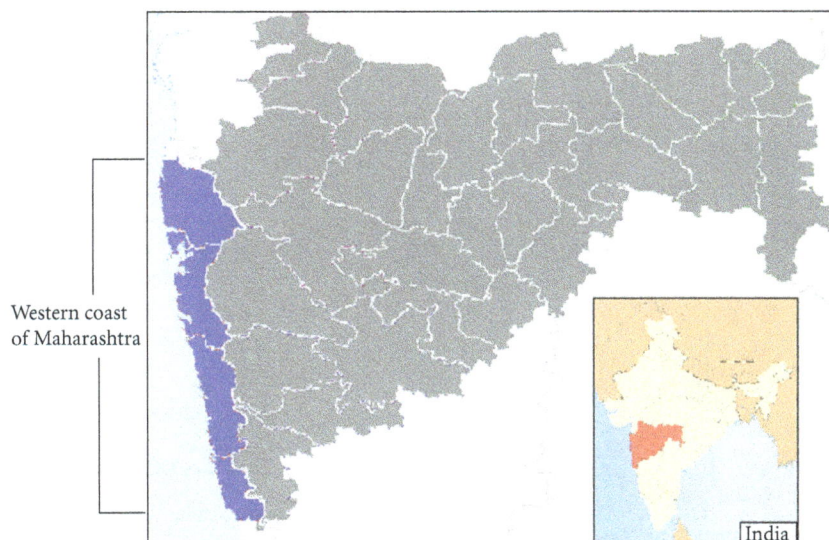

FIGURE 2: Map of western coast region of Maharashtra (INDIA).

bioactive compounds. The next step after isolation of microorganisms is screening. Ideally, selective procedure would allow the detection and isolation of microorganisms producing the desired metabolite. This primary screening should be rapid, inexpensive, predictive, specific, but effective over a broad range and should be applicable to large scale. Sometimes primary screening is time consuming and labour intensive when a large number of isolates have to be screened to identify a few potential ones. However, this is possibly the most critical step since it eliminates the large bulk of unwanted isolates, which are either nonproducers or producers of known compounds.

Screening and isolation method for long chain PUFA by marine protistan was reported but judged unsuitable by Bowles et al. for high throughput screening. The H_2O_2 plate assay method can be applied to a wide range of bacterial samples collected from different regions or from different animal sources. During screening of random hundreds of different marine bacterial samples, we have successfully discovered new strains of marine microorganisms with the special characteristic to produce PUFAs.

Generally, PUFAs are the molecules which are most susceptible to oxygen and reactive oxygen species (ROS) [10].

There are facts that PUFAs are stable when they are in vivo against oxidative stresses caused by ROS. This study was based on application of antioxidative effect of PUFA against ROS (Figure 1), for rapid screening of large number of marine isolates. However, no information regarding the screening of PUFAs producing marine bacteria has been reported. In the following strategy, we have presented qualitative method for rapid screening of PUFA producers.

2. Experimental Procedures

2.1. Materials. All the chemicals and media components used in the present study were AR grade and purchased from Hi Media Ltd, Mumbai, India. Sodium azide was purchased from S.D. Fine Chemicals Limited, Mumbai, India.

2.2. Media and Culture Conditions for Marine Microorganisms

2.2.1. Sample Collection. Various samples were collected from different regions of western coast of Maharashtra, India (Figure 2) and were brought to our laboratory for further investigation.

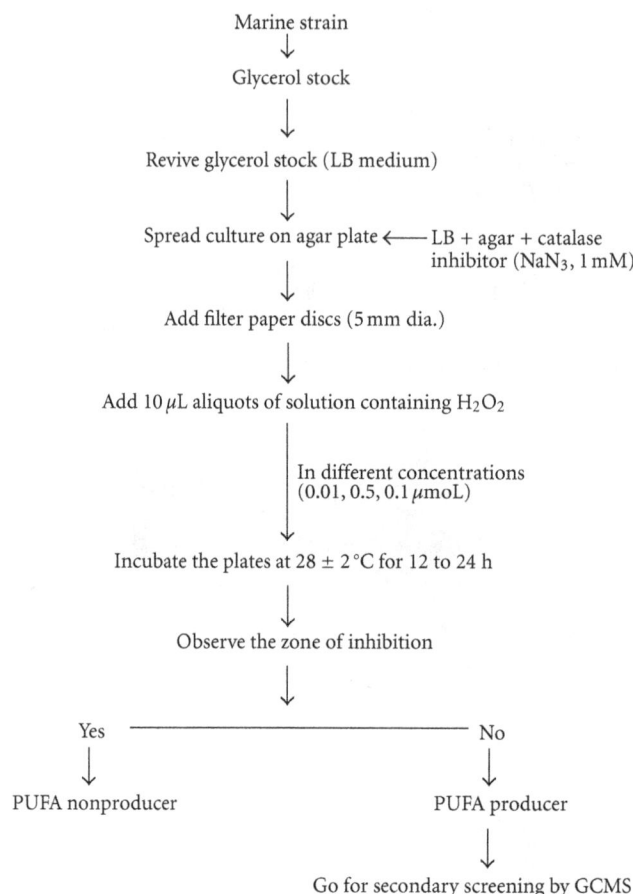

FIGURE 3: Protocol for primary screening of marine isolates.

2.2.2. Isolation of Marine Bacteria. The collected samples were serially diluted and plated over nutrient agar plates. The inoculated plates were incubated at $28 \pm 2°C$ for 24–48 h. After incubation, the selected bacterial colonies were purified and subcultured on nutrient agar medium for further investigation. All the marine isolates were preserved in glycerol stocks under $(-)$ 20°C.

2.3. Primary Screening of Marine Bacterial Isolates by H_2O_2-Plate Assay. About 100 isolated marine strains were randomly selected and screened for PUFA production (Figure 3). All the selected marine cultures were cultivated in Luria-Bertani (LB) medium (1% Tryptone, 0.5% Yeast Extract, 1% NaCl) at $28 \pm 2°C$ for 24 h, 180 rpm. To this medium, 0.5% NaCl was added for proper growth of marine isolates rather than using artificial seawater [11]. To find out response of bacteria to H_2O_2, bacterial culture reaching an optical density at 660 nm of 1.0 was used and spreaded over plate containing LB medium and sodium azide (NaN$_3$, 1 mM). On a surface filter paper discs of diameter 5 mm were placed and ten microliter aliquots of solution containing different concentration of H_2O_2 (0.01, 0.5, 1.0% prepared from 30% stock solution) were added on filter disc. Plates were then

incubated at $28 \pm 2°C$ for 24 h. Zone of inhibition was observed and further confirmation of PUFA producers was done by GCMS.

2.4. Secondary Screening and Confirmation

2.4.1. Media and Cultivation Conditions. Amongst all screened marine bacterial cultures, the positive strains were selected on the basis of no zone of inhibition (PUFA producers) and used further for secondary screening and confirmation for the production of PUFA. A loopful of the bacterial culture grown on previously preserved LB medium slants was transferred into 20 mL of LB broth in a 100 mL Erlenmyer flask and left on a shaker at $28 \pm 2°C$ for 24 h at 180 rpm. One mL of culture was transferred into 50 mL fresh sterilized LB broth and left on shaker at $28 \pm 2°C$ for 24 h at 180 rpm. Cell biomass was harvested from fermented media by centrifugation at 8000 rpm at $25 \pm 2°C$ for 10 min. The biomass was washed thoroughly with distilled water and was finally dried at 50°C overnight.

2.4.2. Direct Lipid Extraction and Esterification of Fatty Acids. A rapid direct extraction and esterification method was

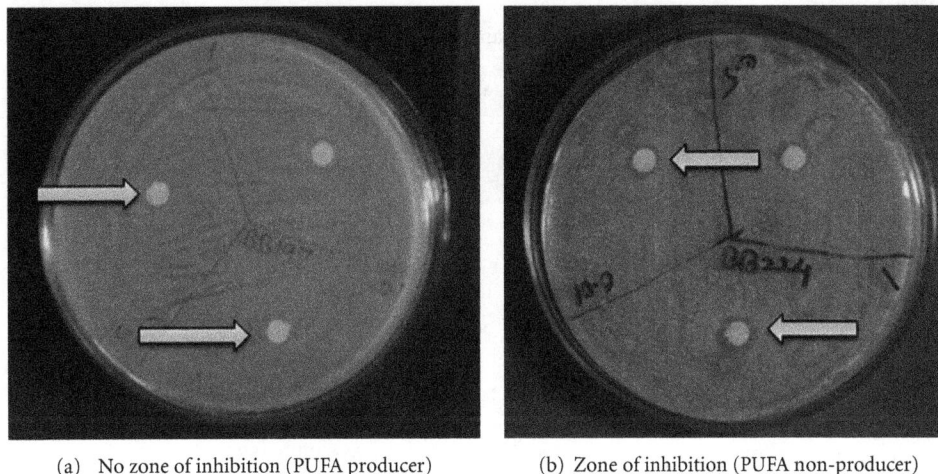

(a) No zone of inhibition (PUFA producer)   (b) Zone of inhibition (PUFA non-producer)

FIGURE 4: H_2O_2-plate assay.

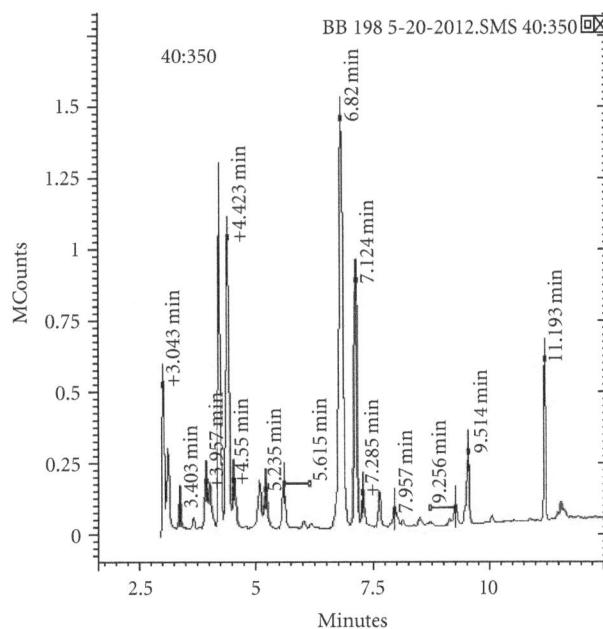

FIGURE 5: Gas chromatogram showing fatty acid profile of a selected marine isolate after primary screening.

developed and applied in this study. Extraction of lipids from biomass was performed according to the modified procedure of Hoshi et al. [12]. Lipids were extracted for 3-4 h with a 3 times volume of chloroform/methanol (2 : 1, v/v) containing 15 mg BHT for prevention of oxidation of PUFAs [13]. Esterification was done by adding 0.2 mL of 20 mM cupric acetate monohydrate in methanol and 1 mL of 0.5 N HCl in methanol was added, and the mixture was left for the specified time (2-3 h) at room temperature or at $28 \pm 2°C$. The reaction was stopped with the addition of 0.4 mL of water. The lower chloroform layer was pooled and then evaporated and was concentrated by rotary evaporator (Buchi Rotavapor) at 35°C. Finally the fatty acid methyl esters (FAME) were dissolved in hexane (1 mL), filtered through Whatman No. 1 filter paper, and analysed by GCMS.

2.4.3. Analysis PUFA by GCMS. Analysis of the FAME was performed by GCMS using slightly modified procedure [14]. All compounds were identified by comparison of their retention times with those of known standards and confirmed by GCMS using Varian 220-MS ion trap mass spectrometer (Varian, Inc. Walnut Creek, CA) connected to Varian 450-GC equipped with CP-SIL 88 capillary column (25 m × 0.25 mm i.d. × 0.39 mm OD, Varian). The injector was maintained at 250°C, and the column oven was programmed to increase from 160 to 220°C at 7°C min and then maintained at 220°C for 10 min. Split ratio was adjusted to 1 : 20. Helium was used as the carrier gas and flow rate was maintained at 1 mL/min. GCMS was operated at an ionization voltage of 70 eV and trap temperature at 220°C with mass range of 40–350 atomic mass units.

TABLE 1: Screening of marine strains using H_2O_2-plate assay method.

(a)

No.	Strain	$H_2O_2^@$ (%)			PUFA*	No.	Strain	$H_2O_2^@$ (%)			PUFA*
		0.1	0.5	1.0	+ve/−ve			0.1	0.5	1.0	+ve/−ve
1	JAY B 8	−	−	−	−ve	26	TH B 9	−	−	−	−ve
2	TAV B 31	−	−	−	−ve	27	DI BB 152	−	−	+	+ve
3	PA BB 8	−	−	−	−ve	28	**TH BB 21**	+	++	++	**+ve**
4	TH BB 31	−	−	−	−ve	29	ALI B 71	−	−	−	−ve
5	DIG BB 113	−	−	−	−ve	30	R BB 11	+	+	−	+ve
6	TH BB 22	−	−	−	−ve	31	BB 82	−	−	−	−ve
7	PA BB 3	−	−	−	−ve	32	JAY B 12	−	−	−	−ve
8	TH BB 15	−	−	−	−ve	33	TH BB 22	−	−	−	−ve
9	P BB 13	−	−	−	−ve	34	MAN BB 358	−	−	−	−ve
10	**DIG BB 4**	+	+	+	**+ve**	35	P BB 9	−	−	−	−ve
11	**M BB 318**	−	+	++	**+ve**	36	TH 33	−	−	−	−ve
12	DI BB 146	−	−	+	+ve	37	R 23	−	−	−	−ve
13	GP B 2	−	+	++	+ve	38	RAT B 18	−	−	−	−ve
14	AH B 53	−	+	++	+ve	39	RA 21	−	−	−	−ve
15	**MD BB 345**	+	+	++	**+ve**	40	MD BB 325	−	−	−	−ve
16	PA BB 8	−	+	++	+ve	41	SHI BB 2	−	−	−	−ve
17	TH BB 5	−	+	+	+ve	42	PA B 12	−	−	−	−ve
18	MAN BB 200	−	−	−	−ve	43	PA BB 4	−	−	−	−ve
19	MAN BB 361	−	−	−	−ve	44	TH BB 27	−	−	−	−ve
20	AL BB 156	−	+	++	+ve	45	TH BB 25	−	−	−	−ve
21	TH BB 20	−	+	++	+ve	46	**BB 227**	+	+	++	**+ve**
22	**TAV B 27**	+	+	++	**+ve**	47	TH BB 30	−	−	−	−ve
23	**ALI B 51**	+	+	++	**+ve**	48	MAN 10	−	−	−	−ve
24	BB 232	−	+	+	+ve	49	BB 72	−	−	−	−ve
25	MAN 353	−	−	+	+ve	50	R 37	−	−	−	−ve

$^@$describes zone of inhibition due to presence H_2O_2; +/++ describes growth of microorganisms or no zone of inhibition due to presence of PUFA.
*PUFA +ve denotes PUFA producer and −ve denotes PUFA nonproducer.
Note: highlighted strains were further screened by secondary analysis.

(b)

No.	Strain	$H_2O_2^@$ (%)			PUFA*	No.	Strain	$H_2O_2^@$ (%)			PUFA*
		0.1	0.5	1.0	+ve/−ve			0.1	0.5	1.0	+ve/−ve
51	**TH BB 19**	++	++	+	**+ve**	76	MD BB 312	−	−	−	−ve
52	**BB 83**	+	+	+	**+ve**	77	MAN BB 63	−	−	−	−ve
53	RAT B 3	−	−	−	−ve	78	TH BB 28	−	−	−	−ve
54	AK BB 189	−	−	−	−ve	79	JAY B 38	−	−	−	−ve
55	BB 201	−	−	−	−ve	80	BB 20	+	−	−	+ve
56	MAN 352	−	−	−	−ve	81	P BB 22	−	−	−	−ve
57	TH 33	+	−	−	+ve	82	RAT B 85	−	−	−	−ve
58	**BB 198**	++	+	+	**+ve**	83	SHI BB 2	−	−	−	−ve
59	R 13	−	−	−	−ve	84	PA BB 33	−	−	−	−ve
60	HAR BB 118	−	−	−	−ve	85	MD BB 122	−	−	−	−ve
61	RO BB 24	+	−	−	+ve	86	MAN BB 229	+	−	−	+ve
62	TH BB 24	−	−	−	−ve	87	TH BB 82	−	−	−	−ve
63	SRK 6	−	−	−	−ve	88	DIG BB 122	−	−	−	−ve
64	BB 245	−	−	−	−ve	89	G BB 89	−	−	−	−ve
65	BB 86	−	−	−	−ve	90	MAN 63	−	−	−	−ve
66	G BB 249	−	−	−	−ve	91	AL BB 148	−	−	−	−ve
67	DIG BB 108	−	−	−	−ve	92	TAV B 47	−	−	−	−ve
68	MD BB 338	−	−	−	−ve	93	R 39	−	−	−	−ve

(b) Continued.

No.	Strain	$H_2O_2^@$ (%)			PUFA*	No.	Strain	$H_2O_2^@$ (%)			PUFA*
		0.1	0.5	1.0	+ve/−ve			0.1	0.5	1.0	+ve/−ve
69	TH BB 166	−	−	−	−ve	94	BB 289	+	−	−	+ve
70	MAN BB 364	−	−	−	−ve	95	JAY B 23	−	−	−	−ve
71	SRK 23	−	−	−	−ve	96	DI BB 278	−	−	−	−ve
72	TH BB 83	−	−	−	−ve	97	TH B 34	−	−	−	−ve
73	BB 39	−	−	−	−ve	98	AK BB 148	−	−	−	−ve
74	TAV B 29	−	−	−	−ve	99	HAR BB 72	−	−	−	−ve
75	SHE BB 29	−	−	−	−ve	100	RO BB 41	−	−	−	−ve

$^@$describes zone of inhibition due to presence H_2O_2; +/++ describes growth of microorganisms or no zone of inhibition due to presence of PUFA.
*PUFA +ve denotes PUFA producer and −ve denotes PUFA nonproducer.
Note: highlighted strains were further screened by secondary analysis.

TABLE 2: Fatty acid profile of marine isolate produced after fermentation in secondary screening.

Fatty acid methyl esters (MUFA + PUFA)	Selected strains for secondary screening and their fatty acid profile									
	M BB 318	BB 83	ALI B 51	BB 227	MD BB 345	BB 198	DIG BB 4	TAV B 27	TH BB 19	TH BB 21
Pentadecanoic acid	+	+	+	+	+	+	−	−	−	−
Hexadecanoic acid	+	+	+	+	+	+	+	+	+	+
14-methylhexadecanoate	+	−	+	+	+	+	−	−	−	−
Margaric acid	+	+	+	+	+	+	−	−	−	−
Oleic acid	+	+	+	+	+	+	−	+	+	+
16-Octadecenoic acid	−	−	−	−	−	−	+	−	−	−
7,10-Octadecadienoic acid	−	−	−	−	−	−	+	−	−	−
Linoleic acid	+	+	+	+	+	+	−	+	+	+
7,10,13-Hexadecatrienoic acid	−	−	−	−	−	−	+	−	−	−
γ-Linolenic acid	−	−	−	−	−	+	−	−	−	−
Linolenic acid	+	+	+	+	+	+	−	+	+	+
Eicosanoic acid	+	−	−	−	−	−	−	−	−	−
Arachidonic acid	−	−	−	−	−	+	−	−	−	−
Eicosapentaenoic acid	−	−	−	−	−	+	−	−	−	−

Note: MUFA- monounsaturated fatty acid.

3. Results and Discussion

3.1. Primary Screening of Marine Bacterial Isolates by H_2O_2-Plate Assay.
In H_2O_2-plate assay method, the cells which are susceptible to externally-added H_2O_2 cannot able to grow suitably and thus shows a zone of inhibition which is dependant on added concentration of H_2O_2 on filter paper disc. The diameter of zone of inhibition is directly proportional to the concentration of added H_2O_2. The contradictory situation was observed for bacterial cells which produce PUFA. These cells able to grow in presence of added H_2O_2 on filter paper disc. As shown in Figure 4(a) the bacteria were grown even in presence of H_2O_2, due to the membrane-shielding effects of PUFAs. In most cases, PUFAs are among the molecules most vulnerable to oxygen and ROS, Okuyama et al. [10]. In Figure 4(b) the bacterial cells which were PUFA deficient or nonproducers were hampered by H_2O_2 and hence could not able to grow where zone of inhibition was directly proportional to the concentration of H_2O_2 added. Higher the concentration of H_2O_2 more will be the zone of inhibition.

To confirm that, growth of bacteria in presence of H_2O_2 is in reality mainly due to presence of PUFA; NaN_3 was added in to the media which is a very powerful inhibitor of catalase. If microorganism is producing catalase enzyme, NaN_3 inhibits catalase enzyme [15] which helps in interfering and promotes the actual interpretation of plate assay. About 1 mM concentration of NaN_3 was used during experimental study. The concentration should be an adequate amount to act as catalase inhibitor at the same time should not be antimicrobial in nature. The concentration of NaN_3 was decided from reported studied by Teixeira and Mota [16]. Out of selected 100 strains, 26 strains were found to give false positive results (Tables I(a) and I(b)). Out of 26, 10 strains were selected which gave false positive results at all H_2O_2 concentrations used during plate assay method. They were screened for further secondary analysis and confirmation.

EPA- and DHA-expressing bacteria were reported to be more resistant to exogenous H_2O_2 [10]. But there were no investigations whether other long chain (LC) PUFAs than EPA and DHA have similar effects. The bacterial cell gets protected by the effect of EPA that has reported earlier [10].

The membrane-shielding effects of n-3 LC-PUFAs have been shown only for bacterial cells producing EPA [17, 18]. From the study we have observed that other than EPA and DHA other PUFA might be responsible for the same protecting effect of exogenous H_2O_2 and our hypothesis was well supported with studied carried out by Okuyama et al. [10].

As shown in Figure 5, the fatty acid profile obtained from the bacterial culture which was able to grow in presence of H_2O_2 shows mainly AA production along with EPA. Hence, from this study not only EPA and DHA but also other PUFAs like AA act as the shield molecules against such oxidative challenges exogenously and endogenously raised in marine environments. There are some reports on protective effect of DHA against the external hydrogen peroxide [19]. It was reported that PUFAs including n-3 PUFAs are the molecules which are most susceptible to oxygen and reactive oxygen species (ROS). But from our study, it was observed that not only n-3 PUFAs are susceptible to oxygen and reactive oxygen species but also n-6 PUFAs like AA may behave in same way [20].

3.2. Secondary Screening and Confirmation. From few selected false positive strains all strains were found to give remarkable response when their lipid extract was injected into GCMS. The mass spectra of all selected strains were found to produce different kind of fatty acids. Some of them were found to produce AA and EPA which are very essential PUFAs from nutritional point of view. The fatty acid profile of 10 selected strains was shown in Table 2 that provides the information about different fatty acids produced by selected marine bacteria after primary screening using H_2O_2-plate assay. Hence this method is a qualitative estimation of PUFAs produced from microorganisms.

4. Conclusion

Thus, the present investigation has clearly revealed the presence of PUFAs in marine bacteria in the marine samples by newly developed novel simplified and rapid plate culture method for screening of PUFAs-producing marine bacteria; collected from different regions of western coast of Maharashtra, India. GCMS analysis studies confirmed the actual production of PUFA in various selected marine bacteria after primary screening. In order to minimise the time required for analysis as well as economic loss, this method gives the suitable solution for large number of samples screening which are abundant in the marine environment.

Acknowledgment

Authors are thankful to the University Grants Commission, Government of India, India, for providing financial assistance during the course of this research.

References

[1] C. A. Drevon, I. Baksaas, and H. E. Krokan, *Omega-3 Fatty Acids: metabolism and Biological Effects*, Birkhäuser, Basel, Switzerland, 1993.

[2] A. P. Simopoulos, A. Leaf, and N. Salem, "Essentiality of and recommended dietary intakes for omega-6 and omega-3 fatty acids," *Annals of Nutrition and Metabolism*, vol. 43, no. 2, pp. 127–130, 1999.

[3] M. A. Crawford, "Placental delivery of arachidonic and docosahexaenoic acids: implications for the lipid nutrition of preterm infants," *American Journal of Clinical Nutrition*, vol. 71, no. 1, pp. 275S–284S, 2000.

[4] S. Bergstrom, H. Danielsson, and B. Samuelsson, "The enzymatic formation of prostaglandin E2 from arachidonic acid," *Biochimica et Biophysica Acta*, vol. 90, pp. 207–210, 1964.

[5] D. A. Van Dorp, R. K. Beerthuis, D. H. Nugteren, and H. Vonkeman, "The biosynthesis of prostaglandins," *Biochimica et Biophysica Acta*, vol. 90, no. 1, pp. 204–207, 1964.

[6] B. A. Jakschik, A. R. Sams, H. Sprecher, and P. Needleman, "Fatty acid structural requirements for leukotriene biosynthesis," *Prostaglandins*, vol. 20, no. 2, pp. 401–410, 1980.

[7] H. Okuyama, T. Kobayashi, and S. Watanabe, "Dietary fatty acids—The N-6/N-3 balance and chronic elderly diseases. Excess linoleic acid and relative N-3 deficiency syndrome seen in Japan," *Progress in Lipid Research*, vol. 35, no. 4, pp. 409–457, 1996.

[8] J. Kobayashi, "Bioactive products from marine micro-and macro-organisms," *Pure and Applied Chemistry*, vol. 71, no. 6, pp. 1123–1126, 1999.

[9] R. D. Bowles, A. E. Hunt, G. B. Bremer, M. G. Duchars, and R. A. Eaton, "Long-chain n-3 polyunsaturated fatty acid production by members of the marine protistan group the thraustochytrids: screening of isolates and optimisation of docosahexaenoic acid production," *Journal of Biotechnology*, vol. 70, no. 1–3, pp. 193–202, 1999.

[10] H. Okuyama, Y. Orikasa, and T. Nishida, "Significance of antioxidative functions of eicosapentaenoic and docosahexaenoic acids in marine microorganisms," *Applied and Environmental Microbiology*, vol. 74, no. 3, pp. 570–574, 2008.

[11] J. D. Oliver and R. R. Colwell, "Extractable lipids of gram negative marine bacteria: fatty acid composition," *International Journal of Systematic Bacteriology*, vol. 23, no. 4, pp. 442–458, 1973.

[12] M. Hoshi, M. Williams, and Y. Kishimoto, "Esterification of fatty acids at room temperature by chloroform methanolic HCl cupric acetate," *Journal of Lipid Research*, vol. 14, no. 5, pp. 599–601, 1973.

[13] M. H. Cheng, T. H. Walker, G. J. Hulbert, and D. R. Raman, "Fungal production of eicosapentaenoic and arachidonic acids from industrial waste streams and crude soybean oil," *Bioresource Technology*, vol. 67, no. 2, pp. 101–110, 1999.

[14] X. Song, X. Zhang, C. Kuang, L. Zhu, and N. Guo, "Optimization of fermentation parameters for the biomass and DHA production of *Schizochytrium limacinum* OUC88 using response surface methodology," *Process Biochemistry*, vol. 42, no. 10, pp. 1391–1397, 2007.

[15] H. C. Lichstein and M. H. Soule, "Studies of the effect of sodium azide on microbic growth and respiration. II. The action of sodiuim azide on bacterial catalase," *Journal of Bacteriology*, vol. 47, no. 3, pp. 231–238, 1944.

[16] J. A. Teixeira and M. Mota, "Determination of catalase activity and its inhibition by a simple manometric method," *Biochemical Education*, vol. 20, no. 3, pp. 174–175, 1992.

[17] T. Nishida, Y. Orikasa, K. Watanabe, and H. Okuyama, "The cell membrane-shielding function of eicosapentaenoic acid for *Escherichia coli* against exogenously added hydrogen peroxide," *FEBS Letters*, vol. 580, no. 28-29, pp. 6690–6694, 2006.

[18] T. Nishida, N. Morita, Y. Yano, Y. Orikasa, and H. Okuyama, "The antioxidative function of eicosapentaenoic acid in a marine bacterium, *Shewanella marinintestina* IK-1," *FEBS Letters*, vol. 581, no. 22, pp. 4212–4216, 2007.

[19] S. Bechoua, M. Dubois, Z. Dominguez et al., "Protective effect of docosahexaenoic acid against hydrogen peroxide-induced oxidative stress in human lymphocytes," *Biochemical Pharmacology*, vol. 57, no. 9, pp. 1021–1030, 1999.

[20] B. Halliwell and J. M. Gutteridge, *Free Radicals in Biology and Medicine*, Oxford University Press, Oxford, UK, 3rd edition, 1998.

18

Isolation and Characterization of Some Phytochemicals from Indian Traditional Plants

Neeharika Srivastava, Aishwarya Singh Chauhan, and Bechan Sharma

Department of Biochemistry, Faculty of Science, University of Allahabad, Allahabad 211002, India

Correspondence should be addressed to Bechan Sharma, sharmabi@yahoo.com

Academic Editor: Ju-Kon Kim

The present study was designed to evaluate relative contribution of different polyphenols (total phenolics, flavonoids, flavonols) and their antioxidants activities in aqueous extracts of different parts of some plants; *Argemone mexicana, Datura metel, Calotropis procera, Thevetia peruviana,* and *Cannabis sativa.* The antioxidants (total phenolics, flavonoids, flavones) were determined by chemical methods. The antioxidant capacities of these extracts were evaluated by FRAP assay. The results demonstrated that phenolic content was maximally present in leaves of *T. peruviana.* This plant exhibited minimum phenolic content in its flower as compared to other plants. The flower of *D. metel* contained maximum phenolic content. The flavonoids were present in highest quantity in leaves of *C. procera* while *T. peruviana* flowers showed maximum flavonoid content. The fruits of *C. sativa* contained maximum quantity of flavonoid as compared to other plants tested. The flower extract of *C. sativa* possessed highest FRAP value followed by *A. mexicana* and fruit of *C. procera.* The values of ratios of different polyphenolic compounds present in plant extracts indicated that flower of *D. metel* contained maximum total flavonoids and minimum phenolics. These results suggested that levels of total phenolics, flavonoids and their FRAP indices exhibited specificity to different plants and their parts.

1. Introduction

The extraction of plant constituents is essential to isolate biologically active compounds and in understanding their role in disease prevention and treatment and in knowing their toxic effects as well. However, meager information is available about the medicinal and pharmacological properties and biological activities of phytochemicals derived from some plants (*Calotropis procera, Datura metal, Cannabis sativa, Argemone Mexicana,* and *Thevetia peruviana*) commonly known to have toxic, narcotic and ornamental properties. The information available on these important plants indicates that not much attention has been paid towards studying their physicochemical properties as well as biological activities towards their potentials as antioxidants. Keeping this information in view, an endeavour has been made in this communication to determine some biochemical constituents and their properties into the aqueous extracts of aforesaid medicinally important plants commonly available in the northern part of India.

Calotropis procera, known as apple of Sodom or mudar, belongs to Apocynaceae family and is found in many countries such as Africa and Western and South Asia, as well as Indochina. It is known for its medicinal and pharmacological properties [1]. The milky sap of this plant is known to contain three toxic glycosides: (i) calotropin, (ii) uscharin, and (iii) calotoxin as well as steroidal heart poisons, known as cardiac aglycones [2]. The crude extract of this plant and its protein fraction possess high fibrinolytic and anticoagulant activity in rabbit and human plasma [3]. Aqueous extracts of different parts of this plants are shown to exert mild diuretic and cardiac as well as respiratory stimulating effects in experimental animals [4].

Datura metal, a well-known traditional Indian plant, is found throughout the warmer parts of the world and contains both the ornamental and medicinal properties. All parts of *Datura* plants contain high levels of tropane alkaloids, which are highly toxic to humans and other animals. This plant is known to possess analgesic [5, 6], antioxidant, and antimicrobial properties [7].

Argemone mexicana, known as Mexican poppy or Mexican prickly poppy, belongs to the species of poppy found in Mexico and now in the United States, India, and Ethiopia. It is poisonous but has been used as a traditional medicine by natives of the western US and parts of Mexico [8]. It possesses the alkaloid sanguinarine reported to be responsible for epidemic dropsy [9, 10]. *A. mexicana* is reported to have antimicrobial activity [11], wound healing capacity in rat [12], larvicidal and chemosterilant activity [13], and nematicidal and allelopathic potential [14].

The root of *Cannabis sativa* is used as an old folk medicine to treat arthritis or joint pain [15]. *C. sativa* contains tetrahydrocannabinol (THC), which is of medical significance [16]. The cannabidiol isolated from *C. sativa* may act as an antipsychotic drug [17]. Terpenes have been detected and isolated from essential oil from flowers, leaves, and roots [18]. The terpenes are responsible for the flavor of different varieties of cannabis [19]. The roots, leaves, stems, pollen, and seeds of *C. sativa* contain piperidine and pyrrolidine, which have been used in different medicinal formulations [20].

Thevetia peruviana, a native in Iran, the Mediterranean regions, and India, produces milky juice. The leaves of this plant are used as a cardiotonic, antibacterial, and diuretic agent and used also to treat cutaneous eruptions and as an antidote against snake venom [21]. Its root is used for curing different types of cancers, ulcers, and leprosy while the root bark is used specifically against ring worm. The aqueous extracts of its leaves, branches, roots, and flowers are toxic to certain insects [22]. The main phytochemicals found in different parts of this plant include glycosides, terpenoids, cardiotonic substances, and steroids.

Oxidation and reduction of molecules are the common reactions in every cell leading to the production of free radicals and these free radicals react with organic substrates, namely, lipids, proteins, and DNA, causing damage to these molecules. Due to altered redox homoeostasis, an imbalance between the production and neutralization of the free radicals within the cell gives rise to oxidative stress (OS). It disturbs their normal functions resulting in the onset of variety of chronic diseases including cancer, heart disease, and other degenerative diseases. A wide variety of sources are present in biological systems producing reactive free radicals and reactive oxygen species. Considering the pathomechanism of free radicals, certain diseases are named as free radical diseases [23]. The organ systems most susceptible to damage are the eyes, brain, pulmonary, circulatory, and the reproductive systems [24–27].

The presence of the excess of oxidants (free radicals and nonradical reactive molecules) derived from free radicals such as reactive oxygen species (ROS): hydroxyl ($OH^•$), superoxide ($O_2^{•-}$), nitric oxide ($NO^•$), thiyl ($RS^•$), and peroxyl ($RO_2^•$)is known to cause oxidative stress. The nonradical ROS includes peroxynitrite ($ONOO^-$), hypochlorous acid (HOCl), hydrogen peroxide (H_2O_2), singlet oxygen ($-^1O_2$), ozone (O_3), and lipid peroxide (LOOH) as well as reactive nitrogen species (RNS) such as nitrous oxide (N_2O), nitrosyl cation (NO^+), peroxynitrite ($OONO^-$), nitrogen dioxide ($NO_2^•$), peroxynitrous acid (ONOOH),

dinitrogen trioxide (N_2O_3), nitroxyl anion (NO^-), nitrous acid (HNO_2), and nitryl chloride (NO_2Cl) [28, 29].

There are evidences which suggest that by quenching the free radicals, antioxidants help reduce the risk of chronic diseases. These antioxidants are either endogenous (internally synthesized) or exogenous (consumed). Nowadays, the application of plant based antioxidants or natural antioxidants is replacing synthetic molecules because of toxicities associated with the later [30, 31]. The phytochemicals like phenolic acids, polyphenols, flavonoids, flavonols, terpenoids vitamin C, vitamin E, carotenes, phenolic acids, phytate, and phytoestrogens scavenge the free radicals activity thus inhibiting the oxidative mechanisms that lead to emergence of various diseases as these molecules are electron rich. They donate electrons to ROS and neutralize these chemical species [32, 33].

In this study we have selected the aforesaid five plants, namely, *Calotropis procera*, *Datura metal*, *Cannabis sativa*, *Argemone Mexicana*, and *Thevetia peruviana* known for their varied properties. *C. procera* has the medicinal and toxic constituents, *D. metal* contains narcotics and toxic substances, *C. sativa* possesses some narcotics, *A. mexicana* causes dropsy, and *T. peruviana* has medicinal and ornamental applications. In our laboratory, some of the properties of the phytochemicals present in aqueous extracts of different parts (leaves, stem, flowers, and fruits) of these plants have been explored. In this paper, the presence of polyphenolic contents (total phenolics, flavonoids, and flavonols) in these aqueous extracts as well as their antioxidant potentials have been demonstrated.

2. Materials and Methods

2.1. Chemicals. 2,4,6-tripyridyl-*s*-triazine (TPTZ), ferrous sulphate, $AlCl_3$, and $FeCl_3$ were purchased from Sisco Research Laboratory; quercetin was purchased from Sigma Chemical Co. (St. Louis, MO, USA); Folin-Ciocalteu's phenol reagent and sodium carbonate were from Merck Chemical Supplies (Darmstadt, Germany). All the other chemicals used including the solvents were of analytical grade.

2.2. Collection and Identification of Plant Materials. Different parts (leaves, stem, flower, and fruits) of some plants such as *Argemone mexicana*, *Datura metal*, *Calotropis procera*, *Thevetia peruviana*, and *Cannabis sativa* were used in this study. These plant samples were collected from Allahabad and adjoining areas during March and April ($38 \pm 1°C$) in the year 2011.

2.3. Preparation of Plant Extracts. The fresh plant parts were collected washed with tap water followed by distilled water. 5.0 g of each was cut into several small pieces, minced well in a pestle and a mortar, and extracted with 50 mL of 50 mM Tris-HCl buffer at (pH 7.0). Freezing and thawing are done twice at the intervals of 2 h each followed by mechanical jerk by grinding in the pestle mortar in order to rupture the plant cell wall. The 10% (w/v) homogenate of each of the plant materials was prepared at 4–6°C. The homogenate was filtered using Whatman's filter paper type

1. The volume of the filtrate was recorded. The filtrate was centrifuged at 1000 xg for 10 min under cooling (4–6°C) conditions. The clear supernatant was used to estimate their antioxidant potential. The difference of the weights of the starting material and the residues was considered as the amount of the plant present in the extract.

2.4. Determination of Total Phenolics. Folin-Ciocalteu method as described elsewhere [34] was employed for estimation of total phenolics in the aqueous plant extracts. An aliquot (100 μL) of the extracts was mixed with 2.5 mL Folin-Ciocalteu reagent (previously diluted with water; 1 : 10 v/v) and 2 mL (75 g/L) of sodium carbonate. The tubes were vortexed for 15 s and allowed to stand for 30 min at 40°C for color development. The optical absorbance was recorded against reagent blank at 765 nm wavelength using the Thermoscientific Spectrascan UV2700 double beam spectrophotometer. The concentration of each plant extract was 0.1 g/mL. Total phenolic contents were expressed as mg/g n-propyl gallate equivalent.

2.5. Determination of Total Flavonoids. Determination of total flavonoid content was done using the method already described elsewhere [35]. In brief, a volume of 0.5 mL of 2% AlCl$_3$ in ethanol solution was added to 0.5 mL of plant extracts. After 1 h incubation at room temperature, the absorbance was measured at 420 nm. Appearance of yellow color indicated the presence of flavonoids. The extract samples were evaluated at a final concentration of 0.1 g/mL. Total flavonoid contents were calculated as quercetin equivalent (mg/g).

2.6. Determination of Total Flavonols. Total flavonols in the plant extracts were estimated by a known method described elsewhere [36]. In brief, 1.0 mL of 2% AlCl$_3$ in ethanol and 1.5 mL sodium acetate (50 g/L) solutions were added in 0.10 mL of extract solution. The absorption at 440 nm was monitored after 2.5 h of incubation at 20°C. The sample extracts were evaluated at a final concentration of 0.1 mg/mL. Total flavonoid content was calculated as quercetin equivalent (mg/g).

2.7. Total Antioxidant Activity Determination by Ferric Reducing Antioxidant Power (FRAP) Assay. The method as described by Benzie and Strain [37] with some modifications was employed for the estimation of antioxidant activity by FRAP assay. The stock solutions included 300 mM acetate buffer (pH 3.6), 10 mM 2,4,6-tripyridyl-s-triazine (TPTZ) solution in 40 mM HCl, and 1 mM FeCl$_3$ · 6H$_2$O solution. TPTZ was dissolved in 40 mM HCl at 50°C in water bath for 30–40 min till it completely dissolves. The fresh working solution was prepared by mixing 10 : 1 : 1 of acetate buffer, TPTZ, and FeCl$_3$ · 6H$_2$O, respectively. The temperature of the working solution was maintained to 37°C before starting the reaction by adding the plant extracts (100 μL) to 2 mL of the FRAP solution. The reaction mixture was incubated for 30 min in the dark condition. The optical absorbance of the colored product (ferrous tripyridyltriazine complex) was recorded at 593 nm. The standard curve was linear between

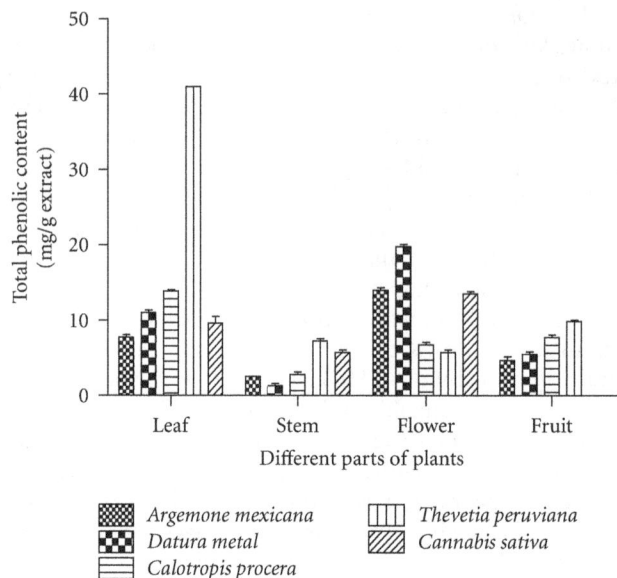

FIGURE 1: Comparative estimates of the analysis of total phenolic content in different parts of *A. mexiacana, D. metal, C. procera, T. peruviana,* and *C. sativa.* The determination of phenolics has been done as described in Section 2. The results indicate average values of three independent experiments.

20 and 100 μM FeSO$_4$ · 7H$_2$O. The results were expressed in μM Fe (II)/g dry mass.

3. Results

3.1. The Evaluation of Phenolics in the Aqueous Extracts of Different Parts of the Plants. The data obtained after analysis of total phenolics as shown in Figure 1 was largely variable not only among the plants but also among their various parts. In case of *A. mexicana,* the highest phenolic content was found in flowers (14 mg/g) followed by leaf, fruit, and stem, with the values being 7.5, 4.62, and 2.5 mg/g, respectively. The aqueous extracts of *D. metal* showed a similar pattern with maximum phenolic content present in flowers (19.75 mg/gm) followed by leaf, fruit, and stem, with the values being 11, 5.5, and 1.25 mg/g, respectively. In *C. procera,* maximum phenolic content was present in leaves (14 mg/g). Other parts of *C. procera* such as fruit, flower, and stem, had values of 7.7, 6.7, and 2.7 mg/g, respectively. Among the different parts of *T. peruviana,* maximum phenolic content was found in leaves (41 mg/g) followed by fruit, stem, and flower, with the values being 9.75, 7.3, and 5.75 mg/g, respectively. The flower of *C. sativa* contained maximum phenolic content (13.5 mg/g) followed by leaf and stem, with the values being 9.62 and 5.7 mg/g, respectively (Figure 1).

Upon comparison of all the five plants, the leaves of *T. peruviana* were found to have maximum phenolic content followed by leaves of *C. procera, D. metal, C. sativa,* and *A. mexicana.* Stems of all the plants did not contain a significant quantity of phenolics (Figure 1). In case of flowers, *D. metal* contained maximum phenolic content followed by that of *A.*

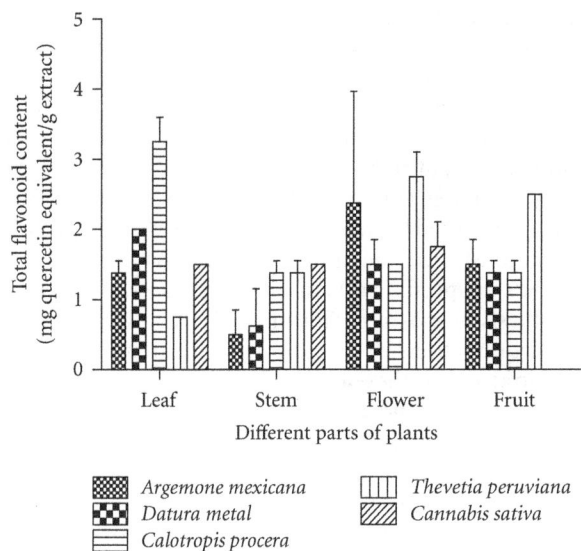

FIGURE 2: Comparative estimates of the analysis of total flavonoid content in different parts of *A. mexiacana, D.* metal, *C. procera, T. peruviana,* and *C. sativa.* The determination of flavonoids has been done as described in Section 2. The results indicate average values of three independent experiments.

mexicana, C. sativa, C. procera, and *T. peruviana.* The trend of total phenolic content in different parts of the selected plants is as the following: leaves: *T. peruviana* > *C. procera* > *D. metal* > *C. sativa* > *A. mexicana;* stem: *T. peruviana* > *C. sativa* > *C. procera* > *A. mexicana* > *D. metal;* flower: *D. metal* > *A. mexicana* > *C. sativa* > *C. procera* > *T. peruviana* and: fruit: *T. peruviana* > *C. procera* > *D. metal* > *A. mexicana* (Figure 1).

3.2. The Analysis of Flavonoids in the Aqueous Extracts of Different Parts of the Plants.
Like the phenolics in the aqueous extracts of aforesaid plants, the analysis of flavonoid was also carried out. As shown in Figure 2, different parts of the plants exhibited the presence of flavonoids, albeit to varying extents. In *A. mexicana,* the highest flavonoid content was found in flowers (2.37 mg/g) followed by fruit, leaf, and stem, with the values being 1.5, 1.37, and 0.5 mg/g, respectively. In *D. metal,* the highest flavonoid content was found in leaves (2 mg/g) followed by flower, fruit, and stem, with the values being 1.5, 1.37, and 0.62 mg/g, respectively. The *C. procera* leaves exhibited highest flavonoid content (3.25 mg/g) whereas its stem, flower, and fruit contained 1.25, 1.5, and 1.75 mg/g flavonoid, respectively. The *T. peruviana,* flowers and fruits were having nearly the same level of flavonoid contents (2.6 and 2.5 mg/g) whereas its stem and leaves contained 1.37 and 0.75 mg/g flavonoid, respectively. The flowers of *C. saiva* displayed the presence of highest flavonoid content (1.75 mg/g). Its leaves and stem had almost the same amount (1.5 mg/g) of flavonoid (Figure 2).

An organwise comparison of the five plants suggested that the maximum flavonoid content was found in the leaves of *C. procera* followed by *D. metal, C. sativa, A. Mexicana,* and *T. peruviana.* In the stem, *C. procera, T. peruviana,* and *C. sativa* had almost same flavonoid content while *A. mexicana*

and *D. metal* contained relatively low level of this molecule. The flowers and fruits of *T. peruviana* were found to contain maximum flavonoid content as compared to other plants tested. The trend of total flavonoid content in different parts of the selected plants is as the following: leaves: *C. procera* > *D. metal* > *C. sativa* > *A. mexicana* > *T. peruviana;* stem: *C. sativa* > *C. procera* = *T. peruviana* > *D. metal* > *A. mexicana;* flower: *T. peruviana* > *A. Mexicana* > *C. sativa* > *D. metal* = *C. procera:* fruit: *T. peruviana* > *A. mexicana* > *C. procera* > *D. metal* (Figure 2).

3.3. The Evaluation of Flavons in the Aqueous Extracts of Different Parts of the Plants.
The analysis of flavon content in these plants preparations indicated its presence in low quantity in *A. mexicana* only. Other plants tested exhibited absence of flavon in their aqueous extracts (data not shown).

3.4. The Evaluation of Ferric Reducing Antioxidant Power (FRAP) in the Aqueous Extracts of Different Parts of the Plants.
When these plant extracts were subjected to FRAP assay, flowers of *C. sativa* showed a significant antioxidant potential ($74.8 \pm 1.93 \mu M\ Fe^{++}g^{-1}$) among all the plants. *A. mexicana* flowers, *D. metal* leaves, fruits from *C. procera,* and *T. peruviana* showed maximum antioxidant capacity, with the values being 69.1 ± 0.28, 24.7 ± 1.13, 41.3 ± 1.20, and $24.2 \pm 0.31 \mu M\ Fe^{++}g^{-1}$, respectively (Table 1).

3.5. The Levels of Ratios of Polyphenolic Compounds in the Aqueous Extracts of Different Parts of the Plants.
The ratios of polyphenolic compounds present in these plant extracts are shown in Table 2. The data indicated varying levels of flavonoid dependent antioxidant activities in the extracts of different parts of the plants. The trend of presence of flavonoid was found to be as follows: leaves of *A. Mexicana* > flowers of *D. metal* > stem of *C. procera* > flowers of *T. peruviana* > stem of *C. sativa* (Table 2), while among the flowers from all the five plants, *D. metal* exhibited maximum levels and leaves of *T. peruviana* exhibited minimum levels.

4. Discussion

It is well known that plant polyphenols, the secondary metabolites, are widely distributed in the plant kingdom and that they are sometimes present in surprisingly high concentrations [38]. Phenolic compounds are characterized by the presence of several phenol groups. By donating a hydrogen atom or an electron they make them very reactive in neutralizing free radicals, chelating metal ions in aqueous solutions [39]. The results of present study indicated that the amount of phenolic contents varied not only plantwise but also from one part of the plant to another. The leaves of *T. peruviana* had maximum phenolic content as compared to the leaves of *C. procera, D. metal, C. sativa,* and *A. mexicana,* while flowers of *D. metal* contained maximum phenolic as compared to the other plants. Stems of all the plants did not contain any significant quantity of phenolics. The aqueous extract of *Datura stramonium* has been reported to exhibit different levels of phenolic contents in different parts of

TABLE 1: Total antioxidant activity of different plant preparations in terms of ferric reducing antioxidant power (FRAP).

S. number	Name of the plant	Plant part used	Extract type	FRAP (μM Fe^{++} g^{-1})
1		Leaf		33.6 ± 0.14
2	Argemone mexicana	Stem		8.40 ± 0.14
3		Flower		69.1 ± 0.28
4		Fruit		11.8 ± 0.07
5		Leaf		24.7 ± 1.13
6	Datura metal	Stem		9.60 ± 0.00
7		Flower		21.3 ± 0.28
8		Fruit		23.1 ± 0.77
9		Leaf		38.7 ± 0.43
10	Calotropis procera	Stem	Aqueous	11.3 ± 0.77
11		Flower		24.5 ± 0.42
12		Fruit		41.3 ± 1.20
13		Leaf		15.8 ± 0.07
14	Thevitia peruviana	Stem		21.5 ± 0.70
15		Flower		19.2 ± 0.21
16		Fruit		24.2 ± 0.31
17		Leaf		34.0 ± 0.02
18	Cannabis sativa	Stem		14.5 ± 0.35
19		Flower		74.8 ± 1.93

The FRAP values for different parts of the plants A. mexicana, D. metal, C. procera, T. peruviana and C. sativa have been determined as described in Section 2. The values are the average of three independent experiments.

the plant, with the values in leaf, seed, whole fruit, and stem, respectively, being 0.397, 0.277, 0.1, and 0.114 mg/g [40]. Liu et al. (2008) have shown total phenolic content into Cannabis fruit to be 0.57 ± 0.002 (mg GAE/g dw) [41].

While screening 70 medicinal plant extracts for their antioxidant capacity and total phenols, Katalinic et al. have reported the presence of phenolic contents in different plants to the varying extents [42]. These compounds act as free radical scavengers and thus help protect cells from oxidative toxicity [43–45]. Some workers have demonstrated the presence of phenolic compounds in the aerial parts of the plants including C. procera, T. peruviana, and C. sativa [46–48] but analysis of these phytoconstituents in the aqueous extracts of different specific parts of these plants has not been worked out.

It is reported that in the Fenton reaction, flavonoids as antioxidants interfere with the biochemical pathways which are involved in the generation of free radicals (ROS), quench them, chelate the transition metals, and make them redox inactive [49]. Commonly flavonoids occur as glycosides in plants and are considered to be very efficient as antioxidants. With different degrees of hydroxylation, oxidation, and substitution, the flavonoids have common diphenylpropane structure ($C_6C_3C_6$) [40, 50].

The results of the present study reflected that the quantity of flavonoids varies from one plant to another and also into different parts of the plants. Maximum flavonoid content was present in the leaves of C. procera and flowers as well as fruits of T. peruviana when compared to D. metal, C. sativa, A. Mexicana, and T. peruviana. The stem of these plants contained low amount of flavonoid. Liu et al. have reported flavonoids content to be absent in Cannabis fruit [41]. Very recently, the levels of flavonoids in different medicinal plants have been reported by de Queiroz Siqueira et al. [51] and they have demonstrated the similar distribution pattern of the flavonoids in specific parts of different plants. According to a hypothesis proposed by Tattini et al. [52], flavonoids have protective functions during drought. Ryan et al. [53] have demonstrated that these molecules impart photoprotection. Likewise, Barceló and Poschenrieder [54] have shown that flavonoids in plants may help ameliorate toxicity of aluminium as they grow in soils contaminated with this heavy metal. Thus in addition to acting as antioxidants, flavonoids are also involved in the regulation of various physiochemical behaviours of plants.

The antioxidant capacity of the plant extract largely depends on both the composition of the extract and the test system [55]. The FRAP assay [37, 56] measures antioxidant power with the help of an oxidant, that is, Fe^{3+}. Reduction of ferric to ferrous ion at low pH produces a coloured ferrous-tripyridyltriazine complex. In the FRAP assay, reductants (antioxidants) present in the sample reduce the Fe (III)/tripyridyltriazine complex to the blue ferrous form. The change in absorbance and FRAP value of the antioxidants is proportional to each other [56]. The FRAP assay in spite of being simple and inexpensive does have few drawbacks too like the antioxidant capacity of certain antioxidants cannot be measured accurately by this assay such as iron (II) and SH group-containing antioxidants [37, 56–58].

The evaluation of FRAP value has been made in the aqueous extracts of different parts of the five plants tested in the present study and the results suggested that flowers of A. mexicana and C. sativa, D. metal leaves, and fruits from C. procera and T. peruviana exhibited maximum antioxidant capacity. The trend of FRAP values obtained from different plant parts having maximum antioxidant potential for being used in various pharmacological preparations is flowers of C. sativa > flowers of A. mexicana> fruits of C. procera > leaves of D. metal > fruits of T. peruviana. These results are in agreement with those reported by Katalinic et al. [42]. In another species of Datura, that is, Datura stramonium, Oseni et al., (2011) have reported aqueous extracts of leaf, seed, whole fruit, and stem to exhibit 68.90, 25.60, 62.70, and 96.69% antioxidant properties, respectively, by using FRAP assay [40]. Ozgen et al. have presented a study on antioxidant properties of medicinal plants belonging to Asclepiadoideae family, in which Calotropis gigantea was reported to have 185.71, 186.13, and 93.07 mmol 100 g^{-1} distilled water, in root, flower, and leaf, respectively, using FRAP assay [57]. Liu et al. have shown FRAP value in Cannabis fruit to be 0.010 ± 0.001 (Fe(II) mmol/g dw) [41].

While evaluating the antioxidant potential of several medicinal plants using FRAP, Katalinic et al. have demonstrated the presence of antioxidant potential of plants to varying degrees. However, the values of FRAP determined

TABLE 2: Ratio of different polyphenolic compounds in the plant extracts (total flavonoids/total phenolics).

S. number	Name of the plant	Plant parts used	Extract type	Ratio of polyphenolic compounds (total flavonoids/total phenolics)
1		Leaf		0.331
2	Argemone mexicana	Stem		0.200
3		Flower		0.169
4		Fruit		0.324
5		Leaf		0.181
6	Datura metal	Stem		0.500
7		Flower		0.750
8		Fruit		0.251
9		Leaf		0.234
10	Calotropis procera	Stem	Aqueous	0.500
11		Flower		0.222
12		Fruit		0.209
13		Leaf		0.018
14	Thevitia peruviana	Stem		0.189
15		Flower		0.478
16		Fruit		0.253
17		Leaf		0.155
18	Cannabis sativa	Stem		0.260
19		Flower		0.129

by them in their tested medicinal plants were much higher than reported into five different plants under the present study [42, 59]. According to Adedapo et al., the extracts from plants, *Bidens pilosa* and *Chenopodium album,* prepared in acetone and methanol showed relatively high FRAP activity in comparison to the FRAP values obtained using aqueous extracts [60]. Tawaha et al. have recorded a large variation in the total antioxidant capacity of the aqueous and methanolic extracts of the selected Jordanian plant species analyzed [61], which could be attributed to substantial differences in the solubility of phytochemicals extracted into organic and aqueous solvents. The strong correlation observed in the present study between antioxidant activity, phenolics, and flavonoid content of different plants suggests a possible use of their partsin making the active ingredients of antioxidant supplement after removing their toxic ingredients, if any.

5. Conclusion

The results from the present study demonstrated that the leaves of *T. peruviana* contained the presence of the maximum phenolic content. This plant exhibited minimum phenolic content in its flower as compared to others. Phenolic contents were maximum in the flowers of *D. metal*. The flavonoids were present in highest quantity in the leaves of *C. procera* while the *T. peruviana* flowers showed maximum flavonoid content. The fruits of *C. sativa* contained maximum quantity of flavonoid as compared to other plants tested. The aqueous extract of the flower of *C. sativa* possessed highest FRAP value followed by the flower of *A. mexicana* and the fruit of *C. procera*. The values of ratios of different polyphenolic compounds present in the plant extracts indicated that the flower of *D. metal* contained maximum flavonoids and minimum phenolics. These results suggested that the levels of total phenolics and flavonoids contents as well their FRAP indices varied not only from one plant to the other but also in their different parts tested. These results indicated that despite the presence of some toxic ingredients, these plants contained high antioxidant activity and sufficient quantity of flavonoids and phenolics in their varying parts, which may be exploited for certain medicinal or pharmacological formulations.

Conflict of Interests

Authors do not have any conflict of interests or a direct financial relation with the commercial identity mentioned in the paper.

Acknowledgments

The authors (N. Srivastava and A. S. Chauhan) express their gratefulness to the University Grants Commission, New Delhi, India, for the financial support in the forms of a Junior Research Fellowship (UGC-JRF, NET) and a Research Fellowship (CRET-University of Allahabad), respectively.

References

[1] V. L. Kumar and S. Arya, "Medicinal uses and pharmacological properties of *Calotropis procera*," in *Recent Progress in Medicinal Plants*, vol. 11, pp. 373–388, Studium Press, Houston, Tex, USA, 2006.

[2] R. H. N. Chaudhuri, "Pharmacognostic studies on the roots of *Calotropis gigantea* R.Br.ex Ait," *Bull Bot Surv India*, vol. 3, pp. 171–173, 1961.

[3] G. N. Srivastava, R. N. Chakravarti, and S. H. Zaidi, "Studies on anticoagulant therapy. III. *In vitro* screening of some Indian plant latices for fibrinolytic and anticoagulant activity," *Indian Journal of Medical Sciences*, vol. 16, pp. 873–877, 1962.

[4] T. Devasari, "Toxic effects of *Calotropic procera*," *Indian Journal of Pharmacology*, vol. 27, pp. 272–275, 1965.

[5] Rajesh and G. L. Sharma, "Studies on antimycotic properties of *Datura metel*," *Journal of Ethnopharmacology*, vol. 80, no. 2-3, pp. 193–197, 2002.

[6] N. N. Wannang, H. C. Ndukwe, and C. Nnabuife, "Evaluation of the analgesic properties of the *Datura metel* seeds aqueous extract," *Journal of Medicinal Plant Research*, vol. 3, no. 4, pp. 192–195, 2009.

[7] F. C. Akharaiyi, "Antibacterial, phytochemical and antioxidant activities of *Datura metel*," *International Journal of PharmTech Research*, vol. 3, no. 1, pp. 478–483, 2011.

[8] R. S. Felger and M. B. Moser, *People of the Desert and Sea*, University of Arizona Press, Tucson, Ariz, USA, 1985.

[9] R. R. Dalvi, "Sanguinarine: its potential as a liver toxic alkaloid present in the seeds of *Argemone mexicana*," *Experientia*, vol. 41, no. 1, pp. 77–78, 1985.

[10] N. N. Sood, M. S. Sachdev, M. Mohan, S. K. Gupta, and H. P. Sachdev, "Epidemic dropsy following transcutaneous absorption of *Argemone mexicana* oil," *Transactions of the Royal Society of Tropical Medicine and Hygiene*, vol. 79, no. 4, pp. 510–512, 1985.

[11] A. A. Izzo, G. Di Carlo, D. Biscardi et al., "Biological screening of Italian medicinal plants for antibacterial activity," *Phytotherapy Research*, vol. 9, no. 4, pp. 281–286, 1995.

[12] G. K. Dash and P. N. Murthy, "Evaluation of *Argemone mexicana* Linn. Leaves for wound healing activity," *Journal of Natural Product and Plant Resources*, vol. 1, no. 1, pp. 46–56, 2011.

[13] M. Sakthivadivel and D. Thilagavathy, "Larvicidal and chemosterilant activity of the acetone fraction of petroleum ether extract from *Argemone mexicana* L. seed," *Bioresource Technology*, vol. 89, no. 2, pp. 213–216, 2003.

[14] S. S. Shaukat, I. A. Siddiqui, G. H. Khan, and M. J. Zaki, "Nematicidal and allelopathic potential of *Argemone mexicana*, a tropical weed: allelopathic and nematicidal potential of *Argemone mexicana*," *Plant and Soil*, vol. 245, no. 2, pp. 239–247, 2002.

[15] C. Bott and D. Bishop, "Frequently asked questions about Cannabis," in *The Eldorado County Chapter of the American Alliance for Medical Cannabis*, 2008.

[16] C. E. Turner, M. A. Elsohly, and E. G. Boeren, "Constituents of *Cannabis sativa* L. XVII. A review of the natural constituents," *Journal of Natural Products*, vol. 43, no. 2, pp. 169–234, 1980.

[17] A. W. Zuardi, J. A. S. Crippa, J. E. C. Hallak, F. A. Moreira, and F. S. Guimarães, "Cannabidiol, a *Cannabis sativa* constituent, as an antipsychotic drug," *Brazilian Journal of Medical and Biological Research*, vol. 39, no. 4, pp. 421–429, 2006.

[18] A. Hazekamp and J. T. Fischedick, "Cannabis-from cultivar to chemovar," *Drug Testing Analysis*, vol. 4, no. 7-8, pp. 660–667, 2012.

[19] M. A. Elsohly, C. E. Turner, C. H. Phoebe, J. E. Knapp, P. L. Schiff, and D. J. Slatkin, "Anhydrocannabisativine, a new alkaloid from *Cannabis sativa* L," *Journal of Pharmaceutical Sciences*, vol. 67, no. 1, article 124, 1978.

[20] I. J. Flores-Sanchez and R. Verpoorte, "Secondary metabolism in cannabis," *Phytochemistry Reviews*, vol. 7, no. 3, pp. 615–639, 2008.

[21] Publications and Information Directorate, Council of Scientific and Industrial Research, *The Wealth of India*, vol. 3, Publications and Information Directorate,Council of Scientific and Industrial Research, New Delhi, India, 1952.

[22] P. S. Vaidyaratnam, *Indian Medicinal Plants. A Compendium of 500 Species*, vol. 4, Orient Longman, Chennai, India, 1994.

[23] R. Kahl and A. G. Hildebrandt, "Methodology for studying antioxidant activity and mechanisms of action of antioxidants," *Food and Chemical Toxicology*, vol. 24, no. 10-11, pp. 1007–1014, 1986.

[24] A. Nadeem, H. G. Raj, and S. K. Chhabra, "Increased oxidative stress and altered levels of antioxidants in chronic obstructive pulmonary disease," *Inflammation*, vol. 29, no. 1, pp. 23–32, 2005.

[25] G. M. Somfai, B. Knippel, É. Ruzicska et al., "Soluble semicarbazide-sensitive amine oxidase (SSAO) activity is related to oxidative stress and subchronic inflammation in streptozotocin-induced diabetic rats," *Neurochemistry International*, vol. 48, no. 8, pp. 746–752, 2006.

[26] P. Studinger, B. Mersich, Z. Lénárd, A. Somogyi, and M. Kollai, "Effect of vitamin E on carotid artery elasticity and baroreflex gain in young, healthy adults," *Autonomic Neuroscience*, vol. 113, no. 1-2, pp. 63–70, 2004.

[27] K. Stadler, V. Jenei, G. von Bölcsházy, A. Somogyi, and J. Jakus, "Increased nitric oxide levels as an early sign of premature aging in diabetes," *Free Radical Biology and Medicine*, vol. 35, no. 10, pp. 1240–1251, 2003.

[28] B. Halliwell, "How to characterize a biological antioxidant," *Free Radical Research Communications*, vol. 9, no. 1, pp. 1–32, 1990.

[29] B. Halliwell and J. M. Gutteridge, *Free Radicals in Biology and Medicine*, Oxford University Press, New York, NY, USA, 1989.

[30] M. Cruz, D. Franco, J. M. Dominguez, J. Senerio, H. Dominguez, and M. J. Nunez Parajao, "Natural antioxidants form residual sources," *Food Chemistry*, vol. 72, no. 2, pp. 145–171, 2001.

[31] T. H. Tseng, E. S. Kao, C. Y. Chu, F. P. Chou, H. W. Lin Wu, and C. J. Wang, "Protective effects of dried flower extracts of *Hibiscus sabdariffa* L. against oxidative stress in rat primary hepatocytes," *Food and Chemical Toxicology*, vol. 35, no. 12, pp. 1159–1164, 1997.

[32] M. I. Gil, F. Ferreres, and F. A. Tomás-Barberán, "Effect of postharvest storage and processing on the antioxidant constituents (flavonoids and vitamin C) of fresh-cut spinach," *Journal of Agricultural and Food Chemistry*, vol. 47, no. 6, pp. 2213–2217, 1999.

[33] J. B. Harborne, H. Baxter, and G. P. Moss, *Phytochemical Dictionary: Handbook of Bioactive Compounds from Plants*, Taylor & Francis, London, UK, 2nd edition, 1999.

[34] K. Wolfe, X. Wu, and R. H. Liu, "Antioxidant activity of apple peels," *Journal of Agricultural and Food Chemistry*, vol. 51, no. 3, pp. 609–614, 2003.

[35] A. A. L. Ordoñez, J. D. Gomez, M. A. Vattuone, and M. I. Isla, "Antioxidant activities of *Sechium edule* (Jacq.) Swartz extracts," *Food Chemistry*, vol. 97, no. 3, pp. 452–458, 2006.

[36] A. Kumaran and R. J. Karunakaran, "*In vitro* antioxidant activities of methanol extracts of five *Phyllanthus* species from

India," *LWT—Food Science and Technology*, vol. 40, no. 2, pp. 344–352, 2007.

[37] I. F. F. Benzie and J. J. Strain, "The ferric reducing ability of plasma (FRAP) as a measure of "antioxidant power": the FRAP assay," *Analytical Biochemistry*, vol. 239, no. 1, pp. 70–76, 1996.

[38] J. B. Harborne, "New naturally occurring plant polyphenols," in *Polyphenolic Phenomena*, A. Scalbert, Ed., INRA, Paris, France, 1993.

[39] S. Petti and C. Scully, "Polyphenols, oral health and disease: a review," *Journal of Dentistry*, vol. 37, no. 6, pp. 413–423, 2009.

[40] O. A. Oseni, F. Igbe, and S. A. Olagboye, "Distribution of anti-nutrients and antioxidant properties in the plant of Thornap-ple (*Datura stramonium L*) *Solanaceae*," *Journal of Agriculture and Biological Sciences*, vol. 2, no. 6, pp. 136–140, 2011.

[41] H. Liu, N. Qiu, H. Ding, and R. Yao, "Polyphenols contents and antioxidant capacity of 68 Chinese herbals suitable for medical or food uses," *Food Research International*, vol. 41, no. 4, pp. 363–370, 2008.

[42] V. Katalinic, M. Milos, T. Kulisic, and M. Jukic, "Screening of 70 medicinal plant extracts for antioxidant capacity and total phenols," *Food Chemistry*, vol. 94, no. 4, pp. 550–557, 2006.

[43] P. K. J. P. D. Wanasundara and F. Shahidi, "Process-induced changes in edible oils," *Advances in Experimental Medicine and Biology*, vol. 434, pp. 135–160, 1998.

[44] J. León, M. A. Lawton, and I. Raskin, "Hydrogen peroxide stimulates salicylic acid biosynthesis in tobacco," *Plant Physiology*, vol. 108, no. 4, pp. 1673–1678, 1995.

[45] R. M. Costa, A. S. Magalhães, J. A. Pereira et al., "Evaluation of free radical-scavenging and antihemolytic activities of quince (*Cydonia oblonga*) leaf: a comparative study with green tea (*Camellia sinensis*)," *Food and Chemical Toxicology*, vol. 47, no. 4, pp. 860–865, 2009.

[46] A. P. Oliveira, P. Valentão, J. A. Pereira, B. M. Silva, F. Tavares, and P. B. Andrade, "*Ficus carica* L.: metabolic and biological screening," *Food and Chemical Toxicology*, vol. 47, no. 11, pp. 2841–2846, 2009.

[47] F. Aqil, I. Ahmad, and Z. Mehmood, "Antioxidant and free radical scavenging properties of twelve traditionally used Indian medicinal plants," *Turkish Journal of Biology*, vol. 30, no. 3, pp. 177–183, 2006.

[48] F. Pourmorad, S. J. Hosseinimehr, and N. Shahabimajd, "Anti-oxidant activity, phenol and flavonoid contents of some selected Iranian medicinal plants," *African Journal of Biotech-nology*, vol. 5, no. 11, pp. 1142–1145, 2006.

[49] I. I. Koleva, T. A. van Beek, J. P. H. Linssen, A. de Groot, and L. N. Evstatieva, "Screening of plant extracts for antioxidant activity: a comparative study on three testing methods," *Phy-tochemical Analysis*, vol. 13, no. 1, pp. 8–17, 2002.

[50] R. K. Sharma, S. Chatterji, D. K. Rai et al., "Antioxidant activ-ities and phenolic contents of the aqueous extracts of some Indian medicinal plants," *Journal of Medicinal Plant Research*, vol. 3, no. 11, pp. 944–948, 2009.

[51] C. F. de Queiroz Siqueira, D. L. V. Cabral, T. J. d. S. P. Sobrinho, E. L. C. de Amorim, J. G. de Melo, T. A. d. S. Araujo et al., "Levels of tannins and flavonoids in medicinal plants: evaluating bioprospecting strategies," *Evidence-Based Complementary and Alternative Medicine*, vol. 2012, Article ID 434782, 7 pages, 2012.

[52] M. Tattini, C. Galardi, P. Pinelli, R. Massai, D. Remorini, and G. Agati, "Differential accumulation of flavonoids and hydrox-ycinnamates in leaves of *Ligustrum vulgare* under excess light and drought stress," *New Phytologist*, vol. 163, no. 3, pp. 547–561, 2004.

[53] K. G. Ryan, E. E. Swinny, K. R. Markham, and C. Winefield, "Flavonoid gene expression and UV photoprotection in trans-genic and mutant *Petunia* leaves," *Phytochemistry*, vol. 59, no. 1, pp. 23–32, 2002.

[54] J. Barceló and C. Poschenrieder, "Fast root growth responses, root exudates, and internal detoxification as clues to the mechanisms of aluminium toxicity and resistance: a review," *Environmental and Experimental Botany*, vol. 48, no. 1, pp. 75–92, 2002.

[55] S. Surveswaran, Y. Z. Cai, J. Xing, H. Corke, and M. Sun, "Antioxidant properties and principal phenolic phytochem-icals of Indian medicinal plants from Asclepiadoideae and Periplocoideae," *Natural Product Research*, vol. 24, no. 3, pp. 206–221, 2010.

[56] F. L. Song, R. Y. Gan, Y. Zhang, Q. Xiao, L. Kuang, and H. B. Li, "Total phenolic contents and antioxidant capacities of selected chinese medicinal plants," *International Journal of Molecular Sciences*, vol. 11, no. 6, pp. 2362–2372, 2010.

[57] M. Ozgen, R. N. Reese, A. Z. Tulio, J. C. Scheerens, and A. R. Miller, "Modified 2,2-azino-bis-3-ethylbenzothiazoline-6-sulfonic acid (ABTS) method to measure antioxidant capacity of selected small fruits and comparison to ferric reduc-ing antioxidant power (FRAP) and 2,2'-diphenyl-1-picryl-hydrazyl (DPPH) methods," *Journal of Agricultural and Food Chemistry*, vol. 54, no. 4, pp. 1151–1157, 2006.

[58] B. Ou, D. Huang, M. Hampsch-Woodill, J. A. Flanagan, and E. K. Deemer, "Analysis of antioxidant activities of common vegetables employing oxygen radical absorbance capacity (ORAC) and ferric reducing antioxidant power (FRAP) assays: a comparative study," *Journal of Agricultural and Food Chem-istry*, vol. 50, no. 11, pp. 3122–3128, 2002.

[59] K. E. Heim, A. R. Tagliaferro, and D. J. Bobilya, "Flavonoid antioxidants: chemistry, metabolism and structure-activity relationships," *Journal of Nutritional Biochemistry*, vol. 13, no. 10, pp. 572–584, 2002.

[60] A. Adedapo, F. Jimoh, and A. Afolayan, "Comparison of the nutritive value and biological activities of the acetone, methanol and water extracts of the leaves of *Bidens pilosa* and *Chenopodium album*," *Acta Poloniae Pharmaceutica—Drug Research*, vol. 68, no. 1, pp. 83–92, 2011.

[61] K. Tawaha, F. Q. Alali, M. Gharaibeh, M. Mohammad, and T. El-Elimat, "Antioxidant activity and total phenolic content of selected Jordanian plant species," *Food Chemistry*, vol. 104, no. 4, pp. 1372–1378, 2007.

19

Recent Advances in the Genetic Transformation of Coffee

M. K. Mishra[1] and A. Slater[2]

[1] Central Coffee Research Institute, Coffee Research Station, Chikmagalur, Karnataka 577117, India
[2] The Biomolecular Technology Group, Faculty of Health and Life Sciences, De Montfort University, Gateway, Leicester LE1 9BH, UK

Correspondence should be addressed to M. K. Mishra, manojmishra.m@gmail.com

Academic Editor: Shengwu Ma

Coffee is one of the most important plantation crops, grown in about 80 countries across the world. The genus *Coffea* comprises approximately 100 species of which only two species, that is, *Coffea arabica* (commonly known as arabica coffee) and *Coffea canephora* (known as robusta coffee), are commercially cultivated. Genetic improvement of coffee through traditional breeding is slow due to the perennial nature of the plant. Genetic transformation has tremendous potential in developing improved coffee varieties with desired agronomic traits, which are otherwise difficult to achieve through traditional breeding. During the last twenty years, significant progress has been made in coffee biotechnology, particularly in the area of transgenic technology. This paper provides a detailed account of the advances made in the genetic transformation of coffee and their potential applications.

1. Introduction

Coffee is one of the most important agricultural commodities, ranking second in international trade after crude oil. The total global production of green coffee is above 134.16 million bags (60 kg capacity) with a retail sales value in excess of $22.7 billion during 2010-11 in the world market [1]. Coffee is grown in about 10.2 million hectares land spanning over 80 countries in the tropical and subtropical regions of the world especially in Africa, Asia, and Latin America. The economics of many coffee growing countries depends heavily on the earnings from this crop. More than 100 million people in the coffee growing areas worldwide derive their income directly or indirectly from the produce of this crop.

Coffee trees belong to the genus *Coffea* in the family Rubiaceae. The genus *Coffea* L. comprises more than 100 species [2], of which only two species, that is, *C. arabica* (arabica coffee) and *C. canephora* (robusta coffee), are commercially cultivated. Another coffee species, *Coffea liberica* is also cultivated in a small scale to satisfy local consumption. Almost all the coffee species are diploid ($2n = 2x = 22$) and generally self-incompatible except *C. arabica* which is a natural allotetraploid ($2n = 4x = 44$) self-fertile species [3]. In the consumer market, *C. arabica* is preferred for its beverage quality, aromatic characteristics, and low-caffeine content compared to robusta, which is characterized by a stronger bitterness, and higher-caffeine content. Arabica contributes towards 65% of global coffee production [4].

C. arabica is mainly native to the highlands of South-western Ethiopia with additional populations in South Sudan (Boma Plateau) and North Kenya (Mount Marsabit) [5–8]. The *C. arabica* varieties grown all over the world are derived from either the "Typica" or "Bourbon" genetic base, which has resulted in low-genetic diversity among cultivated arabicas. In contrast, *C. canephora* has a wide geographic distribution, extending from the western to central tropical and subtropical regions of the African continent, from Guinea and Liberia to Sudan and Uganda with high genetic diversity in the Democratic Republic of Congo [9]. *C. canephora* maintains heterozygosity due to its cross-pollinating nature.

2. Coffee Breeding and Its Limitations

Coffee breeding is largely restricted to the two species, *C. arabica* and *C. canephora*, that dominate world coffee production. However, *C. liberica* and *C. congensis* have contributed useful characters to the gene pool of *C. arabica* and *C. canephora*, respectively, through natural and artificial interspecific hybridisation. In *C. arabica*, initial breeding objectives were to increase productivity and adaptability to local conditions. To achieve these objectives, breeding strategies were directed towards identification of superior plants in the population and their propagation and crossing with existing cultivars. These early breeding efforts, which were carried out from 1920 to 1940, had considerable success in identifying and developing vigorous and productive cultivars. Several of these varieties such as Kents and S.288 from India, Mundo Novo, Caturra and Catuai from Brazil, and Blue Mountain from Jamaica, are still under commercial cultivation. These cultivars are suggested to have a larger degree of genetic variability than the base population [10]. The appearance of coffee leaf rust (*Hemileia vastatrix* Berk and Br) in epidemic scale in Southeast Asia between 1870 and 1900 had a devastating effect on arabica coffee cultivation in several coffee growing countries. This has changed the breeding focus worldwide with emphasis now given to disease resistance. This has resulted in the introduction of other tolerant species, especially *C. canephora,* in many countries. Until now, *C. canephora* has provided the major source of disease and pest resistance traits such as coffee leaf rust (*H. vastatrix*), coffee berry disease (*Colletotrichum kahawae*), and root-knot nematode (*Meloidogyne* spp.) not available in *C. arabica*. Besides, *C. canephora,* other diploid species such as *C. liberica* has been used as source of resistance to leaf rust [11] and *C. racemosa* for imparting resistance to coffee leaf minor [12]. Further, the cultivation of *C. arabica* with other diploid species such as *C. canephora* and *C. liberica* in close proximity has resulted in spontaneous hybrids in many countries. Natural interspecific hybrids such as Hybrido-de Timor (a hybrid between *C. arabica* and *C. canephora* [13] from Timor island), Devamachy (a hybrid between *C. arabica* and *C. canephora*),and S.26 (a hybrid between *C. arabica* and *C. liberica, which* both originated in India [14]) are the main source of resistance to pest and disease and extensively used in *C. arabica* breeding programmes.

Like *C. arabica*, improvement of *C. canephora* was originally aimed at increasing productivity, and improving bean size and liquor quality. The breeding methods adopted for *C. canephora* involvedmass selection and intra- as well as interspecific hybridization. Varieties like Apoata of Brazil, S.274 of India, and Nemaya of Central America were derived through mass selection. The spontaneous diploid interspecific hybrid between *C. canephora var. ugandae* and the *C. congensis* (called Congusta in Indonesia), and the C × R hybrid variety developed through artificial hybridisation between *C. congensis* and *C. canephora* in India are examples of improved robusta cultivars developed through interspecific hybridisation.

Although conventional breeding is mainly used for coffee improvement, it is a long process involving several different techniques, namely selection, hybridization, and progeny evaluation. A minimum of 30 years is required to develop a new cultivar using any of these methods. Further, the long generation time of the coffee tree, the high cost of field trials, the lack of accuracy of the breeding process, the differences in ploidy level between *C. arabica* and other diploid species, and the incompatibility are all major limitations associated with conventional coffee breeding. In addition to these, genetic resistance to coffee white stem borer (*Xylotrechus quadripes*) and coffee berry borer (*Hypothenamus hampei*), drought and cold tolerance, and herbicide resistance are some of the features that are not easily available in the coffee gene pool or are difficult to incorporate using conventional breeding. Another constraint that hinders the arabica coffee improvement programme is the selection of genetically diverse parental lines for hybridization and the identification of hybrids at an early stage of plant growth based on morphological traits. This is because most of the commercial arabica cultivars are morphologically identical and not easily distinguishable from one another. Uniformity of morphological traits in *C. arabica* could be attributed to the origin of the species, its narrow genetic base and self-fertile nature. In coffee, identification of cultivars is mainly based upon phenotypic features, but this approach is not reliable and is subject to environmental influences, mainly because of the long generation time of the coffee trees. In some countries of Asia, Latin America, and Africa, coffee is cultivated under shade in varied agroclimatic conditions and displays remarkably different morphologies in various microclimatic zones. In view of the above, it becomes imperative to develop alternative techniques that are reliable, quick, and efficient for discriminating between coffee cultivars. Among the various markers available for genetic analysis in coffee, molecular markers are more efficient, precise, and reliable than other markers for discriminating closely related species and cultivars. The DNA-based markers have the potential of complimenting coffee breeding and improvement program in form of marker-assisted selection (MAS).

3. Molecular Markers and Coffee Genetic Improvement

Various molecular markers, such as restriction fragment length polymorphism (RFLP), random amplified polymorphic DNA (RAPD), amplified fragment length polymorphism (AFLP), intersimple sequence repeat (ISSR), simple sequence repeats (SSR), and expressed sequence tag derived simple sequence repeats (EST-SSR) have been used in coffee genetic diversity studies [46–51]. In addition to the above, a large number of commercial coffee samples of American, Indian, and African origin were also analyzed using highly polymorphic SSR markers which revealed that Indian cultivars were genetically diverse from the American and African cultivars [52]. More recently, a new type of molecular marker known as a sequence related amplified polymorphism (SRAP) was used in genetic diversity analysis of coffee cultivars and species [53, 54]. SRAP markers were also successfully used to discriminate between parents in hybrid identification [55, 56] and therefore has great potential in coffee breeding programmes. In addition to the above, single-nucleotide polymorphisms (SNPs) and PCR-RFLP markers were used in coffee genome analysis, which revealed that in *C. arabica*, polymorphisms are created by paralogous chromosomes, whereas homozygosity of many genes is maintained by the self-fertile nature of the species [57]. These results further demonstrated that in allopolyploid *C. arabica,* the two parental genomes remain separated and exhibit multiple allelic inheritance patterns, and these findings will be very important for designing strategies and decisions in breeding programmes as well as in sequencing projects. In recent years, concerted efforts have been made by several laboratories across the world, under the International Coffee Genome Network (ICGN) programme, to sequence the coffee genome by using high-throughput sequencing technology which will unravel several key aspects of the coffee genome that may be useful for coffee genetic improvement. In addition to molecular markers, a two-dimensional protein mapping technique was also used to differentiate green coffee samples [58]. A detailed review of the role of various molecular markers in coffee is already available [59] and therefore beyond the scope of the present review.

4. The Need for Genetic Transformation of Coffee

Since its initial application to plants more than 25 years ago, genetic transformation has become an indispensable tool in plant molecular biology and functional genomics research [60]. Genetic transformation technology is considered as an extension of conventional plant breeding technologies [61]. It offers unique breeding opportunities by introducing novel genetic material irrespective of the species barrier and creating phenotypes with desired traits that are not available in the germplasm pool of crop plants. The major objectives for using genetic engineering technique in coffee are to introduce new traits in to elite coffee genotypes, develop new cultivars with desirable traits such as pest and disease resistance, herbicide resistance, drought and frost tolerance, and improved cup quality, which are not possible to incorporate using traditional breeding techniques. The recent developments in coffee transcriptomics and the availability of large amounts of expressed sequence tag (EST) data from both *C. canephora* and *C. arabica* [62–64], as well as the development of coffee bacterial artificial chromosome (BAC) genomic libraries [65, 66], have opened up new possibilities in the area of coffee functional genomics. A key component of most functional genomics approaches is the availability of a highly efficient transformation system useful for designing strategies for gene identification, elucidation of gene functions, regulation and interaction of genes and gene expression analysis to understand the involvement of genes, in coffee biological processes. This will help in precisely targeting the trait of interest using various transformation tools (genes and promoters) with increase probability of success in reducing economic costs.

Genetically modified coffee plants have been produced by different research groups in the world [21, 25, 26, 29, 34, 37, 67]. Despite significant advances over the last 20 years, coffee transformation is far from a routine procedure in many laboratories [35]. The objective of this paper is to provide an update on coffee genetic transformation over the last decade, including the *in vitro* methods used for plant generation.

5. *In Vitro* Plant Regeneration

The establishment of an efficient regeneration system is important for genetic transformation of coffee. Various *in vitro* multiplication methods such as somatic embryogenesis, meristem and axillary bud culture, and induction of adventitious buds have been reported using different types of tissue in various coffee species [68, 69].

5.1. Somatic Embryogenesis. The initiation and development of embryos from somatic tissues without the involvement of sexual fusion are known as somatic embryogenesis. In coffee, induction of somatic embryogenesis and plant regeneration was first reported from the internodal explants of *C. canephora* [70]. In *C. arabica,* calluses were successfully induced from seeds, leaves, and anthers of two different cultivars, that is, Mundo Novo and Bourbon Amarelo [71]. During the last 35 years, a number of protocols for somatic embryogenesis have been developed for various genotypes of coffee [68]. The first protocol to obtain calli with high embryogenic potential from the leaf explants of *C. arabica* cv. Bourbon used two different culture media compositions: a first "conditioning" medium and a second "induction" medium [72, 73]. The availability of auxins is critical for the induction of embryogenic calli [72]. In coffee, both high-frequency somatic embryogenesis (HFSE) and low-frequency somatic embryogenesis (LFSE) were established. 2,4-D strongly increases HFSE in combination in primary cultures where as IBA and NAA combined with K increase LFSE. During somatic embryo induction in *C. arabica* cv. Caturra Rojo, two types of cell clusters, embryogenic and nonembryogenic were observed [74]. The differences in gene expression at both RNA and protein levels were observed between the embryogenic and nonembryogenic cell clusters. Further, it was observed that the number of genes turned off in somatic cells to allow for the change from somatic to embryogenic state is higher than those genes that are turned on [74].

In coffee, somatic embryogenesis follows two distinct developmental patterns: (1) direct somatic embryogenesis, where embryos originate directly from the explants and (2) indirect somatic embryogenesis, where embryos are derived from an embryogenic dedifferentiated tissue (callus). However, both direct and indirect somatic embryos of coffee formed from leaf segments and callus, respectively, have a unicellular origin [75]. Various attempts were made to reduce the time needed for embryogenesis and increase the embryogenesis frequency in coffee. Triacontanol, silver nitrate ($AgNO_3$), salicylic acid, thidiazuron, and 6-(3-methyl-3-butenylamino) purine (2ip) are the widely used growth regulators in coffee embryogenesis. Interestingly, picomolar concentrations of salicylates reported to induce cellular growth and enhance somatic embryogenesis in *C. arabica* tissue culture [76]. Similarly, triacontanol, as well as silver nitrate, at low concentration in combination with indole-3-acetic acid (IAA) and benzyladenine (BA) induced direct somatic embryogenesis in both species of *C. arabica* and *C. canephora* [77, 78]. Additionally, thidiazuron (TDZ) also induced direct somatic embryos from the cultured leaf explants of *C. canephora* cv. C × R [79]. In *C. canephora*, the embryogenic response of the explants has been shown to increase by the addition of polyamines, either alone or in combination with silver nitrate. It has been observed that incorporation in the *in vitro* culture medium of inhibitors of the polyamine biosynthetic pathway such as D,L-.alpha-difloromethylornithine (DFMO) and D,L-.alpha-difloromethylarginine (DFMA) significantly reduced the embryogenic response of the explants in *C. canephora*, indicating the pivotal role played by polyamines in coffee somatic embryogenesis [80]. Besides the polyamines, indoleamines (melatonin and serotonin) as well as calcium and calcium ionophores (A23187) have also been shown to be beneficial in inducing somatic embryogenesis [81]. Apart from exogenous growth hormones, ethylene and dissolved oxygen concentration play a crucial role in coffee somatic embryogenesis [82, 83].

The use of somatic embryos on an industrial scale was achieved by inducing somatic embryos of *C. arabica* in liquid medium using bioreactors [84, 85]. The yield of embryos achieved was about 46,000 embryos/3L Erlenmeyer flask (after 7 weeks of culture). Various other workers also reported the production of somatic embryos for industrial use [86, 87]. Extensive studies were carried out in the use of conventional and temporary immersion system for coffee somatic embryo production [68, 88, 89]. However, to date the major obstacle associated with production of somatic embryos on a commercial scale is synchronisation of embryogenesis and conversion of plantlets.

5.2. Micropropagation. The coffee plant has a single apical meristem with each axil leaf having 4-5 dormant orthotropic buds and two plagiotropic buds. The plagiotropic buds only start development from the 10th to 11th node. For apical meristem culture and the culture of dormant buds, both orthotropic and plagiotropic buds were cultured for obtaining plantlets. Microcuttings or nodal culture comprise a tissue culture approach which involves culturing nodal stem segments carrying dormant auxiliary buds and stimulating them to develop. Each single segment can provide 7–9 microcuttings every eighty days. Most of these studies were carried out during the 1980s, and these topics have been reported in an earlier review [68].

Several studies have been carried out with a view to micropropagating superior coffee genotypes using apical or axillary meristem culture and nodal culture [68]. A maximum of nine shoots was obtained per one shoot explant [90]. Culture of microcuttings in a temporary immersion system resulted in a 6-fold increase in the multiplication rate, in comparison with microcuttings multiplied on solid medium [91, 92]. The field performance of embryo-generated plants was reported and showed a normal response in terms of physiology and yield. The genetic fidelity of

micropropagated plants of *C. canephora* obtained through somatic embryogenesis was assessed in a large-scale field trial [93]. A total number of 5067 trees regenerated from five to 7-month-old embryogenic cell suspension cultures were planted in the Philippines and in Thailand for comparison with control plants derived from auxiliary budding *in vitro*. No significant differences in yield and morphological features were observed between the somatic seedlings and microcutting derived trees [93]. However, in contrast to the above, several studies have clearly demonstrated culture-induced variation and regeneration of somaclonal variants in coffee obtained through direct and indirect somatic embryogenesis [94–96]. Detailed molecular analysis of the plantlets of *C. arabica* derived from high-frequency somatic embryogenesis revealed alterations in both the nuclear and mitochondrial genomes [97]. These reports therefore proposed a critical evaluation of tissue culture-derived plants both at phenotypic and molecular level.

Adventitious shoot development is an alternative method of coffee micropropagation. Shoots originating in tissues located in areas other than leaf axils or shoot tips are subjected to one phase of dedifferentiation followed by differentiation and morphogenesis [68, 98]. Rooting is the most difficult and expensive phase of the micropropagation process, and the success of newly formed plantlets is closely linked to the ability of the root system to adapt to the autotrophic conditions. Several studies have been carried out to improve the rate of rooting of micropropagated plants [68].

6. Development of Transgenic Coffee

Genetic engineering research on coffee has been pursued for the past fifteen years with two major objectives: (1) to elucidate the function, regulation, and interaction of agronomically important genes through a functional genomics approach and (2) to improve coffee genotypes with desirable traits through the introduction of targeted genes.

6.1. Candidate Genes. In recent years, the development in high-throughput sequencing (HTS) technologies has allowed the rapid acquisition of significant amounts of sequence data, and this has increased our understanding of the genomics of a particular species. During the last decade, significant progress has been made in developing an EST database for coffee. Initial efforts in developing ESTs in *Coffea arabica* were initiated by the University of Trieste, Italy, and the EST sequences have been placed in the public domain (http://www.coffee.dna.net/). In a private/public collaboration between Nestle and Cornell University, 47000 ESTs from *C. canephora* were established comprising 13175 unigenes [63]. Subsequently the Brazilian government funded an ambitious coffee genome program, and this has resulted in the establishment of 200 000 ESTs which led to the identification of 30000 genes [64]. Very recently, the Italian group has generated an additional 161 660 ESTs which will be publicly available at the website (http://www.coffeedna.net/ [99]). In parallel with the development of EST database,

BAC libraries of coffee species, *C. arabica* and *C. canephora* were established [65, 66]. Such maps are of central strategic importance for marker-assisted breeding, positional cloning of agronomical important genes, and analysis of gene structure and function.

Due to the concerted efforts on coffee genomics, many candidate genes from coffee have been identified and some of them have been cloned and are currently being characterised. These include a caffeine biosynthesis gene [28, 100], a sucrose synthase gene [66], osmotic stress response genes [101], genes for seeds oil content [102] and several pathogen resistance genes such as *Mex-1* gene [103], *SH₃* gene [104], and *Ck-1* gene to CBD [105]. An efficient genetic transformation protocol is necessary in order to validate the structural and functional aspects of these intrinsic genes. In coffee transformation experiments, genes isolated from coffee as well as from heterologous sources have been used. Some of the genes isolated from coffee used in transformation experiments include a theobromine synthase gene (*CaMXMTI*) for suppressing caffeine biosynthesis [28, 106] and an *ACC oxidase* gene involved in ethylene biosynthesis [34]. The genes introduced to coffee from heterologous sources include a *cry1Ac* gene from *Bacillus thuringiensis* targeted against leaf miner [26], the *α-AII* gene from common bean for imparting resistance to coffee berry borer [30], the *BAR* gene for herbicide tolerance [21, 35], and the homeobox gene *WUSCHEL* (WUS) from *Arabidopsis* responsible for stem cell identity [36]. The functional significance and expression of these transgenes in coffee plants are described subsequently in this paper.

6.2. Promoters.

In most of the coffee transformation experiments reported so far, with few exceptions, the CaMV35S promoter derived from cauliflower mosaic virus is extensively used in transgenic constructs (See Table 1). In a comparative study made earlier, the efficacy of different promoters driving *uidA* transient expression in endosperm, somatic embryos, and leaf explants of *C. arabica* was analyzed using microprojectile bombardment [18, 19]. It was observed that the EF-1α promoter (from *Arabidopsis thaliana* EF-1α translation elongation factor) directed maximum transient expression of the *uidA* gene compared to other promoters [18]. Therefore, this promoter was used subsequently by the same group in *Agrobacterium* gene constructs, driving the *cry1Ac* gene for the control of leaf miner in coffee [26, 33]. The efficacy of the CaMV35S viral promoter was also compared with two coffee promoters (α-tubulin and α-arabicin) which have revealed a similar level of transient *uidA* gene expression [19]. These findings have opened up the possibility of using coffee specific promoter in transformation experiments.

6.3. Reporter Genes.

Reporter genes are used in gene constructs to optimize the transformation procedure. In the majority of coffee transformation experiments, the *uidA* gene is used as a reporter gene (See Table 1). Only recently have the *sgfp* (synthetic green fluorescence protein) and *DsRFP* (Red fluorescent protein) been used in coffee transformation

[28, 31, 37]. However, most transient expression studies have been carried out using the *uidA* reporter gene. In an effort to optimize the *Agrobacterium* transformation protocol in coffee, the expression of the *uidA* gene driven by a CaMV35S promoter was compared in various tissues under different cocultivation conditions [27]. It was observed that the endogenous *GUS* activity was reduced substantially when 20% methanol was added to the *GUS* staining solution. Marked differences in *GUS* activity were observed between the endogenous, and transformed tissue as the transformed plants exhibited a deep blue colour in reaction with X-gluc, while nontransformed plants only exhibited a pale blue coloration. Several factors such as of cocultivation period, preculture of explants, and acetosyringone-influenced *GUS* activity. Furthermore, in a comparative study, it was observed that *GUS* expression in leaf explants was more pronounced in the cut ends of the veins, whereas zygotic and somatic embryos, hypocotyls, and the adjacent region are the main target sites [27]. Recently, in another experiment, the expression of p35S. GUS expression was found to be stronger in the root tip and central vascular system compared to other regions in the root system [43].

Recently, the expression of both *uidA* and *gfp* genes driven by the CaMV35S promoter was monitored using various *Agrobacterium* strains and culture conditions [37]. Expression of the *sgfp-S65T* gene driven by the CaMV35S promoter (signified by green fluorescence) was observed in cocultivated calli just 2 days after cocultivation. Initially, green fluorescence appeared as discrete spots but subsequently, calli that showed green fluorescence increased in size, producing a bright fluorescence mass after 15 days cocultivation. The expression of both *uidA* and *sgfp* was intense in globular- and torpedo-shaped embryos until the development of cotyledonary leaves. In older leaves, green florescence was weak due to the interference of chlorophyll, which emits red fluorescence at the same activating wavelength. The pattern of expression of *GUS* and *gfp* genes driven by the CaMV35S promoter was similar, being much more pronounced in the leaf veins and root tips and in vascular zones. Using the CaMV35S promoter, similar expression patterns were obtained for *DsRFP* in somatic embryos [31] and for *gfp* in roots [44].

7. Transformation Systems

Both direct and indirect DNA delivery systems have been employed to transform coffee by various workers and the details are described below.

7.1. Direct DNA Delivery

7.1.1. Electroporation.

Electroporation is a process through which permeability of the cell plasma membrane is significantly increased by the external application of electrical field. It is usually used in molecular biology as a way of introducing some substance into a cell, such as a drug that can change the cell's function, or a piece of coding DNA. Electroporation was used to integrate foreign

TABLE 1: Summary of transformation studies in *Coffea* sp*.

DNA delivery method	Coffea species	Explants used	Binary vector	Agrobacterium strain	Promoter	Selection marker	Target gene	Results	Reference	Country
Direct delivery	C. arabica	Protoplast	NA	—	pGA472	nptII	uidA	TE	[15]	USA
electroporation	C. arabica	SE	pCAMBIA 3201	—	CaMV35S	bar	uidA	TGI	[16]	Venezuela
	C. canephora	Endosperm	pCAMBIA 1301	—	CaMV35S	hpt	uidA	TE	[17]	India
	C. arabica	Leaf	pPIGK	—	EF-1a	bar	uidA	TE	[18]	France
	C. arabica	ET	pCAMBIA 2301	—	CaMV35S	nptII	uidA	TGI	[19]	Colombia
Microprojectile	C. arabica	ET	pBI-426	—	CaMV35S	nptII	uidA	TGI, PR	[20]	Brazil
bombardment	C. canephora	ET	pCAMBIA 3301	—	CaMV35S	bar	uidA	TGI, PR	[21]	Brazil
	C. arabica	SE	pCAMBIA 2301	—	CaMV35S	nptII	uidA	TE	[22]	Costa Rica
	C. arabica	ET	pBI-426	—	CaMV35S	nptII	uidA	TGI, PR	[23]	Brazil
	C. arabica	Protoplast	pGV2260	NA	CaMV35S	hpt	uidA	TE	[24]	France
	C. canephora	ET	pIGI121-Hm	EHA101		hpt	uidA	TGI, PR	[25]	Japan
	C. arabica and C. canephora	SE	pBIN19	LBA4404	EF-1α	csr-1-1	cry1Ac	TGI, PR	[26]	France
	C. canephora	ZE, SE, H	pBECKS 400	EHA101	CaMV35S	hpt	uidA	TGI,	[27]	India
	C. arabica and C. canephora	ET, SE	pHIBI –IG	EHA101	CaMV35S	hpt	gfp & CaMXMTI	RNAi; PR	[28]	Japan
Indirect delivery	C. canephora	H	pBECKS 400	EHA 101	CaMV35S	hpt	uidA	TGI, PR	[29]	India
A. tumefaciens	C. canephora	ET	pCAMBIA 3301	EHA105	CaMV35S	ppt	uidA	TGI	[30]	Brazil
	C. canephora	ET, Leaf	pER10W-35SRed	C58	CaMV35S	nptII	DsRFP	TGI, PR	[31]	Mexico
	C. canephora	SE	pCAMBIA 1381	A4, EHA 101	CaMV35S	hpt	NMT	RNAi	[32]	India
	C. canephora	SE	pBIN19	LBA 4404	EF-1	csr-1-1	cry1Ac	Field test	[33]	France
	C. arabica	ET	pCAMBIA 3300	EHA 105	CaMV35S	bar	ACC-oxidase	AE	[34]	Brazil
	C. canephora	ET	pCAMBIA 3301	EHA 105	CaMV35S	bar	uidA	TGI, PR	[35]	Brazil
	C. canephora	Embryos	pER10W-35SRed	C58C1	CaMV35S	WUSCHEL	DsRFP	TGI,PR	[36]	Mexico
	C. canephora and C. arabica	ET	pBECKS 2000	EHA 101, EHA 105, LBA 4404, AGL 1	CaMV35S	hpt, visual	uidA, sgfp	TGI, PR	[37]	India
	C. arabica	ET	pMDC32	LBA1119	CaMV35S	hpt	sgfp	PR	[38]	France
	C. arabica and C. canephora	SE	pBIN 19	A4	CaMV35S	hpt	uidA	TGI, PR	[39]	France
A. rhizozenes	C. arabica	Leaf	NA	IFO 14554	NA	NA	NA	TGI, PR	[40]	Japan
	C. canephora	SE	pBIN 19	A4	CaMV35S	csr-1-1	cry1Ac	TGI, PR	[41]	France
	C. canephora	SE	pCAMBIA 1301	A4	CaMV35S	hpt	uidA	TGI, PR	[42]	India
	C. arabica	H	pBIN 19	A4	CaMV35S	visual	uidA	TGI, PR	[43]	France
	C. arabica	H	pCAMBIA 2300	A4	CaMV35S	visual	gfp	TGI, PR	[44]	France
					CaMV35S					

* Updated from Etienne et al. [45]; ET: embryogenic tissue; SE: somatic embryos; H: hypocotyls; TE: transient gene expression; TGI: target gene integration; PR: plant regeneration; RNAi: RNA interference; AE: antisense expression.

DNA into protoplasts of *C. arabica* [15]. Regeneration of transformed embryos and plantlets resistant to kanamycin was obtained but the plantlets did not survive due to a weak root system. In another experiment, various parameters influencing transformation of coffee somatic embryos using electroporation of pCAMBIA 3201 plasmid carrying the *uidA* gene were described [16]. The results showed that the electroporation of somatic embryos at torpedo stage can be a promising approach to coffee transformation since transformed torpedo-shaped embryos produced significantly higher numbers of gus positive secondary embryos in the culture medium. Recently, the expression of *uidA* gene driven by the N-methyltransferase (NMT) promoter was studied by electroporation of coffee endosperm [17]. The results indicated that *uidA* gene expression driven by the NMT promoter is targeted to the external surface of the vacuoles.

7.2. Microprojectile Bombardment. Genetic transformation of coffee via microprojectile bombardment was described for the first time, using a gunpowder driven device and several target explants [18]. The study compared different promoters and demonstrated that the EF-1α promoter from *Arabidopsis thaliana* is more effective than the *CaMV35S* promoter in driving transient *GUS* expression in leaves of microcuttings. The interaction between osmotic preconditioning and physical parameters of helium gun device was studied in *C. arabica* suspension cells. It was observed that four hours of pretreatment of the target tissue with mannitol and sorbitol before bombardment increased the number of cells expressing *GUS* gene without causing cell necrosis [19].

Successful regeneration of transgenic coffee plants (*C. canephora*) by microprojectile bombardment was achieved by using the pCambia3301 plasmid containing the *uidA* and *bar* genes [21]. The study demonstrated the effectiveness of the *bar* gene for selection of transformants *in vitro* and *in vivo* identification of transgenic coffee plants. In *C. arabica*, the plasmid pBI-426 carrying the *nptII* and *uidA* genes was employed in particle bombardment of embryogenic calli, and transformants were selected using kanamycin [20]. The transgenic status of the regenerated plants was confirmed by PCR analysis. Recently, a *C. arabica* suspension culture with a high regenerative capacity for secondary somatic embryogenesis was used for transformation using microprojectile bombardment [22]. However, no transformants could be regenerated due to damage to bombarded tissue. Very recently, successful regeneration of transgenic *C. arabica* was reported using bombardment of embryogenic calli followed by kanamycin selection [23]. The authors reported the normal growth of the transgenic plants and obtained T_1 progeny presenting 3 : 1 segregation of the *uidA* transgene.

7.3. Indirect DNA Delivery. Agrobacterium tumefaciens. The *Agrobacterium tumefaciens* mediated transformation technique has been extensively used for genetic transformation of coffee and is the method of choice for many workers (see Table 1). An initial report of *A. tumefaciens*-mediated transformation of coffee involved the cocultivation of protoplast with different *Agrobacterium* strains carrying *nptII* and *uidA* genes [24]. Transient GUS expression was demonstrated in the callus tissue derived from protoplasts but plant regeneration was not obtained. The first successful *A. tumefaciens* transformation and transgenic plant regeneration were achieved in *C. canephora* [25] and subsequently in *C. arabica* [26]. In *C. canephora*, various parameters that influence T-DNA delivery to coffee tissue were studied using transient GUS gene expression [27]. It was reported that preculture of explants prior to *Agrobacterium* cocultivation, addition of acetosyringone to cocultivation medium and the duration of the cocultivation period significantly influenced T-DNA delivery to coffee tissue. In addition to the culture conditions, the *A. tumefaciens* strains also influence the transformation efficiency in coffee [37]. *A. tumefaciens* strains and their plasmids are classified by the function of the opine genes they carry. These opines, which are synthesized in the infected plant cells, are mainly agropine, nopaline, and octopine. For coffee transformation, various *A. tumefaciens* strains such as LBA 4404 pAL4404 (octopine), C58CI pMP90 (nopaline), EHA 101 pEHA101 (agropine), EHA 105 pEHA101 (agropine), and AGL1 pTiBo542 (agropine) have been used by various workers (see Table 1). Recently, a study was carried out to compare the efficiency of four different *Agrobacterium* strains (LBA 4404, EHA101, EHA105, AGL1) in coffee transformation using pBECKS 2000 vector constructs carrying *uidA* and *gfp* reporter genes [37]. It was observed that EHA 105 and EHA101 were more efficient compared to LBA4404 in T-DNA delivery and transgenic plant regeneration. Based on this improved protocol, mass production of transgenic coffee plants of both *C. arabica* and *C. canephora* was achieved [37].

In order to improve the efficiency of *A. tumefaciens* mediated transformation, sonication and vacuum infiltration methods were incorporated during cocultivation [31, 35]. In most of the *A. tumefaciens* mediated transformation, embryogenic tissues and/or somatic embryos were used as the target material for cocultivation (see Table 1). In *C. canephora*, a highly efficient *A. tumefaciens* transformation and regeneration protocol were established using hypocotyl explants as the target material (Figures 1(a)–1(l)) [29]. In *C. canephora*, the collar region of the hypocotyls was found to be more suitable for *A. tumefaciens* transformation [107]. However, in *C. arabica*, embryogenic calli were found to be more suitable for *A. tumefaciens* transformation [38]. The methodology for genetic transformation of *Coffea* using *Agrobacterium* was described in detail [108]. They were able to transform 20 different genotypes either belonging to *C. arabica* or *C. canephora* by cocultivation of embryogenic calli with *A. tumefaciens*.

A. tumefaciens mediated transformation has also been used for gene silencing using RNAi technology and several genes such as theobromine synthase (*CaMXMTI*) and N-methyltransferase (*NMT*) (both involved in caffeine biosynthesis) and the *ACC oxidase* gene (involved in ethylene biosynthesis) were targeted to coffee tissue in reverse orientation for obtaining stable silencing [28, 32, 34]. Recently, transgenic *C. canephora* plants incorporating a *homeobox* gene *WUSCHEL* (WUS) responsible for stem cell identity were regenerated [36].

Figure 1: *Agrobacterium tumefaciens* mediated transformation and regeneration of *C. canephora* cv. C × R and *C. arabica* genotypes. (b, c, d, f, and k) *C. arabica* (a, e, g, h, i, j, and l) *C. canephora*. (a) Cocultivated hypocotyls of *in vitro* seedlings expressing transient GUS expression; (b) embryogenic calli showing transient GUS expression following cocultivation; (c) initiation of somatic embryos from the transformed calli showing green fluorescence; (d) mass of heart shaped somatic embryo showing green fluorescence; (e) germinating somatic embryo with well-developed cotyledon leaves with bright green fluorescence; (f) Gus expression in the root tips of a germinated transformed plant; (g) strong GFP expression in transgenic root; (h) *in vitro* plant regeneration; (i) Gus staining of the leaf of a transgenic plant; (j) GFP expression in the developing leaf; (k) GUS staining of the regenerated transformed plant; (l) transgenic plant in the soil.

Agrobacterium rhizogenes. Agrobacterium rhizogenes mediated transformation of both *C. canephora* and *C. arabica* species was reported as early as 1993 [39]. Subsequently, many workers have reported successful *A. rhizogenes* mediated transformation and plantlet regeneration in both *C. arabica* and *C. canephora* using different explants [40, 42, 44, 109]. In almost all cases, *Agrobacterium* strain A4 was used except in one case where bacterial strain IFO 14554 has been used (Table 1). Most of the binary constructs used in *A. rhizogenes* transformation are either based on a *pBIN19* or *pCambia* backbone. *A. rhizogenes* transformation is very useful for functional analysis of genes involved in resistance to root knot nematode in coffee.

8. Selection of Transformants

The successful recovery of stable transformants depends upon the choice of a suitable selective agent, its optimal concentration, timing, and frequency of selection. An effective selective agent must allow the growth and development of transformed tissue but simultaneously restrict the proliferation of nontransformed cells in the same culture medium. Therefore, incorporation of the right selectable marker gene in gene constructs is critical to the success of transformation. In coffee transformation, various selection marker genes (*hpt* hygromycin-R, *nptII* kanamycin-R, *csr1-I* chlor and sulfuron-R, *ppt* phosphinothricin-R) were used (Table 1), and their efficacy has been evaluated [18, 27]. The first work in coffee transformation was carried out using kanamycin as the selective agent [39]. However, contradictory results were obtained with regard to the efficacy of kanamycin for selection of coffee transformants. Many workers [15, 39, 41, 110] have reported the development of nontransformed somatic embryos at higher kanamycin concentrations and have attributed this to a poor capacity for transformant selection. However, there have also been several reports of successful selection of transformed coffee plants using kanamycin as the sole selective agent [20, 23, 31]. In many transformation experiments, hygromycin at concentrations of 20–100 mg/L was used successfully for selection of transformants [25, 27–29]. The efficiency of hygromycin as a selective agent was tested in different transformed and nontransformed tissue of *C. canephora* and it was observed that 25 mg/L hygromycin severely checked nontransformed somatic embryo growth and proliferation [27]. Similar results were also obtained by using hygromycin in *A. rhizogenes* mediated coffee transformation [42].

In addition to antibiotics, several other types of selection markers such as herbicide selection and positive selection have also been used in coffee transformation. The reliability of chlorsulfuron (*csr1-I*), phosphinothricin (*ppt*), and ammonium glufosinate (*bar*) as selection markers to regenerate transformed tissue were already confirmed in both *C. arabica* and *C. canephora* using various transformation methods [21, 26, 30]. As an alternative to negative markers such as antibiotics and herbicides, positive selection marker genes such as phosphomannose isomerase (*pmi*) and xylose isomerase (*xylA*) have also been used for coffee transformation, producing transformants able to grow in the presence of mannose and xylose, respectively, without an additional carbohydrate source. The study indicated that compared to mannose, xylose is an effective selective agent for coffee transformation [111].

In recent times, transgenic crops regenerated carrying antibiotic and herbicide resistance genes have generated public disquiet about food safety and environmental impact. This has stimulated research into utilizing visual selection markers instead of using antibiotic and/or herbicide selection markers. In coffee, green fluorescent protein, (*gfp*), and red fluorescent protein (*DsRFP*) were used for visual selection of transformed tissue following *A. tumefaciens* mediated transformation [28, 31, 37]. The regeneration of transformed roots of *C. arabica* using visual selection of green epifluorescence without using any selective agent was achieved through *A. rhizogenes* mediated transformation [44]. Recently transgenic plants of both *C. canephora* and *C. arabica* were regenerated by employing visual selection of green fluorescent protein as the sole screen following *A. tumefaciens* mediated transformation [112].

9. Transgene Expression

9.1. Expression of Reporter Genes. Studies pertaining to expression of transgenes in a perennial crop like coffee are very important. However, such reports are very limited. The expression of *uidA* and *gfp* transgenes driven by a CaMV35S viral promoter was monitored at different stages of plant growth of *C. canephora* following *A. tumefaciens* mediated transformation [27, 37]. It was observed that maximum expression of both *uidA* and *gfp* genes were obtained at the globular- and torpedo-shaped somatic embryos. Following the development of cotyledon leaves, GUS expression was scattered, with pronounced expression shifted to vascular regions in the well-developed leaves. In roots, maximum expression was obtained in the root tips and root hairs compared to the main roots. Recently, the expression of *uidA* gene driven by the double CaMV35S promoter was monitored in the flowers and fruits of *C. arabica* transgenic plants obtained through microprojectile bombardment [23].

9.2. Expression of Insect Resistance Genes. Transgenic coffee (*C. canephora*) plants incorporating synthetic *cry* genes (*cry1Ac*) from *Bacillus thuringiensis* were regenerated [26]. Field assessment demonstrated the resistance of transgenic plants to coffee leaf miner (*Perileucoptera coffeella*) indicating the functional stability of the transgenes [33]. In another study, transgenic plants of *C. canephora* incorporating α-A11 from common bean were produced via *A. tumefaciens* mediated transformation, and bioassays with the insect are underway to confirm functional validation of its protein in coffee [30].

9.3. Expression of Herbicide Resistance Genes. Transgenic coffee plants incorporating the *csr-1-1* gene were produced using *A. tumefaciens* mediated transformation [26]. In this study, plants were selected using chlorsulfuron but several nontransformed escapes were obtained, which suggests that the herbicide is not a very tight selective agent. In another study, transgenic *C. canephora* plants were produced using the *bar* gene [35]. Regenerated plants were sprayed with the herbicide ammonium glufosinate under green house conditions and showed no phytotoxicity effects.

9.4. Modification of Expression of Genes Controlling Biochemical and Physiological Traits. The expression of genes, involved in the caffeine biosynthesis pathway, was modified using RNAi technology to reduce the level of *CaMXMT1* (theobromine synthase) [28]. Transgenic plants expressing antisense *ACC oxidase* (involved in fruit maturation and ethylene production) have also been produced and stable expression of the transgene observed [34].

10. Applications of Transgenic Technology

Genetic transformation technology has potential applications in coffee agriculture by incorporating desirable traits such as disease and insect resistance, drought and frost, tolerance and herbicide resistance. Transgenic technology can also be used to increase nutritional value and improve cup quality, produce varieties with caffeine-free beans, and for production of hybrid crops for molecular farming. Identification of target specific genes is one of the prerequisites for developing transgenic crops. The availability of a large number of EST sequences in coffee and initiation of coffee genome sequencing may speed up the gene discovery and accelerate transgenic research efforts in coffee.

10.1. Insect Resistance. Production of insect resistant coffee plants is one of the major objectives of the breeding programs. The major pests attacking coffee include coffee berry borer (CBB, *Hypothenemus hampei*), white stem borer (WSB, *Xylotrechus quadripes*), leaf miner (*Perileucoptera coffeela*), and root nematodes (*Meloidogyne* spp. and *Pratylenchus* spp.). The CBB is present in almost all the coffee growing countries and considered to the most devastating pest in coffee. To date, there is no reported source of resistance to CBB in the coffee gene pool. Like CBB, WSB is another serious pest in arabica coffee in India and several other

East Asian countries. Both CBB and WSB belong to the order Coleoptera. For India, controlling WSB is the biggest challenge and has therefore become the highest research priority. Robusta is generally resistant to WSB but the interspecific robusta arabica hybrids are susceptible to WSB. Although leaf miner is not yet a serious pest in India and other East Asian countries, it is an economically important pest in East Africa and Brazil.

Effective chemical control of CBB and WSB is difficult due to the nature of their life cycle inside the berry and stem, respectively, as well as environmental concern regarding the use of pesticides. Biological control measures are adapted to combat these insect pests with varying degrees of success. Developing coffee plants resistant to these pests using genetic transformation technology could be one of the alternative strategies to counter pest damage.

For insect resistance, several different classes of proteins from bacterial, plant, and animal sources have been isolated and their insecticidal properties tested against many important pests. Amongst these proteins, Bt toxins are most important and several transgenic crops expressing Bt genes have been commercialized. Coffee transgenic plants carrying a synthetic version of the *cry1Ac* gene have been produced [26, 67]. Indeed, this was the first report that an important agronomic trait has been introduced into a coffee plant. The transgenic plants presented similar features in growth and development compared to normal plants. Transgenic plants highly resistant to leaf minor under greenhouse conditions were tested under field conditions in French Guyana for 4 years for field resistance [33, 113]. From a total of 54 independent transformation events, 70% of the events were resistant to leaf minor. Unfortunately, the field trial was vandalized which led to the termination of the experiment [114].

The effectiveness of *Bt* genes in controlling coleopteran pests is well documented in corn and potato, which indicates that Bt genes might be effective against CBB. The high toxicity of *B. thuringiensis* serovar *israelensis* against first instar larvae of CBB has been demonstrated [115]. In another experiment, an α-amylase inhibitor from *Phaseolus vulgaris* was tested against CBB and found to have an inhibitory effect on its growth and development [116].

In addition to the CBB and WSB, arabica coffee varieties are also susceptible to endoparasitic root-knot nematode (*Meloidogyne* spp.) [117]. So far, 15 species have been reported to be parasites of coffee. Controlling nematodes is extremely difficult and currently seedling grafting with robusta rootstock is followed as one of the control measure. Sources of resistance specific to root-knot nematodes have been identified in coffee trees [118] and the *Mex-1* gene conferring resistance to *M. exigua* in *C. arabica* is in the process of isolation [103]. The functional analysis of nematode resistant genes could be carried out by using the *A. rhizogenes* mediated transformation protocol already developed in coffee (See Table 1). Root-specific promoters could also be used in the vector constructs to drive transgene expression in the root.

Cultivation of insect resistant transgenic coffee should address ecological concerns related to insects and soil microorganisms. Transgene stability also needs to be studied since a coffee plantation can remain for several years without replantation. The constant selection pressure of the transgenic plants on the targeted insect population should be monitored, as there may be a chance of emergence of insect resistant populations.

10.2. Tolerance to Abiotic Stress. In many coffee growing countries, abiotic stress such as drought and frost are the major climatic factors that limit coffee production. Changes in climatic patterns due to global climate change is considered increasingly important for coffee cultivation. Drought induces water stress in plants, which affects vegetative growth and vigour, and triggers floral abnormalities and poor fruit set. It also indirectly increases the incidence of pests and diseases in the plants. Arabica is generally more tolerant to water stress than robusta, partly due to its extensive deep root system. *C. racemosa* is known to be a good source of drought tolerance. In India, hybrids between *C. canephora* and *C. racemosa* have been obtained and are currently under evaluation for drought tolerance.

In many coffee growing countries, coffee is propagated in marginal areas where the annual rainfall is below 1000 mm, with prolonged dry spells of over 4-5 months. In those areas, water shortage and unfavourable temperatures constitute major constraints, and the growth and productivity of robusta coffee are badly affected.

As with drought, periodic frost also affects coffee production in parts of Brazil. The introduction of drought and frost tolerant genes through genetic transformation would be of great importance for alleviating these problems. Research is now being carried out by several groups to identify genes involved in biotic as well as abiotic stress. The most promising approach of genetic engineering for drought tolerance includes the use of functional or regulatory genes as well as the transfer of transcription factors. In recent years, plants tolerant to high temperature and water stress have been the subject of intense research [119–121]. For achieving drought tolerance, genes that have been targeted include those encoding enzymes involved in detoxification or osmotic response metabolism, enzymes active in signalling, proteins involved in the transport of metabolites, and regulating the plant energy status [119–121]. The dissection of molecular mechanisms related to signal transduction and transcriptional regulation might help in engineering drought tolerance in coffee.

10.3. Disease Resistance. Coffee leaf rust (CLR) caused by the fungus *Hemileia vastatrix* is the most important disease in coffee with substantial loss to coffee production and productivity in all the coffee growing countries. In addition to leaf rust, coffee berry disease (CBD) caused by fungus *Colletotrichum kahawae* can be a devastating anthracnose causing substantial crop loss in Africa. Several other fungal and bacterial diseases may also affect coffee, causing economic damage to a small extent. Arabica is more susceptible to many diseases than robusta coffee. Though most of the disease control measures rely upon chemical control, they are more expensive and labour intensive. The long-term solution is the breeding of resistant varieties, which is the focus of many breeding programmes. However, breeding for disease resistant varieties is time consuming due to the perennial nature of coffee, with its long gestation period. India has a long history of arabica coffee breeding especially for leaf rust resistance and is the first country to demonstrate the existence of multiple races of leaf rust. Resistance to CLR is conditioned primarily by a number of major (S_H) genes, and coffee genotypes are classified into different resistance groups based on their interaction with different rust races pathogen [122]. Currently *C. canephora* provides the main source of resistance to pests and diseases including CLR (*H. vastatrix*) and CBD (*C. kahawae*) and is therefore used in breeding programs. Other diploid species like *C. liberica* and *C. racemosa* have been used as a source of resistance to coffee leaf rust and coffee leaf miner, respectively [12, 123].

The development of coffee varieties resistant to major fungal diseases such as CLR and CBD using transgenic technology will benefit the coffee industry immensely. During the last 15 years, significant progress has been made in the area of host-pathogen interactions [124, 125] and many resistance genes involved in recognizing invading pathogens have been identified and cloned [126]. A number of signalling pathways, which are induced following pathogen infection, have been dissected [127]. Many antifungal compounds that are synthesized by plants to combat fungal infection have been identified [128]. Understanding the specific induction of targeted pathways and identification of specific pathways responsible for particular fungal resistance is important in order to employ this strategy in transgenic technology. The recent investigation of gene expression during coffee leaf rust infection could give an insight into the defence pathways operating in coffee [129, 130].

Efforts have been made to identify and clone resistance genes from coffee for achieving durable resistance. Recently, the genetic and physical map of two resistance genes, that is, the S_H 3 gene conferring resistance to rust [104] and the *Ck1* gene conferring resistance to *C. kahawae* CBD [105] have been established. These genes could be used for molecular marker assisted breeding programmes.

10.4. Production of Low-Caffeine Coffee.
Low-caffeine and decaffeinated coffee represent around 10% of the coffee sales around the world [35]. The industrial process for coffee decaffeination can be expensive and affects the original flavour and aroma in coffee [131].

Transgenic coffee plants with suppressed caffeine synthesis using RNA interference (RNAi) technology have been obtained [28, 106]. Specific sequences in the 3′ untranslated region of the theobromine synthase gene (*CaMXMT1*) were selected for construction of RNAi short and long fragments. The caffeine and theobromine content of the transgenic plants reduced by up to 70% compared to the untransformed plants. In *C. canephora*, RNAi technology has also been employed to silence the N-methyl transferase gene involved in caffeine biosynthesis [17]. Recently the promoter of an N-methyltransferase (*NMT*) gene involved in caffeine biosynthesis was cloned [100], which will be very useful for studying the regulation of caffeine biosynthesis.

10.5. Improvement in Cup Quality.
Improvement in coffee cup quality requires elaborate knowledge of the chemical constituents as well as the metabolic pathways involved in the elaboration of quality. The constituents of coffee beans include minerals, proteins, carbohydrates caffeine, chlorogenic acids (CGA), glycosides, lipids, and many volatile compounds that give flavour to coffee by roasting. Among these, the role of three major constituents: sucrose, CGA, and trigonelline have been studied in coffee. The sucrose content of coffee bean is associated with coffee flavour; the higher the sucrose content in green beans, the more intense will be the cup flavour [132, 133]. The sucrose content of *C. arabica* (8.2-8.3%) is higher than *C. canephora* (3.3–4.0%). The sucrose:amino acids ratio in green beans determines the profile of volatile compounds. Manipulating sucrose content in coffee bean is therefore important in improving cup quality. Recently, the sucrose synthase gene (*CcSUS2*) from *C. canephora* has been cloned and sequenced [66]. This provides an opportunity to manipulate the sucrose content in coffee.

Chlorogenic acids are products of phenylpropanoid metabolism. Chlorogenic acids are a group of hydroxycinnamoyl quinic acids (HQA) formed by esterification between caffeic acids, coumaric acids, and quinic acids [134]. They are present in relatively large quantities in the coffee bean and are the precursor of phenolic compounds in roasted beans. Robusta beans contain higher CGA (10%) compared to arabica beans (6-7%). CGA are known to have antioxidant properties as well as being associated with disease resistance [135]. Genetic manipulation of genes involved in CGA synthesis can serve either of these purposes by up- or downregulating the pathway.

Phenylalanine ammonia-lyase (PAL) catalyzes the first step of the phenylpropanoid pathway leading to the synthesis of a wide range of chemical compounds including flavonoids, coumarins, hydroxycinnamoyl esters, and lignins [136]. Recently, the full length cDNA and corresponding genomic sequences of PAL from *C. canephora* was isolated, characterized, and functionally validated [137, 138]. This has opened up new possibilities for manipulating the level of the PAL enzyme in coffee which in turn will be useful for cup quality improvement and manipulating antioxidant properties in coffee.

10.6. Fruit Ripening.
Uniformity during fruit ripening is decisively related to cup quality in coffee, and consequently to the value of the product. Fruits at the ideal ripening stage produce the best organoleptic characteristics for coffee. The presence of overripened or green fruits changes the acidity, the bitterness and consequently the cup quality. In order to maximise uniform ripening of coffee fruits, it is essential to control the action of genes involved in the last step of maturation process. Ethylene is known to trigger ripening, and increasing ethylene biosynthesis is associated with various stages of ripening process [139]. To control coffee fruit maturation, two of the major genes involved in ethylene biosynthesis, namely, ACC synthase and ACC oxidase, have been cloned [139, 140]. Introduction of the *ACC oxidase* gene in antisense orientation has been achieved

in both *C. arabica* and *C. canephora* [34]. The effect of the transgene on ethylene production and fruit maturation has yet to be reported. The inhibition of genes downstream to the initial ethylene burst is also an option to control coffee fruit maturation [141].

11. GM Coffee and Consumer Approval

Considering that the technology for coffee transformation is available, and given the rapid progress in gene discovery, it may not be very long before transgenic coffee hits the market. Since transgenic coffee is at the initial stages of commercial development and needs to be integrated into the main breeding programmes for evaluation, it will take at least 15–20 years for field release of GM coffee. However, the main obstacle will be consumer approval and acceptance of genetically modified coffee. With a rapid increase in the cultivation of various transgenic crops around the globe, consumer perception towards transgenic coffee may be more positive than it is today. Undoubtedly, GM coffee must undergo rigorous testing on both health and environmental effects before it is released for commercial cultivation.

12. Conclusion and Future Perspectives

During the last 15 years, transgenic crops became an integral part of the agricultural landscape. The number of transgenic crops and the area under cultivation is rapidly increasing in many parts of the world. This has been made possible by the application of genetic transformation technology and its integration with plant breeding programmes. Despite significant advances made over the last 15 years, coffee transformation is still time consuming and laborious. In addition, a genotype independent transformation protocol is not yet achieved in coffee. Genetic transformation of coffee has two major applications: (1) a tool for the validation of gene function and (2) production of transgenic crops with agronomically important traits. In order to achieve these goals, a simple, efficient, genotype-independent routine transformation protocol needs to be developed for coffee. Public concern regarding the use of antibiotic marker genes in transgenic technology need to be addressed. In this regard, coffee transformation should be based on several strategies such as use of positive selection markers and GFP for transformant selection, and on the *cre/lox* system for elimination of selectable marker genes. Development of a clean gene transgenic technology for coffee based on the *gfp* gene for visual selection and the *cre/lox* vector system is currently in progress. All coffee transformation programs should address the expression of transgenes in appropriate tissues, for which tissue specific promoters need to be used. The stable expression of transgenes should be monitored at every stage of plant growth and development. In addition, genomic technologies such as transgenics, molecular marker assisted breeding, genomics, proteomics, and metabolomics should complement traditional breeding efforts for hastening the genetic improvement of coffee.

Further coffee transformation programmes and investments should involve public and private companies. At the same time, researchers must make an effort to educate the public and help them understand the real advantages and risks associated with the use of genetically modified coffee. This is the only way to address irrational fears about the transgenic crops, and this will pave the way to the use of transgenic technology for coffee improvement.

Acknowledgments

M. K. Mishra thanks the Department of Biotechnology, Government of India for an overseas Associateship to De Montfort University, UK and Coffee Board, India for deputation.

References

[1] ICO. International Coffee Organization, "ICO Annual Review 2010," 2010, http://www.ico.org/.

[2] A. P. Davis, R. Govaerts, D. M. Bridson, and P. Stoffelen, "An annotated taxonomic conspectus of the genus *Coffea* (Rubiaceae)," *Botanical Journal of the Linnean Society*, vol. 152, no. 4, pp. 465–512, 2006.

[3] A. Charrier and J. Berthaud, "Botanical classification of coffee," in *Coffee, Botany, Biochemistry and Production of Beans and Beverage*, M. N. Clifford and K. C. Wilson, Eds., pp. 13–47, Croom Helm, London, UK, 1985.

[4] A. Lécolier, P. Besse, A. Charrier, T. N. Tchakaloff, and M. Noirot, "Unraveling the origin of *Coffea arabica* "Bourbon pointu" from la Réunion: a historical and scientific perspective," *Euphytica*, vol. 168, no. 1, pp. 1–10, 2009.

[5] T. W. M. Gole, T. Demel, M. Denich, and T. Bosch, "Diversity of traditional coffee production systems in Ethiopia and their contribution to conservation of genetic diversity," in *Proceedings of the International Agricultural Research for Development*, Deutscher Tropentag, Bonn, Germany, 2001.

[6] A. S. Thomas, "The wild arabica coffee on the Boma plateau of Anglo-Egyptian Sudan," *Empirical Journal of Experimental Agriculture*, vol. 10, pp. 207–212, 1942.

[7] F. Anthony, J. Berthaud, J. L. Guillaumet, and M. Lourd, "Collecting wild coffee species in Kenya and Tanzania," *Plant Genetic Resources Newsletter*, vol. 69, pp. 23–29, 1987.

[8] E. W. Githae, M. Chuah-Petiot, J. K. Mworia, and D. W. Odee, "A botanical inventory and diversity assessment of Mt. Marsabit forest, a sub-humid montane forest in the arid lands of Northern Kenya," *African Journal of Ecology*, vol. 46, no. 1, pp. 39–45, 2008.

[9] M. F. Carneiro, "Coffee biotechnology and its application in genetic transformation," *Euphytica*, vol. 96, no. 1, pp. 167–172, 1997.

[10] H. A. M. van der Vossen, "Coffee selection and breeding," in *Coffee: Botany, Biochemistry and Production of Beans and Beverage*, M. N. Clifford and K. C. Wilson, Eds., pp. 48–96, Croom Helm, London, UK, 1985.

[11] K. H. Srinivasan and R. L. Narasimhaswamy, "A review of coffee breeding work done at the Government coffee experiment station, Balehonnur," *Indian Coffee*, vol. 34, pp. 311–321, 1975.

[12] O. Guerreiro Filho, M. B. Silvarolla, and A. B. Eskes, "Expression and mode of inheritance of resistance in coffee to leaf miner *Perileucoptera coffeella*," *Euphytica*, vol. 105, no. 1, pp. 7–15, 1999.

[13] A. Bettencourt, *Considerações gerais sobre o 'Híbrido de Timor'*, Instituto Agronomico de Campinas. Circular, no. 31, Instituto Agronomico de Campinas, 1973.

[14] "Coffee Guide," Central Coffee Research Institute, Coffee Board, Bangalore, India, 2000.

[15] C. R. Barton, T. L. Adams, and M. A. Zarowitz, "Stable transformation of foreign DNA into *Coffea arabica* plants," in *Proceedings of the 14th International Conference on Coffee Science (ASIC '91)*, pp. 460–464, San Francisco, Calif, USA, 1991.

[16] R. Fernandez-Da Silva and A. Menéndez-Yuffá, "Transient gene expression in secondary somatic embryos from coffee tissues electroporated with the genes *GUS* and *BAR*," *Electronic Journal of Biotechnology*, vol. 6, no. 1, pp. 29–35, 2003.

[17] V. Kumar, K. V. Satyanarayana, A. Ramakrishna, A. Chandrashekar, and G. A. Ravishankar, "Evidence for localization of *N*-methyltransferase (MMT) of caffeine biosynthetic pathway in vacuolar surface of *Coffea canephora* endosperm elucidated through localization of GUS reporter gene driven by NMT promoter," *Current Science*, vol. 93, no. 3, pp. 383–386, 2007.

[18] J. van Boxtel, M. Berthouly, C. Carasco, M. Dufour, and A. Eskes, "Transient expression of beta-glucuronidase following biolistic delivery of foreign DNA into coffee tissues," *Plant Cell Reports*, vol. 14, no. 12, pp. 748–752, 1995.

[19] A. G. Rosillo, J. R. Acuna, A. L. Gaitan, and M. de Pena, "Optimised DNA delivery into *Coffea arabica* suspension culture cells by particle bombardment," *Plant Cell, Tissue and Organ Culture*, vol. 74, no. 1, pp. 45–49, 2003.

[20] W. G. Cunha, F. R. B. Machado, G. R. Vianna, J. B. Teixeira, and E. V. S. Albuquerque, *Obtencao De Coffea Arabica Geneticamente Modificadas Por Bombardmento De Calos Embriogenecos*, vol. 73 of *Boletim de Pesquisa e desenvolvimento*, Embrapa, Brasilia, Brazil, 2004.

[21] A. F. Ribas, A. K. Kobayashi, L. F. P. Pereira, and L. G. E. Vieira, "Genetic transformation of *Coffea canephora* by particle bombardment," *Biologia Plantarum*, vol. 49, no. 4, pp. 493–497, 2005.

[22] A. M. Gatica-Arias, G. Arrieta-Espinoza, and A. M. Espinoza Esquivel, "Plant regeneration via indirect somatic embryogenesis and optimisation of genetic transformation in *Coffea arabica* L. cvs. Caturra and Catuaí," *Electronic Journal of Biotechnology*, vol. 11, no. 1, pp. 1–11, 2008.

[23] E. V. S. Albuquerque, W. G. Cunha, A. E. A. D. Barbosa et al., "Transgenic coffee fruits from *Coffea arabica* genetically modified by bombardment," *In Vitro Cellular & Developmental Biology—Plant*, vol. 45, no. 5, pp. 532–539, 2009.

[24] J. Spiral and V. Petiard, "Protoplast culture and regeneration in coffee species," in *Proceedings of the 14th International Conference on Coffee Science (ASIC '91)*, pp. 383–391, San Francisco, Calif, USA, 1991.

[25] T. Hatanaka, Y. E. Choi, T. Kusano, and H. Sano, "Transgenic plants of coffee *Coffea canephora* from embryogenic callus via *Agrobacterium tumefaciens*-mediated transformation," *Plant Cell Reports*, vol. 19, no. 2, pp. 106–110, 1999.

[26] T. Leroy, A. M. Henry, M. Royer et al., "Genetically modified coffee plants expressing the *Bacillus thuringiensiscry1Ac* gene for resistance to leaf miner," *Plant Cell Reports*, vol. 19, no. 4, pp. 382–389, 2000.

[27] M. K. Mishra, H. L. Sreenath, and C. S. Srinivasan, "*Agrobacterium*-mediated transformation of coffee: an assessment of factors affecting gene transfer efficiency," in *Proceedings of the 15th Plantation Crops Symposium, K. Sreedharan, P. K.

[28] S. Ogita, H. Uefuji, M. Morimoto, and H. Sano, "Application of RNAi to confirm theobromine as the major intermediate for caffeine biosynthesis in coffee plants with potential for construction of decaffeinated varieties," *Plant Molecular Biology*, vol. 54, no. 6, pp. 931–941, 2004.

[29] M. K. Mishra and H. L. Sreenath, "High-efficiency *Agrobacterium*-mediated transformation of coffee (*Coffea canephora* Pierre ex. Frohner) using hypocotyl explants," in *Proceedings of the 20th International Conference on Coffee Science (ASIC '04)*, pp. 792–796, Bangalore, India, October 2004.

[30] A. R. R. Cruz, A. L. D. Paixao, F. R. Machado et al., *Obtencao de plantas transformadas de Coffea canephora por co-cultivo de calos embriogenicos com A. Tumefaciens*, vol. 73 of *Boletim de pesquiza e Desenvolvimento*, Embrapa, Brasilia, Brazil, 2004.

[31] R. L. R. Canche-Moo, A. Ku-Gonzalez, C. Burgeff, V. M. Loyola-Vargas, L. C. Rodríguez-Zapata, and E. Castaño, "Genetic transformation of *Coffea canephora* by vacuum infiltration," *Plant Cell, Tissue and Organ Culture*, vol. 84, no. 3, pp. 373–377, 2006.

[32] V. Kumar, K. V. Sathyanarayana, S. Saarala Itty, P. Giridhar, A. Chandrasekhar, and G. A. Ravishankar, "Post transcriptional gene silencing for down regulating caffeine biosynthesis in *Coffea canephora* P. ex Fr," in *Proceedings of the 20th International Conference on Coffee Science (ASIC '04)*, pp. 769–774, Bangalore, India, 2004.

[33] B. Perthuis, J. L. Pradon, C. Montagnon, M. Dufour, and T. Leroy, "Stable resistance against the leaf miner *Leucoptera coffeella* expressed by genetically transformed *Coffea canephora* in a pluriannual field experiment in French Guiana," *Euphytica*, vol. 144, no. 3, pp. 321–329, 2005.

[34] A. F. Ribas, R. M. Galvao, L. F. P. Pereira, and L. G. E. Vieira, "Transformacao de *Coffea arabica* com o gene da ACC oxidase em orientacao antinsenso," in *Proceedings of the 50th Congreso Brasileiro de Genetica*, p. 492, Sao Pulo, Brazil, September 2005.

[35] A. F. Ribas, A. K. Kobayashi, L. F. P. Pereira, and L. G. E. Vieira, "Production of herbicide-resistant coffee plants (*Coffea canephora* P.) via *Agrobacterium tumefaciens*-mediated transformation," *Brazilian Archives of Biology and Technology*, vol. 49, no. 1, pp. 11–19, 2006.

[36] A. Arroyo-Herrera, A. Ku Gonzalez, R. Canche Moo et al., "Expression of *WUSCHEL* in *Coffea canephora* causes ectopic morphogenesis and increases somatic embryogenesis," *Plant Cell, Tissue and Organ Culture*, vol. 94, no. 2, pp. 171–180, 2008.

[37] M. K. Mishra, H. L. Sreenath, Jayarama et al., "Two critical factors: *Agrobacterium* strain and antibiotics selection regime improve the production of transgenic coffee plants," in *Proceedings of the 22nd International Association for Coffee Science (ASIC '08)*, pp. 843–850, Campinas, Brazil, 2008.

[38] A. F. Ribas, E. Dechamp, A. Champion et al., "*Agrobacterium*-mediated genetic transformation of *Coffea arabica* (L.) is greatly enhanced by using established embryogenic callus cultures," *BMC Plant Biology*, vol. 11, article 92, 2011.

[39] J. Spiral, C. Thierry, M. Paillard, and V. Petiard, "Obtention de plantules de *Coffea canephora* Pierre (Robusta) transformées par *Agrobacterium rhizogenes*," *Comptes Rendus de l' Academie des Sciences de Paris*, vol. 316, no. 1, pp. 1–6, 1993.

[40] M. Sugiyama, C. Matsuoka, and T. Takagi, "Transformation of *Coffea* with *Agrobacterium rhizogenes*," in *Proceedings of the

[—] V. Kumar, Jayarama, and B. M. Chulaki, Eds., pp. 251–255, Mysore, India, December 2002.

16th International Conference on Coffee Science (ASIC '95), pp. 853–859, Kyoto, Japan, 1995.

[41] C. A. Giménez, A. Menéndez-Yuffá, and E. de García, "Efecto del antibiótico kanamicina sobre diferentes explantes del híbrido de café (*Coffea* sp.) Catimor," *Phyton*, vol. 59, pp. 39–46, 1996.

[42] V. Kumar, K. V. Satyanarayana, S. Sarala Itty et al., "Stable transformation and direct regeneration in *Coffea canephora* P ex. Fr. by *Agrobacterium rhizogenes* mediated transformation without hairy-root phenotype," *Plant Cell Reports*, vol. 25, no. 3, pp. 214–222, 2006.

[43] E. Alpizar, E. Dechamp, B. Bertrand, P. Lashermes, and H. Etienne, "Transgenic roots for functional genomics of coffee resistance genes to root-knot nematodes," in *Proceedings of the 21st International Conference on Coffee Science (ASIC '06)*, pp. 653–659, Montpellier, France, September 2006.

[44] E. Alpizar, E. Dechamp, S. Espeout et al., "Efficient production of *Agrobacterium rhizogenes*-transformed roots and composite plants for studying gene expression in coffee roots," *Plant Cell Reports*, vol. 25, no. 9, pp. 959–967, 2006.

[45] H. Etienne, P. Lashermes, A. Menendez-Yuffa, Z. de Guglielmo-Croquer, E. Alpizar, and H. L. Sreenath, "Coffee," in *Compendium of Transgenic Crop Plants, Transgenic Plantation Crops, Ornamentals and Turf Grasses*, C. Kole and T. Hall, Eds., pp. 57–84, Wiley Blackwell Publishers, London, UK, 2008.

[46] M. C. Combes, S. Andrzejewski, F. Anthony et al., "Characterization of microsatellite loci in *Coffea arabica* and related coffee species," *Molecular Ecology*, vol. 9, no. 8, pp. 1178–1180, 2000.

[47] D. L. Steiger, C. Nagai, P. H. Moore, C. W. Morden, R. V. Osgood, and R. Ming, "AFLP analysis of genetic diversity within and among *Coffea arabica* cultivars," *Theoretical and Applied Genetics*, vol. 105, no. 2-3, pp. 209–215, 2002.

[48] L. E. C. Diniz, N. S. Sakiyama, P. Lashermes, E. T. Caixeta, A. C. B. Oliveira, Zambolim et al., "Analysis of AFLP marker associated to the *Mex-1* resistance locus in Icatu progenies," *Crop Breeding and Applied Biotechnology*, vol. 5, pp. 387–393, 2005.

[49] M. P. Maluf, M. Silvestrini, L. M. C. Ruggiero, O. Guerreiro Filho, and C. A. Colombo, "Genetic diversity of *Coffea arabica* inbreed lines assessed by RAPD, AFLP and SSR marker system," *Scientia Agricola*, vol. 62, no. 4, pp. 366–373, 2005.

[50] P. S. Hendre, R. Phanindranath, V. Annapurna, A. Lalremruata, and R. K. Aggarwal, "Development of new genomic microsatellite markers from robusta coffee (*Coffea canephora* Pierre ex A. Froehner) showing broad cross-species transferability and utility in genetic studies," *BMC Plant Biology*, vol. 8, article 51, 2008.

[51] R. F. Missio, E. T. Caixeta, E. M. Zambolim, L. Zambolim, C. D. Cruz, and N. S. Sakiyama, "Polymorphic information content of SSR markers for *Coffea* spp.," *Crop Breeding and Applied Biotechnology*, vol. 10, no. 1, pp. 89–94, 2010.

[52] P. Tornincasa, R. Dreos, B. de Nardi, E. Asquini, J. Devasia, M. K. Mishra et al., "Genetic diversity of commercial coffee (*C. arabica* L) from America, India and Africa assessed by simple sequence repeats (SSRs)," in *Proceedings of the 21st International Association for Coffee Science (ASIC '06)*, pp. 778–785, Montpellier, France, 2006.

[53] M. K. Mishra, S. Nishani, and Jayarama, "Genetic relationship among indigenous coffee species from India using RAPD, ISSR and SRAP markers," *Biharean Biologists*, vol. 5, no. 1, pp. 17–24, 2011.

[54] M. K. Mishra, S. Nishani, and Jayarama, "Molecular identification and genetic relationships among coffee species (*Coffea* L.) inferred from ISSR and SRAP marker analyses," *Archives of Biological Sciences*, vol. 63, no. 3, pp. 667–679, 2011.

[55] M. K. Mishra, N. Suresh, A. M. Bhat et al., "Genetic molecular analysis of *Coffea arabica* (Rubiaceae) hybrids using SRAP markers," *Revista de Biologia Tropical*, vol. 59, no. 2, pp. 607–617, 2011.

[56] M. K. Mishra, A. M. Bhat, N. Suresh et al., "Molecular genetic analysis of arabica coffee hybrids using SRAP marker approach," *Journal of Plantation Crops*, vol. 39, no. 1, pp. 41–47, 2011.

[57] M. K. Mishra, P. Tornincasa, B. de Nardi, E. Asquini, R. Dreos, L. Del Terra et al., "Genome organization in coffee as revealed by EST PCRRFLP, SNPs and SSR analysis," *Journal of Crop Science and Biotechnology*, vol. 14, no. 1, pp. 25–37, 2011.

[58] M. T. Gil-Agusti, N. Campostrini, L. Zolla, C. Ciambella, C. Invernizzi, and P. G. Righetti, "Two-dimensional mapping as a tool for classification of green coffee bean species," *Proteomics*, vol. 5, no. 3, pp. 710–718, 2005.

[59] P. S. Hendre, R. K. Aggarwal, and D. N. A. Markers, "Development and applications for genetic improvement of coffee," in *Genomics Assisted Crop Improvement*, R. K. Varshney and R. Tuberosa, Eds., vol. 2 of *Genomics Application in Crops*, pp. 399–434, Springer.

[60] M. de Block, L. Herrera-Estrella, M. van Montagu, J. Schell, and P. Zambryski, "Expression of foreign genes in regenerated plants and in their progeny," *The EMBO Journal*, vol. 3, no. 8, pp. 1681–1689, 1984.

[61] G. Y. Zhong, "Genetic issues and pitfalls in transgenic plant breeding," *Euphytica*, vol. 118, no. 2, pp. 137–144, 2001.

[62] A. Pallavicini, L. Del Terra, M. R. Sondahl et al., "Transcriptomics of resistance response in *Coffea arabica* L," in *Proceedings of the 20th International conference on coffee science (ASIC '04)*, pp. 66–67, Bangalore, India, October 2004.

[63] C. Lin, L. A. Mueller, J. M. Carthy, D. Crouzillat, V. Pétiard, and S. D. Tanksley, "Coffee and tomato share common gene repertoires as revealed by deep sequencing of seed and cherry transcripts," *Theoretical and Applied Genetics*, vol. 112, no. 1, pp. 114–130, 2005.

[64] L. G. E. Vieira, A. C. Andrade, C. Colombo et al., "Brazillian coffee genome project: an EST-based genomic resource," *Brazillian Journal of Plant Physiology*, vol. 18, no. 1, pp. 95–108, 2006.

[65] S. Noir, S. Patheyron, M. C. Combes, P. Lashermes, and B. Chalhoub, "Construction and characterisation of a BAC library for genome analysis of the allotetraploid coffee species (*Coffea arabica* L.)," *Theoretical and Applied Genetics*, vol. 109, no. 1, pp. 225–230, 2004.

[66] T. Leroy, P. Marraccini, M. Dufour et al., "Construction and characterization of a *Coffea canephora* BAC library to study the organization of sucrose biosynthesis genes," *Theoretical and Applied Genetics*, vol. 111, no. 6, pp. 1032–1041, 2005.

[67] J. Spiral, T. Leroy, M. Paillard, and V. Petiard, "Transgenic coffee (*Coffea* sp.)," in *Biotechnology in Agricultulture and Forestry*, Y. P. S. Bajaj, Ed., pp. 55–76, Springer, Heidelberg, Germany, 1999.

[68] M. F. Carneiro, "Advances in coffee biotechnology," *AgBiotechNet*, vol. 1, pp. 1–14, 1999.

[69] V. Kumar, M. M. Naidu, and G. A. Ravishankar, "Developments in coffee biotechnology—*in vitro* plant propagation and crop improvement," *Plant Cell, Tissue and Organ Culture*, vol. 87, no. 1, pp. 49–65, 2006.

[70] G. Staritsky, "Embryoid formation in callus tissues of coffee," *Acta Botanica Neerlandica*, vol. 19, pp. 509–514, 1970.

[71] W. R. Sharp, L. S. Caldas, O. J. Crocomo, L. C. Monaco, and A. Carvalho, "Production of *Coffea arabica* callus of three ploidy levels and subsequent morphogenesis," *Phyton*, vol. 31, pp. 67–74, 1973.

[72] M. R. Sondahl and W. Sharp, "High frequency induction of somatic embryos in cultured leaf explants of *Coffea arabica* L," *Zeitschrift für Pflanzenphysiologie*, vol. 81, pp. 395–408, 1977.

[73] M. R. Sondahl, D. Spahlinger, and W. R. Sharp, "A histological study of high frequency and low frequency induction of somatic embryos in cultured leaf explants of *Coffea arabica* L," *Zeitschrift für Pflanzenphysiologie*, vol. 94, no. 2, pp. 101–108, 1979.

[74] F. R. Quiroz-Figueroa, C. F. J. Fuentes-Cerda, R. Rojas-Herrera, and V. M. Loyola-Vargas, "Histological studies on the developmental stages and differentiation of two different somatic embryogenesis systems of *Coffea arabica*," *Plant Cell Reports*, vol. 20, no. 12, pp. 1141–1149, 2002.

[75] F. R. Quiroz-Figueroa, M. Méndez-Zeel, F. Sánchez-Teyer, R. Rojas-Herrera, and V. M. Loyola-Vargas, "Differential gene expression in embryogenic and non-embryogenic clusters from cell suspension cultures of *Coffea arabica*," *Journal of Plant Physiology*, vol. 159, no. 11, pp. 1267–1270, 2002.

[76] F. R. Quiroz-Figueroa, M. Méndez-Zeel, A. Larqué-Saavedra, and V. M. Loyola-Vargas, "Picomolar concentrations of salicylates induce cellular growth and enhance somatic embryogenesis in *Coffea arabica* tissue culture," *Plant Cell Reports*, vol. 20, no. 8, pp. 679–684, 2001.

[77] P. Giridhar, E. P. Indu, G. A. Ravishankar, and A. Chandrasekar, "Influence of triacontanol on somatic embryogenesis in *Coffea arabica* L. and *Coffea canephora* P. ex Fr," *In Vitro Cellular & Developmental Biology—Plant*, vol. 40, no. 2, pp. 200–203, 2004.

[78] P. Giridhar, E. P. Indu, K. Vinod, A. Chandrashekar, and G. A. Ravishankar, "Direct somatic embryogenesis from *Coffea arabica* L. and *Coffea canephora* P ex Fr. under the influence of ethylene action inhibitor-silver nitrate," *Acta Physiologiae Plantarum*, vol. 26, no. 3, pp. 299–305, 2004.

[79] P. Giridhar, V. Kumar, E. P. Indu, A. Chandrasekar, and G. A. Ravishankar, "Thidiazuron induced somatic embryogenesis in *Coffea arabica* L. and *Coffea canephora* P ex Fr," *Acta Botanica Croatica*, vol. 63, no. 1, pp. 25–33, 2004.

[80] V. Kumar, P. Giridhar, A. Chandrashekar, and G. A. Ravishankar, "Polyamines influence morphogenesis and caffeine biosynthesis in *in vitro* cultures of *Coffea canephora* P. ex Fr," *Acta Physiologiae Plantarum*, vol. 30, no. 2, pp. 217–223, 2008.

[81] A. Ramakrishna, P. Giridhar, M. Jobin, C. S. Paulose, and G. A. Ravishankar, "Indoleamines and calcium enhance somatic embryogenesis in *Coffea canephora* P ex Fr," *Plant Cell, Tissue and Organ Culture*, vol. 108, no. 2, pp. 267–278, 2012.

[82] T. Hatanaka, E. Sawabe, T. Azuma, N. Uchida, and T. Yasuda, "The role of ethylene in somatic embryogenesis from leaf discs of *Coffea canephora*," *Plant Science*, vol. 107, no. 2, pp. 199–204, 1995.

[83] M. de Feria, E. Jiménez, R. Barbón, A. Capote, M. Chávez, and E. Quiala, "Effect of dissolved oxygen concentration on differentiation of somatic embryos of *Coffea arabica* cv.

Catimor 9722," *Plant Cell, Tissue and Organ Culture*, vol. 72, no. 1, pp. 1–6, 2003.

[84] A. Zamarripa, J. P. Ducos, H. Tessereau, H. Bollon, A. B. Eskes, and V. Pétiard, "Développement d'un procédé demultiplication en masse du caféier par embryogenèse somatique en milieu liquide," in *Proceedings of the 14th International Scientific Colloquium on Coffee (ASIC '91)*, pp. 392–402, San Francisco, Calif, USA, 1991.

[85] A. Zamarripa, *Etude et development de l'embryogenese en millieu liquid du cafeier (Coffea canephora P., Coffea arabica L. Et Hybrid Arabusta) [Ph.D. thesis]*, Ecolo National Superieure Agronomique, Rennes, France, 1993.

[86] J. P. Ducos, A. Zamarripa, A. B. Eskes, and V. Petiard, "Production of somatic embryos of coffee in a bioreactor," in *Proceedings of the 15th International Conference on Coffee Science (ASIC '93)*, pp. 89–96, Montpellier, France, 1993.

[87] C. Noega and M. R. Sondahl, "Arabica coffee micropropagation through somatic embryogenesis via bioreactors," in *Proceedings of the 15th International Conference on Coffee Science (ASIC '93)*, pp. 73–81, Montpellier, France, 1993.

[88] H. Etienne and M. Berthouly, "Temporary immersion systems in plant micropropagation," *Plant Cell, Tissue and Organ Culture*, vol. 69, no. 3, pp. 215–231, 2002.

[89] J. Albarrán, B. Bertrand, M. Lartaud, and H. Etienne, "Cycle characteristics in a temporary immersion bioreactor affect regeneration, morphology, water and mineral status of coffee (*Coffea arabica*) somatic embryos," *Plant Cell, Tissue and Organ Culture*, vol. 81, no. 1, pp. 27–36, 2005.

[90] M. F. Carneiro and T. M. O. Ribeiro, "*In vitro* meristem culture and plant regeneration in some genotypes of *Coffea arabica*," *Broteria Genetica*, vol. 85, pp. 127–138, 1989.

[91] M. Berthouly, D. Alvarad, C. Carrasco, and C. Teisson, "*In vitro* micropropagation of *coffee sp.* By temporary immersion," in *Abstracts Eighth International Congress of Plant Tissue and Cell Culture*, p. 162, Florence, Italy, 1994.

[92] C. Teisson, D. Alvard, M. Berthouly, F. Cote, J. V. Escalant, and H. Etienne, "Culture *in vitro* par immersion temporaire un nouveau récipient," *Plantations, Recherche, Développement*, vol. 2, no. 5, pp. 29–31, 1995.

[93] J. P. Ducos, R. Alenton, J. F. Reano, C. Kanchanomai, A. Deshayes, and V. Pétiard, "Agronomic performance of *Coffea canephora* P. trees derived from large-scale somatic embryo production in liquid medium," *Euphytica*, vol. 131, no. 2, pp. 215–223, 2003.

[94] M. R. Sondahl, W. R. Romig, and A. Bragin, "Induction and selection of somaclonal variation in coffee," US patent 5436395, 1995.

[95] V. M. Loyola-Vargas, C. Fuentes, M. Monforte-Gonzalez, M. Mendez-Zeel, R. Rojas, and J. Mijangos-Cortes, "Coffee tissue culture as a new modelfor the study of somaclonal variation," in *Proceedings of the 18th International Conference on Coffee Science (ASIC '99)*, pp. 302–307, Paris, France, 1999.

[96] L. F. Sanchez-Teyer, F. R. Quiroz-Figueroa, V. M. Loyola-Vargas, and D. Infante-Herrera, "Culture-induced variation in plants of *Coffea arabica* cv. Caturra rojo, regenerated by direct and indirect somatic embryogenesis," *Molecular Biotechnology*, vol. 23, no. 2, pp. 107–115, 2003.

[97] V. Rani, K. P. Singh, B. Shiran et al., "Evidence for new nuclear and mitochondrial genome organizations among high-frequency somatic embryogenesis-derived plants of allotetraploid *Coffea arabica* L. (Rubiaceae)," *Plant Cell Reports*, vol. 19, no. 10, pp. 1013–1020, 2000.

[98] D. Ganesh and H. L. Sreenath, "Embryo culture in coffee: technique and applications," *Indian Coffee*, vol. 4, pp. 7–9, 1999.

[99] G. de Moro, M. Modonut, E. Asquini, P. Tornincasa, A. Pallavicini, and G. Graziosi, "Development and analysis of an EST databank of *Coffea arabica*," in *Proceedings of the 6th Solanaceae Genome Workshop*, p. 127, New Delhi, India, 2009.

[100] K. V. Satyanarayana, V. Kumar, A. Chandrashekar, and G. A. Ravishankar, "Isolation of promoter for *N*-methyltransferase gene associated with caffeine biosynthesis in *Coffea canephora*," *Journal of Biotechnology*, vol. 119, no. 1, pp. 20–25, 2005.

[101] C. Hinniger, V. Caillet, F. Michoux et al., "Isolation and characterization of cDNA encoding three dehydrins expressed during *Coffea canephora* (Robusta) grain development," *Annals of Botany*, vol. 97, no. 5, pp. 755–765, 2006.

[102] A. J. Simkin, T. Qian, V. Caillet et al., "Oleosin gene family of *Coffea canephora*: quantitative expression analysis of five oleosin genes in developing and germinating coffee grain," *Journal of Plant Physiology*, vol. 163, no. 7, pp. 691–708, 2006.

[103] S. Noir, F. Anthony, B. Bertrand, M. C. Combes, and P. Lashermes, "Identification of a major gene (*Mex*-1) from *Coffea canephora* conferring resistance to *Meloidogyne exigua* in *Coffea arabica*," *Plant Pathology*, vol. 52, no. 1, pp. 97–103, 2003.

[104] N. S. Prakash, D. V. Marques, V. M. P. Varzea, M. C. Silva, M. C. Combes, and P. Lashermes, "Introgression molecular analysis of a leaf rust resistance gene from *Coffea liberica* into *C. arabica* L," *Theoretical and Applied Genetics*, vol. 109, no. 6, pp. 1311–1317, 2004.

[105] E. K. Gichuru, M. C. Combes, E. W. Mutitu et al., "Characterization and genetic mapping of a gene conferring resistance to coffee berry disease (*Colletotrichum kahawae*) in Arabica coffee (*Coffea arabica*)," in *Proceedings of the 21st International Conference on Coffee Science (ASIC '06)*, pp. 786–793, Montpellier, France, 2006.

[106] S. Ogita, H. Uefuji, Y. Yamaguchi, N. Koizumi, and H. Sano, "RNA interference: producing decaffeinated coffee plants," *Nature*, vol. 423, no. 6942, p. 823, 2003.

[107] V. Sridevi, P. Giridhar, P. S. Simmi, and G. A. Ravishankar, "Direct shoot organogenesis on hypocotyl explants with collar region from *in vitro* seedlings of *Coffea canephora* Pierre ex. Frohner cv. C × R and *Agrobacterium tumefaciens*-mediated transformation," *Plant Cell, Tissue and Organ Culture*, vol. 101, no. 3, pp. 339–347, 2010.

[108] T. Leroy and M. Dufour, "*Coffea* spp. Genetic transformation," in *Transgenic Crops of the World: Essential Protocols*, I. S. Curtis, Ed., pp. 159–170, Kluwer Academic Publishers, Dordrecht, The Netherlands, 2004.

[109] T. Leroy, M. Royer, M. Paillard et al., "Introduction de genes d'interet agronomique dans l'espece *Coffea canephora* Pierre par transformation avec *Agrobacterium sp*," in *Proceedings of the 17th International Conference on Coffee Science (ASIC '97)*, pp. 439–446, Nairobi, Kenya, 1997.

[110] J. van Boxtel, A. Eskes, and M. Berthouly, "Glufosinate as an efficient inhibitor of callus proliferation in coffee tissue," *In Vitro Cellular & Developmental Biology—Plant*, vol. 33, no. 1, pp. 6–12, 1997.

[111] N. P. Samson, C. Campa, M. Noirot, and A. de Kochko, "Potential use of D-xylose for coffee transformation," in *Proceedings of the 20th International Conference on Coffee Science (ASIC '04)*, pp. 707–713, Bangalore, India, 2004.

[112] M. K. Mishra, S. Devi, A. McCormac et al., "Green fluorescent protein as a visual selection marker for coffee transformation," *Biologia*, vol. 65, no. 4, pp. 639–646, 2010.

[113] B. Perthuis, T. Leroy, M. Dufour et al., "Variability in the insecticidal protein concentration within transformed *Coffea canephora* observed in a field experiment," in *Proceedings of the 21st International Cconference on Coffee Science (ASIC '06)*, pp. 1390–1393, Montpellier, France, September 2006.

[114] C. Montagnon, "Genetically modified coffees- the experience of CIRAD," in *ICO seminar on Genetically Modified Coffee Workshop*, May 2005, http://dev.ico.org/event_pdfs/gm/presentations/Christophe%20Montagnon.pdf.

[115] I. Méndez-López, R. Basurto-Ríos, and J. E. Ibarra, "*Bacillus thuringiensis* serovar *israelensis* is highly toxic to the coffee berry borer, *Hypothenemus hampei* Ferr. (Coleoptera: Scolytidae)," *FEMS Microbiology Letters*, vol. 226, no. 1, pp. 73–77, 2003.

[116] M. F. Grossi de Sa, R. A. Pereira, E. V. S. A. Barros et al., "Uso de inibidores de alfa-amilases no controle da broca-do-cafe," in *Anais do workshop Internacional de Manejo da Broca-do-cafe*, Paraná, Brazil, 2004.

[117] V. P. Campos, P. Sivapalan, and N. C. Gnanapragasam, "Nematode parasites of coffee, cocoa and tea," in *Plant Parasitic Nematodes in Subtropical and Tropical Agriculture*, M. Luc, R. A. Sikora, and J. Bridge, Eds., pp. 113–126, CAB International, Wallingford, UK, 1990.

[118] B. Bertrand, F. Anthony, and P. Lashermes, "Breeding for resistance to *Meloidogyne exigua* in *Coffea arabica* by introgression of resistance genes of *Coffea canephora*," *Plant Pathology*, vol. 50, no. 5, pp. 637–643, 2001.

[119] M. M. Chaves, J. P. Maroco, and J. S. Pereira, "Understanding plant responses to drought—from genes to the whole plant," *Functional Plant Biology*, vol. 30, no. 3, pp. 239–264, 2003.

[120] M. M. Chaves and M. M. Oliveira, "Mechanisms underlying plant resilience to water deficits: prospects for water-saving agriculture," *Journal of Experimental Botany*, vol. 55, no. 407, pp. 2365–2384, 2004.

[121] I. Coraggio and R. Tuberosa, "Molecular basis of plant adaptation to abiotic stress and approaches to enhance tolerance to hostile environment," in *Hand Book of Plant Biotechnology*, P. Christou and H. Klee, Eds., pp. 413–466, John Wiley & Sons, West Sussex, UK, 2004.

[122] H. A. M. van der Vossen, "Agronomy I: coffee breeding practices," in *Coffee: Recent Developments*, R. J. Clarke and O. G. Vitzthum, Eds., pp. 184–201, Blackwell Science, London, UK, 2001.

[123] M. S. Sreenivasan, "Breeding for leaf rust resistance in India," in *Coffee Rust: Epidemiology, Resistance and Management*, A. C. Kushalappa and A. B. Eskes, Eds., pp. 316–323, CRC Press, Boca Raton, Fla, USA, 1989.

[124] B. J. Staskawicz, F. M. Ausubel, B. J. Baker, J. G. Ellis, and J. D. G. Jones, "Molecular genetics of plant disease resistance," *Science*, vol. 268, no. 5211, pp. 661–667, 1995.

[125] B. J. Feys and J. E. Parker, "Interplay of signaling pathways in plant disease resistance," *Trends in Genetics*, vol. 16, no. 10, pp. 449–455, 2000.

[126] F. L. W. Takken and M. H. A. J. Joosten, "Plant resistance genes: their structure, function and evolution," *European Journal of Plant Pathology*, vol. 106, no. 8, pp. 699–713, 2000.

[127] X. Dong, "SA, JA, ethylene, and disease resistance in plants," *Current Opinion in Plant Biology*, vol. 1, no. 4, pp. 316–323, 1998.

[128] M. P. Does and B. J. C. Cornelissen, "Emerging strategies to control fungal diseases using transgenic plants," in *in proceedings of International Crop Science Congress*, V. L. Chopra, R. B. Singh, and A. Verma, Eds., pp. 233–244, Oxford and IBH, New Delhi, India, 1998.

[129] D. Fernandez, P. Santos, C. Agostini et al., "Coffee (*Coffea arabica* L.) genes early expressed during infection by the rust fungus (*Hemileia vastatrix*)," *Molecular Plant Pathology*, vol. 5, no. 6, pp. 527–536, 2004.

[130] B. de Nardi, R. Dreos, L. Del Terra et al., "Differential responses of *Coffea arabica* L. leaves and roots to chemically induced systemic acquired resistance," *Genome*, vol. 49, no. 12, pp. 1594–1605, 2006.

[131] K. David, "Fun without the buzz: decaffeination process and issues," 2002, http://www.virtualcoffee.com/sept_2002/deca-fe.html.

[132] M. N. Clifford, "Chemical and physical aspects of green coffee and coffee products," in *Coffee: Botany Biochemistry and Production of Beans and Beverages*, M. N. Clifford and K. C. Williamson, Eds., pp. 304–374, AVI, Westport, Conn, USA, 1985.

[133] C. A. B. de Maria, L. C. Trugo, F. R. Aquino Neto, R. F. A. Moreira, and C. S. Alviano, "Composition of green coffee water-soluble fractions and identification of volatiles formed during roasting," *Food Chemistry*, vol. 55, no. 3, pp. 203–207, 1996.

[134] M. N. Clifford, "Chlorogenic acids and other cinnamates—nature, occurrence and dietary burden," *Journal of the Science of Food and Agriculture*, vol. 79, no. 3, pp. 362–372, 1999.

[135] C. Campa, M. Noirot, M. Bourgeois et al., "Genetic mapping of a caffeoyl-coenzyme A 3-0-methyltransferase gene in coffee trees. Impact on chlorogenic acid content," *Theoretical and Applied Genetics*, vol. 107, no. 4, pp. 751–756, 2003.

[136] K. Hahlbrock and D. Scheel, "Physiology and molecular biology of phenyl-propanoid metabolism," *Annual Review of Plant Physiology and Plant Molecular Biololgy*, vol. 40, pp. 347–369, 1989.

[137] V. Mahesh, J. J. Rakotomalala, L. L. Gal et al., "Isolation and genetic mapping of a *Coffea canephora* phenylalanine ammonia-lyase gene (*CcPAL1*) and its involvement in the accumulation of caffeoyl quinic acids," *Plant Cell Reports*, vol. 25, no. 9, pp. 986–992, 2006.

[138] G. Parvatam, V. Mahesh, G. A. Ravishankar, C. Campa, and A. de Kochko, "Functional validation of *Coffea PAL* genes using genetic engineering," in *Proceedings of the 21st International Conference on Coffee Science (ASIC '06)*, pp. 702–705, Montpellier, France, September 2006.

[139] L. F. Protasio Pereira, R. M. Galvão, A. K. Kobayashi, S. M. B. Cação, and L. G. Esteves Vieira, "Ethylene production and ACC oxidase gene expression during fruit ripening of *Coffea arabica* L," *Brazilian Journal of Plant Physiology*, vol. 17, no. 3, pp. 283–289, 2005.

[140] K. R. Neupane, S. Moisyadi, and J. Stiles, "Cloning and characterization of fruit expressed ACC synthase and ACC oxidase from coffee," in *In Proceedings of the 18th International Conference on Coffee Science (ASIC '99)*, pp. 322–326, Helsinki, Finland, 1999.

[141] A. F. Ribas, L. F. P. Pereira, and L. G. E. Vieira, "Genetic transformation of coffee," *Brazilian Journal of Plant Physiology*, vol. 18, no. 1, pp. 83–94, 2006.

20

Biodegradation of Used Motor Oil in Soil Using Organic Waste Amendments

O. P. Abioye,[1,2] P. Agamuthu,[1] and A. R. Abdul Aziz[3]

[1] Institute of Biological Sciences, University of Malaya, 50603 Kuala Lumpur, Malaysia
[2] Department of Microbiology, Federal University of Technology, PMB 65, Minna 920281, Nigeria
[3] Department of Chemical Engineering, University of Malaya, 50603 Kuala Lumpur, Malaysia

Correspondence should be addressed to O. P. Abioye, bisyem2603@yahoo.com

Academic Editor: Goetz Laible

Soil and surface water contamination by used lubricating oil is a common occurrence in most developing countries. This has been shown to have harmful effects on the environment and human beings at large. Bioremediation can be an alternative green technology for remediation of such hydrocarbon-contaminated soil. Bioremediation of soil contaminated with 5% and 15% (w/w) used lubricating oil and amended with 10% brewery spent grain (BSG), banana skin (BS), and spent mushroom compost (SMC) was studied for a period of 84 days, under laboratory condition. At the end of 84 days, the highest percentage of oil biodegradation (92%) was recorded in soil contaminated with 5% used lubricating oil and amended with BSG, while only 55% of oil biodegradation was recorded in soil contaminated with 15% used lubricating oil and amended with BSG. Results of first-order kinetic model to determine the rate of biodegradation of used lubricating oil revealed that soil amended with BSG recorded the highest rate of oil biodegradation (0.4361 day^{-1}) in 5% oil pollution, while BS amended soil recorded the highest rate of oil biodegradation (0.0556 day^{-1}) in 15% oil pollution. The results of this study demonstrated the potential of BSG as a good substrate for enhanced remediation of hydrocarbon contaminated soil at low pollution concentration.

1. Introduction

Contamination of soil by used lubricating oil is rapidly increasing due to global increase in the usage of petroleum products [1]. Environmental pollution with petroleum and petrochemical products has attracted much attention in recent decades. The presence of different types of automobiles and machinery has resulted in an increase in the use of lubricating oil. Spillage of used motor oils such as diesel or jet fuel contaminates our natural environment with hydrocarbon [2]. Hydrocarbon contamination of the air, soil, and freshwater especially by PAHs attracts public attention because many PAHs are toxic, mutagenic, and carcinogenic [3–5].

Prolonged exposure to high oil concentration may cause the development of liver or kidney disease, possible damage to the bone marrow, and an increased risk of cancer [6–8]. In addition, PAHs have a widespread occurrence in various ecosystems that contribute to the persistence of these

compounds in the environment [9]. The illegal dumping of used motor oil is an environmental hazard with global ramifications [10]. Used motor oil contains metals and heavy polycyclic aromatic hydrocarbons (PAHs) that could contribute to chronic hazards including mutagenicity and carcinogenicity [11, 12].

Lack of essential nutrients such as nitrogen and phosphorus is one of the major factors affecting biodegradation of hydrocarbon by microorganisms in soil and water environment. Therefore, the addition of inorganic or organic nitrogen-rich nutrients (biostimulation) is an effective approach to enhance the bioremediation process [13–15]. Positive effects of nitrogen amendment on microbial activity and/or petroleum hydrocarbon degradation have been widely demonstrated by various authors [16–19].

Concentration of petroleum hydrocarbon determines to a greater extent the rate of breakdown of the hydrocarbons from soil environment. High concentration of hydrocarbon can be inhibitory to microorganisms, and concentration at which inhibition occurs varied with the compound. Ijah and Antai [20] reported high degradation of hydrocarbons in soil contaminated with 10% and 20% crude oil compared to those contaminated with 30 and 40% crude oil which experienced partial degradation of hydrocarbons within a period of 12 months. Rahman et al. [21] reported that percentage of degradation by mixed bacterial consortium decreased from 78% to 52%, as the concentration of crude oil increased from 1 to 10%. High concentrations of hydrocarbons can be associated with heavy, undispersed oil slicks in water, causing inhibition of oil biodegradation due to oxygen limitation or through toxic effects exerted by volatile hydrocarbons on microorganisms.

The objectives of this study are to determine the potential of banana skin, brewery spent grain, and spent mushroom compost for enhanced biodegradation of used lubricating oil in soil, as an alternative to the use of inorganic fertilizers. These organic materials are widely available as wastes in our environment. The study also aimed to determine the effects of oil concentration on biodegradation of used lubricating oil.

2. Methods

2.1. Collection of Samples. The soil sample used was collected from the Nursery Section of the Asia-European Institute, University of Malaya, Kuala Lumpur, Malaysia, in a sack and transported to the laboratory for analysis. Used lubricating oil was collected from the Perodua Car Service Centre, Petaling Jaya, while the organic wastes were collected from different locations; banana skins (BS) were collected from the IPS Canteen, University of Malaya, brewery spent grains (BSG) were collected from Carlsberg Brewery, Shah Alam, Selangor, and spent mushroom compost (SMC) was the collected from Gano Mushroom Farm, Tanjung Sepat, Selangor.

2.2. Bioremediation Setup. 1.5 kg of soil (sieved with 2 mm mesh size) was placed in plastic vessels with a volume of about 3000 cm^3, and 5% and 15% (w/w) used lubricating oil was added separately, thoroughly mixed, and left undisturbed for 48 hours to allow the volatilization of toxic components of the oil. After two days, 10% of each organic waste (ground dry banana skin (BS), brewery spent grain (BSG), and spent mushroom compost (SMC)) were individually introduced into each oil-polluted soil and thoroughly mixed. The moisture was adjusted to 60% water holding capacity and incubated at room temperature (28 ± 2°C). Treatment with only soil and used lubricating oil served as control. Additional control was also set up which contained autoclaved soil poisoned with 0.5% (w/w) sodium azide to monitor nonbiological loss of oil in the oil-contaminated soil. The content of each vessel was tilled twice a week for aeration and the moisture maintained at 60% water holding capacity by the addition of sterile distilled water. The experiment was set up in triplicate. Periodic sampling from each vessel was carried out at 14-day intervals for 84 days. Composite samples were obtained by mixing 5 g of soil collected from four different areas of the plastic vessels for

isolation and enumeration of hydrocarbon utilizing bacteria and determination of total petroleum hydrocarbon.

2.3. Physicochemical Analysis of Soil and Organic Wastes. Nitrogen contents of soil used for bioremediation and organic wastes were determined using the Kjeldahl method, while phosphorus and carbon contents were determined using ICP-QES and furnace method, respectively. pH was determined with pH meter (HANNA HI 8424) on 1 : 2.5 (w/v) soil/distilled water after 30minute equilibration. Triplicate determinations were made.

2.4. Total Petroleum Hydrocarbon (TPH) Determination. Residual hydrocarbon contents of the soil samples were determined by toluene cold extraction method of Adesodun and Mbagwu [22]. 10 g of soil sample was weighed into 50 mL flask, and 20 mL of toluene (AnalaR grade) added. After shaking for 30 minutes on an orbital shaker (model N-Biotek-101M), the liquid phase of the extract was measured at 420 nm using DR/4000 Spectrophotometer. The total petroleum hydrocarbon (TPH) in soil was estimated with reference to a standard curve derived from fresh used lubricating oil diluted with toluene. TPH data were fitted to the first-order kinetics model:

$$C = C_o e^{-kt}, \tag{1}$$

where C is the hydrocarbon content in soil (g kg^{-1}) at time t, C_o is the initial hydrocarbon content in soil (g kg^{-1}), k is the biodegradation rate constant (d^{-1}), and t is time (d).

2.5. Enumeration and Identification of Bacteria. Three replicate samples from each oil-polluted soil were withdrawn every 14 days for the enumeration of hydrocarbon utilizing bacteria (HUB). 0.1 mL of serially diluted samples were plated on oil agar prepared from mineral salt medium of Zajic and Supplisson [23] (1.8 g K$_2$HPO$_4$, 4.0 g NH$_4$Cl, 0.2 g MgSO$_4 \cdot$7H$_2$O, 1.2 g KH$_2$PO$_4$, 0.01 g FeSO$_4 \cdot$7H$_2$O, 0.1 g NaCl, 20 g agar, 1% used lubricating oil in 1000 mL distilled water, pH 7.4). Triplicate plates were incubated at 30°C for 5 days before the colonies were counted and randomly picked; pure isolates were obtained by repeated subculturing on nutrient agar (Oxoid). The bacterial isolates were characterized using microscopic techniques and biochemical tests and further confirmed by using API 20NE for Gram-negative bacteria, and BBL Crystal rapid identification kit for Gram-positive bacteria. For Gram-positive bacterial identification, colonies of pure culture of bacteria were introduced into the BBL inoculums fluid with the aid of sterile wire loop and vortexed for 10–15 seconds. The turbidity was adjusted to the equivalent of McFarland no. 0.5 standard; the entire inoculum was poured into the BBL base that contains different wells. The inoculum was gently rolled with both hands to ensure that all the wells are filled. The wells containing the inoculums were later covered with BBL lid that contained 29 dehydrated biochemical and enzymatic substrates and a fluorescence control on tips of plastic prongs. The inoculated panels were incubated for 18–24 hours at 35–37°C; at the end of incubation period the wells were examined for

TABLE 1: Physicochemical properties of soil and organic wastes used for bioremediation.

Parameter	Soil	Organic wastes		
		BSG	BS	SMC
pH	6.12 ± 0.23	6.66 ± 0.49	7.04 ± 0.29	5.64 ± 0.25
Nitrogen (%)	0.4 ± 0.02	1.02 ± 0.1	0.4 ± 0.01	0.5 ± 0.03
Phosphorus (mg/kg)	21.8 ± 1.5	20.6 ± 2.0	21.2 ± 1.4	22.5 ± 1.8
Organic C (%)	10.3 ± 1.1	10.9 ± 0.91	10.5 ± 1.3	10.2 ± 1.1
Moisture (%)	7.0 ± 0.3	71.84 ± 3.5	38.5 ± 2.86	62.3 ± 4.12
Sand (%)	37.5 ± 2.6	—	—	—
Silt (%)	18.75 ± 1.95	—	—	—
Clay (%)	43.75 ± 2.75	—	—	—
HUB (CFU/g)	6.2×10^3	7.4×10^2	2.1×10^2	4.5×10^2
Texture	Clayey	—	—	—

BSG: Brewery spent grain, BS: banana skin, SMC: spent mushroom compost, HUB: hydrocarbon utilizing bacteria.

colour change or presence of fluorescence that resulted from metabolic activities of the microorganisms. The resulting patterns of the 29 reactions were converted into a ten-digit profile number that were used as the basis for identification. The resulting profile number derived from different colour changes and cell morphology were entered into PC in which the BBL Crystal MIND Software has been installed to obtain the bacterial identification.

Gram-negative bacterial isolates were identified using API 20 NE. Pure culture colonies of bacterial sample were transferred into an ampoule of API NaCl 0.85% medium (2 mL) with the aid of inoculating wire loop to prepare a suspension with a turbidity equivalent to 0.5 McFarland standard. Tests of NO_3 to PNPG in the API panel were inoculated by distributing the saline suspension into the tubes using sterile pipette. $200 \mu L$ of the remaining suspension was added into an ampoule of API AUX medium and homogenized. The cupules tests GLU to PAC were filled with the suspension from API AUX medium followed by addition of mineral oil to the test cupule-labeled GLU, ADH, and URE until a convex meniscus was formed. The incubation box was closed and incubated at $29°C \pm 2°C$. At the end of the incubation period, the results were read based on colour changes and converted into numerical profile. The identification was performed by using the database (V7.0) with the analytical profile index which was earlier installed into the PC.

2.6. Germination Toxicity Test of Remediated Soil. Toxicity of the remediated soils was assessed using germination test. Lettuce was used in this study owing to its sensitivity to hydrocarbon in soil [24, 25]. The germination test was conducted over a 5-day test period. Seeds of lettuce were obtained commercially. For each soil sample, 150 g of thoroughly mixed remediated soil was placed in 100 × 15 mm Petri dish. Ten viable seeds of lettuce (*Lactuca sativa* L.) were placed evenly throughout each petri dish and covered with 10 g of dry sand. Three replicates of the samples were prepared. The moisture of the soil was maintained at 80% water holding capacity. The Petri dishes were placed in a room with 16 hours light and 8 hours darkness for 5 days.

At the end of 5 days, the number of seedlings that emerged from the surface of the sand was counted and recorded.

Germination index of lettuce seed on the remediated soil was calculated using the formula of Millioli et al. [26]:

$$\text{Germination index (\%)} = \frac{(\% \text{ SG}) \times (\% \text{ GR})}{100}, \quad (2)$$

$$\% \text{ SG} = (\% \text{ EG}/\% \text{ CG}) \times 100, \quad (3)$$

$$\% \text{ GR} = (\text{GERm}/\text{GERCm}) \times 100, \quad (4)$$

where % SG = seed germination, % GR = growth of the root, % EG = germination on contaminated soil, % CG = germination on control soil, GERm = elongation of root on contaminated soil, GERCm = elongation of root on control soil.

2.7. Statistical Analysis. Statistical analysis of data was carried out using Analysis of Variance (ANOVA).

3. Results and Discussion

3.1. Physicochemical Properties of Soil and Organic Wastes. The physicochemical properties of soil and organic wastes used for the bioremediation studies are shown in Table 1. The soil used for bioremediation had C : N ratio of 25.7; this is a low C : N ratio for effective biodegradation of oil in the soil, hence the need for addition of organic wastes as a source of nutrients (N and P). BSG had the highest N content among the three organic wastes used; this is one of the most important limiting nutrient for effective bioremediation to take place [27, 28]. The moisture contents of BSG (71.8%) were as well higher than those of BS (38.5%) and (62.3%); this might enable the BSG to harbor some important microorganisms that will contribute positively to the biodegradation of oil in the soil. The pH of SMC (5.6) was slightly acidic; the reason for this might be because it was used to grow fungi (mushroom) which grow better in an acidic environment. Therefore, the initial substrate of SMC might be slightly acidic in nature.

3.2. Biodegradation of Used Lubricating Oil. The percentage of oil biodegradation in the soil contaminated with 5% and 15% used lubricating oil is shown in Figures 1 and 2, respectively. The results revealed rapid and high (between 79% and 92%) biodegradation of the used lubricating oil at the end of 84 days in soil contaminated with 5% oil. Soil amended with different organic wastes recorded the highest rate of oil mineralization compared to unamended polluted soil. The reason for this relatively high and progressive biodegradation in all the soil contaminated with 5% used lubricating oil might be due to low concentration of oil in the soil which does not pose serious challenge to the metabolic activities of soil microrganisms. It could also be due to the presence of organic waste amendments which likely supply nutrient to the microbial population present in the contaminated soil, thereby enabling them to degrade almost completely the oil contaminant. The result is in agreement with the findings of Rahman et al. [21] who reported increase in the rate of biodegradation of crude oil, as the concentration of oil reduced.

At the end of 28 days in soil contaminated with 15% oil, there were 17%, 24%, and 5% total petroleum hydrocarbon (TPH) degradation in soil amended with BSG, BS, and SMC, respectively. The reason for the low percentage of oil degradation within the first 28 days might be attributed to the toxicity of the oil on the microbial flora of the soil, due to high concentration of oil which might likely had negative effects on the biodegradative activities of the microbial population in the contaminated soil. This initial trend of low biodegradation due to high oil concentration has been reported by different authors [20, 21] who argued that high concentration of hydrocarbon can be inhibitory at the initial stage to the indigenous microorganisms in the soil. At the end of 84 days, 55%, 49%, and 36% oil biodegradation were recorded in soil contaminated with 15% oil amended with BSG, BS, and SMC, respectively. In soil contaminated with 5% oil, 92% oil biodegradation was recorded in soil amended with BSG, followed by 84% degradation in soil treated with BS, and 79% in soil amended with SMC at the end of 84 days. The results are in contrast with the findings of Adesodun and Mbagwu [22] who reported 30% and 42% biodegradation in soil contaminated with 5% spent lubricating oil and amended with cow dung and piggery wastes within the period of three months. The differences in these results might be due to different composition of used lubricating oil utilized for the studies or differences in the organic wastes used. It might as well be due to differences in the soil composition used for the studies.

BSG-amended soil recorded highest percentage biodegradation (92% and 55%) throughout the 84 days period in 5% and 15% oil-contaminated soil, respectively. This might be due to high N and P contents present in BSG. N and P are known as the most important nutrients needed by hydrocarbon-utilizing bacteria to carry out effective and efficient biodegradative activities of xenobiotics in the soil environment [27–29]. 8% and 5% of oil degradation in 5% and 15% oil-polluted soil might be due to nonbiological factors such as evaporation or photodegradation. This was recorded in poisoned controlled soil, that is, autoclaved contaminated

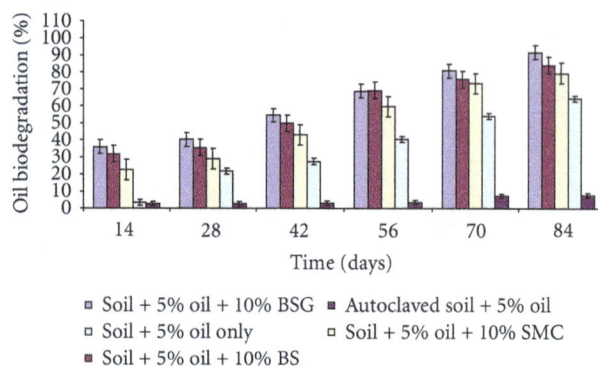

FIGURE 1: Biodegradation of petroleum hydrocarbon in soil contaminated with 5% used lubricating oil and amended with 10% organic wastes.

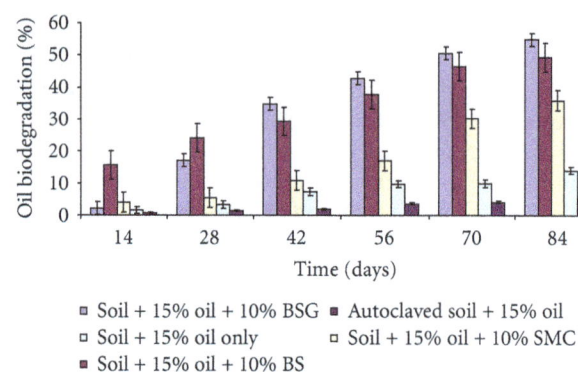

FIGURE 2: Biodegradation of petroleum hydrocarbon in soil contaminated with 15% used lubricating oil and amended with 10% organic wastes.

soil treated with 0.5% sodium azide. This was in sharp contrast to the findings of Palmroth et al. [30], who recorded as high as 70% diesel oil loss within 28 days of study in sodium azide-treated soil. The differences in these results might be because poisoned control in this study was an autoclaved soil mixed with 0.5% sodium azide, whereas Palmroth et al. [30] used only 0.5% sodium azide without autoclaving the soil; thus the sodium azide effect possibly could not completely sterilize the soil.

3.3. Biodegradation Rate. First-order kinetics was used to determine the rate of biodegradation of used lubricating oil in the various treatments as shown in Table 2. In soil contaminated with 5% used lubricating oil, BSG-amended soil recorded the highest biodegradation rate of 0.4361 day^{-1}. The biodegradation rates of soil amended with BS and SMC were 0.410 day^{-1} and 0.3100 day^{-1}, respectively. Unamended and autoclaved contaminated soil recorded biodegradation rates of 0.1886 day^{-1} and 0.0079 day^{-1}, respectively. However, in 15% used lubricating-oil-contaminated soil, BS-amended soil recorded highest biodegradation rate of 0.0556 day^{-1}. The biodegradation rates of soil amended with BSG and SMC were 0.0479 day^{-1} and 0.0216 day^{-1}, respectively. High biodegradation rate recorded in BS-amended

TABLE 2: Biodegradation rates of hydrocarbon in used lubricating-oil-contaminated soil.

Treatment	Biodegradation constant (k) day^{-1}
Soil + 5% oil + BS	0.4010[b]
Soil + 5% oil + BSG	0.4361[b]
Soil + 5% oil + SMC	0.3100[b]
Soil + 5% oil	0.1886[a]
Autoclaved soil + 5% oil	0.0079[a]
Soil + 15% oil + BS	0.0556[b]
Soil + 15% oil + BSG	0.0479[a]
Soil + 15% oil + SMC	0.0216[b]
Soil + 15% oil	0.0092[a]
Autoclaved soil + 15% oil	0.0033[a]

Values followed by letter b indicate significant difference at $P < 0.05$ level, while values followed by "a" are not different significantly at $P < 0.05$ level.

soil above that of BSG might be due to initial rapid loss of used lubricating oil in the first 28 days of study in BS-amended soil than those of BSG- and SMC-amended soil. This is however different from the results of Adesodun and Mbagwu [22], who reported highest biodegradation rate in oil-contaminated soil amended with piggery wastes, which had highest percentage of biodegradation throughout the study period.

The results show significant relationships between the rate of biodegradation and concentration of oil in the contaminated soil. From the results, higher biodegradation rates were recorded in soil contaminated with 5% oil; this high biodegradation rate could be attributed to increase in the activity of soil microbes in this oil pollution level [22]. Bossert and Bartha [31] stated that sensitivity of soil microflora to petroleum hydrocarbons is a factor of quantity and quality of oil spilled and previous exposure of the native soil microbes to oil. Schaefer and Juliane [32] also concluded that bioremediation is a useful method of soil remediation if pollutant concentrations are moderate.

3.4. Microbial Counts. Count of hydrocarbon utilizing bacteria (HUB) in soil contaminated with 5% used lubricating oil and amended with organic wastes is shown in Figure 3. The count of HUB in soil amended with BSG was about 8% higher than those amended with BS and SMC. HUB count in BSG amended soil ranged from 47.0×10^6 CFU/g to 146.0×10^6 CFU/g while those amended with BS and SMC ranged from 42×10^6 CFU/g to 120×10^6 CFU/g and 12.0×10^6 CFU/g to 51.0×10^6 CFU/g, respectively, within 84 days of study. The count of HUB in 15% used lubricating-oil-contaminated soil amended with BSG was about 3% higher than those amended with BS and SMC. HUB count in BSG-amended soil ranged from 24.0×10^5 CFU/g to 210.0×10^5 CFU/g, while those amended with BS and SMC ranged from 15.0×10^5 CFU/g to 167×10^5 CFU/g, and 3.0×10^5 CFU/g and 38.0×10^5 CFU/g respectively (Figure 4). However, the HUB count in unamended control soil was extremely (2.0×10^5 CFU/g to 14.0×10^5 CFU/g) lower than those amended with organic wastes.

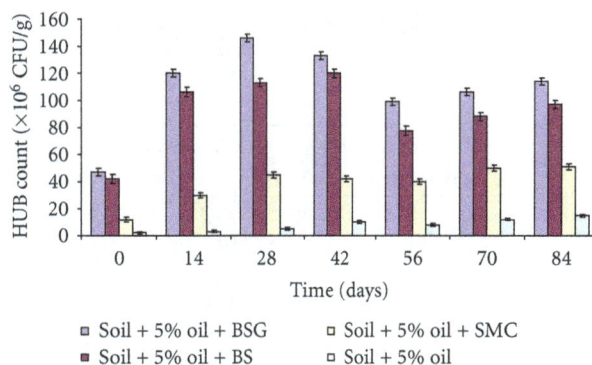

FIGURE 3: Hydrocarbon-utilizing bacteria (HUB) in soil contaminated with 5% used lubricating oil and amended with organic wastes.

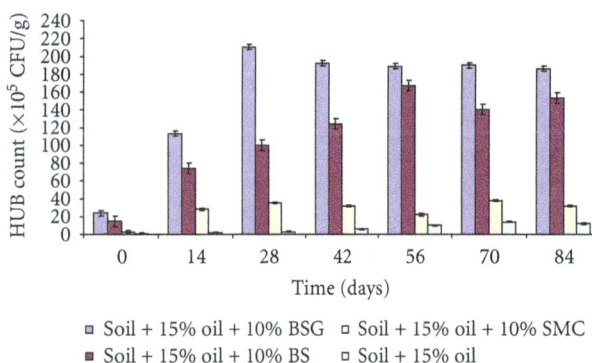

FIGURE 4: Hydrocarbon-utilizing bacteria (HUB) in soil contaminated with 15% used lubricating oil and amended with organic wastes.

The counts of hydrocarbon utilizing bacteria (HUB) in all the soil amended with organic wastes were appreciably higher compared to those of unamended and poisoned control soil. The reason for higher counts of bacteria in amended soil might be as a result of presence of appreciable quantities of nitrogen and phosphorus in the organic wastes, especially high nitrogen content in BSG, which are necessary nutrients for bacterial biodegradative activities [20, 22, 33–35]. The reason for increased biodegradation of oil in amended soil as compared to the unamended soil might also be due to the presence of organic wastes in the soil which helps to loosen the compactness of the soil making sufficient aeration available for the indigenous bacteria present in the soil, thereby enhancing their metabolic activities in the contaminated soil. It might as well be due to the ability of these organic wastes (mostly BSG that recorded higher counts) to neutralize the toxic effects of the oil on the microbial population by rapid improvement of the soil physicochemical properties [16].

The HUB isolated from the used lubricating-oil-contaminated soil were identified as species of Acinetobacter, Micrococcus, Pseudomonas aeruginosa, Nocardia, Bacillus mega-terium, Bacillus sp., and Corynebacterium. These bacterial species had been implicated in hydrocarbon degradation by different authors [9, 36–40].

Table 3: Toxicity test based on seed germination (%).

Percentage of Oil pollution	Treatments					
	A	B	C	D	E	F
5	80 ± 6.0	100	80 ± 6.0	40 ± 6.0	20 ± 0	100
15	40 ± 5.8	40 ± 6.0	20 ± 0	10 ± 0	0	100

A = Soil + Oil + BS, B = Soil + Oil + BSG, C = Soil + Oil + SMC, D = Soil + Oil, E = Autoclaved soil + Oil + NaN$_3$, F = Uncontaminated soil.

Table 4: Seed germination toxicity index (%).

Percentage of Oil pollution	Germination toxicity index (%)				
	A	B	C	D	E
5	40.00	83.33	33.34	13.33	3.27
15	6.53	13.33	5.00	1.65	0.00

A = Soil + Oil + BS, B = Soil + Oil + BSG, C = Soil + Oil + SMC, D = Soil + Oil, E = Autoclaved soil + Oil + NaN$_3$.

3.5. Germination Toxicity. Lettuce (*Lactuca sativa*) is an important agricultural crop, and it is fairly sensitive to toxic chemicals (mostly petroleum contaminants), which led to its wide use for toxicity tests [41, 42]. The results of germination toxicity test conducted after 84 days of remediation for soil contaminated with 5% and 15% used lubricating oil and amended organic wastes are shown in Table 3. The results reveal 100%, 80%, and 80% germination in soil contaminated with 5% oil and amended with BSG, BS, and SMC, respectively. However, 40%, 40%, and 20% seed germination were recorded in soil contaminated with 15% oil and amended with BSG, BS, and SMC, respectively. 100% germination was recorded in uncontaminated control soil, while only 20% and 0% were recorded in poisoned controlled soil in soil contaminated with 5% and 15% used lubricating oil, respectively. The result shows positive correlation between loss of oil in the remediated soil and seed germination. It also revealed that remediation of soil contaminated with high concentration of petroleum hydrocarbons needs a longer period of time possibly with increased quantity of organic wastes amendment to be completely restored into a state suitable for agricultural purposes. The results are in agreement with the findings of Banks and Schultz [41] and Millioli et al. [26], who recorded decrease in number of germinated seeds with increased quantities of petroleum concentration in the soil.

3.6. Seed Germination Index. Germination index of lettuce seed on the remediated soil was calculated using the formula of Millioli et al. [26]. Table 4 shows the results of seed germination index in soil contaminated with 15% and 5% used lubricating oil and amended with different organic wastes. Soil treated with BSG recorded the highest germination index (83.33%, & 13.33%) in all the treatments with organic wastes amendments; this result further proved the effectiveness of BSG in enhancing biodegradation of hydrocarbon in oil-contaminated soil. The result is similar to the finding of Molina-Barahona et al. [43] and Oleszczuk [42], who reported that composted wastewater sludge reduced phytotoxicity of diesel oil to the germination of *Lepidium sativum*

after composting the sludge for 76 days. The negative effect of hydrocarbons on the germination index may be attributed to their inherent toxicity or to the perturbations they cause in soil and plants due to their hydrophobic properties [44, 45]. Hydrocarbons may coat root surface, preventing or reducing gas and water exchange and nutrient absorption. They may also enter the seeds and alter the metabolic reactions or kill the embryo by direct, acute toxicity after penetrating the plant tissues. Hydrocarbons damage cell membranes and reduce the metabolic transport and respiration rate [44, 46]. But, a more likely reason for the inhibitory effect of hydrocarbons on germination is its physical water-repellent property. The film of hydrocarbons around the seeds may act as a physical barrier, preventing or reducing both water and oxygen from entering the seeds. This would inhibit the germination response [44].

4. Conclusion

Amendment of soil contaminated with used lubricating oil with organic wastes positively enhanced the rate of biodegradation of used lubricating oil in soil within the period of 84 days. The results of the studies in soil contaminated with 5% and 15% used lubricating oil amended with organic wastes (BS, BSG, and SMC) show low (55%) oil biodegradation in soil contaminated with 15% oil compared with 92% oil biodegradation recorded in 5% oil pollution, thus, showing that level of oil contamination influenced the rate of oil biodegradation in soil environment. Contaminated soil amended with BSG recorded highest rate of oil biodegradation and counts of hydrocarbon utilizing bacteria compared to soil amended with BS and SMC in both 5% and 15% oil pollution. Results of germination toxicity test carried out on the remediated soil showed less toxicity to lettuce in 5% oil-contaminated soil compared to those of 15% oil-contaminated soil. Therefore, brewery spent grain, which is a waste from brewery, can be utilized effectively to reclaim soil contaminated with used lubricating oil.

Acknowledgments

The authors would like to acknowledge the support of University of Malaya, IPPP grant PS 244/2008C, and FRGS Grant FP014/2010A. Also, they would like to thank the management of the Carlsberg Brewery for providing brewery spent grain and the Ganofarm Sdn. Bhd for the provision of spent mushroom compost.

References

[1] T. Mandri and J. Lin, "Isolation and characterization of engine oil degrading indigenous microorganisms in Kwazulu-Natal," *African Journal of Biotechnology*, vol. 6, no. 1, pp. 23–27, 2007.

[2] A. Husaini, H. A. Roslan, K. S. Y. Hii, and C. H. Ang, "Biodegradation of aliphatic hydrocarbon by indigenous fungi isolated from used motor oil contaminated sites," *World Journal of Microbiology and Biotechnology*, vol. 24, no. 12, pp. 2789–2797, 2008.

[3] J. A. Bumpus, "Biodegradation of polycyclic aromatic hydro-carbons by Phanerochaete chrysosporium," *Applied and Environmental Microbiology*, vol. 55, no. 1, pp. 154–158, 1989.

[4] A. R. Clemente, T. A. Anazawa, and L. R. Durrant, "Biodegradation of polycyclic aromatic hydrocarbons by soil fungi," *Brazilian Journal of Microbiology*, vol. 32, no. 4, pp. 255–261, 2001.

[5] C. E. Cerniglia and J. B. Sutherland, "Bioremediation of polycyclic aromatic hydrocarbons by ligninolytic and non-ligninolytic fungi," in *Fungi in Bioremediation*, G. M. Gadd, Ed., pp. 136–187, Cambridge University Press, Cambridge, UK, 2001.

[6] S. Mishra, J. Jyot, R. C. Kuhad, and B. Lal, "Evaluation of inoculum addition to stimulate in situ bioremediation of oily-sludge-contaminated soil," *Applied and Environmental Microbiology*, vol. 67, no. 4, pp. 1675–1681, 2001.

[7] T. L. Propst, R. L. Lochmiller, C. W. Qualls, and K. McBee, "In situ (mesocosm) assessment of immunotoxicity risks to small mammals inhabiting petrochemical waste sites," *Chemosphere*, vol. 38, no. 5, pp. 1049–1067, 1999.

[8] A. C. Lloyd and T. A. Cackette, "Diesel engines: environmental impact and control," *Journal of the Air and Waste Management Association*, vol. 51, no. 6, pp. 809–847, 2001.

[9] J. D. Van Hamme, A. Singh, and O. P. Ward, "Recent advances in petroleum microbiology," *Microbiology and Molecular Biology Reviews*, vol. 67, no. 4, pp. 503–549, 2003.

[10] W. C. Blodgett, "Water-soluble mutagen production during the bio-remediation of oil—contaminated soil," *Florida Scientist*, vol. 60, no. 1, pp. 28–36, 2001.

[11] L. S. Hagwell, L. M. Delfino, and J. J. Rao, "Partitioning of polycyclic aromatic hydrocarbons from diesel fuel into water," *Environmental Science and Technology*, vol. 26, no. 11, pp. 2104–2110, 1992.

[12] S. Boonchan, M. L. Britz, and G. A. Stanley, "Degradation and mineralization of high-molecular-weight polycyclic aromatic hydrocarbons by defined fungal-bacterial cocultures," *Applied and Environmental Microbiology*, vol. 66, no. 3, pp. 1007–1019, 2000.

[13] J. Hollender, K. Althoff, M. Mundt, and W. Dott, "Assessing the microbial activity of soil samples, its nutrient limitation and toxic effects of contaminants using a simple respiration test," *Chemosphere*, vol. 53, no. 3, pp. 269–275, 2003.

[14] K. T. Semple, N. M. Dew, K. J. Doick, and A. H. Rhodes, "Can microbial mineralization be used to estimate microbial availability of organic contaminants in soil?" *Environmental Pollution*, vol. 140, no. 1, pp. 164–172, 2006.

[15] J. Walworth, A. Pond, I. Snape, J. Rayner, S. Ferguson, and P. Harvey, "Nitrogen requirements for maximizing petroleum bioremediation in a sub-Antarctic soil," *Cold Regions Science and Technology*, vol. 48, no. 2, pp. 84–91, 2007.

[16] K. S. Jørgensen, J. Puustinen, and A. M. Suortti, "Bioremediation of petroleum hydrocarbon-contaminated soil by composting in biopiles," *Environmental Pollution*, vol. 107, no. 2, pp. 245–254, 2000.

[17] R. Margesin and F. Schinner, "Bioremediation (Natural Attenuation and Biostimulation) of diesel-oil-contaminated soil in an alpine glacier skiing area," *Applied and Environmental Microbiology*, vol. 67, no. 7, pp. 3127–3133, 2001.

[18] R. Riffaldi, R. Levi-Minzi, R. Cardelli, S. Palumbo, and A. Saviozzi, "Soil biological activities in monitoring the bioremediation of diesel oil-contaminated soil," *Water, Air, and Soil Pollution*, vol. 170, no. 1–4, pp. 3–15, 2006.

[19] R. Margesin, M. Hämmerle, and D. Tscherko, "Microbial activity and community composition during bioremediation of diesel-oil-contaminated soil: effects of hydrocarbon concentration, fertilizers, and incubation time," *Microbial Ecology*, vol. 53, no. 2, pp. 259–269, 2007.

[20] U. J. J. Ijah and S. P. Antai, "The potential use of chicken-drop micro-organisms for oil spill remediation," *Environmentalist*, vol. 23, no. 1, pp. 89–95, 2003.

[21] K. S. M. Rahman, J. Thahira-Rahman, P. Lakshmanaperumalsamy, and I. M. Banat, "Towards efficient crude oil degradation by a mixed bacterial consortium," *Bioresource Technology*, vol. 85, no. 3, pp. 257–261, 2002.

[22] J. K. Adesodun and J. S. C. Mbagwu, "Biodegradation of waste-lubricating petroleum oil in a tropical alfisol as mediated by animal droppings," *Bioresource Technology*, vol. 99, no. 13, pp. 5659–5665, 2008.

[23] J. E. Zajic and B. Supplisson, "Emulsification and degradation of "Bunker C" fuel oil by microorganisms," *Biotechnology and Bioengineering*, vol. 14, no. 3, pp. 331–343, 1972.

[24] K. Vaajasaari, A. Joutti, E. Schultz, S. Selonen, and H. Westerholm, "Comparisons of terrestrial and aquatic bioassays for oil-contaminated soil toxicity," *Journal of Soils and Sediments*, vol. 2, no. 4, pp. 194–202, 2002.

[25] G. Płaza, G. Nałęcz-Jawecki, K. Ulfig, and R. L. Brigmon, "The application of bioassays as indicators of petroleum-contaminated soil remediation," *Chemosphere*, vol. 59, no. 2, pp. 289–296, 2005.

[26] V. S. Millioli, E. L. C. Servulo, L. G. S. Sobral, and D. D. Carvalho, "Bioremediation of crude oil-bearing soil: evaluating the effect of rhamnolipid addition to soil toxicity and to crude oil biodegradation efficiency," *Global Nest Journal*, vol. 11, no. 2, pp. 181–188, 2009.

[27] I. O. Okoh, "Biodegradation alternative in the cleanup of petroleum hydrocarbon pollutants," *Biotechnology and Molecular Biology Reviews*, vol. 1, no. 2, pp. 38–50, 2006.

[28] S. J. Kim, D. H. Choi, D. S. Sim, and Y. S. Oh, "Evaluation of bioremediation effectiveness on crude oil-contaminated sand," *Chemosphere*, vol. 59, no. 6, pp. 845–852, 2005.

[29] C. Frederic, P. Emilien, G. Lenaick, and D. Daniel, "Effects of nutrient and temperature on degradation of petroleum hydrocarbons in contaminated sub-Antarctic soil," *Chemosphere*, vol. 58, no. 10, pp. 1439–1448, 2005.

[30] M. R. T. Palmroth, J. Pichtel, and J. A. Puhakka, "Phytoremediation of subarctic soil contaminated with diesel fuel," *Bioresource Technology*, vol. 84, no. 3, pp. 221–228, 2002.

[31] I. Bossert and R. Bartha, "The fate of petroleum in soil ecosystems," in *Petroleum Microbiology*, R. M. Atlas, Ed., Macmillan, New York, NY, USA, 1984.

[32] M. Schaefer and F. Juliane, "The influence of earthworms and organic additives on the biodegradation of oil contaminated soil," *Applied Soil Ecology*, vol. 36, no. 1, pp. 53–62, 2007.

[33] K. Nakasaki, H. Yaguchi, Y. Sasaki, and H. Kubota, "Effects of C/N ratio on thermophilic composting of garbage," *Journal of Fermentation and Bioengineering*, vol. 73, no. 1, pp. 43–45, 1992.

[34] H. S. Joo, C. G. Phae, and J. Y. Ryu, "Comparison and analysis on characteristics for recycling of multifarious food waste," *Journal of KOWREC*, vol. 9, pp. 117–124, 2001.

[35] H. S. Joo, M. Shoda, and C. G. Phae, "Degradation of diesel oil in soil using a food waste composting process," *Biodegradation*, vol. 18, no. 5, pp. 597–605, 2007.

[36] U. J. J. Ijah, "Studies on relative capabilities of bacterial and yeast isolates from tropical soil in degrading crude oil," *Waste Management*, vol. 18, no. 5, pp. 293–299, 1998.

[37] Y. H. Ahn, J. Sanseverino, and G. S. Sayler, "Analyses of polycyclic aromatic hydrocarbon-degrading bacteria isolated from

contaminated soils," *Biodegradation*, vol. 10, no. 2, pp. 149–157, 1999.

[38] F. M. Bento, F. A. O. Camargo, B. C. Okeke, and W. T. Frankenberger, "Comparative bioremediation of soils contaminated with diesel oil by natural attenuation, biostimulation and bioaugmentation," *Bioresource Technology*, vol. 96, no. 9, pp. 1049–1055, 2005.

[39] K. Das and A. K. Mukherjee, "Crude petroleum-oil biodegradation efficiency of Bacillus subtilis and Pseudomonas aeruginosa strains isolated from a petroleum-oil contaminated soil from North-East India," *Bioresource Technology*, vol. 98, no. 7, pp. 1339–1345, 2007.

[40] A. Roldán-Martín, G. Calva-Calva, N. Rojas-Avelizapa, M. D. Díaz-Cervantes, and R. Rodríguez-Vázquez, "Solid culture amended with small amounts of raw coffee beans for the removal of petroleum hydrocarbon from weathered contaminated soil," *International Biodeterioration and Biodegradation*, vol. 60, no. 1, pp. 35–39, 2007.

[41] M. K. Banks and K. E. Schultz, "Comparison of plants for germination toxicity tests in petroleum-contaminated soils," *Water, Air, and Soil Pollution*, vol. 167, no. 1–4, pp. 211–219, 2005.

[42] P. Oleszczuk, "Phytotoxicity of municipal sewage sludge composts related to physico-chemical properties, PAHs and heavy metals," *Ecotoxicology and Environmental Safety*, vol. 69, no. 3, pp. 496–505, 2008.

[43] L. Molina-Barahona, L. Vega-Loyo, M. Guerrero et al., "Ecotoxicological evaluation of diesel-contaminated soil before and after a bioremediation process," *Environmental Toxicology*, vol. 20, no. 1, pp. 100–109, 2005.

[44] G. Adam and H. Duncan, "Influence of diesel fuel on seed germination," *Environmental Pollution*, vol. 120, no. 2, pp. 363–370, 2002.

[45] I. A. Ogboghodo, E. K. Iruaga, I. O. Osemwota, and J. U. Chokor, "An assessment of the effects of crude oil pollution on soil properties, germination and growth of maize (zea mays) using two crude types—forcados light and escravos light," *Environmental Monitoring and Assessment*, vol. 96, no. 1– 3, pp. 143–152, 2004.

[46] V. Labud, C. Garcia, and T. Hernandez, "Effect of hydrocarbon pollution on the microbial properties of a sandy and a clay soil," *Chemosphere*, vol. 66, no. 10, pp. 1863–1871, 2007.

Permissions

The contributors of this book come from diverse backgrounds, making this book a truly international effort. This book will bring forth new frontiers with its revolutionizing research information and detailed analysis of the nascent developments around the world.

We would like to thank all the contributing authors for lending their expertise to make the book truly unique. They have played a crucial role in the development of this book. Without their invaluable contributions this book wouldn't have been possible. They have made vital efforts to compile up to date information on the varied aspects of this subject to make this book a valuable addition to the collection of many professionals and students.

This book was conceptualized with the vision of imparting up-to-date information and advanced data in this field. To ensure the same, a matchless editorial board was set up. Every individual on the board went through rigorous rounds of assessment to prove their worth. After which they invested a large part of their time researching and compiling the most relevant data for our readers. Conferences and sessions were held from time to time between the editorial board and the contributing authors to present the data in the most comprehensible form. The editorial team has worked tirelessly to provide valuable and valid information to help people across the globe.

Every chapter published in this book has been scrutinized by our experts. Their significance has been extensively debated. The topics covered herein carry significant findings which will fuel the growth of the discipline. They may even be implemented as practical applications or may be referred to as a beginning point for another development. Chapters in this book were first published by Hindawi Publishing Corporation; hereby published with permission under the Creative Commons Attribution License or equivalent.

The editorial board has been involved in producing this book since its inception. They have spent rigorous hours researching and exploring the diverse topics which have resulted in the successful publishing of this book. They have passed on their knowledge of decades through this book. To expedite this challenging task, the publisher supported the team at every step. A small team of assistant editors was also appointed to further simplify the editing procedure and attain best results for the readers.

Our editorial team has been hand-picked from every corner of the world. Their multi-ethnicity adds dynamic inputs to the discussions which result in innovative outcomes. These outcomes are then further discussed with the researchers and contributors who give their valuable feedback and opinion regarding the same. The feedback is then collaborated with the researches and they are edited in a comprehensive manner to aid the understanding of the subject.

Apart from the editorial board, the designing team has also invested a significant amount of their time in understanding the subject and creating the most relevant covers. They scrutinized every image to scout for the most suitable representation of the subject and create an appropriate cover for the book.

The publishing team has been involved in this book since its early stages. They were actively engaged in every process, be it collecting the data, connecting with the contributors or procuring relevant information. The team has been an ardent support to the editorial, designing and production team. Their endless efforts to recruit the best for this project, has resulted in the accomplishment of this book. They are a veteran in the field of academics and their pool of knowledge is as vast as their experience in printing. Their expertise and guidance has proved useful at every step. Their uncompromising quality standards have made this book an exceptional effort. Their encouragement from time to time has been an inspiration for everyone.

The publisher and the editorial board hope that this book will prove to be a valuable piece of knowledge for researchers, students, practitioners and scholars across the globe.

List of Contributors

Francisco Fábio Cavalcante Barros, Ana Paula Resende Simiqueli, Cristiano José de Andrade and Gláucia Maria Pastore
Department of Food Science, Faculty of Food Engineering, University of Campinas, P.O. Box 6121, 13083-862 Campinas, SP, Brazil

Secil Berna Kuzu and Hatice Korkmaz Guvenmez
Biotechnology & Molecular Biology Division, Department of Biology, Cukurova University, 01330 Adana, Turkey

Aziz Akin Denizci
Research Institute for Genetic Engineering and Biotechnology, The Scientific and Technical Research Council of Turkey (TUBITAK), Marmara Research Center Campus, Gebze-Kocaeli 41470, Turkey

R. J. Mohd Salim and M. I. Adenan
Natural Product Division, Drug Discovery Centre (DDC), Forest Research Institute Malaysia (FRIM), Jalan Kepong, Selangor Darul Ehsan, 52109 Kepong, Malaysia
Malaysian Institute of Pharmaceuticals and Nutraceuticals, Ministry of Science, Technology and Innovation, USM, 10 Persiaran Bukit Jambul, 11900 Bukit Jambul, Malaysia

M. H. Jauri and A. S. Sued
Natural Product Division, Drug Discovery Centre (DDC), Forest Research Institute Malaysia (FRIM), Jalan Kepong, Selangor Darul Ehsan, 52109 Kepong, Malaysia

A. Amid
Department of Biotechnology Engineering, International Islamic University Malaysia, Gombak, P.O. Box 10, 50728 Kuala Lumpur, Malaysia

C. Karunanithy, Y.Wang and K.Muthukumarappan
Department of Agricultural and Biosystems Engineering, South Dakota State University, Brookings, SD 57007, USA

S. Pugalendhi
Department of Bioenergy, Tamil Nadu Agricultural University, Coimbatore 641003, India

Maria Soledad Diaz, Lorena Palacio, Ana Cristina Figueroa and Marta Ester Goleniowski
CEPROCOR, Science and Technology Ministry, Arenales 230, X5004APP Cordoba, Argentina

Ndatsu Yakubu and Amuzat Aliyu Olalekan
Department of Biochemistry, Ibrahim Badamasi Babangida University, Niger State, Lapai, Nigeria

Ganiyu Oboh
Department of Biochemistry, Federal University of Technology, Ondo State, Akure, Nigeria

Natalia N. Pozdnyakova
Institute of Biochemistry and Physiology of Plants and Microorganisms, Russian Academy of Sciences, 13 Prospekt Entuziastov, Saratov 410049, Russia

P. Suganya Devi and M. Saravana Kumar
P.G. Department of Biotechnology, Dr. Mahalingam Centre for Research and Development, N.G.M.College, Pollachi 642001, India

S. Mohan Das
Kaamadhenu Arts and Science College, Sathyamangalam 638503, India

Nanda Kumar Yellapu and Bhaskar Matcha
Division of Animal Biotechnology, Department of Zoology, Sri Venkateswara University, Tirupati, Andhra Pradesh 517502, India

Kalpana Kandlapalli
Department of Biochemistry, Sri Venkateswara Institute of Medical Sciences, Tirupati, Andhra Pradesh 517507, India

Koteswara Rao Valasani
Department of Pharmacology and Toxicology, University of Kansas, Lawrence, KS 66047, USA

P. V. G. K. Sarma
Department of Biotechnology, Sri Venkateswara Institute of Medical Sciences, Tirupati, Andhra Pradesh 517507, India

Ivana A. Cavello, Roque A. Hours and Sebastian F. Cavalitto
Research and Development Center for Industrial Fermentations (CINDEFI) (UNLP, CONICET La Plata), Calle 47 y 115, B1900ASH La Plata, Argentina

Javier Alonso Iserte, Betina Ines Stephan, Sandra Elizabeth Goñi, Cristina Silvia Borio and Mario Enrique Lozano
LIGBCM- Area Virosis Emergentes y Zoon´oticas, Universidad Nacional de Quilmes, B1876BXD Buenos Aires, Argentina

Pablo Daniel Ghiringhelli
LIGBCM- Area Virosis de Insectos, Universidad Nacional de Quilmes, B1876BXD Buenos Aires, Argentina

Giuseppe Vassalli
Department of Cardiology, Centre Hospitalier Universitaire Vaudois (CHUV), Avenue du Bugnon, 1011 Lausanne, Switzerland
Molecular Cardiology Laboratory, Fondazione Cardiocentro Ticino, via Tesserete 48, 6900 Lugano, Switzerland

Marc-Estienne Roehrich
Department of Cardiology, Centre Hospitalier Universitaire Vaudois (CHUV), Avenue du Bugnon, 1011 Lausanne, Switzerland

Nadimpalli Ravi S. Varma and Ali S. Arbab
Cellular and Molecular Imaging Laboratory, Department of Radiology, Henry Ford Hospital, Detroit, MI 48202, USA

Haryanti Toosa
Institute of Bioscience, Faculty of Biotechnology and Biomolecular Sciences, Universiti Putra Malaysia, 43400 Serdang, Selangor, Malaysia

Hooi Ling Foo
Institute of Bioscience, Faculty of Biotechnology and Biomolecular Sciences, Universiti Putra Malaysia, 43400 Serdang, Selangor, Malaysia
Department of Bioprocess Technology, Faculty of Biotechnology and Biomolecular Sciences, Universiti Putra Malaysia, 43400 Serdang, Selangor, Malaysia

Raha Abdul Rahim and Noorjahan Banu Mohamed Alitheen
Institute of Bioscience, Faculty of Biotechnology and Biomolecular Sciences, Universiti Putra Malaysia, 43400 Serdang, Selangor, Malaysia
Department of Cell and Molecular Biology, Faculty of Biotechnology and Biomolecular Sciences, Universiti Putra Malaysia, 43400 Serdang, Selangor, Malaysia

Mariana Nor Shamsudin
Institute of Bioscience, Faculty of Biotechnology and Biomolecular Sciences, Universiti Putra Malaysia, 43400 Serdang, Selangor, Malaysia
Department of Medical Microbiology and Parasitology, Faculty of Medicine and Health Sciences, 43400 Serdang, Selangor, Malaysia

Khatijah Yusoff
Institute of Bioscience, Faculty of Biotechnology and Biomolecular Sciences, Universiti Putra Malaysia, 43400 Serdang, Selangor, Malaysia
Department of Microbiology, Faculty of Biotechnology and Biomolecular Sciences, Universiti Putra Malaysia, 43400 Serdang, Selangor, Malaysia

María A. Martos and Emilce R. Zubreski
Facultad de Ciencias Exactas, Qu´ımicas y Naturales, Universidad Nacional de Misiones, Felix de Azara 1552, N3300LQH Posadas, Argentina

Oscar A. Garro
Universidad Nacional del Chaco Austral, Comandante Fern´andez 755, H3700LGO Presidencia Roque Saenz Pena, Argentina

Roque A. Hours
Centro de Investigación y Desarrollo en Fermentaciones Industriales (CINDEFI, UNLP, CONICET La Plata), Facultad de Ciencias Exactas, Universidad Nacional de la Plata, Calle 47 y 115, B1900ASH La Plata, Argentina

Chukwuma S. Ezeonu and Richard Tagbo
Industrial Biochemistry and Environmental Biotechnology Unit, Chemical Sciences Department, Godfrey Okoye University, P.M.B. 01014, Enugu, Nigeria

Ephraim N. Anike
Pure and Industrial Chemistry Unit, Chemical Sciences Department, Godfrey Okoye University, P.M.B. 01014, Enugu, Nigeria

Obinna A. Oje and Ikechukwu N. E. Onwurah
Pollution Control and Biotechnology Unit, Department of Biochemistry, University of Nigeria, Nsukka, Enugu State, Nigeria

Elena Dehnavi, Mojtaba Ahani Azari, Saeed Hasani, Leila Shahmohamadi and Soheil Yousefi
Department of Animal Science, Gorgan University of Agricultural Sciences and Natural Resources, P.O. Box 4913815739, Gorgan, Iran

Mohammad Reza Nassiry
Department of Animal Science, Ferdowsi University of Mashhad, P.O. Box 9177948978, Mashhad, Iran

Mokhtar Mohajer
Golestan Agriculture Jahad, P.O. Box 49174, Gorgan, Iran

Alireza Khan Ahmadi
Department of Animal Science, Faculty of Agricultural Science and Natural Resources, Gonbad University, P.O. Box 4971799151, Gonbad, Iran

Ashwini Tilay and Uday Annapure
Food Engineering and Technology Department, Institute of Chemical Technology, Matunga, Mumbai 400019, India

Neeharika Srivastava, Aishwarya Singh Chauhan and Bechan Sharma
Department of Biochemistry, Faculty of Science, University of Allahabad, Allahabad 211002, India

M. K.Mishra
Central Coffee Research Institute, Coffee Research Station, Chikmagalur, Karnataka 577117, India

A. Slater
The Biomolecular Technology Group, Faculty of Health and Life Sciences, De Montfort University, Gateway, Leicester LE1 9BH, UK

O. P. Abioye
Institute of Biological Sciences, University of Malaya, 50603 Kuala Lumpur, Malaysia
Department of Microbiology, Federal University of Technology, PMB 65, Minna 920281, Nigeria

P. Agamuthu
Institute of Biological Sciences, University of Malaya, 50603 Kuala Lumpur, Malaysia

A. R. Abdul Aziz
Department of Chemical Engineering, University of Malaya, 50603 Kuala Lumpur, Malaysia

www.ingramcontent.com/pod-product-compliance
Lightning Source LLC
Chambersburg PA
CBHW080645200326
41458CB00013B/4741